Lecture Notes on Data Engineering and Communications Technologies

Volume 124

Series Editor

Fatos Xhafa, Technical University of Catalonia, Barcelona, Spain

The aim of the book series is to present cutting edge engineering approaches to data technologies and communications. It will publish latest advances on the engineering task of building and deploying distributed, scalable and reliable data infrastructures and communication systems.

The series will have a prominent applied focus on data technologies and communications with aim to promote the bridging from fundamental research on data science and networking to data engineering and communications that lead to industry products, business knowledge and standardisation.

Indexed by SCOPUS, INSPEC, EI Compendex.

All books published in the series are submitted for consideration in Web of Science.

More information about this series at https://link.springer.com/bookseries/15362

Ngoc Hoang Thanh Dang · Yu-Dong Zhang ·
João Manuel R. S. Tavares · Bo-Hao Chen
Editors

Artificial Intelligence in Data and Big Data Processing

Proceedings of ICABDE 2021

 Springer

Editors
Ngoc Hoang Thanh Dang
Department of Information Technology
College of Technology and Design
University of Economics Ho Chi Minh City
(UEH)
Ho Chi Minh City, Vietnam

João Manuel R. S. Tavares ⓘ
Department of Mechanical Engineering
Faculty of Engineering
University of Porto
Porto, Portugal

Yu-Dong Zhang
Department of Informatics
University of Leicester
Leicester, UK

Bo-Hao Chen
Department of Computer Science
and Engineering
Yuan Ze University
Taoyuan, Taiwan

ISSN 2367-4512   ISSN 2367-4520   (electronic)
Lecture Notes on Data Engineering and Communications Technologies
ISBN 978-3-030-97612-5   ISBN 978-3-030-97610-1   (eBook)
https://doi.org/10.1007/978-3-030-97610-1

This Springer imprint is published by the registered company Springer Nature Switzerland AG
The registered company address is: Gewerbestrasse 11, 6330 Cham, Switzerland

This book is gratefully dedicated to all the researchers in the fields of Artificial Intelligence, Machine Learning, Deep Learning, Data and Big Data Processing and Analytics, in both theory and application.

About ICABDE 2021

Digital transformation is an important step to reduce administrative procedures, improve labor efficiency, augment the ability of human–machine interaction, increase customers' experience, and more. Digital transformation has been applied widely in many fields of life from technology, business, engineering, medicine, and to other social fields. In the digital transformation, it is necessary to apply advanced technologies such as Artificial Intelligence (AI) and Big Data. Big Data contains large collections of various Data types such as plain text, Data records, media, and others. To discover knowledge from Data and Big Data, it is necessary to have strong and efficient tools like Artificial Intelligence, Machine Learning, and Deep Learning. With the theme focusing on Artificial Intelligence, Data and Big Data Processing and Analytics, The 2021 International Conference on "Artificial Intelligence and Big Data in Digital Era" (ICABDE 2021) is a prestigious venue for bringing together scientists, researchers, experts, and practitioners from academia, business, and industry, who are engaged in the different research domains related to AI, Data, Big Data, and other relevant fields.

Key topics of the conference:

- AI, Machine Learning, and Deep Learning: Theory and Application.
- Data and Big Data Processing and Analytics.
- Signal, Image, and Natural Language Processing and Analyzing.
- Expert Systems and Intelligent Systems.
- Smart University and Smart City.
- Data Mining and Knowledge Discovery.
- High-Performance Computing and Cloud Computing.

During the call for papers, ICABDE 2021 received 140 submissions from various countries and territories over the world such as USA, UK, France, Russia, Turkey, Italy, Czechia, China, India, Mexico, Singapore, Philippines, Thailand, Indonesia, Malaysia, Taiwan, Egypt, Morocco, Nigeria, Algeria, Nepal, Iran, Iraq, and Vietnam. Each paper was evaluated by at least three reviewers. Only 57 papers were selected to publish in the conference proceedings.

ICABDE 2021 Organizing Committee

General Chairs

Prof. Su Dinh Thanh, University of Economics Ho Chi Minh City, Vietnam
Dr. Bui Quang Hung, University of Economics Ho Chi Minh City, Vietnam
Assoc. Prof. Nguyen Phong Nguyen, University of Economics Ho Chi Minh City,
Vietnam

Honorary Chairs

Prof. Yu-Dong Zhang, University of Leicester, UK
Prof. Kostas Nikolopoulos, Durham University, UK
Prof. Wilfried Philips, Ghent University, Belgium

Organizing Chairs

Dr. Thai Kim Phung, University of Economics Ho Chi Minh City, Vietnam
M.Sc. Phan Hien, University of Economics Ho Chi Minh City, Vietnam

Program Chairs

Prof. João Manuel R. S. Tavares, University of Porto, Portugal
Prof. Sergey Dvoenko, Tula State University, Russia
Dr. Nguyen Quoc Hung, University of Economics Ho Chi Minh City, Vietnam

Dr. Ngo Tan Vu Khanh, University of Economics Ho Chi Minh City, Vietnam
Assoc. Prof. Bo-Hao Chen, Yuan Ze University, Taiwan

Publication Chairs

Dr. Dang Ngoc Hoang Thanh, University of Economics Ho Chi Minh City, Vietnam
Dr. Bui Thanh Hieu, University of Economics Ho Chi Minh City, Vietnam

Publicity Chairs

M.Sc. Truong Viet Phuong, University of Economics Ho Chi Minh City, Vietnam
M.Sc. Nguyen Manh Tuan, University of Economics Ho Chi Minh City, Vietnam
M.Sc. Bui Xuan Huy, University of Economics Ho Chi Minh City, Vietnam

Secretaries and Finance Managers

Ms. Ngo Mai Thuy Quyen, University of Economics Ho Chi Minh City, Vietnam
M.Sc. Do Thi Bich Le, University of Economics Ho Chi Minh City, Vietnam
M.Sc. Tran Trung Nguyen, University of Economics Ho Chi Minh City, Vietnam
M.Sc. Ho Thi Thanh Tuyen, University of Economics Ho Chi Minh City, Vietnam

Technical Program Committee (TPC)

Assoc. Prof. Alvaro Rodriguez, University of A Coruña, Spain
Dr. Amirreza Mahbod, Medical University of Vienna, Austria
Dr. Brendan Morris, University of Nevada, USA
Assoc. Prof. Chinthaka Premachandra, Shibaura Institute of Technology, Japan
Dr. Cong Wang, Northwestern Polytechnical University, China
Dr. David Van Hamme, Ghent University, Belgium
Prof. Davide Moroni, Italian National Research Council, Italy
Dr. Dawood Al Chanti, Central School of Nantes, France
Assoc. Prof. Diaa Salama Abd Elminaam, Benha University, Egypt
Prof. Diego Reforgiato Recupero, University of Cagliari, Italy
Assoc. Prof. Essam H. Houssein, Minia University, Egypt
Dr. Eugene Shalnov, Lomonosov Moscow State University, Russia
Assoc. Prof. Fu Bo, Liaoning Normal University, China
Dr. Garg Bharat, Thapar University, India

Dr. Ha Kieu-Phung, Hanoi University of Science and Technology, Vietnam
Assoc. Prof. Haidi Ibrahim, Universiti Sains Malaysia (USM), Malaysia
Prof. Hiep Quang Luong, Ghent University, Belgium
Dr. Huynh Van Duc, University of Economics Ho Chi Minh City, Vietnam
Prof. Jia-Li Yin, Fuzhou University, China
Dr. Jinming Duan, University of Birmingham, UK
Dr. Ke Gu, Beijing University of Technology, China
Assoc. Prof. Khamparia Aditya, Babasaheb Bhimrao Ambedkar University, India
Dr. Le Ngoc Thanh, University of Economics Ho Chi Minh City, Vietnam
Dr. Le Zou, Hefei University, China
Dr. Luca Scimeca, University of Cambridge, UK
Prof. Manuel Mazzara, Innopolis University, Russia
Dr. Muhammad Ahmad, National University of Computer and Emerging Sciences, Pakistan
Dr. Ngo Duc Thanh, Vietnam National University of Ho Chi Minh City (UIT), Vietnam
Dr. Ngo Thi Lua, Korea University, South Korea
Dr. Nguyen An Te, University of Economics Ho Chi Minh City, Vietnam
Dr. Nguyen Dang Cao, University of Economics Ho Chi Minh City, Vietnam
Dr. Nguyen Gia Tri, FPT University, Vietnam
Dr. Nguyen Hoang Hai, University of Danang (VKU), Vietnam
Dr. Nguyen Thanh Le Vi, Edith Cowan University, Australia
Dr. Nguyen The Kien, Vietnam National University (UEB), Vietnam
Assoc. Prof. Nguyen Van Huan, Thai Nguyen University, Vietnam
Assoc. Prof. Nitish Pathak, Guru Gobind Singh Indraprastha University, India
Dr. Omar Elharrouss, Qatar University, Qatar
Dr. Omid Mahdi Ebadati E., Kharazmi University, Iran
Dr. Ong Chi Toan, University of Economics Ho Chi Minh City, Vietnam
Assoc. Prof. Pham Van Hai, Hanoi University of Science and Technology, Vietnam
Assoc. Prof. Phung Trung Nghia, Thai Nguyen University, Vietnam
Dr. Prayag Tiwari, Aalto University, Finland
Dr. Ricardo Silva Peres, NOVA University Lisbon, Portugal
Dr. Sachin Kumar, South Ural State University, Russia
Dr. Sema Candemir, Ohio State University, USA
Prof. Sergey Ablameyko, Belarusian State University, Belarus
Dr. Simon Graham, University of Warwick, UK
Dr. Stavroula Mougiakakou, University of Bern, Switzerland
Dr. Sudeb Majee, Umeå University, Sweden
Dr. Surya Prasath, University of Cincinnati, USA
Assoc. Prof. Thoai Nam, Vietnam National University of Ho Chi Minh City (UT), Vietnam
Assoc. Prof. Tran Duc Tan, Phenikaa University, Vietnam
Dr. Tran Hong Thai, University of Economics Ho Chi Minh City, Vietnam
Assoc. Prof. Tran Thi-Thao, Hanoi University of Science and Technology, Vietnam
Prof. Tudor Barbu, Romanian Academy, Romania

Dr. Ugur Erkan, Karamanoğlu Mehmetbey University, Turkey
Dr. Vivek Kumar, University of Cagliary, Italy
Dr. Vu Thi Phuong Lan, University of Economics Ho Chi Minh City, Vietnam
Dr. Xiaodong Jia, University of Cincinnati, USA
Prof. Zhi-Feng Pang, Henan University, China
Assoc. Prof. Zuherman Rustam, University of Indonesia, Indonesia

Preface I

With the theme focusing on Artificial Intelligence, Data and Big Data Processing and Analytics, The 2021 International Conference on "Artificial Intelligence and Big Data in Digital Era" (ICABDE 2021) is a prestigious venue for bringing together scientists, researchers, experts, and practitioners from academia, business, and industry, who are engaged in the different research domains related to Artificial Intelligence (AI), Data, Big Data, and other relevant fields. ICABDE 2021 was organized by the College of Technology and Design, University of Economics Ho Chi Minh City (UEH), Vietnam, during December 18–19, 2021. During two active working days, ICABDE 2021 welcomed about 200 participants, including authors, listeners, and other participants. It proved the great attraction of the conference. The papers contributed to ICABDE 2021 are about the theory and application of AI, Machine Learning, and Deep Learning; Data and Big Data Analytics in economics, industry, agriculture, medicine, and other relevant fields such as Knowledge Discovery, Sentiment Analysis, Image Processing, Computer Vision, and Intelligent Systems.

On behalf of ICABDE 2021 organizing committee, we would like to express our sincere gratitude to all contributors including the authors, the keynote speakers, the chairpersons, the reviewers, and other supporters and participants. The conference cannot succeed without your active support.

Ho Chi Minh City, Vietnam

Dr. Ngoc Hoang Thanh Dang
Publication Chair of ICABDE 2021

Preface II

This conference proceedings volume contains the papers accepted and presented at ICABDE 2021: The 2021 International Conference on Artificial Intelligence and Big Data in Digital Era held on December 18–19, 2021, in Ho Chi Minh City, Vietnam. The conference provided a setting to let authors present and discuss their research results. The papers are divided into eight sessions: one session about Data and Big Data Analytics; two sessions about AI, Machine Learning, and Deep Learning; one session about Image Processing; 2 sessions about Computer Vision; and one session about Signal Processing, Intelligent Systems, and Networking.

Papers accepted for presentation at the conference were passed the strict peer-review rules and evaluated by at least three independent referees. Only 57 papers out of 140 papers were selected to publish in the conference proceedings, equivalent to 40.7.

Also, ICABDE 2021 welcomed two leading scientists in the field of AI as keynote speakers: Prof. Konstantinos Nikolopoulos, Professor in Business Information System, Durham University, UK, with a topic "Long-Term Forecasting for Policymaking with Structured Analogies" and Prof. Yu-Dong Zhang, Professor in Knowledge Discovery and Machine Learning, University of Leicester, UK, with a topic "Artificial Intelligence and Big Data in COVID-19 Diagnosis".

On behalf of ICABDE 2021 organizing committee, we would like to express our sincere gratitude to all authors for their contributions to this book.

Ho Chi Minh City, Vietnam   Dr. Ngoc Hoang Thanh Dang
Leicester, UK   Prof. Yu-Dong Zhang
Porto, Portugal   Prof. João Manuel R. S. Tavares
Taoyuan, Taiwan   Assoc. Prof. Bo-Hao Chen

Acknowledgments

ICABDE 2021 organizer would like to acknowledge the great work of the conference committees. On behalf of the organizer, the editors, and the authors, we would like to thank keynote speakers for their interesting and attractive presentations at the conference. We would like to acknowledge the active support of referees in the peer-review procedure. Especially, we would like to thank the ICABDE 2021 organizing committee for their hard work. We also would like to thank the University of Economics Ho Chi Minh City (UEH), Vietnam, for the financial support to organize the conference. We also would like to thank all the chairpersons and all the members of the conference committees for their great support. Many thanks to all persons who helped and supported this conference. Finally, we would like to thank the Springer publisher for the active support in the guidance and production of this volume.

Machines take me by surprise with great frequency.

Alan Turing

Contents

Editors and Contributors

About the Editors

Dr. Ngoc Hoang Thanh Dang graduated B.Sc. and M.Sc. majoring in Applied Mathematics, and Ph.D. in Computer Science. He is currently Assistant Professor in the Department of Information Technology, UEH College of Technology and Design, University of Economics Ho Chi Minh City (UEH), Vietnam. He has about 100 works on international peer-reviewed journals and conference proceedings, ten chapters, two books, and one European Patent. His research interests include Image Processing, Computer Vision, Machine Learning, Data classification, computational mathematics, fuzzy mathematics, and optimization. He is Member of the scientific organization INSTICC, Portugal, ACM, USA, and IAENG, Taiwan. He is also Member of international conferences committees, such as the IEEE ICCE 2018, Vietnam, IWBBIO, Spain, the IEEE ICIEV, USA, the IEEE ICEEE, Turkey, ICIEE, Japan, ICoCTA, Australia, and ICMTEL, UK. He is now Associate Editor of *The Journal of Engineering* (IET and Wiley), and he also served as Guest Editor of the *Current Medical Imaging journal, Frontiers in Applied Mathematics and Statistics.*

Prof. Yu-Dong Zhang serves as Professor in the Department of Informatics, University of Leicester, UK. His research interests include Deep Learning and medical image analysis. Prof. Zhang is Fellow of IET (FIET) and Senior Member of IEEE and ACM. He was included in "Most Cited Chinese Researchers (Computer Science)" by Elsevier from 2014 to 2018. He was 2019 Recipient of "Highly Cited Researcher" by Web of Science. He won "Emerald Citation of Excellence 2017" and "MDPI Top 10 Most Cited Papers 2015". He was included in "Top Scientist" in Guide2Research. He is Author of over 200 peer-reviewed articles, including more than 30 "ESI Highly Cited Papers" and two "ESI Hot Papers". His citation reached 17,158 in Google Scholar and 10,183 in Web of Science. He has conducted many successful industrial projects and academic grants from NSFC, NIH, Royal Society, EPSRC, MRC, and British Council.

Prof. João Manuel R. S. Tavares graduated in Mechanical Engineering at the Universidade do Porto, Portugal, in 1992. He also earned his M.Sc. degree and Ph.D. degree in Electrical and Computer Engineering from the Universidade do Porto in 1995 and 2001 and attained his Habilitation in Mechanical Engineering in 2015. He is Senior Researcher at the Instituto de Ciência e Inovação em Engenharia Mecânica e Engenharia Industrial (INEGI) and Associate Professor at the Department of Mechanical Engineering (DEMec) of the Faculdade de Engenharia da Universidade do Porto (FEUP). João Tavares is Co-editor of more than 60 books, Co-author of more than 50 chapters, 650 articles in international and national journals and conferences, and three international and three national patents. He has been Committee Member of several international and national journals and conferences, is Co-founder and Co-editor of the book series "Lecture Notes in Computational Vision and Biomechanics" published by Springer, Founder and Editor-in-Chief of the journal *Computer Methods in Biomechanics and Biomedical Engineering: Imaging & Visualization* published by Taylor & Francis, Editor-in-Chief of the journal *Computer Methods in Biomechanics and Biomedical Engineering* published by Taylor & Francis, and Co-founder and Co-chair of the international conference series: CompIMAGE, ECCOMAS VipIMAGE, ICCEBS, and BioDental. Additionally, he has been (co-)Supervisor of several M.Sc. and Ph.D. theses and Supervisor of several postdoc projects, and has participated in many scientific projects both as Researcher and as Scientific Coordinator. His main research areas include computational vision, medical imaging, computational mechanics, scientific visualization, human–computer interaction, and new product development.

Dr. Bo-Hao Chen (Member, IEEE) received the Ph.D. degree in Electronic Engineering from the National Taipei University of Technology, Taipei, Taiwan, in 2014. He is currently Associate Professor in the Department of Computer Science and Engineering, Yuan Ze University, Taoyuan, Taiwan. His current research interests are in Computer Vision, closely integrated with Deep Learning, autonomous vehicles, knowledge representation/reasoning, and human–computer interaction. Dr. Chen was Recipient for involvement in research with the Department of Cybernetics, Tula State University, Tula, Russia, in 2014. He received the Best Student Paper Award from the IEEE International Symposium on Multimedia in 2013, the Best Paper Award from the ACM International Conference on Big Data and Advanced Wireless Technologies in 2016, the First Paper Award from the IEEE International Conference on Applied System Innovation in 2017, the Young Scholar Research Award from the Yuan Ze University in 2017, the Best Ph.D. Dissertation Award from the IEEE Taipei Section and the Taiwan Institute of Electrical and Electronic Engineering in 2014, and the Gold Medal Award, the Bronze Medal Award, and the Silver Medal Award at the Taiwan Innotech Expo Invention Contest in 2019, 2020, and 2021, respectively. He has been serving as Associate Editor for the *Journal of Engineering* (IET) since 2021. He has also been on the Editorial Board of *Mathematics* (MDPI) and *Foundations* (MDPI) since 2020.

Contributors

Mohd Zaid Abdullah School of Electrical and Electronic Engineering, Engineering Campus, Universiti Sains Malaysia, Nibong Tebal, Penang, Malaysia

Adeel Ahmad University of Littoral Côte d'Opale, UR 4491, LISIC, Laboratoire d'Informatique Signal et Image de la Côte d'Opale, Calais, France

Kamal Al-Barznji Department of Computer Science, University of Raparin, Ranya, Kurdistan Region, Iraq

Aakash Bhandari Deerwalk Institute of Technology, Tribhuvan University (TU), Kirtipur, Nepal

Mikhail Bogatyrev Tula State University, Tula, Russia

Mourad Bouneffa University of Littoral Côte d'Opale, UR 4491, LISIC, Laboratoire d'Informatique Signal et Image de la Côte d'Opale, Calais, France

Gregory Bourguin University of Littoral Côte d'Opale, UR 4491, LISIC, Laboratoire d'Informatique Signal et Image de la Côte d'Opale, Calais, France

Ngoc-Mai Bui University of Information Technology, Ho Chi Minh City, Vietnam; Vietnam National University, Ho Chi Minh City, Vietnam

Nguyen Tan Cam Hoa Sen University, Ho Chi Minh City, Vietnam; University of Information Technology, Ho Chi Minh City, Vietnam; Vietnam National University, Ho Chi Minh City, Vietnam

Qian-qian Chen School of Artificial Intelligence and Big Data, Hefei University, Hefei, Anhui, China

Tianzhen Chen Dalian Neusoft University of Information, Dalian, China; Neusoft Institute of Modern Industry, Meizhouwan Vocational Technology College, Putian, China

Dongfei Cui Northeast Electric Power University, Jilin, China

Minh-Thu Dao Faculty of Information Technology, VNU-University of Engineering and Technology, Hanoi, Vietnam

Thi-Kien Dao Fujian Provincial Key Lab of Big Data Mining and Applications, Fujian University of Technology Fujian, Fujian, China; School of Computer Science and Mathematics, Fujian University of Technology, Fuzhou, China

Viet-Hang Dao Institute of Gastroenterology and Hepatology, Hanoi Medical University Hospital, Hanoi, Vietnam

Nguyen Tan Dat Vietnam National University, Ho Chi Minh City, Vietnam; University of Technology, Ho Chi Minh City, Vietnam; OLLI Technology JSC, Ho Chi Minh City, Vietnam

Dinh Khanh Nguyen Diep FPT University Can Tho, Can Tho, Vietnam

Ze-sheng Ding School of Artificial Intelligence and Big Data, Hefei University, Hefei, Anhui, China

Dien Dinh University of Science-VNU-HCM, Ho Chi Minh City, Vietnam; Faculty of Information Technology, University of Science, Ho Chi Minh City, Vietnam; Vietnam National University, Ho Chi Minh City, Vietnam

Hoang Phu Dinh School of Electrical and Electronic Engineering, Hanoi University of Science and Technology, Hanoi, Vietnam

Le Do Thi Department of Electronics and Telecommunications, Dalat University, Lam Dong, Viet Nam

Huong-Giang Doan Control and Automation Faculty, Electric Power University, Hanoi, Vietnam

Duc Do Vietnam National University, Ho Chi Minh City, Vietnam

Ly Anh Do Falculty of Engineering and Technology, Nguyen Tat Thanh University, Ho Chi Minh City, Vietnam

Thao Do University of Bath, Bath, UK

Trong-Hop Do University of Information Technology, Ho Chi Minh City, Vietnam; Vietnam National University, Ho Chi Minh City, Vietnam

Tuong Thanh Do Can Tho University, Can Tho, Vietnam

Van-Chieu Do Haiphong University of Management and Technology, Haiphong, Vietnam

Long Pham Duc Faculty of Computer Science, Phenikaa University, Hanoi, Vietnam

Quang Vu Duc Thai Nguyen University of Education, Thai Nguyen, Vietnam; National Central University, Taoyuan, Taiwan

Nguyen Duc Dung Vietnam National University, Ho Chi Minh City, Vietnam; University of Technology, Ho Chi Minh City, Vietnam

Ninh Duong-Bao College of Computer Science and Electronic Engineering, Hunan University, Changsha, China; Faculty of Mathematics and Informatics, Dalat University, Lam Dong, Viet Nam

Hoang Thanh Duong Tra Vinh University, Tra Vinh, Viet Nam

Thuong Ta Duy Faculty of Computer Science, Phenikaa University, Hanoi, Vietnam

Sergey Dvoenko Tula State University, Tula, Russia

Hamid Reza Ebadati E. Department of Computer Engineering, Islamic Azad University, Tehran, Iran

Omid Mahdi Ebadati E. Department of Mathematics and Computer Science, Kharazmi University, Tehran, Iran

Tamer Z. Emara Faculty of Computer Science and Artificial Intelligence, Pharos University in Alexandria, Alexandria, Egypt

Serdar Enginoğlu Department of Mathematics, Faculty of Arts and Sciences, Çanakkale Onsekiz Mart University, Çanakkale, Turkey

Uğur Erkan Department of Computer Engineering, Faculty of Engineering, Karamanoğlu Mehmetbey University, Karaman, Turkey

Andrei Filin Institute of Applied Mathematics and Computer Science, Tula State University, Tula, Russia

Bharat Garg Thapar Institute of Engineering and Technology, Patiala, Punjab, India

Moncef Garouani University of Littoral Côte d'Opale, UR 4491, LISIC, Laboratoire d'Informatique Signal et Image de la Côte d'Opale, Calais, France;
CCPS Laboratory, ENSAM, University of Hassan II, Casablanca, Morocco;
Study and Research Center for Engineering and Management, HESTIM, Casablanca, Morocco

Inessa Gracheva Institute of Applied Mathematics and Computer Science, Tula State University, Tula, Russia

Zhuang Guanqun Electric Power Research Institute, State Grid Jilin Electric Power Co.Ltd, Changchun, China

Hoang Thi Thanh Ha DaNang University of Economics, The University Of DaNang, DaNang, Vietnam

Zhang Haifeng Electric Power Research Institute, State Grid Jilin Electric Power Co.Ltd, Changchun, China

Mohamed Hamlich CCPS Laboratory, ENSAM, University of Hassan II, Casablanca, Morocco

Rostam Affendi Hamzah Faculty of Electrical and Electronic Engineering Technology, Universiti Teknikal Malaysia Melaka, Melaka, Malaysia

Tri Handhika Centre for Computational Mathematics Studies, Gunadarma University, Depok, Indonesia

Jing He College of Computer Science and Electronic Engineering, Hunan University, Changsha, China

Zhi-huang He School of Artificial Intelligence and Big Data, Hefei University, Hefei, Anhui, China

Nghia Dang Hieu Can Tho University, Can Tho, Vietnam

Nguyen Huu Hoa Can Tho University, Can Tho City, Vietnam

Viet Nguyen Hoang Can Tho University, Can Tho, Vietnam

Viet Tran Hoang Can Tho University, Can Tho, Vietnam

Hanh Hong-Phuc Vo University of Information Technology, Ho Chi Minh City, Vietnam;
Vietnam National University, Ho Chi Minh City, Vietnam

Bui Thanh Hung Faculty of Information Technology, Ton Duc Thang University, Ho Chi Minh City, Vietnam

Phat Nguyen Huu Hanoi University of Science and Technology (HUST), Hanoi, Vietnam

Dinh Thi Hong Huyen Quynhon University, Quynhon, Binhdinh, Vietnam

Haidi Ibrahim School of Electrical and Electronic Engineering, Engineering Campus, Universiti Sains Malaysia, Nibong Tebal, Penang, Malaysia

Yunyun Jiang School of Computer and Information Technology, Liaoning Normal University, Dalian, China

Zhan Jiao Liaoning Vocational College of Light Industry, Dalian, China

Jugal Kalita University of Colorado, Colorado Springs, USA

Nghi Hoang Khoa University of Information Technology, Ho Chi Minh City, Vietnam;
Vietnam National University, Ho Chi Minh City, Vietnam

Andrei Kopylov Institute of Applied Mathematics and Computer Science, Tula State University, Tula, Russia

Vivek Kumar Marie Sklodowska-Curie Researcher, University of Cagliari, Cagliari, Italy

Mikhail Kurbakov Tula State University, Tula, Russia

Chi Qin Lai Peninsula College Malaysia, Kampung Batu Kawan, Penang, Malaysia

Khang Nhut Lam Can Tho University, Can Tho, Vietnam

E. V. Larkin Tula State University, Tula, Russia

Binh Le Faculty of Information Technology, University of Science, Ho Chi Minh City, Vietnam;
Vietnam National University, Ho Chi Minh City, Vietnam

Ngoc-Bich Le School of Biomedical Engineering, International University, Ho Chi Minh City, Vietnam;
Vietnam National University Ho Chi Minh City, Linh Trung Ward, Ho Chi Minh City, Vietnam

Ngoc-Yen Le University of Science-VNU-HCM, Ho Chi Minh City, Vietnam

Thanh-Hai Le Department of Mechatronics, Faculty of Mechanical Engineering, Ho Chi MinhCity University of Technology (HCMUT), Ho Chi Minh City, Vietnam;
Vietnam National University Ho Chi Minh City, Linh Trung Ward, Ho Chi Minh City, Vietnam

Jun Leng Electric Power Research Institute, State Grid Jilin Electric Power Co., Ltd, Changchun, China

Arnaud Lewandowski University of Littoral Côte d'Opale, UR 4491, LISIC, Laboratoire d'Informatique Signal et Image de la Côte d'Opale, Calais, France

Naiyu Liu Electric Power Research Institute, State Grid Jilin Electric Power Co., Ltd, Changchun, China

An-Vinh Luong University of Science, Ho Chi Minh City, Vietnam;
Vietnam National University, Ho Chi Minh City, Vietnam;
Saigon Technology University, Ho Chi Minh City, Vietnam

Thi Hong Thu Ma Tan Trao University, Tuyen Quang, Vietnam

Nurulfajar Abd Manap Faculty of Electronic and Computer Engineering, Universiti Teknikal Malaysia Melaka, Melaka, Malaysia

Samet Memiş Department of Computer Engineering, Faculty of Engineering and Natural Sciences, İstanbul Rumeli University, İstanbul, Turkey

Xiangdong Meng Electric Power Research Institute, State Grid Jilin Electric Power Co., Ltd, Changchun, China

Khoi Tran Minh Can Tho University, Can Tho, Vietnam

Quang Tran Minh Hanoi University of Science and Technology (HUST), Hanoi, Vietnam

Quang Tran Minh Faculty of Computer Science and Engineering, Ho Chi Minh City University of Technology (HCMUT), Ho Chi Minh City, Vietnam;
Vietnam National University Ho Chi Minh City (VNU-HCM), Ho Chi Minh City, Vietnam

Murni Centre for Computational Mathematics Studies, Gunadarma University, Depok, Indonesia

Quoc-Tao Ngo Institute of Information Technology, Academy of Science and Technology, Hanoi, Vietnam

Truong-Giang Ngo Faculty of Computer Science and Engineering, Thuyloi University, Hanoi, Vietnam

Huy Nguyen G. Faculty of Computer Science and Engineering, Ho Chi Minh City University of Technology (HCMUT), Ho Chi Minh City, Vietnam;
Vietnam National University Ho Chi Minh City (VNU-HCM), Ho Chi Minh City, Vietnam

Tu-Anh Nguyen-Hoang University of Information Technology, Ho Chi Minh City, Vietnam;
Vietnam National University, Ho Chi Minh City, Vietnam

Khanh Nguyen-Huu Department of Electronics and Telecommunications, Dalat University, Lam Dong, Viet Nam

Bao Yen Nguyen Thai Nguyen University of Education, Thai Nguyen, Vietnam

Bich-Ngan T. Nguyen Faculty of Information Technology, Ho Chi Minh city University of Food Industry, Ho Chi Minh, Vietnam;
Faculty of Electrical Engineering and Computer Science, VŠB-Technical University of Ostrava, Ostrava, Czech Republic

Binh Nguyen Faculty of Information Technology, University of Science, Ho Chi Minh City, Vietnam;
Vietnam National University, Ho Chi Minh City, Vietnam

Ho-Si-Hung Nguyen University of Science and Technology, The University of Danang, Danang, Vietnam

Hoang Dung Nguyen School of Electrical and Electronic Engineering, Hanoi University of Science and Technology, Hanoi, Vietnam

Huy Hoang Nguyen University of Information Technology, Ho Chi Minh City, Vietnam;
Vietnam National University, Ho Chi Minh City, Vietnam

Huy Quang Duy Nguyen FPT University, Can Tho, Vietnam

Khanh Duy Nguyen Tra Vinh University, Tra Vinh, Viet Nam

Long Nguyen Faculty of Information Technology, University of Science, Ho Chi Minh City, Vietnam;
Vietnam National University, Ho Chi Minh City, Vietnam

Mai Nguyen Thai Nguyen University of Education, Thai Nguyen, Vietnam

Phuong-Thao Nguyen School of Electrical and Electronic Engineering, Hanoi University of Science and Technology, Hanoi, Vietnam

Q. H. Nguyen University of Economics Ho Chi Minh City, Ho Chi Minh, Vietnam

Thai Son Nguyen Tra Vinh University, Tra Vinh, Viet Nam

Thanh Nguyen Faculty of Computer Science and Engineering, Ho Chi Minh City University of Technology (HCMUT), Ho Chi Minh City, Vietnam;
Vietnam National University Ho Chi Minh City (VNU-HCM), Ho Chi Minh City, Vietnam

Thi-Xuan-Huong Nguyen University of Management and Technology Haiphong, Haiphong, Vietnam

Thu Nguyen University of Information Technology, Ho Chi Minh City, Vietnam;
Vietnam National University, Ho Chi Minh City, Vietnam

Thu Hien Nguyen Thai Nguyen University of Education, Thai Nguyen, Vietnam

Trinh-Dong Nguyen University of Information Technology, VNU-HCM, Ho Chi Minh City, Vietnam;
Vietnam National University, Ho Chi Minh City, Vietnam

Trong-The Nguyen University of Information Technology, VNU-HCM, Ho Chi Minh City, Vietnam;
Vietnam National University, Ho Chi Minh City, Vietnam;
School of Computer Science and Mathematics, Fujian University of Technology, Fuzhou, China;
Haiphong University of Management and Technology, Haiphong, Vietnam;
Fujian Provincial Key Lab of Big Data Mining and Applications, Fujian University of Technology Fujian, Fuzhou, China

V. S. Nguyen Military Weapon Institute, Hanoi, Vietnam

Van Thuan Nguyen School of Electrical and Electronic Engineering, Hanoi University of Science and Technology, Hanoi, Vietnam

Van Truong Nguyen Thai Nguyen University of Education, Thai Nguyen, Vietnam

Vinh-Tiep Nguyen University of Information Technology, VNU-HCM, Ho Chi Minh, Vietnam;
Vietnam National University, Ho Chi Minh City, Vietnam

Vinh Dinh Nguyen FPT University Can Tho, Can Tho, Vietnam

A Nguyen Thi Yen Nhi University of Information Technology, Ho Chi Minh City, Vietnam;
Vietnam National University, Ho Chi Minh City, Vietnam

Michel Occello Grenoble Alpes University, Grenoble, France

Dmitry Orlov Tula State University, Tula, Russia

Canh V. Pham ORLab, Faculty of Computer Science, Phenikaa University, Hanoi, Vietnam

Duc-Huy Pham Department of Mechatronics, Faculty of Mechanical Engineering, Ho Chi MinhCity University of Technology (HCMUT), Ho Chi Minh City, Vietnam; Vietnam National University Ho Chi Minh City, Linh Trung Ward, Ho Chi Minh City, Vietnam

H. V. Pham Hanoi University of Science and Technology, Hanoi, Vietnam

Nguyet-Hue Thi Pham Can Tho University, Can Tho, Vietnam

Phuong N. H. Pham Faculty of Information Technology, Ho Chi Minh city University of Food Industry, Ho Chi Minh, Vietnam; Faculty of Electrical Engineering and Computer Science, VŠB-Technical University of Ostrava, Ostrava, Czech Republic

Thi-Thu-Hien Pham School of Biomedical Engineering, International University, Ho Chi Minh City, Vietnam; Vietnam National University Ho Chi Minh City, Linh Trung Ward, Ho Chi Minh City, Vietnam

Van-Hau Pham University of Information Technology, Ho Chi Minh City, Vietnam; Vietnam National University, Ho Chi Minh City, Vietnam

Viet Pham Faculty of Information Technology, University of Science, Ho Chi Minh City, Vietnam; Vietnam National University, Ho Chi Minh City, Vietnam

Anh-Cang Phan Vinh Long University of Technology Education, Vinh Long, Vietnam

Thuong-Cang Phan Can Tho University, Ninh Kieu district, Can Tho, Vietnam

Van Thinh Phan School of Electrical and Electronic Engineering, Hanoi University of Science and Technology, Hanoi, Vietnam

Nguyen Thinh Phu Falculty of Engineering and Technology, Nguyen Tat Thanh University, Ho Chi Minh City, Vietnam

Trang Phung Thai Nguyen University, Thai Nguyen, Vietnam

Jeeya Prakash Thapar Institute of Engineering and Technology, Patiala, India

A. N. Privalov Tula State Lev Tolstoy Pedagogical University, Tula, Russia

Hao Qiankun Northeast Electric Power University, Jilin, China

Loc Duc Quan FPT University, Can Tho, Vietnam

Le Ngoc Quoc Falculty of Engineering and Technology, Nguyen Tat Thanh University, Ho Chi Minh City, Vietnam

Dhananjay Raina Thapar Institute of Engineering and Technology, Patiala, India

Shuo-yi Ran School of Artificial Intelligence and Big Data, Institute of Applied Optimization, Hefei University, Hefei, Anhui, China

Ahmad Sabri Department of Informatics, Gunadarma University, Depok, Indonesia

Oleg Seredin Institute of Applied Mathematics and Computer Science, Tula State University, Tula, Russia

Archit Sethi Thapar Institute of Engineering and Technology, Patiala, Punjab, India

Václav Snášel Faculty of Electrical Engineering and Computer Science, VŠB-Technical University of Ostrava, Ostrava, Czech Republic

Shreyansh Soni Thapar Institute of Engineering and Technology, Patiala, India

Shahrel Azmin Suandi School of Electrical and Electronic Engineering, Engineering Campus, Universiti Sains Malaysia, Nibong Tebal, Penang, Malaysia

Wei Sun Dalian Neusoft University of Information, Dalian, China;
Neusoft Institute of Modern Industry, Meizhouwan Vocational Technology College, Putian, China

Egor Surkov Tula State University, Tula, Russia

Thu-Thuy Ta University of Information Technology, Ho Chi Minh City, Vietnam;
Vietnam National University, Ho Chi Minh City, Vietnam

Nguyen Thai-Nghe Can Tho University, Can Tho City, Vietnam

K. P. Thai School of Business Information Technology, University of Economics, Ho Chi Minh City, Vietnam

Dang N. H. Thanh University of Economics Ho-Chi-Minh-City (UEH), Ho-Chi-Minh-City, Vietnam

Luong Nguyen Thi Faculty of Information Technology, Dalat University, Lam Dong, Viet Nam

Chuong Nguyen Thien Nha Trang University, Nha Trang, Vietnam

Pham Thi Thien Huong University of Economics Ho-Chi-Minh-City (UEH), Ho-Chi-Minh-City, Vietnam

Nguyen Vuong Thinh University of Information Technology, Ho Chi Minh City, Vietnam;
Vietnam National University, Ho Chi Minh City, Vietnam

Thi Hong Thu Ma Tan Trao University, Tuyen Quang, Vietnam

Manh Tran X. Tula State University, Tula, Russia

Anh-Thu Tran University of Science-VNU-HCM, Ho Chi Minh City, Vietnam

Chinh Trung Tran Faculty of Computer Science, Phenikaa University, Hanoi, Vietnam

Dinh-Khoa Tran University of Science and Technology, The University of Danang, Danang, Vietnam

Ho-Dat Tran Vinh Long University of Technology Education, Vinh Long, Vietnam

Huy Q. Tran Falculty of Engineering and Technology, Nguyen Tat Thanh University, Ho Chi Minh City, Vietnam

Loi H. Tran Faculty of Information Technology, Ho Chi Minh city University of Food Industry, Ho Chi Minh, Vietnam

Phuc Hoang Tran FPT University, Can Tho, Vietnam

T. A. H. Tran School of Business Information Technology, University of Economics, Ho Chi Minh City, Vietnam

Thai-Son Tran University of Science-VNU-HCM, Ho Chi Minh City, Vietnam

Thanh-Hai Tran School of Electrical and Electronic Engineering, Hanoi University of Science and Technology, Hanoi, Vietnam

Thi-Hoang-Giang Tran University of Science and Technology, The University of Danang, Danang, Vietnam

Ngo Thanh Tri Can Tho University, Can Tho City, Vietnam

Thanh-Ngoan Trieu Université de Bretagne Occidentale, Brest, France; Can Tho University, Ninh Kieu district, Can Tho, Vietnam

Thanh Trinh Faculty of Computer Science, Phenikaa University, Hanoi, Vietnam; Phenikaa Research and Technology Institute (PRATI), A&A Green Phoenix Group JSC, Hanoi, Vietnam

Thong Duc Trinh FPT University Can Tho, Can Tho, Vietnam

Hieu Le Trung Faculty of Computer Science and Engineering, Ho Chi Minh City University of Technology (HCMUT), Ho Chi Minh City, Vietnam; Vietnam National University Ho Chi Minh City (VNU-HCM), Ho Chi Minh City, Vietnam

Lam Quang Tuong Vietnam National University, Ho Chi Minh City, Vietnam; University of Science, Ho Chi Minh City, Vietnam; OLLI Technology JSC, Ho Chi Minh City, Vietnam

Trung Vu-Thanh College of Computer Science and Electronic Engineering, Hunan University, Changsha, China; VNU School of Interdisciplinary, Vietnam National University, Ha Noi, Viet Nam

Hai Vu School of Electrical and Electronic Engineering, Hanoi University of Science and Technology, Hanoi, Vietnam

Hefei Wang School of Computer and Information Technology, Liaoning Normal University, Dalian, China

Ruizi Wang Liaoning Vocational College of Light Industry, Dalian, China

Xiao-feng Wang School of Artificial Intelligence and Big Data, Hefei University, Hefei, Anhui, China

Yun-sheng Wei School of Energy Materials and Chemical Engineering, Hefei University, Hefei, Anhui, China

Han Wenqi Electric Power Research Institute, State Grid Jilin Electric Power Co.Ltd, Changchun, China

Zhi-ze Wu School of Artificial Intelligence and Big Data, Institute of Applied Optimization, Hefei University, Hefei, Anhui, China

Rana Pratap Yadav Thapar Institute of Engineering and Technology, Patiala, Punjab, India

Zhang Yifu Electric Power Research Institute, State Grid Jilin Electric Power Co.Ltd, Changchun, China

Zhang Yu Electric Power Research Institute, State Grid Jilin Electric Power Co.Ltd, Changchun, China

Madiha Zahari Faculty of Electronic and Computer Engineering, Universiti Teknikal Malaysia Melaka, Melaka, Malaysia

Haifeng Zhang Electric Power Research Institute, State Grid Jilin Electric Power Co., Ltd, Changchun, China

Guanqun Zhuang Electric Power Research Institute, State Grid Jilin Electric Power Co., Ltd, Changchun, China

Le Zou School of Artificial Intelligence and Big Data, Hefei University, Hefei, Anhui, China

AI, Machine Learning, and Deep Learning: Theory and Application

Conceptual Graphs Clustering with Evolutionary Algorithms

Mikhail Bogatyrev and Dmitry Orlov

Abstract Conceptual graphs have been applied as a powerful and computationally affordable model of semantic contents of natural language texts. As for clustering natural language texts, as for clustering conceptual graphs, there exists the problem of semantic interpretation of derived clusters. It is shown that the application of evolutionary algorithms in clustering procedures expands the possibilities of semantic interpretations. It happens due to applying sets of suboptimal solutions in evolutionary algorithms and due to the valid assumption that semantically correct solution of clustering problem is not necessarily associated with the global extreme of fitness function of evolutionary algorithm. Several measures of similarity between conceptual graphs, including conceptual and relational measures were investigated. The proposed technique was tested on the standard Information Retrieval problems of named entity recognition and relations extraction. Experiments were made on the BioNLP texts corpus.

Keywords Conceptual graph · Conceptual clustering · Evolutionary computation · Genetic algorithm

1 Introduction

Clustering is one of the instruments of Natural Language Processing (NLP). It has been applied for solving many Text Mining tasks: finding texts similarities, text rubricating, etc. [1]. The stable tendency in NLP for Text Mining is replacing direct text processing by processing text representation models. This tendency is also actual for clustering due to the following reasons. There are two well-known problems in clustering: the general problem of calculating a similarity between objects to be clustered and the particular NLP problem of semantic interpretation of final clusters. In direct text processing, texts similarity is often calculated by using keywords. This way is valid for many tasks but it has internal restrictions since it could not represent

M. Bogatyrev (✉) · D. Orlov
Tula State University, Tula, Russia
e-mail: okkambo@mail.ru

© The Author(s), under exclusive license to Springer Nature Switzerland AG 2022
N. H. T. Dang et al. (eds.), *Artificial Intelligence in Data and Big Data Processing*,
Lecture Notes on Data Engineering and Communications Technologies 124,
https://doi.org/10.1007/978-3-030-97610-1_1

texts semantics. Replacing text with a model which somehow represents semantics is the way to solve the problem of semantic interpretation of clustering results. Here the problem of calculating texts similarity is replaced by the problem of calculating the similarity between text's model objects. Among models being applied for text representation for clustering, conceptual models play a significant role. Conceptual clustering was established as a special field of clustering and realized by several approaches [2, 3].

Conceptual models realize graph-based semantic representations and the simplest such model is a conceptual graph [4]. A Conceptual graph is a kind of semantic network representing the logical structure of its elements: concepts and conceptual relations. Although conceptual graphs always have been illustrated on texts, their elements, concepts, and conceptual relations may have other than textual origins. This caused the existence of many applications of conceptual graphs and their generalizations in the area of Information Retrieval [5]. Using conceptual graphs in Text Mining is conditioned by a solution to the problem of acquiring conceptual graphs from natural language text. This problem has several solutions [6, 7] but it is still open.

There are several approaches to conceptual graphs clustering [2, 3, 5]. Works [2, 3] are based on the traditional approach, and in the work [5] special kind of similarity between two conceptual graphs is proposed.

The objective of this study is to apply a new approach to the problem of conceptual graphs clustering based on Evolutionary computation. The new evolutionary algorithm for clustering conceptual graphs is proposed. Its main feature is that several solutions to the clustering problem may exist representing various semantic interpretations of clusters. The proposed approach of clustering was tested on the standard Information Retrieval problems of Named Entity Recognition and Relation Extraction. Experiments were made on BioNLP textual corpus.

2 Conceptual Graphs and Their Semantics

Conceptual graphs belong to *conceptual structures*, which are one of the formal representations of knowledge [2]. A conceptual graph is a bipartite directed graph consisting of two types of nodes: concepts and conceptual relations which are connected by binary relations. Figure 1 shows the conceptual graph for the sentence *"Symbiosis is a tremendously creative force in evolution, and bacteria, in particular, are critical partners with eukaryotes."*

This graph was created by our system SemText [8] working on BioNLP textual corpus [9]. It has as standard conceptual relations known from linguistics as semantic roles ("agent", "patient", "attribute", etc.) as specific conceptual relations ("with") which can be acquired from the text when the mode of adding new relations is allowed in SemText. There are two subgraphs on Fig. 1 which are corresponded to two sub sentences in the processed sentence and are marked by the "and" between them.

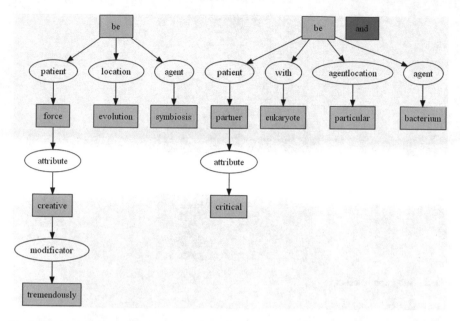

Fig. 1 Conceptual graph for the sentence "Symbiosis is a tremendously creative force in evolution, and bacteria, in particular, are critical partners with eukaryotes"

The semantics of conceptual graph is defined by its concepts, conceptual relations, and patterns of connections between concepts and relations. Concepts may be represented by types and values. If acquiring types is supported in a system producing conceptual graphs then the concept "bacterium" on Fig. 1 may be replaced by the concept "Living organism: bacterium", where "Living organism" is the type and "bacterium" is the value. To support types in conceptual graphs, data from external resources (e.g. ontology) must be processed together with texts [7].

If types of concepts are supported then measuring similarity between conceptual graphs will be organized with external resources such as WordNet. Important elements of the semantics of conceptual graphs are structural patterns in them represented as subgraphs. They are useful for solving fact extraction problems and may be considered as facts. For example in the sentence "*Xylella fastidiosa is a Gram-negative, fastidious, xylem-limited bacterium*" the fact is that bacterium named *Xylella fastidiosa* is Gram-negative. So subgraph shown in Fig. 2 must be treated as one complete object, the fact.

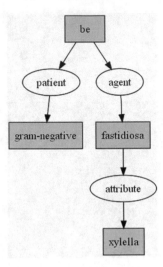

Fig. 2 Subgraph is a fact

In the example in Fig. 2 SemText system has divided the name of bacterium into two concepts: "*xylella*" and "*fastidiosa*". The reason for it is that probably not all bacteria named *Xylella* are fastidious and they have names with another second word. In other cases the pattern [concept 1] → (attribute) → [concept 2] may be treated as one *named* entity.

3 Clustering of Conceptual Graphs

In cluster analysis, conceptual clustering implies the interpretation of clusters as concepts. Each such concept has a certain semantics.

The semantics of conceptual graphs is determined by the semantics of their concepts and the type of relations linking the concepts. In Text Mining, the standard similarity measure of text documents is known as Dice coefficients [10]. There are several variants of Dice coefficients. The general principle of their calculation is to find the ratio of the common elements of the objects under study (sets, texts, or documents) to the total volume of these objects. For conceptual graphs in which there are two types of nodes, this principle is implemented as follows.

For conceptual graphs G_1 and G_2, there are two similarity measures. The conceptual similarity measure is set by the expression

$$s_c = \frac{2n(G_c)}{n(G_1) + n(G_2)} \tag{1}$$

where $G_c = G_1 \cup G_2$, $n(G)$ is the number of conceptual nodes of graph G. The relational similarity measure is calculated as follows:

$$s_r = \frac{2m(G_c)}{m_{G_c}(G_1) + m_{G_c}(G_2)} \tag{2}$$

where $m_{G_c}(G_1)$ is the number of relational nodes of a conceptual graph G_c, $m_{G_c}(G)$ is the number of relational nodes of conceptual graph G, at least one of which belongs to graph G_c.

We use similarity measures of conceptual graphs based on these coefficients. Some improvements for measures (1), (2) had been proposed in [11] and used in the SemText system.

3.1 Evolutionary Approach to Conceptual Graphs Clustering

In all known approaches to conceptual clustering [2, 3, 5] classical clustering algorithms of k—means and hierarchical clustering have been applied. When texts or conceptual graphs are clustered their similarity measure is always very relatively represents semantic or another kind of nearness of clustering objects. It is known that the clustering problem solution will be correct if clustering objects represented by their measure of nearness belong to metric space [12]. Metric is a mapping that assigns the real number $\rho(x, y) \in R$ to each ordered pair of elements of the set M $x, y \in M$ such that the following properties are true.

A metric is positive: $\rho(x, y) \geq 0$, $\forall x, y \in M$, $\rho(x, y) = 0 \Leftrightarrow x = y$, a metric has symmetry property: $\rho(x, y) = \rho(x, y)$, $\forall x, y \in M$ and it meets the demand of triangle axiom: $\rho(x, y) \leq \rho(x, y) + \rho(x, z)$, $\forall x, y, z \in M$.

Unfortunately, conceptual graphs similarity measures may not correspond to these conditions and therefore the solution of the clustering problem may not be correct. Most clustering algorithms work in that they stop processing data when a certain function $f(\rho(x, y))$ defined on the used measure of nearness reaches an extreme. If clustering algorithm works with quasi metric nearness then very often $f(\rho(x, y))$ is multimodal function and algorithm may stop in local extreme instead of stopping in the global one. That means that there may be several different results of clustering corresponding to various local extremes $f(\rho(x, y))$.

Semantic nuances. We always need to interpret the results of clustering. *In Text Mining, if there are several results of clustering objects having certain semantics then these results may be interpreted as semantic nuances.* This is a very interesting hypothesis, which needs to be verified. It is hard to do it with classical clustering algorithms since all of them produce a single solution to the clustering problem.

Evolutionary algorithms are those algorithms that are applied for solving optimization problems with multimodal optimization functions [13]. They also initially use a set (population) of solutions and the final optimal solution may be also presented as a set.

Evolutionary Clustering Algorithm. The evolutionary clustering algorithm can be presented in general form by the following. Let G be a set of objects to be clustered, P is a set of solutions of clustering problem, i.e. each $p \in P$ belongs to the distinct set of clusters. Every clustering solution is characterized by the value of *fitness function* $f(p)$. Then optimal clustering solution p^* is the following:

$$p^* = \arg\max_{p^* \in P} f(p) \tag{3}$$

Since the nature of a set P is complicated it is needed a unified way to represent elements of P to be able to manipulate them. This way is named as *encoding scheme* or *encoding* and it is the mapping $\varphi : P \to S$ where a set S contains objects which encode elements from P i.e. clustering solutions. These objects are named *chromosomes*. Chromosomes consist of *genes*.

Many evolutionary algorithms have been realized as *genetic algorithms* [13]. In the genetic algorithm, every value $p \in P$ is represented as a binary string.

We use non-binary encoding which is actual for clustering. For mapping $\varphi : S \to P$ there necessarily exists an inverse mapping $\varphi^{-1} : S \to S$, so for every $s \in S$ there exists $p \in P$.

For the given encoding scheme, the following evolutionary algorithm solves problem (3) for conceptual graphs.

Algorithm 1: Evolutionary clustering algorithm

Input: the set of n chromosomes as encoded conceptual graphs;
Output: the set of clusters.
1 Randomly generate a population of clustering solutions as chromosomes $S_0 = \{S_k^0\}$ **while** *stopping criterion is false* **do**
2   **for** *all* n **do**
3   evaluate fitness function $f[\varphi^{-1}(s_k)] \to f(S_k)$;
4   *Selection of parent chromosomes: $S_k \to S_{k+1}/2$;*
5   *Crossover on chromosomes: $S_{k+1}/2 \to S_{k+1}$;*
6   *Mutation on chromosomes;*

The evolution of the populations representing several clustering solutions starts with initial population S_0. Then the population evolves to its final state under the action of *selection, crossover,* and *mutation* operators. Evolution finishes by the stopping criterion. The stopping criterion is the immutability of the fitness function values over several steps of evolution or the achievement of the maximum number of steps of evolution.

At each step of evolution, evolutionary operators of *selection, mutation,* and *crossover* are performed on the population, and they are implemented as functions.

The selection operator performs selection of chromosomes using *proportional, random universal, tournament,* and *truncation* selection methods [13] realized in the algorithm.

The mutation operator performs a random change of genes in certain positions of chromosomes, choosing new values from the coding alphabet.

The crossover operator works on pairs of chromosomes and performs a random exchange of code fragments of two chromosomes. There are single and multipoint crossover options.

The action of crossover and mutation operators on a set of chromosomes is equivalent to random walks in the search space for optimal solutions. By controlling the parameters of the algorithm, it is possible to achieve its convergence in the neighborhood of the global extremum of the fitness function. The algorithm finds an approximation to the global extremum regardless of the type of fitness function, which is its important advantage over other global search algorithms.

Encoding chromosomes is a crucial problem in Evolutionary computation and genetic algorithms [13]. It is also actual for evolutionary clustering. For conceptual graphs clustering, we use our chain-encoding scheme [14] which is not redundant as some known schemata [15]. This means that the same clustering variant will not be repeated in several different chromosomes.

4 Experimental Data and Results

The constantly growing number of publications in Bioinformatics has caused the urgency of the tasks of clustering text data. Ananiadou and Artech [16], Zweigenbaum et al. [17]. *Biomedical Natural Language Processing* (BioNLP) [18] is the new area in Bioinformatics where methods of analyzing text data are being investigated and applied, for example, of the texts of articles in the well-known PubMed repository [19].

In our experiments of clustering, we used BioNLP texts corpus [9] consisted of textual descriptions of biotopes of bacteria. Biotope is an area of uniform environmental conditions providing a living place for plants, animals, or any living organism.

There are Bacteria Biotope (BB) tasks on corpus [9] which can be formulated as Text Mining tasks of Named Entity Recognition and Relations Extraction.

The task of Named Entity Recognition has a direct solution with conceptual graphs. As is illustrated above (Fig. 2) conceptual graphs can represent names of bacteria as named entities (concepts).

Clustering conceptual graphs acquired from textual data from corpus [9] was made by applying Dice coefficients (1), (2) in the task of Named Entity Recognition. Similar to the example in Fig. 2 we have found that it is possible to gather in one cluster all variants of the names of bacteria. For example in Fig. 2, it was found that it is only one bacterium of *Xylella* in the corpus: *Xylella fastidiosa*.

Relations Extraction is a more complicated task formulated on the corpus. It is needed to find a habitat of bacteria as the fact of presence relation (connection) between the names of bacteria and subgraphs representing biotopes. This relation is not presented directly as a graph as it takes place for Named Entity Recogni-

M. Bogatyrev and D. Orlov

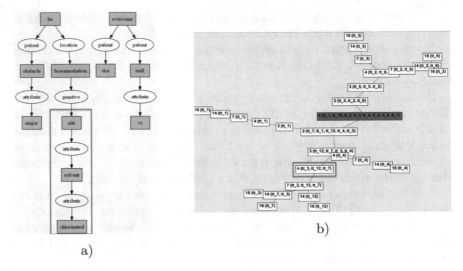

a)

b)

Fig. 3 Example of subgraph denoting biotope (**a**) and the fragment of diagram used for visualizing clustering process (**b**)

tion. Information about the names of bacteria and their biotopes may be in different sentences and so in different conceptual graphs.

We have supposed that using conceptual similarity (2) with our modifications [11] we will have the possibility to gather in clusters subgraphs denoting biotopes. These clusters together with others constitute sub-clusters of some parent cluster which describes bacteria more generally. That hierarchical clustering constitutes the principal solution of the task of Relations Extraction.

Figure 3 illustrates this solution. The subgraph denoting biotope "chlorinated solvent sites" is in the graph corresponding to the sentence "Overcoming the DCE and VC stall is a major obstacle in bioremediation of chlorinated solvent sites" which is shown in Fig. 3a, b shows the fragment of a special diagram of clusters used for visualizing the clustering process. The subgraph "chlorinated solvent sites" is presented in all three conceptual graphs in the cluster $\{tt\_3, tt\_12, tt\_7\}$ shown in Fig. 3b. The parent cluster for this cluster is the cluster $\{tt\_12, tt\_7, tt\_3, tt\_4\}$ containing conceptual graph tt_4 which directly describes bacterium *Dehalococcoides*. The assertion that chlorinated solvent sites is the biotope for *Dehalococcoides* is verisimilar from clustering results and it is true in fact. The evolutionary genetic algorithm for this experiment has the following parameters: numeric system base is 1000, population size is 10, selection type is proportional, one-point crossover, and the absence of mutation. The algorithm was used for clustering 10 sentences, their conceptual graphs are encoded using 10 chromosomes and form 7 clusters shown in Fig. 3b. Using our chromosome encoding scheme, allows us to work with the small-sized chromosome populations.

Using a non-binary and big valued numeric system base is needed for clustering. Also, the small size of the population with the absence of mutation is typical in our task of conceptual graphs clustering.

5 Conclusion and Future Work

In this paper, the solution of two Text Mining tasks of Named Entity Recognition and Relations Extraction is achieved by using clustering of conceptual graphs by the evolutionary algorithm. The application of evolutionary algorithms in clustering procedures expands the possibilities of semantic interpretations of results of clustering. The semantic nuances described above in Sect. 3.1 can be analyzed on clustering variants represented by a population of chromosomes. The classical clustering algorithms produce a single solution, so it is impossible to analyze variants of clustering. We can control decisions at every step of the evolution performed by the genetic algorithm, and the various options obtained on the steps for combining sentences into clusters allow us to structure texts more flexibly.

Future research work is planned in the following directions.

1. Incorporating this technology into a fact extraction system from biomedical texts. In such a system, conceptual graphs are used as semantic models of text data to be queried. The query text is also may be transformed into a conceptual graph, and the fact extraction task is formulated as graph search.
2. Clustering with evolutionary algorithms can be used in the tagging text corpora. Conceptual graphs themselves may serve as tags representing the semantics of sentences. Clusters built on conceptual graphs can serve as thematic tagging.

Acknowledgements The reported study was funded by the Russian Foundation of Basic Research, the research projects No- 19-07-01178, No- 20-07-00055, and RFBR and Tula Region research project No- 19-47-710007.

References

1. Kao A, Poteet SR (eds) (2007) Natural language processing and text Mining. Springer-Verlag, London
2. Mineau G, Godin R (1995) Automatic structuring of knowledge bases by conceptual clustering. IEEE Trans Knowl Data Eng 7:824–828
3. Bournaud I, Ganascia J (1996) Conceptual clustering of complex objects: a generalization space based approach, LNAI, 954. Springer, pp 118-131
4. Sowa J (1999) Conceptual graphs: draft proposed American national standard. In: International Conference on Conceptual Structures ICCS-99, Lecture Notes in Artificial Intelligence 1640, Springer
5. Montes-y-Gomez M, Gelbukh A, Lopez-Lopez M (2002) Text mining at detail level using conceptual graphs. Lect Notes Comput Sci 2393:122–136

6. Boytcheva S, Dobrev P, Angelova G (2001) CGExtract: towards extraction of conceptual graphs from controlled English. Lecture Notes in Computer Science No- 2120, Springer Verlag (2001)
7. Hensman S, Dunnion J (2004) Using linguistic resources to construct conceptual graph representation of texts. Lect Notes Comput Sci 3206:81–88
8. Bogatyrev MY, Vakurin VS (2013) Conceptual modeling in research of biomedical data. Math Biol Bioinformat 8(1):340–349 (2013). Russia
9. Bossy R, Jourde J, Manine A-P, Veber P, Alphonse E, van de Guchte M (2012) BioNLP shared task—the bacteria track. BMC Bioinformatics 13(Supp. 11)
10. Montes-y-Gomez G, Lopez-Lopez, B-Y (2001) Flexible comparison of conceptual graphs. Lecture Notes in Computer Science 2113. Springer-Verlag
11. Bogatyrev MY, Terekhov AP (2009) Framework for evolutionary modelling in text mining—proceedings of the SENSE'09—conceptual structures for extracting natural language semantics. In: Workshop at 17th International Conference on Conceptual Structures (ICCS'09)—Moscow, Russia, pp 26–37
12. Gan G, Ma C, Wu J (2007) Data clustering theory, algorithms, and applications. SIAM, Philadelphia
13. Goldberg DE (1989) Genetic algorithms in search optimization and machine learning. Addison-Wesley, Reading, MA, USA
14. Bogatyrev M, Orlov D, Shestaka T (2021) Multimodal clustering with evolutionary algorithms. Proceeding of the 9th International Workshop What can FCA do for Artificial Intelligence? Co-located with the 30th International Joint Conference on Artificial Intelligence (IJCAI 2021). Montréal, Québec, Canada, pp 71–86
15. Hruschka ER, Campello RJGB, Freitas AA, de Carvalho ACPLF (2009) A survey of evolutionary algorithms for clustering. IEEE Trans Syst Man, Cybernet Part C: Appl Rev 39(2):133–155
16. Ananiadou S, Artech MJ (eds) (2006) Text mining for biology and biomedicine. House Books
17. Zweigenbaum P, Demner-Fushman D, Yu H, Cohen KB (2007) Frontiers of biomedical text mining: current progress. Briefings Bioinformat 8(5):358–375
18. Proceedings of the Workshop on Biomedical Natural Language Processing. 15 September, Hissar, Bulgaria (2011)
19. U.S. National Library of Medicine. http://www.ncbi.nlm.nih.gov/pubmed

An Updated IoU Loss Function for Bounding Box Regression

Thi-Hoang-Giang Tran, Dinh-Khoa Tran, and Ho-Si-Hung Nguyen

Abstract In object detection, bounding box regression (BBR) is a crucial portion that optimizes size of an object and determines the object localization performance. To improve convergence and performance in object detection, many researchers have modified and proposed Intersection over Union (IoU) loss functions. In existing researches, the loss functions have some main drawbacks. Firstly, the IoU-based loss functions are inefficient enough to perform the detection of object in BBR. Secondly, the loss functions ignore the imbalance issues in BBR that the large number of anchor boxes which have small overlaps with the target boxes contribute most to the optimization of BBR. Thirdly, loss functions own redundant parameters which make for extending process. Therefore, to solve these problems, we propose a new method by using an Updated-IoU loss function. Three geometric factors are considered in the proposed function including: (i) the overlap area; (ii) the distances; (iii) the side length. Especially, the Updated-IoU loss function is extremely concentrates on the overlap areas, and predicted object localization to obtain the higher position accuracy performance. In this way, the proposal allows for optimizing and relocating an anchor box in order that the box closes to the ground truth in the training process. The proposal is performed on MS COCO and VOC Pascal dataset. The results are compared to existing IoU models and show that the proposal improves detection ability of object for bounding box regression.

Keywords Updated IoU · Bounding box regression · Object detection

T.-H.-G. Tran · D.-K. Tran · H.-S.-H. Nguyen (✉)
University of Science and Technology, The University of Danang, Danang, Vietnam
e-mail: nhshung@dut.udn.vn

T.-H.-G. Tran
e-mail: tthgiang@dut.udn.vn

D.-K. Tran
e-mail: tdkhoa@dut.udn.vn

© The Author(s), under exclusive license to Springer Nature Switzerland AG 2022
N. H. T. Dang et al. (eds.), *Artificial Intelligence in Data and Big Data Processing*,
Lecture Notes on Data Engineering and Communications Technologies 124,
https://doi.org/10.1007/978-3-030-97610-1_2

13

1 Introduction

Object detection and instance segmentation play fundamental roles in computer vision. In recent years, many real-world applications related to this topic have been proposed. In fact, object detection problems have attracted the attention of many researchers due to its applicability in practice such as visual tracking [13], face detection [14], road scene segmentation [11], etc. Because of the prosperous development of deep learning, many object detection models had been successfully studied. Depending on what additional modules detector needs to create candidate bounding boxes, the popular solutions include three flavors: one-stage [1, 5, 17, 21], two-stage [3, 7, 9, 10], and multi-stage [6] detector. In general, object detection is done in two works including classification and localization. The classification task directly recognizes the object in a given box, while the location task focuses on predicting the precise bounding box of the object.

Two-stage detectors arise a restricted mount of bounding box by using category-independent method. Some typical works of this category include R-CNN [8], Fast-RCNN [7] and etc. On the other hand, one-stage detector takes advantage of the densely pre-defined candidate boxes (anchors). Popular works are known like RefineDet [22], Yolov4 [1] and etc. From pieces of information in AP-loss article [2] point out that one-stage detectors generally computes much faster than two-stage detectors. So that, they could be easily scrutinized about notable holes in the accuracy. One of the main problems leading to decay accuracy is imbalance between the foreground and background region. The background can be seen as the hardness of identification when the number of overwhelm foreground ones. Most of them can be easily detected from foreground objects while few of them are hard to be classified.

To deflate that disadvantage, until now, many solutions were proposed for the purpose to write off the hamper. Zhang et al. proposed a novel single-shot based detector (RefineDet) based on a feed-forward convolutional neural network that produces a fixed number of bounding boxes and the scores indicating the presence difference in classes of those bounding boxes follow by non-maximum suppression to produce the final result. Another state-of-the-art object detection is YOLOv5 [18]. The structure of the model is divided into two parts: The backbone part includes bottleneckCSP, SPP, and other modules. The head part consists of PANet and Detect modules. The Focus modules function is used to reduce the calculation amount of the model, which helps the model in processing image subsampling and does not cause loss of information. This prevents background noise inside the foreground in training process.

In framework of object detection, to evaluate the accuracy of bounding box regression, researchers have given the concept of Intersection over Union (IoU) [20]. IoU is the ratio of the intersection and union of two bounding boxes. Many studies have been proposed IoU models to improve the accuracy in object detection such as IoU+ [19], Generalized IoU (GIoU) [20], Distance IoU (DIoU) [23] and Complete IoU (CIoU) [23] and etc. In our study, we also propose a new IoU, named Updated IoU(UIoU), to focus on the space in the true bounding box and enhance geometric factors of bounding box regression into the inference of deep models for object detection. The UIoU is a promising solution to improve convergence speed and accuracy for bounding box regression.

The paper is organized as follow: after the introduction section, we continue to describe related works and propose an UIoU loss function by taking into account complete geometric factors. Section 3 gives experimental results. Finally, some conclusions and discussions about the obtained results are presented in Sect. 4.

2   Related Work

2.1   The Overlap Area

In recent years, the overlap area in BBR is usually thought of as IoU loss function [19].

$$L_{iou} = 1 - IoU \tag{1}$$

This function faces much trouble in training regression such as weakness of optimization, instability, box regression speed, and so on. Lu et al. [16] had figured out a new highlight formula to fix these issues. The new measurement is defined as:

$$L_{Co} = \gamma * IoU + (1 - \gamma) * Co \tag{2}$$

where γ is a positive trade-off parameter, which determines the weight of intersection over union and intersection over ground true. The formula contributes for an increase in the focusing foreground. In fact, the intersection over union have to be seen more significant than intersection over ground true. From the reasons, in the study of Lu et al. [16],γ should be set in [0.75, 0.8, 0.85] to get a good result of validation period in the training duration . Hence, γ has been defaulted as 0.8. Co is a new generation of the IoU and is defined as follows:

$$Co = \frac{B^{pd} \cap B^{gt}}{B^{gt}} \tag{3}$$

where B^{gt} is the true area and B^{pd} is the predicted area.

2.2   The Distances and the Side Length

The CIoU [24] and Lo-IoU [16] were proposed to optimize the route between the prediction box and the true box. Their weakness that need to be addressed is how decrease the distances between two vertices of two boxes and change the length of

Fig. 1 Schematic diagram
of vertice distance

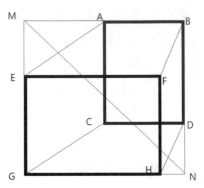

edges of predicted boxes in order to close to the length of true boxes gradually. To
solve this problem, a distance loss function is proposed as below:

$$L_{distances} = \frac{\rho^2(\mathbf{b}_{pl}, \mathbf{b}_{gl}) + \rho^2(\mathbf{b}_{pr}, \mathbf{b}_{gr})}{MN^2} = \frac{AE^2 + DH^2}{MN^2} \tag{4}$$

where $\rho(.) = \| . \|_2$ indicates the Euclidean distance. $\mathbf{b}_{gr}, \mathbf{b}_{gl}$ denote right bottom
corner and left top corner of the true box, while $\mathbf{b}_{pr}\mathbf{b}_{pl}$ denote right bottom corner and
left top corner of the predicted box. This is more clearly that AE denotes the distances
between left top vertex of two boxes, DH denotes the distances between right bottom
vertex of two boxes and MN is the diagonal length of the smallest enclosing box
covering two boxes as Fig. 1. It should be noted that black rectangle is defined the
real box and the red rectangle is the prediction box. If the true box region completely
covers the predicted box region, the $L_{distances} \to 0$. It is a expected condition like
the best result in object detection method. On the other hand, the total parameter of
$L_{distances}$ is not too much. Hence, it not a burden of GPU, as well as not a cause of
time-consuming.

One of the best formulas to degenerate IoU loss function [19] was CIoU[24],
it was a technique to regress scale of width of the true box w^{gt} and width of the
predicted box w toward 1 as well as height of the true box h^{gt} and height of the
predicted box h toward 1. The formula was defined as:

$$v = \frac{4}{\pi^2} \left(arctan \frac{w^{gt}}{h^{gt}} - arctan \frac{w}{h} \right)^2 \tag{5}$$

Value of v is unidentified. Hence, v need be restricted in [0, 1] to fit the condition
of the loss function in object detection method [12]. The trade-off parameter helped
normalize $\mathbf{v}$ to [0, 1], which was defined as:

$$\alpha = \begin{cases} 0, & \text{if } IoU < 0.5; \\ \frac{v}{(1-IoU)+v} & \text{if } IoU \geqslant 0.5; \end{cases} \tag{6}$$

The aspect ratio is a partial improvement of CIoU [24] over DioU [23]. When $IoU \geqslant 0.5$, αv will contribute to the accuracy improvement of loss function. However, it hold back the convergence speed of CIoU [24]. The reduction of convergence speed can be explained by:

$$\frac{\partial v}{w} = -\frac{8}{\pi^2} \left(arctan\frac{w^{gt}}{h^{gt}} - arctan\frac{w}{h} \right) * \frac{h}{w^2 + h^2} \tag{7}$$

$$\frac{\partial v}{h} = \frac{8}{\pi^2} \left(arctan\frac{w^{gt}}{h^{gt}} - arctan\frac{w}{h} \right) * \frac{w}{w^2 + h^2} \tag{8}$$

$$\rightarrow \frac{\partial v}{w} = -\frac{h}{w}\frac{\partial v}{h} \tag{9}$$

Equation 9 indicates that the convergence speed of CIoU loss function [24] is not effective enough. Assuming that if one of two variables (w, h) rise, the other is going to come down such as $w < w^{gt}$ and $h > h^{gt}$ or $w > w^{gt}$ and $h < h^{gt}$. But these two variables must increase or decrease in the same time according to Eq. 9, it points out that the equation is unreasonable.

2.3 The Proposed UIoU Loss Function

In this subsection, an UIoU loss function is proposed to possess the results with higher precision in term of classification and localization. Three geometric factors are considered in the proposed function including the overlap area, the distances and the side length. The formula was defined as:

$$L_{UIoU} = 1 - L_{co} + L_{distance} \tag{10}$$

$$= 1 - \gamma IOU - (1 - \gamma)Co + \frac{AE^2 + DH^2}{MN^2} \tag{11}$$

According to theory provided by Kosub [12], $1 - L_{co}$ is defined as a part of L_{UIoU}. This definition plays the main role of the classification task. By that way, the calculation process is going to increase the rate of union between the ground true area and predicted area. Based on CIoU [24], the quality of loss function can not completely climb up the peak if a shortage of the aspect ratio is occurred. To addressed this problem, the proposed $L_{distance}$ presented in Sect. 2.2 is used to obtain more well performance. $L_{distance}$ not only adjusts the rate between each edge length of the true box and predicted box (toward approximately 1), but also the calculation obtains a considerably improved localization accuracy without a complex operation.

The stability level of L_{UIoU} can be checked by checking it's ingredient. Each of parts must follow properties: (1) $0 \leqslant 1 - L_{co} \leqslant 1$, (2) $0 \leqslant L_{distance} < 2$. Hence, it confidently ensures that the value of L_{UIoU} always limits in range of [0, 3). So that, this evidence points out the good stability of the formula. L_{UIoU} significantly reduces risks of overwhelmed computation.

3   Experimental Results

In this section, the experimental datasets are two popular benchmarks being MS COCO [15] and PASCAL VOC dataset [4]. PASCAL VOC dataset has 20 classes. We use VOC2012 containing 11,530 for train/val. MS COCO is large image recognition/classification, object detection, segmentation, and captioning dataset. We use the COCO train-2017 split (115 k images) for training and report the ablation studies on the 2017 split(5 k images). The dataset has 80 different object categories with 5 captions per image.

3.1   Base Model

YOLOv5: YOLOv5 is selected for the based model due to its highlighting quality. Model is trained with default parameters for three stages Train > Val > Test in turn. We choose version S of YOLOv5 due to the simple structure of the model, little BottleneckCSP will decrease time-consuming in the training process. Then, we get the results after 12 iterations, 0.6 confidence score for VOC PASCAL dataset. By that way, the prediction ability of different IoU models is resulted in Figs. 2 and 3.

In Figs. 2 and 3, there are significantly different in the prediction ability of UIoU and others. Considering the results in scenario of high density objects, the UIoU has an advantage over others in term of both IoU score and how many objects were detected. In addition, the pictures above clearly show that the UIoU keep the wrong predicted objects to minimum.

Table 1 presents obtained results when the proposed UIoU is compared to the existing models in object detection. The primary baseline of these determinants had been applied on the two validation datasets of MS COCO and PAS-CAL VOC. The

(a) IoU   (b) CIoU   (c) UIoU

Fig. 2 The prediction ability of different IoU models with threshold score of mAP being 0.7

(a) IoU   (b) CIoU   (c) UIoU

Fig. 3 The prediction ability of different IoU models with threshold score of mAP being 0.1

Table 1 Mean average precision(mAP) of different IoU models

$Method$	$mAP_{0.5}(VOC)$	$mAP_{0.5:0.95}(VOC)$	$mAP_{0.5}(COCO)$	$mAP_{0.5:0.95}(COCO)$
IoU	0.78031	0.50398	0.53934	0.34755
GIoU	0.77705	0.50671	0.53934	0.34755
DIoU	0.77744	0.50297	0.54037	0.34844
CIoU	0.78301	0.50892	0.53823	0.34716
UIoU(our)	0.78699	0.514	0.54047	0.34993

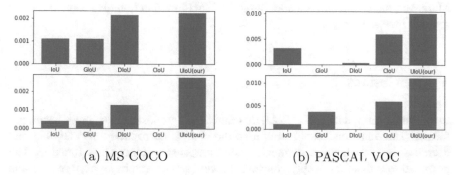

(a) MS COCO   (b) PASCAL VOC

Fig. 4 The change in average precision (mAP) of each method in tests with the MS COCO and the PASCAL VOC datasets

experimental results implemented on the validation dataset of PAS-CAL and VOC prove that the proposed UIoU loss is better accuracy than the others as shown in Table 1. The performance is significantly improved for object categories with a large ratio of background to object pixel like Fig. 2.

From Table 1, the minimum value on each column is always determined. Based on that, the deviations between other values and the minimum value on the same column are determined and shown by bar graphs as Fig. 4.

Figure 4 shows the mAP score of IoU models at the final iteration. The UIoU proves its extraordinary ability when its performance is more outstanding than other IoU models. Hence, the UIoU substantially improves the accuracy after a permanent training time.

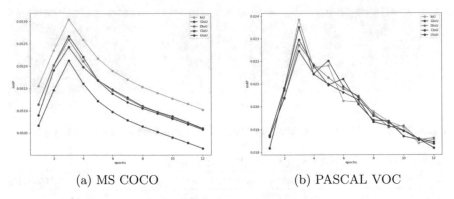

(a) MS COCO   (b) PASCAL VOC

Fig. 5 The classification loss score implemented over the MS COCO and PASCAL VOC dataset

Figure 5 shows that the ample classification ability of the UIoU completely pass over others. This is obviously that the contribution of the UIoU in the base model does faster convergence with non-convex optimization in a large amount of dataset. Moreover, the ample classification ability of IoU loss function also force the mAP score to increase sharply.

4   Conclusions

In this paper, an Update-IoU (UIoU) loss function is proposed to enhances geometric factors. This function helps to improve the computing capacity of YOLOv5 and decrease the classification so much when compared to others loss function. The proposed loss function deeply concentrates on overlap area between true area and predicted area to precisely identify foreground. On the other hand, the UIoU loss function also considers the others geometric factors such as the distances and the side length. As shown in experimental results, the UIoU loss function is able to improve speed replacement of wrong label to get right labels in training period when compared to existing models in computing the distance. In the other words, the predicted box reaches to the true box more quickly than other methods. Besides, the results also confirm that the proposed UIoU loss notably contribute to object detection in term of accuracy. In the further works, we will use this proposal in some realistic applications such as face recognition, object tracking, passenger tracking etc.

Acknowledgements This work was supported by The University of Danang, University of Science and Technology, code number of Project: T2020-02-38.

References

1. Bochkovskiy A, Wang CY, Liao HYM (2020) Yolov4: optimal speed and accuracy of object detection. arXiv preprint arXiv:2004.10934 (2020)
2. Chen K, Lin W, See J, Wang J, Zou J et al (2020) Ap-loss for accurate one-stage object detection. IEEE Trans Pattern Anal Mach Intell (2020)
3. Dai J, Li Y, He K, Sun J (2016) R-fcn: object detection via region-based fully convolutional networks. In: Advances in neural information processing systems, pp 379–387
4. Everingham M, Van Gool L, Williams CKI, Winn J, Zisserman A (2012) The PASCAL Visual Object Classes Challenge 2012 (VOC2012) Results. http://www.pascal-network.org/challenges/VOC/voc2012/workshop/index.html
5. Farhadi A, Redmon J (2018) Yolov3: an incremental improvement. Computer vision and pattern recognition. Springer, Berlin/Heidelberg, Germany, pp 1804–02767
6. Gidaris S, Komodakis N (2015) Object detection via a multi-region and semantic segmentation-aware cnn model. In: Proceedings of the IEEE international conference on computer vision, pp 1134–1142 (2015)
7. Girshick R (2015) Fast r-cnn. In: Proceedings of the IEEE international conference on computer vision, pp 1440–1448 (2015)
8. Girshick R, Donahue J, Darrell T, Malik J (2014) Rich feature hierarchies for accurate object detection and semantic segmentation. In: Proceedings of the IEEE conference on computer vision and pattern recognition, pp 580–587
9. He K, Gkioxari G, Dollár P, Girshick R (2017) Mask r-cnn. In: Proceedings of the IEEE international conference on computer vision, pp 2961–2969
10. He K, Zhang X, Ren S, Sun J (2015) Spatial pyramid pooling in deep convolutional networks for visual recognition. IEEE Trans Pattern Anal Mach Intell 37(9):1904–1916
11. Jung S, Lee J, Gwak D, Choi S, Choo J (2021) Standardized max logits: a simple yet effective approach for identifying unexpected road obstacles in urban-scene segmentation. ArXiv preprint arXiv:2107.11264 (2021)
12. Kosub S (2019) A note on the triangle inequality for the Jaccard distance. Pattern Recogn Lett 120:36–38
13. Li W, Xiong Y, Yang S, Deng S, Xia W (2020) Smot: single-shot multi object tracking. ArXiv preprint arXiv:2010.16031
14. Li X, Lai S, Qian X (2021) Dbcface: towards pure convolutional neural network face detection. IEEE Trans Circ Syst Video Technol
15. Lin TY, Maire M, Belongie S, Hays J, Perona P, Ramanan D, Dollár P, Zitnick CL (2014) Microsoft coco: common objects in context. In: European conference on computer vision. Springer, pp 740–755
16. Lu Z, Liao J, Lv J, Chen F (2021) Relocation with coverage and intersection over union loss for target matching. In: VISIGRAPP (4: VISAPP), pp 253–260
17. Redmon J, Farhadi A (2017) Yolo9000: better, faster, stronger. In: Proceedings of the IEEE conference on computer vision and pattern recognition, pp 7263–7271
18. Wang L, Zhang H, Yang T, Zhang J, Cui Z, Zhu N, Liu Y, Zuo Y (2021) Optimized detection method for Siberian crane (Grus leucogeranus) based on yolov5. Tech. rep, EasyChair
19. Yu J, Jiang Y, Wang Z, Cao Z, Huang T (2016) Unitbox: an advanced object detection network. In: Proceedings of the 24th ACM international conference on Multimedia, pp 516–520
20. Zhai H, Cheng J, Wang M (2020) Rethink the iou-based loss functions for bounding box regression. In: 2020 IEEE 9th joint international information technology and artificial intelligence conference (ITAIC), vol 9. IEEE, pp 1522–1528
21. Zhang S, Chi C, Yao Y, Lei Z, Li SZ (2020) Bridging the gap between anchor-based and anchor-free detection via adaptive training sample selection. In: Proceedings of the IEEE/CVF conference on computer vision and pattern recognition, pp 9759–9768
22. Zhang S, Wen L, Bian X, Lei Z, Li SZ (2018) Single-shot refinement neural network for object detection. In: Proceedings of the IEEE conference on computer vision and pattern recognition, pp 4203–4212

23. Zheng Z, Wang P, Liu W, Li J, Ye R, Ren D (2020) Distance-iou loss: faster and better learning for bounding box regression. In: Proceedings of the AAAI conference on artificial intelligence, vol 34, pp 12993–13000 (2020)
24. Zheng Z, Wang P, Ren D, Liu W, Ye R, Hu Q, Zuo W (2020) Enhancing geometric factors in model learning and inference for object detection and instance segmentation. ArXiv preprint arXiv:2005.03572

Exhaustive Search for Weighted Ensemble Classifiers to Improve Performance on Imbalanced Dataset

Tri Handhika⊚, Ahmad Sabri⊚, and Murni⊚

Abstract We compare performance of six single classifiers trained on German credit dataset, an imbalanced dataset of 1000 instances with binary-valued dependent variable. To improve the performance, we consider resampling the dataset and ensembling the classifiers. The benchmarks are taken from the best performance among six considered classifiers. Resampling the dataset includes oversampling and undersampling. The performance of ensemble classifiers are then analyzed and examined. The experimental results provide three benchmarks, i.e. SVM trained on plain dataset, NB trained on plain dataset, and SVM trained on undersampled dataset. Furthermore, ensemble of kNN, LDA and SVM outperforms the first benchmark for all metrics used in this research, i.e. recall 92.71%, precision 79.14%, F1 84.73%, AUC 79.96%, and accuracy 76.88%. The ensemble of LR, SVM and NB and the ensemble of LDA, SVM, and NB outperforms the second and third benchmark, respectively.

Keywords Ensemble classifiers · Imbalanced dataset · Resampling

This document is the results of the research project funded by Direktorat Riset dan Pengabdian Masyarakat Deputi Bidang Penguatan Riset dan Pengembangan Kementerian Riset dan Teknologi/Badan Riset dan Inovasi Nasional (Grant No. 234/SP2H/LT/DRPM/2021).

T. Handhika (✉) · Murni
Centre for Computational Mathematics Studies, Gunadarma University, Depok 16424, Indonesia
e-mail: trihandika@staff.gunadarma.ac.id

Murni
e-mail: murnipskm@staff.gunadarma.ac.id

A. Sabri
Department of Informatics, Gunadarma University, Depok 16424, Indonesia
e-mail: sabri@staff.gunadarma.ac.id

1 Introduction

Classifiers trained on an imbalanced dataset might have low performance. Because of unequal distribution of classes in the dataset, the classifiers might not have enough knowledge to identify the minority class. This unequal distribution of the classes happens in dataset such as the people having a certain sickness compared to those who are healthy, the number of non eligible credit applicants compared to those who are eligible applicants, etc.

By oversampling the minority, undersampling the majority, or combination of both oversampling and undersampling, the balanced dataset can be achieved. Resampling is done by creating synthetic instances for the minority, and/or removing certain instances of the majority with appropriate rules or randomly chosen. According to [1], combining oversampling and undersampling gives better accuracy than plain undersampling.

In this research, we apply weighted ensemble voting classifier on an imbalanced dataset and search exhaustively for weight sets that improves performance compared to those of six considered single classifiers. We test this method on the German credit dataset [2], an imbalanced dataset consisting of 1000 instances of credit applicants with 20 recorded features, where the dependent variables are classification whether an applicant is eligible or not.

The benchmarks are taken from the best performance among six single classifiers, that are LR, kNN, LDA, DT, Gaussian NB, and SVM, on the same dataset. The goal is to obtain ensemble models that perform better than those single models in terms of recall, precision, F1 measurement, Area Under Curve (AUC), and accuracy.

The search for weight set that gives maximum performance adopts the generating algorithm given in [3, 4]. It is done by generating a combination of weights, continued by fitting ensemble classifier with those weights to the dataset. This procedure runs recursively until all possible weight sets are considered. We call this procedure as *exhaustive weight generating algorithm*.

2 Literature Review

In terms of classifiers, research [5] revealed that the Support Vector Machines (SVM) model outperforms the Artificial Neural Network (ANN) model in terms of generalization (i.e., giving better performance on the test set). This result was then confirmed in [6], that SVM outperforms Logistic Regression (LR) and ANN on their dataset. In [7] was shown that SVM performs better than LR, Linear Discriminant Analysis (LDA) and k-nearest neighbors (kNN), while [8] shows that ANN is the best among five other classifiers: LDA, LR, kNN [9], Naïve Bayes (NB) and Decision Tree (DT) [10, 11], to predict the probability of credit default.

To improve classifier performance, there are several approaches such as by ensembling several classifiers [12–14], by feature selection [15, 16], or combination of both ensembling and feature selection [14]. In ensembled classifiers, the decisions are combined in some way, typically by weighted or unweighted voting, when classifying new examples [17]. In most cases the ensemble classifiers produce more accurate predictions than the base classifiers [18].

3 Methodology

3.1 General Approach

Our general approach is divided into two steps (see Fig. 1). The first step is to set the benchmarks based on the best metric performance of single classifiers. The second step is to conduct weighted ensemble voting classifier that consists of the best

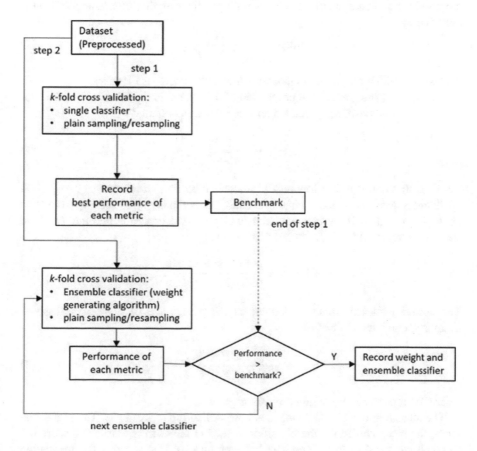

Fig. 1 General scheme

classifiers based on the previous step (in this work, we employ 2 or 3 classifiers). The goal is to obtain ensemble classifiers along with their optimal weight.

3.2 Evaluating the Model

The dataset is preprocessed by applying one-hot encoding to all categorical columns and standardizing all numerical columns. In the dependent column, the labels 1 and 2, each represents "good" and "bad", are transformed to 1 and 0, respectively. In this context, the value 1 is considered as positive case, while 0 as negative case.

Based on such labeling, it follows that the false positives (FP) case, i.e., actual bad applicant predicted as good applicant, is more harmful than the opposite case (false negatives (FN)). This happens more frequent if the classifier is trained on an imbalanced dataset. Thus, beside accuracy, we consider *precision* to measure the proportion of true positives (TP) over all positives predictions. The precision is obtained by:

$$\text{precision} = \frac{TP}{TP + FP}. \tag{1}$$

A classifier with high precision indicates low occurrences of FP cases.

We also consider *recall*, the proportion of TP over actual positives, i.e., the proportion of actual positives predicted as positives. It is obtained by:

$$\text{recall} = \frac{TP}{TP + FN}. \tag{2}$$

Recall is less harmful, but low recall indicates high FN, which means many good applicants classified as bad. This might cause the company loses the opportunity to get higher revenue. To convey the balance between the precision and the recall we can use F1 score which is formulated as follows:

$$F1 = 2 \times \frac{\text{precision} \times \text{recall}}{\text{precision} + \text{recall}}. \tag{3}$$

The accuracy measurement is defined as the percentage of correctly classified instances which is obtained by:

$$\text{accuracy} = \frac{TP + TN}{TP + FP + FN + TN}, \tag{4}$$

where TN represents the number of true negatives.

The measurement of AUC indicates how skilled the classifier is. The higher the score, the more ability for the classifier to predict the correct class. The score 0.5 means it has no skill to differentiate between classes. The score < 0.5 means the classifier predicts against the correct class, i.e., actual 1 predicted as 0 and vice versa.

Due to the imbalancement of the dataset, our investigation involves oversampling the minority and undersampling majority to obtain a more balanced data. The following sampling scheme is applied:

1. Plain sampling, using the original dataset with 1000 instances where the ratio majority: minority = 7:3.
2. Oversampling 100%. This is to achieve 1:1 ratio between majority and minority. By generating synthetically 400 additional minority instances. The resampled dataset contains 1400 instances.
3. Undersampling 100%. This is an alternative way to achieve 1:1 ratio between majority and minority. It is done by reducing the number of majority instances to the same as that of minority. The resampled dataset contains 600 instances.
4. Oversampling 50%. The resampled dataset contains 1050 instances.

We consider six single classifiers as mentioned earlier. Oversampling and undersampling is done by Synthetic Minority Oversampling Technique (SMOTE) [1] and by random technique, respectively.

Evaluation uses repeated stratified 5-fold cross-validation. Performance measured are recall, precision, F1, AUC and accuracy. Oversampling and undersampling are only applied to the folds that contains training data. Figure 2 shows the methods for evaluating classifier.

3.3   The Exhaustive Weight Generating Algorithm

The ensemble voting classifier is expressed by $\sum_{i=1}^{n} w_i C_i$ where: C_i is the binary label in $\{0, 1\}$ predicted by i-th classifier, and w_i is an integer belongs to $\{1, 2, \cdots, n\}$ that represents the weight of i-th classifier. The threshold is $T = \frac{1}{2}\sum_{i=1}^{n} w_i$, which means the predicted label is 1 if $\sum_{i=1}^{n} w_i C_i \geq T$, and 0 otherwise. We denote by $(C_1, C_2, \ldots, C_n)$ the ensemble of n classifiers $C_1, C_2, \cdots, C_n$. The sequence of weights that correspond to $(C_1, C_2, \cdots, C_n)$ is denoted by n-tuple $(w_1, w_2, \ldots, w_n)$ where w_i is the weight of C_i.

The weight sequences are generated exhaustively by recursive generating algorithm called *weight*, as shown in Algorithm 1. The algorithm is adapted from that of [3, 4], and it generates the sequence exactly once for every possible combination. The recursive scheme is preferred rather than using ordinary fixed block loop (e.g. for loop) because of its ability to generate sequences of any length depending to the number of classifiers ensembled, while using fixed block loop needs additional physical loop block for any additional weight.

For convenience, the initial weight is set to $W = (w_1, w_2, \ldots, w_n) = (0, 0, \ldots, 0)$. Every recursive call in level k, weight(k), assigns a weight value for w_k. Recursive calls are initiated by first call weight(1). When $k = n$, a weight sequence of length n is generated. Then the ensemble classifiers with current weight sequence is fitted to

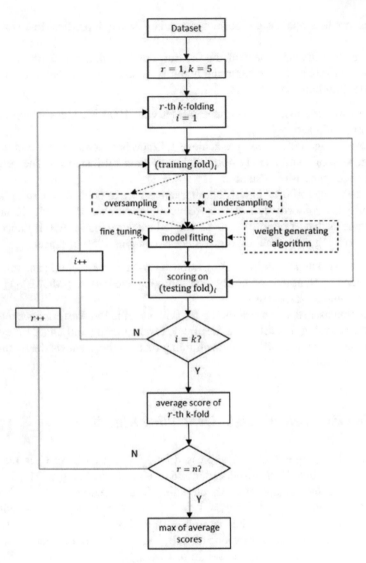

Fig. 2 The flow for evaluating the model

the training data and validated by testing data. This process goes continuously until all n^n possible sets are generated. The weight sets that give best performance are recorded. Figure 3 shows generating tree of the weight sequences for ensemble of 3 classifiers, generated by algorithm weight.

Algorithm 1 Algorithm to construct weighted ensembled classifier.

1: Input: $(C_1, C_2, \cdots, C_n)$
2: Output: $(w_1, w_2, \cdots, w_n)$
3: function weight(k):
4:   if k > length of W then
5:     ensemble_classifier = $w_1 C_1 + w_2 C_2 + \cdots + w_n C_n$
6:     fit ensemble_classifier to the training set
7:     calculate score
8:   else
9:     for $i = 1$ to length of W
10:       $w_k = i$
11:       weight(k+1)

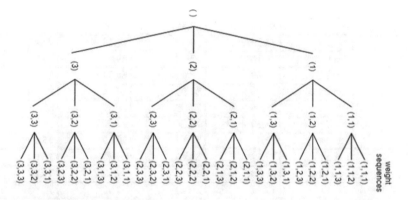

Fig. 3 Generating tree produced by algorithm weight for ensemble of 3 classifiers. The classifiers then is fitted for each generated set of weights

3.4 Selecting Classifiers to Ensemble

We define the following steps to select classifiers to ensemble:

1. Take the best p classifiers on each metric performance.
2. Count the total occurrences of those p classifiers.
3. For a $q \leq p$, select q classifiers with frequent occurences.
4. Construct all possible ensemble of q classifiers using these selected classifiers.

We conjecture that if two classifiers are ensembled, then the performance of the ensembled classifiers are at least the same as the smallest performance among them.

4 Experiment and Results

Table 1 shows the score performance of the six single models in plain sampling, over-sampling 100%, undersampling 100%, and oversampling 50%. In plain sampling,

Table 1 Performance of single models with plain and resampling scheme (the results in bold give the best score)

Model	Metric	Scores (%)			
		Plain	Oversampling 100%	Undersampling 100%	Oversampling 50%
LR	Recall	86.91	73.14	69.69	84.61
	Precision	79.68	84.70	85.51	80.98
	F1	83.11	78.45	76.73	82.71
	AUC	78.73	78.36	77.93	78.54
	Accuracy	75.30	71.92	70.48	75.28
kNN	Recall	91.63	65.91	72.00	85.80
	Precision	76.27	81.89	82.91	78.25
	F1	83.23	72.96	77.00	81.81
	AUC	73.39	70.63	74.13	73.42
	Accuracy	74.16	65.87	69.95	73.32
LDA	Recall	86.69	72.07	69.84	84.57
	Precision	79.96	84.95	85.68	81.00
	F1	83.16	77.93	76.89	82.71
	AUC	78.39	78.34	77.80	78.34
	Accuracy	75.45	71.47	70.69	75.28
DT	Recall	77.11	75.19	62.54	76.20
	Precision	77.20	77.52	79.62	77.28
	F1	77.11	76.28	69.98	76.68
	AUC	61.92	62.11	63.41	61.97
	Accuracy	68.00	67.34	62.53	67.66
NB	Recall	61.91	54.86	63.13	60.44
	Precision	**84.25**	85.12	83.53	84.86
	F1	69.54	65.61	69.99	68.91
	AUC	73.69	72.68	72.92	73.62
	Accuracy	64.98	61.65	64.92	64.59
SVM	Recall	**92.34**	79.83	68.17	88.66
	Precision	77.79	82.03	**86.70**	79.53
	F1	**84.42**	80.87	76.26	83.82
	AUC	**79.36**	77.14	78.89	78.02
	Accuracy	**76.16**	73.60	70.36	76.06

the highest score for recall, F1, AUC, and accuracy is achieved by SVM, and the highest precision is given by NB. Resampling scheme is only superior in precision score, which is given by SVM in undersampling 100% scheme.

Table 2 Benchmark models

Metrics	Benchmark (%)		
	I	II	III
Recall	92.34	61.91	68.17
Precision	77.79	84.25	86.70
F1	84.42	69.54	76.26
AUC	79.36	73.69	78.89
Accuracy	76.16	64.98	70.36

In addition, we observed some regularities on this experiment:

- kNN and SVM give recall score greater than 90% in plain dataset.
- In general, recall scores drops significantly (greater than 10%) in oversampling 100% (except for DT), and get worst in undersampling 100%, except for kNN and NB where the score in undersampling 100% is better than in oversampling 100%.
- In general, precision score is better in undersampling 100% than the other resampling schemes.
- Plain sampling gives better accuracy.

Based on the results in Table 1, we define three benchmarks as shown on the Table 2. Benchmark I is taken from performance of SVM trained on plain dataset, since it gives the best scores of four metrics out of five, benchmark II taken from NB, also trained on plain dataset, since it gives the best precision among all single classifiers scores, and benchmark III is taken from SVM trained on undersampled dataset, since it gives the best precision among all considered classifiers score in both plain and resampling scheme.

4.1 Outperforming Benchmark I

Here we construct ensemble of two or three classifiers. First, we select which classifiers to ensemble by using method described in Sect. 3.4. Referring to performance in Table 1, the best three classifiers in each metric are as follows (starting from the best):

–Recall	:	SVM, kNN, LR
–Precision	:	NB, LR, LDA
–F1	:	SVM, KNN, LDA
–AUC	:	SVM, LR, LDA
–Accuracy	:	SVM, LDA, LR

Next, we count the total occurences of each classifier in all metrics above. Each of SVM, LR, and LDA has four occurences, while kNN has two and NB has one. We decide to discard NB and select SVM, LR, LDA, and kNN to form ensemble of three classifiers. Accordingly, to form ensemble of two classifiers, one can consider the best two classifiers. In this case, SVM has four occurences, kNN and LR have two, LDA and NB have one. We select SVM, kNN, LR, and discard LDA and NB. Ensemble classifiers that go to the experiment are:

–Two classifiers	:	(LR, SVM), (LR, kNN), (kNN, SVM)
–Three classifiers	:	(LR, kNN, SVM), (LR, kNN, LDA)
		(kNN, LDA, SVM), (SVM, LR, LDA)

The experiment results are shown in Table 3. It follows that ensemble (kNN, LDA, SVM) has all five scores above benchmark I. However, those are achieved in different weight sequences. The model (LR, SVM) and (LR, kNN, SVM) almost have the same performance, but they do not outperform benchmark I in recall scores. The results are shown in Table 4. It follows that our considered weighted ensemble voting models outperform benchmark I in all metrics except recall.

4.2 Outperforming Benchmark II

From Table 3, the performance of (LR, SVM) outperforms benchmark II except for the precision. By ensembling NB with (LR, SVM), we expect (LR, SVM, NB) gives better precision than that of NB. On the contrary, (LR, SVM, NB) probably gives lower performance on four other metrics (recall, F1, AUC, accuracy) compared to that of (LR, SVM). This expectation is confirmed by the experiment results in Table 5. The results show that the performance of (LR, SVM, NB) is in between that of NB and of (LR, SVM). For particular weight set, the ensemble of (LR, SVM, NB) with weights (1, 3, 3) outperforms all metrics in benchmark II as shown on Table 6.

4.3 Outperforming Benchmark III

Ensembling SVM with LDA and NB outperforms all the metrics of benchmark III. See the general results given in Table 7. For particular weight set, the ensemble of (LDA, SVM, NB) with weights (1, 1, 3) outperforms all metrics in benchmark III as shown on Table 8.

5 Conclusions

We give a method to determine classifiers to ensemble. It needs further works to have more general insights on this method. The use of weight generating algorithm gives the weight set that maximizes performance of the weighted ensemble voting classifiers. As our experiment shows, this weight set probably different for each metric. The weighted ensemble voting classifier might increase the performance

Table 3 Maximum of each metric performance of weighted ensemble voting classifier consisting of two and three classifiers, along with their weights (the results in bold outperform the corresponding score of SVM plain)

Ensemble model	Metric	Max mean (%)	Standard deviation	Weights
(LR, SVM)	Recall	91.36	0.0275	(1, 2)
	Precision	**79.50**	0.0164	(2, 1)
	F1	**84.62**	0.0145	(1, 2)
	AUC	**80.06**	0.0218	(2, 2)
	Accuracy	**76.77**	0.0209	(1, 2)
(LR, kNN)	Recall	92.21	0.0241	(1, 1)
	Precision	**78.44**	0.0149	(2, 1)
	F1	84.29	0.0147	(1, 1)
	AUC	79.18	0.0206	(2, 1)
	Accuracy	76.10	0.0212	(1, 1)
(kNN, SVM)	Recall	**93.05**	0.0225	(1, 1)
	Precision	77.44	0.0146	(1, 2)
	F1	84.31	0.0147	(1, 2)
	AUC	78.03	0.0232	(1, 2)
	Accuracy	75.90	0.0218	(1, 2)
(LR, kNN, SVM)	Recall	92.29	0.0234	(1, 3, 1)
	Precision	**79.40**	0.0154	(3, 1, 3)
	F1	**84.54**	0.0153	(1, 2, 2)
	AUC	**79.95**	0.0208	(3, 1, 3)
	Accuracy	**76.78**	0.0212	(3, 1, 3)
(LR, kNN, LDA)	Recall	91.90	0.0246	(1, 3, 1)
	Precision	**79.52**	0.0175	(3, 1, 3)
	F1	84.24	0.0146	(1, 2, 1)
	AUC	79.22	0.0205	(3, 2, 3)
	Accuracy	76.05	0.0208	(1, 2, 1)
(kNN, LDA, SVM)	Recall	**92.71**	0.0219	(3, 1, 1)
	Precision	**79.14**	0.0155	(1, 3, 1)
	F1	**84.73**	0.0157	(1, 2, 3)
	AUC	**79.96**	0.0202	(1, 3, 3)
	Accuracy	**76.88**	0.0229	(1, 2, 3)
(SVM, LR, LDA)	Recall	90.52	0.0296	(1, 1, 3)
	Precision	**79.58**	0.0219	(3, 3, 1)
	F1	84.31	0.0183	(1, 1, 2)
	AUC	**80.19**	0.0237	(2, 1, 3)
	Accuracy	**76.50**	0.0268	(1, 1, 2)

Table 4 Performance of ensembled models trained on plain dataset that beat benchmark I in at least 4 metrics (the results in bold outperform their corresponding benchmark)

Ensemble model	Weights	Metric	Mean (%)	Standard deviation	Difference with benchmark I
(LR, SVM)	(1, 2)	Recall	91.36	0.0275	−0.99
		Precision	**78.85**	0.0158	**1.07**
		F1	**84.62**	0.0145	**0.20**
		AUC	**80.04**	0.0225	**0.68**
		Accuracy	**76.77**	0.0209	**0.61**
(LR, kNN, SVM)	(3, 1, 3)	Recall	90.31	0.0281	−2.03
		Precision	**79.40**	0.0154	**1.61**
		F1	**84.47**	0.0151	**0.06**
		AUC	**79.95**	0.0208	**0.59**
		Accuracy	**76.78**	0.0212	**0.62**
(kNN, LDA, SVM)	(1, 2, 3)	Recall	91.67	0.0285	−0.68
		Precision	**78.82**	0.0175	**1.03**
		F1	**84.73**	0.0157	**0.31**
		AUC	**79.90**	0.0208	**0.54**
		Accuracy	**76.88**	0.0229	**0.72**

Table 5 The best performance of (LR, SVM, NB) (all scores outperform their counterparts in benchmark II)

Ensemble model	Weights	Metric	Mean (%)	Standard deviation
(LR, SVM, NB)	(3, 1, 3)	Recall	86.62	0.0341
	(1, 3, 3)	Precision	84.42	0.0332
	(3, 1, 3)	F1	83.48	0.0184
	(3, 1, 3)	AUC	80.00	0.0239
	(3, 1, 3)	Accuracy	76.03	0.0253

Table 6 Ensemble model (LR, SVM, NB) with weights (1, 3, 3) trained in resampled dataset that beats benchmark II.

Ensemble model	Weights	Metric	Mean (%)	Standard deviation	Difference with benchmark II
(LR, SVM, NB)	(1, 3, 3)	Recall	71.74	0.1135	9.82
		Precision	84.42	0.0332	0.17
		F1	76.82	0.0668	7.29
		AUC	78.98	0.0254	5.29
		Accuracy	70.60	0.0537	5.62

Table 7 The best performance of (LDA, SVM, NB) (all scores outperform their counterparts in benchmark III).

Ensemble model	Weights	Metric	Mean (%)	Standard deviation
(LR, SVM, NB)	(3, 1, 3)	Recall	69.45	0.0448
	(1, 1, 3)	Precision	86.77	0.0246
	(3, 1, 3)	F1	76.97	0.0289
	(1, 1, 3)	AUC	79.05	0.0243
	(3, 1, 3)	Accuracy	71.02	0.0292

Table 8 Ensemble model (LDA, SVM, NB) with weights (1, 1, 3) trained in resampled dataset that beats benchmark III

Ensemble model	Weights	Metric	Mean (%)	Standard deviation	Difference with benchmark III
(LDA, SVM, NB)	(1, 1, 3)	Recall	68.95	0.0580	0.78
		Precision	86.77	0.0246	0.07
		F1	76.67	0.0357	0.41
		AUC	79.05	0.0243	0.16
		Accuracy	70.83	0.0338	0.47

compared to that given by single classifier. The experiments also show that the performance of ensemble classifiers are at least the same as the weakest performance of the involved classifier. On the contrary, its best performance might outperform the best performance of the involved classifiers. In general, resampling the German credit dataset by oversampling 100%, undersampling 100%, or oversampling 50%, decrease the performance except for precision. It needs further investigation to find the ratio that maximizes the performance. For future work, the exhaustive weight generating algorithm can also be implemented in other single classifiers for credit scoring proposed by Lessmann et al. [19].

References

1. Chawla NV, Bowyer KW, Hall LO, Kagelmeyer WP (2002) SMOTE: synthetic minority over-sampling technique. J Artif Intell Res 16:321–357
2. Dua D, Graff C (2019) UCI machine learning repository, Irvine, University of California, School of Information and Computer Science. http://archive.ics.uci.edu/ml
3. Sabri A, Vajnovszki V (2015) Two reflected Gray code based orders on some restricted growth sequences. Comput J 58(5):1099–1111
4. Sabri A, Vajnovszki V (2019) On the exhaustive generation for ballot sequences in lexicographic and Gray code order. Pure Math Appl 28(1):109–119
5. Li ST, Shiueb W, Huang MH (2006) The evaluation of consumer loans using support vector machines. Expert Syst Appl 30(4):772–782

6. Khemakhem S, Boujelbene Y (2017) Artificial intelligence for credit risk assessment: artificial neural network and support vector machines. ACRN Oxford J Fin Risk Perspect 6(2):1–17
7. Bellotti T, Crook J (2009) Support vector machines for credit scoring and discovery of significant features. Expert Syst Appl 36(2):3302–3308
8. Yeh IC, Lien CH (2009) The comparisons of data mining techniques for the predictive accuracy of probability of default of credit card clients. Exp Syst Appl 36(2):2473–2480
9. Henley WE, Hand DJ (1996) A k-nearest-neighbour classifier for assessing consumer credit risk. J Royal Statist Soc: Ser D (The Statistician) 45(1):77–95
10. Davis RH, Edelman DB, Gammerman AJ (1992) Machine-learning algorithms for credit-card applications. IMA J Manage Math 4(1):43–51
11. Frydman H, Altman EI, Kao DL (1985) Introducing recursive partitioning for financial classification: the case of financial distress. J Fin 40(1):269–291
12. Handhika T, Fahrurozi A, Zen RIM, Lestari DP, Sari I (2019) Murni: modified average of the base-level models in the hill-climbing bagged ensemble selection algorithm for credit scoring. Proc Comput Sci 157:229–237
13. Hung C, Chen JH (2009) A selective ensemble based on expected probabilities for bankruptcy prediction. Expert Syst Appl 36(3):5297–5303
14. Singh NP, Dahiya S, Handa SS (2015) Credit scoring using ensemble of various classifiers on reduced feature set. Industrija 43(4):163–172
15. Waad B, Ghazi BM, Mohamed L (2013) A three-stage feature selection using quadratic programming for credit scoring. Appl Artif Intell 27(8):721–742
16. Sang HV, Nam NH, Nhan ND (2016) A novel credit scoring prediction model based on feature selection approach. Ind J Sci Technol 9(20):1–6
17. Kuncheva LI (2014) Combining pattern classifiers: methods and algorithms, 2nd edn. Wiley, Hoboken
18. Dietterich TG (1997) Machine-learning research: four current directions. AI Magaz 18(4):97–136
19. Lessmann S, Baesens B, Seow H-V, Thomas LC (2015) Benchmarking state-of-the-art classification algorithms for credit scoring: an update of research. Eur J Operat Res 247:124–136

Automatic Generation of Course Schedules Using Genetic Algorithm

Thai Son Nguyen, Hoang Thanh Duong, and Khanh Duy Nguyen

Abstract Course schedule plays an important role in college and university education because it allows students and lecturers to organize their academic year. It establishes a routine that informs the student of their responsibilities throughout the semester. Creating a well-constructed course schedule takes a long time and a lot of human effort when managers have to put up subjects, classes, lecturers into constrained duration. To solve these issues, we tried to apply a Genetic algorithm in order to make automatically course schedules. In the proposed scheme, both hard and soft constraints are constructed to obtain more accurate results. The experimental results showed that using a Genetic algorithm can take short time to generate a course schedule with high accuracy while maintaining free human errors.

Keywords Genetic algorithm · Course scheduling · Constraints · Chromosomes

1 Introduction

Course schedules can impact a wide range of student outcomes including student satisfaction, graduation rates, and time to graduation. To receive a bachelor's degree from Tra Vinh University (TVU), students must complete eight semesters with 4 years of duration [1]. In each semester, teaching and studying managers arrange a number of subjects and select lecturers for these subjects, which are then assigned to traditional classes to help students know what they must do during the semester. So, course schedules can be seen as study plans of student life. Creating a course schedule will actually be time-consuming and difficult task if there are too many classes in each semester. The planing of course timetable is one of the most complex and stressful jobs due to the fact that works done by hand can lead to variety of errors.

T. S. Nguyen (✉) · H. T. Duong · K. D. Nguyen
Tra Vinh University, Tra Vinh, Viet Nam
e-mail: thaison@tvu.edu.v

H. T. Duong
e-mail: dhthanh@mail.fcu.edu.tw

© The Author(s), under exclusive license to Springer Nature Switzerland AG 2022
N. H. T. Dang et al. (eds.), *Artificial Intelligence in Data and Big Data Processing*,
Lecture Notes on Data Engineering and Communications Technologies 124,
https://doi.org/10.1007/978-3-030-97610-1_4

Nowadays, the advantage of information technology (IT) has been applied at college and university administration. For example, lecturers can organize their class by using Google Classroom or structure a time table by using Google Calender, Microsoft Excel. However, most of these tools or software do not supply automaticaly creating course schedules. The managers must manually arrange lecturers, subjects, and classes into a schedule and the software are solely used to keep track of time, send notifications or sometimes design a timetable. To overcome these problems, Gore et al. [2] used ant colony optimization (ACO) to develop a heuristic algorithm to solve the University timetabling problem. In 2009, the automata-based approximation algorithm we proposed for solving the minimum vertex coloring problem [3]. Runa Ganguli and Siddhartha in 2017 propose solving course timetable scheduling using a graph coloring approach [4], Also in this year, Yazdani et al. [5] developed the mathematical model to solve the problem of source scheduling. He mentioned genetic algorithm concept in his work. The genetic algorithm is the most popularly used for tackling optimization problems such as generating exam timetables or train and bus schedules, etc. In the study of José Joaquim Moreira the author propose to make a system of automatic construction of exam timetable by genetic algorithm [6]. In the study of Chowdhary et al. The researchers mentioned that genetic algorithm is utilized to make timetable system [7]. Al-Jarrah and Al Sawalqah proposed an automated course timetabling system using a genetic algorithm [8]. In 2019, Ansari [9] and coauthors also applied the genetic algorithm concept on course scheduling at Informatic Study Program, Muhammadiyah Banjarmasin University. Although the result is still a conflict on some chromosome, it can be able to generate a lecture schedule.

Genetic algorithm is a meta-heuristic method based on the natural selection process which is commonly used to generate high-quality solutions to optimization and search problems [10]. To save time and to avoid most complex and stressful jobs of planning the course timetable in university, in this paper, the new scheme based on Genetic algorithm is proposed to generate automatically the course schedules. The proposed algorithm aids solving the timetabling problem while giving importance to lecturer's availability. The proposed system took various inputs like the number of subjects, the number of lecturers, the limited number of subjects that could be taught by each lecturer, preference value for each subject given by each lecturer, etc. Then, course scheduling will be created after satisfying basically two types of constraints, soft constraints, and hard constraints to find out the optimal solution. Experimental results showed that the proposed scheme obtained the shortest average time to create a course schedule when population size 10 is used.

In this paper is organized as follows: Sect. 2 described the genetic algorithm. Section 3 explained the proposed approach and worked on the proposed approach. Section 4 illustrated the result and discussion and Sect. 5 described the conclusion and future scope.

2 Genetic Algorithm

In 1962, John Holland [11] proposed genetic algorithm for combinatorial optimization problems by simulating the structure of natural populations. Unlike many heuristic schemes, which have only one optimal solution at any time, the Genetic algorithm maintains many individual solutions in the form of population. Individuals (parents) are chosen from the population and then formed a new individual (child). The child will be mutated to introduce diversity into the population.

Genetic algorithm operators

Chromosome representation. Chromosome, a set of parameters, is considered as a proposed solution to solve the problem by GA. It can be depicted as a simple sequence (example: 11,011,011), the fitness of a chromosome depends upon how well that chromosome solves the problem at hand.

Initial population. In the functioning of a GA, it generates an initial population in the first step. With the initial population, according to the fitness function, each individual is evaluated and assigned a fitness value.

Fitness function. The fitness function evaluates the fitness of each chromosome. The chromosome with the highest level of fitness is closest to the required solution.

Selection. Selection (or reproduction) is used to make more copies of better sequences in a new population. Selection methods rate the fitness of each individual and preferentially select the best solution.

Mutation. A Mutation is doing minor random changes in the chromosome from a population that creates a new chromosome (example: the chromosome 111,111 when a mutation in the third element, the new chromosome will be created is 110111).

Recombination. By mixing more than one chromosome. For example, two chromosomes are chosen and some of their parts are swapped.

A Structure diagram for genetic algorithm

A simple genetic algorithm describes as followings: (Fig. 1)
    Step 1: Randomly generating n chromosomes from the initial population;
    Step 2: Evaluating each individual of the population;
    Step 3: Verifying the termination criteria;
    Step 4: If termination criterion verified—ending the cycle;
    Step 5: Selecting n/2 pairs of chromosomes for a crossover;
    Step 6: Reproducing chromosomes with recombination and mutation;
    Step 7: New population of chromosomes called new generation;
    Step 8: Go to step 2.

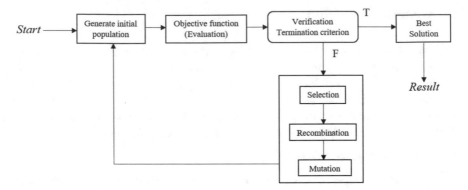

Fig. 1 Basic structure of the genetic algorithm

3 Proposed Scheme

In order to study the computational effort involved in solving the course scheduling problem through GA, the following mathematical model is proposed. The course scheduling problem is defined the following sets to be used in the proposed model: (Fig. 2)

Input Data on course scheduling

GV—Total number of available lecturers.
M—Total number of subjects.
C—Total number of available classes.
TP—An array that gives information about preference values for each subject given by the lecturer.

Constraints on course scheduling

System constraints are divided into hard constraints and soft constraints:

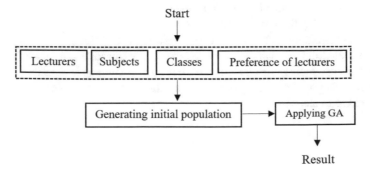

Fig. 2 General observation for automatic course scheduling generator

Hard Constraints. In our scheme, three types of hard constraints have to be satisfied by a valid solution, including: (1) Lecturer must teach subjects that are relevant to their major. (2) Lecturers must teach their favorite subjects. (3) Each lecturer must teach at least one subject during a semester.

Soft Constraints. Such constraints are not much concern. However, they are still taken into contemplation to satisfy that if conditions permit, thereby improving the feasibility of the solution [12]. The soft constraints are following. (1) Summary teaching hours of each lecturer that is suitable with their major. (2) Each lecturer must teach at least three subjects in a semester.

For each university, different soft constraints are evaluated according to their own situation. On the basis of fulfilling the hard conditions, this scheme aims to generate a suitable course schedule with a combination of the soft constraints without serious conflict or error to improve the working efficiency of the lecturer.

Fitness function. The fitness function value of each solution by

$$F(x) = H_1 + H_2 + H_3 + S_1 + S_2 \tag{1}$$

where H_1, H_2, H_3 is penalty value of each hard constraint and S_1, S_2 is penalty value of each soft constraint.

Example Course Scheduling

In the example about course scheduling is used to real data about five lecturers and four classes, their subjects will be held in each class during the semester, population have 100 individuals.

M—Set of all subjects in semester. $M = \{1, 2, 3 \dots n\}$, in this example, $M = \{1, 2, 3 \dots 12\}$.

GV—Set of all lecturers, $GV = \{1, 2, 3 \dots n\}$, in this example $GV = \{1, 2, 3, 4, 5\}$.

C—Set of all classes, $L = \{1, 2, 3 \dots n\}$, in this example $L = \{1, 2, 3, 4\}$. When create population include 100 chromosomes after that take randomly one chromosome.

Where n is the maximum number of subjects, lecturers, or classes (Fig. 3).

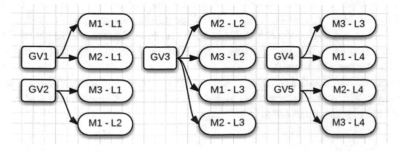

Fig. 3 The structure of a chromosome

Table 1 Result of course scheduling

	L1	L2	L3	L4
M1	GV1	GV2	GV3	GV4
M2	GV1	GV3	GV3	GV5
M3	GV2	GV3	GV4	GV5

When chromosome was created and arranged into table. Table 1 illustrates that lecturer GV1 can teach subjects M1 and M2 in class L1. Similarly, lecturer GV2 can teach subject M3 in class L1 and subject M1 in class L2.

4   Experimental Results

The course scheduling generation code was executed several times to get an optimal result by using genetic algorithm. In the genetic algorithm, the crossover and mutation rates, the population size and the number of iterations both affect the experimental results. The iteration number parameters, namely, $T = 10$, $T = 30$, $T = 50$, $T = 70$ and $T = 100$, are tested in the experiment. The teaching hours of a lecturer are divided into two types: high limit and low limit. The high limit, which is 450, is the maximum number of teaching hours of a lecturer per semester. The low limit, which is 270, is the minimum number of teaching hours of a lecturer per semester.

The inputs of the algorithm are data from the Department of Information Technology in Tra Vinh University that is about 23 lecturers and 123 subjects.

Table 2 gives the information about preference values for each subject given by each lecturer for one particular semester and there are 10 subjects on this semester. Lecturers are represented by GV1 to GV10, whilst subjects are represented by M1 to M10.

For example, let's consider the first column of the reference table described above. It shows the references value given by lecturer GV1 for subjects from M1 to M10. The references value for subjects M1, M3, M4, M6, M8, M9 is 1. So, lecturer GV1 can teach subjects M1, M3, M4, M6, M8, and M9.

Corresponding tests are conducted when the population sizes are $M = 10$, $M = 30$, $M = 50$, $M = 70$ and $M = 100$. The algorithm converges, and the test result is shown in Table 3. It can be seen in Table 3, the fitness value for course scheduling is shortest when $M = 100$, and the time consumption for scheduling is shortest when $M = 10$ as shown in Table 4.

In general, GA showed the relationships of the population size and fitness value are inversely proportional but the relationships between the population size and running time of the algorithm are ratio.

Table 2 Preference table is given by each lecturer for a particular subject

	M1	M2	M3	M4	M5	M6	M7	M8	M9	M10
GV1	1	0	1	1	0	1	0	1	1	0
GV2	0	1	0	0	1	0	1	0	1	0
GV3	0	1	0	1	0	1	1	0	1	0
GV4	0	0	0	1	1	1	0	0	1	1
GV5	1	1	1	0	0	0	1	1	0	0
GV6	0	0	1	1	0	0	1	1	0	0
GV7	1	1	0	0	1	1	0	0	1	1
GV8	0	0	0	1	1	1	0	0	1	1
GV9	1	1	1	0	0	0	1	1	0	0
GV10	1	1	1	1	1	0	0	0	0	0
GV11	0	1	0	1	1	0	1	1	0	1
GV12	0	1	0	1	1	0	1	1	1	1
GV13	0	1	1	1	0	0	1	1	1	1
GV14	0	1	1	0	0	0	0	1	1	1
GV15	0	0	1	0	0	0	0	0	1	0
GV16	0	0	0	0	1	1	0	0	0	0
GV17	0	0	0	0	1	1	0	1	0	0
GV18	1	0	0	1	1	1	0	0	0	0
GV19	1	0	0	1	0	1	1	0	1	1
GV20	0	1	1	0	0	0	1	0	1	1
GV21	0	1	1	0	0	0	1	1	1	1
GV22	0	0	1	1	0	1	1	1	1	1
GV23	0	1	1	1	1	0	1	0	1	0

Table 3 Relationship of population scale and fitness value

Group number	M = 10	M = 30	M = 50	M = 70	M = 100
1	10	6	8	8	7
2	9	6	8	7	7
3	6	8	6	6	5
4	8	7	8	4	5
5	7	5	8	7	5
6	9	5	8	8	7
7	9	8	5	8	7
8	8	6	8	6	4
9	111	9	8	5	4
10	10	5	6	8	5
Average fitness value	18.7	6.5	7.3	6.3	5.6

Table 4 Relationship of population scale and running time

Group number	M = 10	M = 30	M = 50	M = 70	M = 100
1	43	70	100	121	181
2	50	66	113	173	219
3	36	58	117	168	253
4	47	75	100	114	222
5	43	78	88	142	181
6	37	59	90	120	189
7	31	57	88	123	175
8	56	62	141	126	209
9	55	66	115	185	204
10	33	54	173	130	250
Average time	43.1	64.5	112	140.2	208.3

5 Conclusions

According to the actual situation of colleges and universities, the genetic algorithm was applied to generate a suitable course schedule with their data including the lecturers, subjects, and classes. In this paper, applying the genetic algorithm to course scheduling was scientific and feasible. On the basis of experimental results, it can be concluded that:

- The average time to create a course schedule is the shortest when the population size is 10.
- When the population is 100, the average scheduling fitness value is the highest.
- The relationships between the fitness value of scheduling and scheduling time are inversely proportional.

This study discussed the application of the genetic algorithm in our university course scheduling. Course scheduling is a complex combinatorial optimization that will be optimization better in the future by an improved genetic algorithm.

References

1. "Vietnam Government Scholarship" (2021). http://studyinvietnam.edu.vn/detail-university/tra-vinh-university-332.html Accessed 6 Dec 2021
2. Gore P, Sonawane P, Potdar S (2017) Timetable generation using ant colony optimization algorithm. Int J Innov Res Comput Commun Eng 5(3):6033–6039
3. Torkestani JA, Meybodi M (2009) Graph coloring problem based on learning automata. pp 718–722
4. Ganguli R, Roy S (2017) A study on course timetable scheduling using graph coloring approach. Int J Comput Appl Math 12(2):469–485

5. Yazdani M, Naderi B, Zeinali E (2017) Algorithms for university course scheduling problems. Tehnicki vjesnik/Technical Gazette 24
6. Jose JM (2008) A system of automatic construction of exam timetable using genetic algorithms. pp 4200–465
7. Chowdhary A, Kakde P, Dhoke S, Ingle S, Rushiya R, Gawande D (2014) Timetable generation system, pp 5
8. Al-Jarrah MA, Al-Sawalqah AA, Al-Hamdan SF (2017) Developing acourse timetable system for academic departments using genetic algorithm. Jordanian J Comput Info Technol (JJCIT) 3(1)
9. Ansari R, Saubari N (2020) Application of genetic algorithm concept on course scheduling. In: IOP conference series: materials science and engineering, vol 821(1). pp 012043. https://doi.org/10.1088/1757-899x/821/1/012043
10. Mitchell M (1998) In: An introduction to genetic algorithms. MIT Press
11. Holland JH (1992) In: Adaptation in natural and artificial systems
12. Alves SSA, Oliveira SAF, Rocha Neto AR (2015) A novel educational timetabling solution through recursive genetic algorithms. In: Latin America Congress on Computational Intelligence (LA-CCI) 2015. pp 1–6

Integrating Nôm Language Model Into Nôm Optical Character Recognition

Anh-Thu Tran, Ngoc-Yen Le, Dien Dinh, and Thai-Son Tran

Abstract To help preserve Nôm script—an ancient ideographic writing system of Viet Nam, we work on the task of building Nôm Optical Character Recognition (OCR) model for Nôm woodblock-print images. Among top-N results of a character prediction, we pick out a single and more accurate result using Nôm language model (LM). An OCR-LM weighting mechanism is introduced to integrate the language model into the OCR baseline as a post-processing stage. Our OCR baseline's top-1 achieves 98.91% mAP on training dataset *Truyện Kiều* (1871), 71.80% mAP on testing dataset *Truyện Kiều* (1902) and 72.96% mAP on testing dataset *Truyện Lục Vân Tiên*. The integration of the Nôm language model (whose training corpus is limited) helps boost top-1 performance by a small margin: +0.04% on an already high result of the training dataset, +2.72 and +0.14% on the two testing datasets, respectively.

Keywords Nôm OCR · Nôm Language Model · OCR-LM Integration

1 Introduction

Nôm script had been used for centuries in the history of Viet Nam (from the 10th to the 20th century). Since researching Nôm materials is the key to have a better understand about our past, these materials must be preserved at all costs. According to Di san Hán Nôm Việt Nam-thu mục dề yếu, in 1993 [1] the Institute of Sino-Nôm Studies

A.-T. Tran (✉) · N.-Y. Le · D. Dinh · T.-S. Tran
University of Science-VNU-HCM, 72711 Ho Chi Minh City, Vietnam
e-mail: ttathu@apcs.fitus.edu.vn

N.-Y. Le
e-mail: lpnyen@apcs.fitus.edu.vn

D. Dinh
e-mail: ddien@fit.hcmus.edu.vn

T.-S. Tran
e-mail: ttson@fit.hcmus.edu.vn

© The Author(s), under exclusive license to Springer Nature Switzerland AG 2022
N. H. T. Dang et al. (eds.), *Artificial Intelligence in Data and Big Data Processing*,
Lecture Notes on Data Engineering and Communications Technologies 124,
https://doi.org/10.1007/978-3-030-97610-1_5

47

had collected 5,038 books and around 30,000 documents of topics like Literature, History, Religion, Education. Although the number of documents in preservation has been increased [2], there can be more Nôm resources scattered across the country that have yet to be reported.

The fact that many Nôm documents have been severely damaged due to disasters, wars, or natural deformation of papers poses a challenge to Nôm digitization progress. The conventional Nôm digitization usually requires a Nôm knowledgeable workforce re-typing Nôm text on a computer. This task is extremely time- and resource-consuming because it is hard for human to recognize text in badly damaged pages and current Input Method Editor for Nôm has yet to cover wide range of Nôm variants. In recent year, however, with the development of machine learning and deep learning, many advanced Optical Character Recognition (OCR) models have managed to recognize characters of Chinese-Japanese-Korean (CJK) system on which Nôm script was based.

In this research, we first create a Nôm OCR baseline. Then, we further improve the recognition rate by using Nôm language model (RoBERTa's Mask Language Model (MLM) [3]). The characteristic of the OCR results that we are trying to exploit is the time-series nature of language in which a word-level language model can be incorporated to help pick out most linguistically probable results from the top-N outcomes of OCR.

Although language models can be applied to OCR, it is recommended that they should only be adapted to OCR systems that are not fully established [4]. Therefore, incorporating a language model into Nôm OCR system is reasonable as it is built with very limited available annotated imagery data. Compared to [5, 6], our model conducts a different type of incorporation where the Nôm language model acts externally in a post-processing stage to finalize the OCR results. In the OCR model and language model integration stage (shortly referred to as *OCR-LM Integration*), we introduce a weight-learning mechanism to decide whether to accept top-1 results

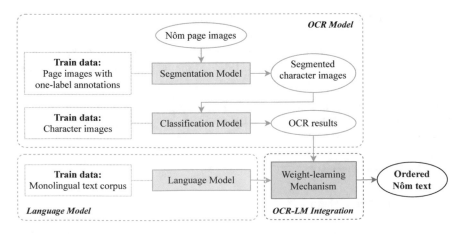

Fig. 1 Proposed model pipeline

offered by the OCR model or top-i results selected by the language model. The overall pipeline of our proposed method is shown in Fig. 1 and the demo video can be accessed at https://youtu.be/1XdP8sxxDxM.

2 Related Works

2.1 Nôm OCR

Phan et al. [7] proposed a segmentation method for Nôm historical documents that uses area Voronoi diagram and recursive X-Y cut method. The method achieved 86.97% recall and 84.60% precision. Their Nôm recognition system proposed in 2016 [8] uses a common approach to recognize Chinese and Japanese characters including a series of nonlinear normalization, feature extraction, coarse classification, and fine classification. The recognition rate of the system on their true Nôm test set averages to 66.92% for top-1 and 78.34% for top-10. However, this is computed to the exclusion of 865 unknown characters. In 2017, Nguyen et al. [9] replace the previous model's classifiers with CNN-based models. The reported result, although of different testing set other than the True Nôm pages, shows a significant improvement in top-1 recognition rate (from 69.08 to 81.73%).

2.2 OCR and Language Model Integration

Tasks like Automatic Speech Recognition and Machine Translation commonly have language model to improve performance for the main system [10, 11]. However, there are several arguments that language model is usually not a performance gain for (Latin) OCR system often due to inadequate integration strategies [4]. In [4], the author uses the Google Book N-gram language model and addresses the integration issue by using a threshold function that decides between OCR and language model results. The issue of using threshold is that the distribution of OCR confidence score can vary across different set of input images.

3 Method

3.1 CNN as an Approach to OCR

OCR problem is an instance of the object detection problem where the target object class is tokens (character, word) and the bounding boxes imply the token-order of the sequence. Therefore, object detection models can also be adapted for OCR purpose in which the returned sequences also have linguistic properties such as semantics and word order for which a sequence modeling can be applied. In our case, the order of writings in Nôm historical documents is right-to-left and top-to-bottom. To create a Nôm OCR baseline, we experiment two common types of object detection models: (1) end-to-end model, using YOLO (version 5, referred to as YOLOv5 [12]); (2) two-stage model, using YOLOv5 for character image segmentation and a CNN-based classifier for character classification, where the classifiers in consideration are AlexNet [13] (a simpler architecture) and ResNet-101 [14] (a more complex architecture).

3.2 Nôm Language Model with RoBERTa's Masked Language Model

Usually for written text, the meaning of a word does not necessarily depend only on its preceding words [15]. Based on this philosophy, Devlin et al. [16] proposed Bidirectional Encoder Representations from Transformers (BERT)—a novel deep bidirectional representation of language by conditioning on the context from both sides of a word. Consequently, BERT-based MLM as a language model does not strictly follow the discrete Markov chain model [17]. However, as pointed out by Wang et al. [18], to use BERT-based MLM as an approximate language model that assigns a score to a sequence, one can choose to compute the pseudo-perplexity (PPL) of the sequence based on the token probabilities modelled by MLM. Inspired by the methods proposed by Salazar et al. [19] and Shin et al. [20], we choose RoBERTa's MLM—an improved version of BERT that allows better masked language modeling [3], to model a sequence:

$$PPL(S) = \exp\left(-\frac{1}{|S|} \sum_{x_i \in S} pLog(x_i)\right) \tag{1}$$

where $pLog(x_i)$ is the log probability of the softmax score of the token x_i with respect to an input sequence S.

3.3 OCR-LM Integration

As illustrated in Fig. 1, OCR-LM Integration uses Nôm language model to pick out one result from top-N OCR outcomes for each segmented character image. A major benefit in using bidirectional character-level language model to post-process the OCR results is that the scoring of the first token of a sentence can rely on its succeeding context.

OCR Results Selection The final hypothesis of the two sequential OCR model and language model scoring follows Bayes's rule, in which the input of the language model depends on the available detection offered by the OCR model. Inspired by the incorporation of a language model into the Automatic Speech Recognition system [21], and machine translation [22, 23], we model our OCR-LM Integration theory as followed:

- Let Y_{OCR} denote a set of sequences of characters proposed by the OCR model. Let Y denote the set of final sequences y (consisting of characters c) that are chosen from Y_{OCR} using a language model-based weighting function $f_W(y_{OCR})$.
- $P(Y|Y_{OCR})$ is a conditional probability that the given y_{OCR} resulting from the OCR stage may have a chance to be selected to become a y with respect to parameter W. While the inverse $P(Y_{OCR}|Y)$ can be interpreted as a function, when force to output a y, would select y_{OCR}.
- The aim of the weighting function is to maximize $P(Y|Y_{OCR})$, so based on Bayes's rule we have:

$$P(Y|Y_{OCR}) \propto P(Y_{OCR}|Y)P(Y) \tag{2}$$

where $P(Y_{OCR}|Y)$ and $P(Y)$ can be maximized separately. In this case, $P(Y_{OCR}|Y)$ is given by the OCR model and $P(Y)$ is given by the language model.
- Motivated by Salazar et al. [19], the multiplication of the two probability terms can be approximated using a log-linear model as follows:

$$\log P(Y|Y_{OCR}) = w_0 + w_1 \log P(Y_{OCR}|Y) + w_2 \log P(Y) \tag{3}$$

Since MLM is a type of fixed-length language model, each time it can only score a fixed-length sequence. Hence, y_{OCR} must be broken into sub-sequences s with respect to a window size r. Given that the OCR result for each document page image consists of a list of M bounding box detection, each detection comes with N possibilities (top-N possible character classes), we denote i as the index to a bounding box detection ($i = 1 \ldots M$) and j as the index to one of its possible classes which are ranked descendingly by class confidence score ($j = 1 \ldots N$). Then, the notation $c_{i,j}$ refers to the character c at top jth of the ith detection.

Let the set of sub-sequences, denoted as S_t, represent the combinations created from the top-N OCR characters of r consecutive detection at sliding time t:

$$S_t = \{s_t | s_t = \{c_{i,j} | i = t \ldots t + r - 1, j = 1 \ldots N \\ \text{and no two characters share the same } i\}\} \tag{4}$$

Then the $\log P(S_t | S_{tOCR})$ can be approximated by the sum of all conditional log-probabilities of every token $c_{i,j}$ within the sequence s_t.

$$\log P(s_t | s_{tOCR}) = \sum_{c_{i,j} \in s_t} \text{pLog}(c_{i,j}) \tag{5}$$

where:

$$\text{pLog}(c_{i,j}) = w_0 + w_1 \log (P_{\text{OCR}}(c_{i,j})) \\ + w_2 \log (P_{\text{MLM}}(c_{i,j} | \forall c \in s_t, c \neq c_{i,j})) \tag{6}$$

w_0 is the bias, w_1 and w_2 are weights that control the decision whether to emphasize OCR result or language model result.

- As we use RoBERTa's MLM to score a sequence s_t, we apply the following formula to calculate its pseudo-perplexity, just like formula (1):

$$\text{PPL}(s_t) = \exp \left(-\frac{1}{|s_t|} \log P(s_t | s_{t\text{OCR}}) \right) \tag{7}$$

with the criteria that: the smaller the value of PPL score of s_t is, the higher the quality of being probable for the s_t and hence, the higher the chance of a token within that sequence to be chosen in the result.

Weight Learning Mechanism Because smaller $\text{PPL}(s_t)$ indicates s_t is a better choice among a set of windowed sub-sequences, our target is to maximize $\log P(s_t | s_{t\text{OCR}})$. Since it is the sum of pseudo-log-likelihood of each character, we need to maximize each term $\text{pLog}(c_{i,j})$. As confidence score returned by OCR model and language model is probability within the range [0..1], its log-probability is negative. So, to maximize $\text{pLog}(c_{i,j})$, it is ideal to suppress one of the weighted terms. Therefore, parameter W should learn how to emphasize one term and suppress the other. While this parameter is usually used as hyperparameter [19], it would fail to generalize for results returned by weak OCR model (trained on low-resource language) since the OCR confidence score distribution can vary greatly depending on different input.

The criteria of the choosing which term to emphasize or suppress are that: (1) since OCR model works directly with the target data, its top-1 prediction should be of first choice; and (2) the sign of a failed OCR detection is a low confidence score in top-1, for this case, the OCR-LM Integration stage must accept whatever the language model selects from top-N candidates.

To achieve this goal for a log-linear model, we use linear regression method whose input X is a 2D matrix where x_{OCR} is $\log (P_{\text{OCR}}(c_{i,j}))$ and x_{MLM} is $\log (P_{\text{MLM}}(c_{i,j} | \forall c \in s_t \text{ and } c \neq c_{i,j}))$. The label Y is the value chosen between x_{MLM} or x_{OCR} using the followings criteria:

- Each time a token $c_{i,j}$ in s_t being scored, we let $y_{c_{i,j}}$ equals to: a) $\log (P_{\text{OCR}}(c_{i,j}))$ if this token $c_{i,j}$ is of top-1 OCR and matches the ground-truth of the bounding box; or b) $\log (P_{\text{MLM}}(c_{i,j} | \forall c \in s_t \text{ and } c \neq c_{i,j}))$ if token $c_{i,j}$ matches the ground-truth but is not top-1.
- We do not derive training data for cases where token $c_{i,j}$ is not top-1 OCR and does not match ground-truth.

To compute the intercept, we let the first column of X to contain all 1's. Then W is calculated using least squares estimation: $\hat{W} = (X^T X)^{-1} X^T Y$. Our proposed weight learning mechanism applies for two scenarios as follows:

1. To improve pre-defined weight w_0.
   In this case, the mechanism initially starts with a pre-defined set of weights w_0. And in the first scoring iteration, the training data for linear regression model is derived. This data is then used to learn w_1 which is next used in the second scoring iteration. Intuitively, we observe that there are many times that the OCR model predicts correctly, so we try to first start with a set of pre-defined weight that emphasizes the OCR term, such as 0.85 for OCR result and 0.15 for language model result.

2. To pre-learn weight w_0 and apply it for just one scoring iteration.
   In contrast to the first case, in this case we try to eliminate the use of pre-defined weight by directly learning w_0 from the given OCR results. The training data for the linear regression model is derived from the process where we score every possible combination given by the OCR results. This pre-learned weight w_0 is then applied to the scoring process as usual. Since the weight is pre-learned from the actual data, it may not need any further iteration for adjustment. However, since all combinations are considered, this would take quite long to compute, which might be an overhead to start the OCR-LM Integration stage.

4 Experiments and Results

4.1 Dataset

Dataset for OCR Model As shown in Table 1, we will refer data created for OCR model from *Truyện Kiều* (1871) [24], *Truyện Kiều* (1902) [25], and *Truyện Lục Vân Tiên* [26] as K1871, K1902, and LVT, respectively. The chosen training and testing datasets are as follows: Training set and parameter tuning is K1871. Testing set 1 is K1902. Testing set 2 is LVT. For segmentation model, we use augmented page-level K1871 and TKH-MTH dataset (version 2) [27] as training dataset for YOLOv5. For classification model, augmented character-level datasets including HWDB1.1 [28], K1871, K1902, and LVT are used to create training and testing datasets. In particular, images in HWDB1.1 and K1871 are used as training sets, while K1902 and LVT are combined to create NomTest as a testing set.

Dataset for Language Model We use ten literary works, with an average sentence length of 7. To comply with the OCR model, we preserve text data of *Truyện Kiều* 1902 and *Truyện Lục Vân Tiên* for test corpus while the eight remaining text data are used as train corpus. The dataset statistics is shown in Table 2. This corpus is considered as small (low-resource) since there are up to 6,133 distinct characters while each character only occurs about 19–20 times on average, and it is estimated that up to one fourth of the vocabulary are 1-frequency characters.

Table 1 Dataset statistics for OCR

	Number of images	Number of sentences	Number of characters	Number of vocabularies
Truyện Kiều (1871)	136	3194	22,808	2841
Truyện Kiều (1902)	163	3665	23,757	2917
Truyện Lục Vân Tiên	105	2060	14,450	2222
Entire dataset	404	8919	61,015	4294

Table 2 Dataset statistics for language model

	Number of sentences	Number of characters	Number of vocabularies	Type
Bài Ca Răn Cờ' Bạc	107	756	409	tr
Bu'ố'm Hoa tân truyện	403	2828	877	tr
Chinh Phụ Ngâm Khúc (Nôm)	412	2885	1098	tr
Gia Huấn Ca	802	5613	1613	tr
Hồ Xuân Huong thi tập	645	4198	1521	tr
Truyện Lục Vân Tiên	2060	14,450	2218	ts
Truyện Kiều (1870)	3255	22,806	2854	tr
Truyện Kiều (1871)	3194	22,808	2838	tr
Truyện Kiều (1902)	3665	23,757	2914	ts
Ngọc Kiều Lê tân truyện	3014	21,041	2743	tr
Entire dataset	17,557	121,142	6133	

tr train, *ts* test

4.2 Results

Nôm OCR Baseline The target of baseline OCR models is to detect Nôm characters on page images, the final validation on Nôm datasets is listing in Table 3. The end-to-end YOLO detector performs poorly since the number of labels is too large (5,568 labels) which is one of the limitations of YOLO itself [29]. When using as one-label segmentation module, YOLO performs exceptionally well. The classifier AlexNet

Table 3 Final top-1 and top-5 validation (mAP@0.5) on Nôm datasets of the three OCR baselines

Model	NomTrain	NomTest	
	K1871	K1902	LVT
End-to-end YOLO	0.0618	0.062	
YOLO-AlexNet	98.47–99.14	59.02–68.76	59.78–70.60
YOLO-ResNet101	98.91–99.32	71.80–80.54	72.96–82.36

uses Stochastic Gradient Descent (SGD) optimizer with a learning rate of $1e\text{-}3$ for the first 10 epochs and reduced to $1e\text{-}4$ for the latter epochs. The AlexNet fails to improve after 16 epochs. Meanwhile, although the ResNet101 classifier takes more epochs with longer time to train, it converges better than AlexNet. In conclusion, YOLO-ResNet101 has better performance than YOLO-Alexnet, therefore, we choose to it for further experiments with language model integration.

The prediction shown in Fig. 2a is an example for the results on the training dataset in which there are some wrong captures at non-existed character areas such as bounding boxes numbered 26 and 50. The character class at the 6th bounding box is 又 , denotes a repetition of the character preceding it. The prediction is correct, however, the font Nom Na Tong we used to write on image is incapable of rendering it. The prediction shown in Fig. 2b, 2c are examples for the results on the two testing datasets. The dataset *Truyện kiều* (1902) contains irregular comment texts in the middle area of the page which affects the sorting order of the OCR outcomes. For the testing dataset *Truyện Lục Vân Tiên*, we pick the page that is unevenly lit to illustrate. The results show that the model manages to predict correctly in such extreme conditions with just some wrong detection such as bounding boxes numbered 132 and 141.

OCR-LM Integration Our purpose in choosing the testing dataset *Truyện Kiều* (1902) used in OCR stage to be the OCR-LM Integration stage's training dataset because its OCR confidence score top-1 and top-N range is representative for unseen data and also its text is quite similar to that of the version 1871 which exists in the text corpus used to train the language model. The configuration for our experiments is explained in details as below:

- *beam size*: The number of beams that are in considered: 1, 2 and 3.
- *sliding window*: The window size r_t of each time step t.

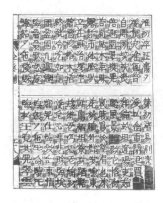  

| (a) *Truyện Kiều* 1871 (page 115) | (b) *Truyện Kiều* 1902 (page 10) | (c) *Truyện Lục Vân Tiên* (page 94) |

Fig. 2 Baseline OCR results

- w_0, w_1, w_2: Components of weight W.
- *weight init*: There are two modes, pre-defined (pd) or pre-learned (pl) W_0.
- *weight combo*: This string value lists the dataset(s) whose generated data is used to derive the next W.
- *accept beam-1 lm*: This boolean value governs when to accept language model's selection as the new top-1 OCR results for next iterations.

There are four experiments namely Experiment-1, Experiment-2, Experiment-3 and Experiment-4, each contains some iterations, denoted with format: *<experiment>.<iteration>* e.g.: *1.1* means Experiment-1 at iteration 1. With multiprocessing, it takes about 4 to 6 h training one iteration of the OCR-LM Integration stage on the two versions of *Truyện Kiều* simultaneously; and about 4 h testing on *Truyện Lục Vân Tiên*. The results are listed in Table 4.

1. Experiment-1:
   This experiment deals with *sliding window* of 4, 1 beam, a pre-learned W_0 using the same *sliding window* on dataset *Truyện Kiều* (1902), and the subsequent weight updates based only on this dataset. This experiment also tests the case where the recognition of the language model's selection in iteration 1.2 is not applied to iteration 1.3. The first iteration already lifts the mAP of top-1 choices of the three datasets by a small margin (+0.01% for *Truyện Kiều* 1871, +0.53% for *Truyện Kiều* 1902 and +0.03% for *Truyện Lục Vân Tiên*). The second and third iterations boost the performance on the OCR-LM Integration stage's train datasets, but not on its testing dataset.

2. Experiment-2:
   Compared to Experiment-1, this experiment has a *sliding window* of 5 and runs 3 beams. Since the context used for each time step is now longer, the language model shows more effect on the PLL of each character. As a result, more correct characters are selected. This first iteration achieved the highest result regarding the testing dataset *Truyện Lục Vân Tiên* which is +0.14% from the OCR top-1 mAP.

3. Experiment-3:
   This experiment uses *sliding window* of 5 and 1 beam. Compared to Experiment-1 and Experiment-2, it uses a pre-defined W_0 on dataset *Truyện Kiều* 1902. The first iteration 3.1 which applies this W_0 performs poorly. However, in iteration 3.2, its performance is drastically improved where the mAP on *Truyện Kiều* 1902 exceeds its OCR top-1 baseline by 2.33% while the mAP on *Truyện Lục Vân Tiên* is comparable to its OCR baseline (a decrease of only 0.01%). This experiment tests the case where the selection of the language model in iteration 3.2 is applied to iteration 3.3 and from iteration 3.3 to iteration 3.4. The highest performance achieved on the training dataset *Truyện Kiều* 1902 is 74.52%, which is +2.72% compared to the OCR top-1 result while the performance on the testing dataset remains stable.

4. Experiment-4:
   The performance using pre-learned W_0 is equal to or higher than the OCR baseline for all three datasets. However, when entering the second loop, it is not stably

Table 4 Experiment results (mAP@0.5)

Order	Parameter								Truyê Kiêu (1871)	Truyê Kiêu (1902)	Truyê Lục Vân Tiên
	Beam size	Sliding window	w_0	w_1	w_2	Weight init	Weight combo	Accept beam-1 lm			
1.1	1	4	23.99	10.8702	0.1025	pl w4	1902	False	**98.92%**	**72.33%**	**72.99%**
1.2	1	4	3.6385	2.8233	0.0779	pl w4	1902	False	**98.94%**	**73.40%**	72.71%
1.3	1	4	3.6686	2.8331	0.0819	pl w4	1902	False	**98.94%**	**73.48%**	72.68%
2.1	3	5	23.99	10.8702	0.1025	pl w4	1902	False	–	**72.21%**	<u>**73.10%**</u>
2.2	3	5	2.1109	2.0969	0.0660	pl w4	1902	False	–	**73.21%**	72.69%
3.1	1	5	0	0.85	0.15	pd	1902	False	93.50%	65.11%	44.87%
3.2	1	5	3.5202	2.7749	0.0859	pd	1902	False	<u>**98.95%**</u>	**74.13%**	72.95%
3.3	1	5	3.2286	2.6633	0.0715	pd	1902	True	–	**74.24%**	72.95%
3.4	1	5	3.1055	2.6080	0.0617	pd	1902	True	–	<u>**74.52%**</u>	72.95%
4.1	2	3	20.5112	10.7774	0.0691	pl w3	1871+1902	False	98.91%	72.08%	**73.06%**
4.2	2	3	2.1071	2.1202	0.0510	pl w3	1871+1902	False	98.91%	72.46%	72.88%
4.3	2	4	2.1316	2.1318	0.0521	pl w3	1871+1902	False	**98.93%**	**72.83%**	**73.01%**
Baseline: top-1 OCR results (from Sect. 4.2)									98.91%	71.80%	72.96%

increasing or decreasing on the testing dataset. Therefore, in iteration 4.3 we try to make a change in the *sliding window* in which it is upgraded from value 3 to value 4. The resulting mAPs on the three datasets in iteration 4.3 are all increased by a small margin.

5. Experiment summary

Our experiments confirm the argument that pre-learned weight brings about improvement on the first iteration and learned weight gives higher chance of improvement than using pre-defined one. Regarding our additional experiments to learn the weight iteratively, our results suggest two ways to do so: (1) Using larger sliding window so that the language model can receive more contexts; and (2) Iteratively replacing top choice of the OCR-LM Integration stage to be the OCR top-1 result for the next iteration could preserve some correct selection of the current iteration. However, in testing time and application scenarios, it is unforeseen when to apply this manner.

5 Conclusion

We conclude that OCR model and language model integration using pre-learned weight and learning weight gives higher performance than using pre-defined weight. Although applying language model to post-process OCR results manages to boost performance only by a small margin of mAP, this method still leaves room for further improvements in which the foremost tasks to do are to expand the size and quality of the training datasets; and improve the pipeline's run time. The correcting ability of the language model in our method depends greatly on the performance of the OCR models since it can only choose among the sequences that are proposed by the OCR model. Therefore, in the case that there are many wrong OCR predictions for input, it can lead to declination and even a decrease in the final results.

6 Future Work

For future works, we plan to collect and build more Nôm literary works to increase corpus size so that language model can converge better. We would like to conduct more experiments with our current set of parameters and try different weight learning mechanism to pick out the better method that could boost post-processing performance. Besides, since our character-level language model can only choose from the results given by the OCR model, for cases where the top-N results are all incorrect, we think that a lower-level Nôm language model such as radical-level or stroke-level may offer improvement by suggesting a different character other than the given top-N OCR results. Consequently, the number of Nôm vocabularies to be covered will be extended more than which of the OCR model.

Acknowledgements The research is financially supported by Ho Chi Minh City Department of Science and Technology.

References

1. Tran N Di san Hán Nôm Viet Nam - thu' mục đe yeu, vol I, pp 15–47. Social Sciences, Ha Noi (1993). http://www.hannom.org.vn/default.asp?CatID=462
2. http://www.hannom.org.vn/default.asp?CatID=8
3. Liu Y, Ott M, Goyal N, Du J, Joshi M, Chen D, Levy O, Lewis M, Zettlemoyer L, Stoyanov V (2019) Roberta: a robustly optimized BERT pretraining approach. CoRR Arxiv: 1907:11692
4. Smith R (2011) Limits on the application of frequency-based language models to ocr. In: ICDAR. pp 538–542. https://doi.org/10.1109/ICDAR.2011.114, won Best Industrial Paper Award
5. Wang QF, Yin F, Liu CL (2012) Improving handwritten chinese text recognition by unsupervised language model adaptation. In: 2012 10th IAPR international workshop on document analysis systems, pp 110–114 (2012). https://doi.org/10.1109/DAS.2012.46
6. Su T, Ma P, Wei T, Liu S, Deng S (2013) Exploring mpe/mwe training for chinese handwriting recognition. In: 2013 12th International conference on document analysis and recognition, pp 1275–1279. https://doi.org/10.1109/ICDAR.2013.258
7. Van Phan T, Zhu B, Nakagawa M (2011) Development of nom character segmentation for collecting patterns from historical document pages. In: Proceedings of the 2011 workshop on historical document imaging and processing. Association for computing machinery, New York, NY, USA, pp 133—139. https://doi.org/10.1145/2037342.2037365
8. Van Phan T, Nguyen KC, Nakagawa M (2016) A nom historical document recognition system for digital archiving. Int J Docum Anal Recog (IJDAR) 19:49–64. Springer (2016). https://doi.org/10.1007/s10032-015-0257-8
9. Nguyen CK, Nguyen CT, Masaki N (2017) Tens of thousands of nom character recognition by deep convolution neural networks. In: Proceedings of the 4th international workshop on historical document imaging and processing. Association for Computing Machinery, New York, NY, USA, pp 37—41.. https://doi.org/10.1145/3151509.3151517
10. Baziotis C, Haddow B, Birch A (2020) Language model prior for low-resource neural machine translation. Arxiv: 2004.14928
11. Raju A, Filimonov D, Tiwari G, Lan G, Rastrow A (2019) Scalable multi corpora neural language models for asr. Arxiv: 1907.01677
12. Jocher G, Stoken A, Borovec J, NanoCode012, Chaurasia A, TaoXie, C, L V, Laughing A, tkianai, yxNONG, Hogan A, lorenzomammana, AlexWang1900, Hajek J, Diaconu L, Marc, Kwon Y, oleg, wanghaoyang0106, Defretin Y, Lohia A, ml5ah, Milanko B, Fineran B, Khromov D, Yiwei D, Doug D, Ingham F (2021) ultralytics/yolov5: v5.0 - YOLOv5-P6 1280 models, AWS, Supervise.ly and YouTube integrations. https://doi.org/10.5281/zenodo.4679653
13. Krizhevsky A, Sutskever I, Hinton GE (2017) Imagenet classification with deep convolutional neural networks. Commun ACM 60(6):84–90. https://doi.org/10.1145/3065386
14. He K, Zhang X, Ren S, Sun J (2015) Deep residual learning for image recognition. http://arxiv.org/abs/1512.03385
15. Taylor WL (1953) "cloze procedure": a new tool for measuring readability. J Quarterly 30(4):415–433. https://doi.org/10.1177/107769905303000401
16. Devlin J, Chang M, Lee K, Toutanova K (2018) BERT: pre-training of deep bidirectional transformers for language understanding. CoRR, Arxiv: 1810:04805
17. Shannon CE (1948) A mathematical theory of communication. The Bell System Tech J 27(3):379–423. https://doi.org/10.1002/j.1538-7305.1948.tb01338.x
18. Wang A, Cho K (2019) BERT has a mouth, and it must speak: BERT as a markov random field language model. CoRR. Arxiv 1902:04094

19. Salazar J, Liang D, Nguyen TQ, Kirchhoff K (2019) Masked language model scoring CoRR. https://doi.org/10.14659
20. Shin J, Lee Y, Jung K (2019) Effective sentence scoring method using bert for speech recognition. In: Lee, WS, Suzuki T (eds) Proceedings of The eleventh Asian conference on machine learning. Proceedings of machine learning research, vol 101, pp 1081–1093. PMLR, Nagoya, Japan (17–19 Nov 2019). http://proceedings.mlr.press/v101/shin19a.html
21. Büler D, Minker W, Elciyanti A (2005) Using language modelling to integrate speech recognition with a flat semantic analysis. In: Proceedings of the 6th SIGdial workshop on discourse and dialogue, pp 212–216
22. Brown PF, Della Pietra SA, Della Pietra VJ, Mercer RL (1993) The mathematics of statistical machine translation: parameter estimation. Comput Linguistics 19(2), 263–311. https://aclanthology.org/J93-2003
23. Gulcehre C, Firat O, Xu K, Cho K, Bengio Y (2017) On integrating a language model into neural machine translation. Comput Speech Lang 45(C):137–148. https://doi.org/10.1016/j.csl.2017.01.014
24. http://nomfoundation.org/nom-project/tale-of-kieu/tale-of-kieu-version-1871
25. http://nomfoundation.org/nom-project/tale-of-kieu/tale-of-kieu-version-1902
26. http://nomfoundation.org/nom-project/Luc-Van-Tien/Luc-Van-Tien-Text
27. Ma W, Zhang H, Jin L, Wu S, Wang J, Wang Y (2020) Joint layout analysis, character detection and recognition for historical document digitization. ICFHR 2020
28. Liu CL, Yin F, Wang DH, Wang QF (2011) Casia online and offline chinese handwriting databases. In: 2011 International Conference on Document Analysis and Recognition, pp 37–41 https://doi.org/10.1109/ICDAR.2011.17. http://www.nlpr.ia.ac.cn/databases/handwriting/Download.html
29. Redmon J, Divvala SK, Girshick RB, Farhadi A (2015) You only look once: unified, real-time object detection. CoRR. Arxiv: 1506:02640

Calib-StyleSpeech: A Zero-Shot Approach in Voice Cloning of High Adaptive Text to Speech System with Imbalanced Dataset

Nguyen Tan Dat, Lam Quang Tuong, and Nguyen Duc Dung

Abstract In this article, we propose Calib-StyleSpeech, a novel end-to-end Text-to-Speech (TTS) system, which achieves a Zero-shot learning approach in cloning a new voice without using the meta-learning method. Our model uses a block to extract the style from the reference mel-spectrogram, embedding this style vector into the acoustic model to condition the hidden features. And then, training it with a multi-speaker dataset that still contains noise and resonant utterances. We also extract a content vector from this mel-spectrogram simultaneously with the style vector. This vector will imitate the hidden vector extracted from the phoneme sequence by the encoder of the acoustic model. The extracted content vector takes responsibility on calibrate the style vector by the Mutual Information (MI) constraint in order to eliminate the dependency between content and style representation space. Calib-StyleSpeech can synthesize a high fidelity voice with only 36 utterances (about 3 min) in training data. Furthermore, the experimental results prove that model can learn to speak in a new voice with only one small reference record (about 5 s) of the target voice without any fine-tune stage and get a competitive similarity score with another state-of-the-art while maintaining naturalness and intelligence.

Supported by OLLI Technology JSC.

N. T. Dat · L. Q. Tuong · N. D. Dung (✉)
Vietnam National University, Ho Chi Minh City, Vietnam
e-mail: nddung@hcmut.edu.vn

N. T. Dat
e-mail: dat.nguyen_bk@hcmut.edu.vn

L. Q. Tuong
e-mail: tuong@olli-ai.com

N. T. Dat · N. D. Dung
University of Technology, Ho Chi Minh City, Vietnam

L. Q. Tuong
University of Science, Ho Chi Minh City, Vietnam

N. T. Dat · L. Q. Tuong
OLLI Technology JSC, Ho Chi Minh City, Vietnam

Keywords Speech synthesis · Text-to-speech · Zero-shot voice cloning ·
Sequence-to-sequence · Low-resource language

1 Introduction

The Text-to-Speech system is an essential component in smart agents and has become
more and more familiar to people in daily life. Many researches, which created huge
improvement in naturalness, intelligence, fidelity, realtime synthesizing, such as [1–
4], contribute on that developments. Along with that rapid development, personal
voice are demanded for many potential applications of personalized TTS systems
include entertainment, personal assistant, etc. This has been attracted researchers on
extended the acoustic model to synthesize more than one voice. Wang et al. [5], Ping
et al. [6], Ren et al. [4] proposed exciting methods for multi-speaker TTS, which
still widely influence recent research on many problems of TTS. These methods try
to extract the high level prosodic features, e.g. duration, pitch, energy, etc., from the
voice as style vectors to condition the reconstruction stage. But, these models require
a large amount of new data from a speaker to synthesize his/her voice. Because of
the limitation of the corpora, especially in low-resource language, a high adaptive
TTS model is a low-cost solution to due with that problem.

The most important requirement of a high adaptive TTS model is to clone a voice
with a small number of records from the target speaker while maintaining the naturalness and intelligence of the output audio. The more decrease on amount of data the
TTS system needs to achieve the new voice, the more convenient, easy, and cheap it
gets for deployment. However, there are two main distinctive challenges on that problem: (1) The mismatch in the recording condition of different speakers, which also
involve the unexpected noise and resonant on the utterances. In the adaptation stage,
data is often recorded with diverse speaking prosodies, styles, emotions, accents,
and other recording environment factors. It cause the difficulty on generalization and
lead to poor performance on adaptation. (2) The trade-off between the fine-tuning
parameters and voice quality when the model needs to adapt to a new voice. The more
parameters update in adaptation stage, the better voice quality, which also increases
the memory storage and lead to the increasement of serving cost.

A popular approach to handle this challenge is transfer learning. Transfer learning approach in voice adaption require a good pre-train model (relative task and
knowledge) in a large and diverse dataset from many speakers [1, 7, 8]. However,
this approach ask new records of the target voice, and also hundreds of fine-tuning
steps, which limits its applicability to real-world scenarios. Another approach is to
extract the latent style vector from a reference voice sample to control the reconstruction stage [5]. This approach is able to clone voices from new speakers without
fine-tuning. However, they rely a lot on the diversity and cleaning of the source
dataset and hence often show poor adaptation performance on low-resource and
noisy datasets. Besides, because of the mutual entangled of the linguistic feature and
speaker identity features in mel-spectrogram, some linguistic information often be

included in style vectors. But, the linguistic feature is a shared space overall speakers in the dataset. So, a good captured style vector of a new voice with least linguistic information promise a well performance on multi-speaker TTS. Recently, there are many works try to create a well capture on style from the speech (e.g., [9] try to disentangle style representation space and content representation space, [10] employs an ASR-liked block to learn style of speakers).

In this paper, we propose Calib-StyleSpeech—a high adaptive multi-speaker TTS model training on an imbalanced Vietnamese dataset. The method is a one-time training for multi-time customizing without any fine-tuning step. After the training stage, our model can well perform on multi-speaker task (e.g., our model can synthesize a voice of a low-resource speaker who has only 36 utterances in training dataset). Besides, our proposed adaptive TTS model has a high performance and efficient in voice customization. Calib-StyleSpeech needs only one small reference voice from the target speaker to be able to synthesize the target voice while still maintain high fidelity. Our work employs the acoustic model of the [4], and further several works on preprocessing the dataset and model architecture design. We also guarantee the naturalness and intelligence of synthesized voice even the model was trained with a assembled dataset that still contains noise and resonant.

2 Related Works

2.1 Neural Text to Speech

With the development of the Deep Neural Network (DNN), several current works apply it in all steps of a TTS system and achieve many improvements on fidelity of synthesized voice. DeepVoice [1] is one of popular model that still follow the three-component statistical parametric TTS model with corresponding neural network. To simplify text analysis modules and directly synthesize speech from grapheme or phoneme, furthermore, reduce the compound error of multi-component of old architecture, end-to-end TTS models have more concentration and so significant improvement in TTS research. Tacotron1/2 [2, 3] and many Tacotron-based TTS models, which are considered as autoregressive models, get text-level feature as input and use RNN-based network with attention mechanism to reconstruct the mel-spectrogram features. FastSpeech1/2 [4, 11], FastPitch, [12], TalkNet2 [13] are some popular state-of-the-art (SOTA) non-autoregressive TTS models. These models are non-RNN networks also can capture some prosody features on speech. Hence, these model architectures reduce synthesis time, potential for multispeaker and high adaptive TTS system. Our proposed build the acoustic module base on FastSpeech 2 because of above advantages.

2.2   Speaker Adaptation

Speaker adaptation researches have been promoted by the demand for custom voices. A popular approach is to fine-tune all the models with new dataset [1] or only a part of the model [8, 14]. Another approach is Tacotron-based model using GST to capture the high level features in speech such as [5] or using GST with another speaker identity features such as pitch [15]. These methods achieve some significant improvements in speaker adaptation, but the entangled of the linguistic feature in style information limit its performance. There is also another popular approach is to apply VAE-based network to encode the utterance into a content feature and style feature, and then using the style feature as one condition on the reconstruction stage [9]. However, these models still have poor similarity even much data have to be collected from the new speaker. This work proposed a method to enrich the style information included in style vector. Which can improve the adaptive model to clone a new voice with only 5 s of target voice are required.

3   Proposed Method

Without loss of generality, we assume that we have a dataset $\mathcal{D} = \cup_{u=1}^{S} \mathcal{D}_u$ where $\mathcal{D}_u$ contains N_u pairs $\boldsymbol{x}_i^u$ and $\boldsymbol{t}_i^u$ which are utterance and its text content of speaker $u \in \mathcal{S} = \{u_i\}_{i=1}^{M}$. Our proposed method encodes each voice input $\boldsymbol{x}_i^u$ of pair $(\boldsymbol{x}_i^u, \boldsymbol{t}_i^u) \in \mathcal{X}$ into two lower-dimensional speaker-related embedding $\boldsymbol{s} = E_s(\boldsymbol{x}_i^u)$ and content-related embedding $\boldsymbol{c} = E_c(\boldsymbol{x}_i^u)$. Where $E_s(\cdot)$ and $E_c(\cdot)$ are style and content learnable function, respectively. In order to synthesize a random text content $\boldsymbol{t}$ utterance with the style of speaker u, we first extract the style $\boldsymbol{s}_i^u$ from voice sample $\boldsymbol{x}_i^u$ without knowledge about its content. Then, we combine it with phoneme encoder $E_p(\boldsymbol{t}, \boldsymbol{s}_i^u)$ to generate a hidden vector $\boldsymbol{h}$. And then add predicted pitch, energy and regularize the length by predicted duration, the hidden vector will be decoded into the utterance $\hat{\boldsymbol{x}}_i^u$ by the decoder $D(\boldsymbol{h}, \boldsymbol{s}_i^u)$. The content of this utterance is $\boldsymbol{t}$ and its voice is belong to speaker u.

3.1   Calibrated Style Extraction

Disentangling Style and Content A mel-spectrogram generated from $\boldsymbol{x}$ contains many informations, and two most importance for a multispeaker TTS system are speaker identity and content. The style embedding $\boldsymbol{s}$ is a private features of speaker voices, such as prosody, tone, timbre, etc., which distinguish this speaker with others in the voice domain $\mathcal{S}$. Opposite, the latent content representation $\boldsymbol{c}$ include the shared linguistic features overall speaker domain and not determine the different in voice of speakers. Unfortunately, these extracted vectors often capture unexpected

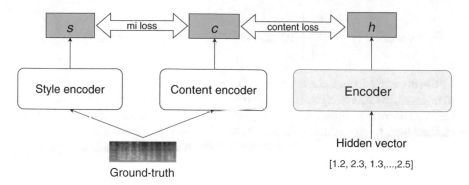

Fig. 1 Ground-truth mel-spectrogram is extracted into style embedding s and content embedding c. This step use a MI-based loss to reduce the dependency between these vectors. Meanwhile, the output hidden vector of the Encoder of Acoustic model will calibrate the vector c by a content loss function. This is why we call it Calib-StyleSpeech (Calibrated Style)

information and the worst behavior in our case is that the style vector contains content information, which is shared with all speakers in the dataset. We build a **Content encoder** Block, which is the same with **Style encoder** Block, to encode mel spectrogram to a content vector. Both content vector and style vector are extracted from mel-spectrogram and have the same shape. The important thing in our method is that we connect them by two loss functions. We use a MSE loss to minimize the discrepancy between the content vector and the phoneme encoded vector. This makes the content vector tend to contain more informations of input text. Besides, we use MI loss function to maximize the independent between the content vector and style vector. Thank that, that style vector tend to have only style information and eliminate the linguistic information. This is the key to robust the model of our method.

We propose an MI-based framework that disentangles the style with content by minimizing the dependency between style encoder $E_s(x)$ output and content encoder $E_c(x)$ output. Our framework also use the hidden state from $E_p(t, s)$ to calibrate the content which extract from the $E_c(\cdot)$ as depicted in Fig. 1. In our proposed, we try to minimize an upper bound of MI [16]. But, the problem here is that the ground-truth conditional distribution $p(s|c)$ is unknown. In our work, we approximate this conditional distribution by a simple neural network to get the closed form $q_\theta(s|c)$. The objective of this neural network is to maximize:

$$\mathcal{L}(\theta_{mi}) = \sum_{u=1}^{M} \sum_{i=1}^{N_u} \log q_{\theta_{mi}}(s_i^u | c_i^u) \qquad (1)$$

At training time, the change of weights of $E_s(\cdot)$ and $E_c(\cdot)$ lead to the change of the distribution $p(s|c)$. Hence, the $q_\theta(s|c)$ also need to updated in order to approximate the current $p(s|c)$. Then, with the closed form distribution q_θ, the MI minimization objective can be formulate as:

$$\mathcal{L}_{mi}(\theta_{model}) = \frac{1}{N} \sum_{u=1}^{M} \sum_{i=1}^{N_u} \left[\log q_{\theta_{mi}} \left(s_i^u \mid c_i^u \right) - \frac{1}{N} \sum_{v=1}^{M} \sum_{j=1}^{N_v} \log q_{\theta_{mi}} \left(s_i^u \mid c_j^v \right) \right] \quad (2)$$

Calibrate Style Even though we try to disentangle the style and content representative spaces, the E_c and E_s still do not know exactly the value or any information about the ground-truth to robust the learning progress except the gradient. Our proposed method here is approximating c with h. In other words, we force the $E_p(t, s)$ and $E_c(x)$ to generate the same output, the middle abstract level of linguistic features. By this work, when minimizing the dependency between the embedding space of s and c, we can make sure that the model is learning to disentangle the style information from content information.

$$E_c(x; \theta_{model}) \approx E_s(t, s; \theta_{model}) \implies p_{\theta_{model}}(c \mid x) \approx p_{\theta_{model}}(h \mid t, s) \quad (3)$$

So, minimization the mutual information between c and s can be considered as minimization the mutual information between h (the rich-linguistic hidden vector) with s (the style vector need to be eliminate content information). We use a content loss $\mathcal{L}_{content}$ to implement above idea. The objective of our model is to minimize the different between c and h:

$$\mathcal{L}_{content}(\theta_{model}) = \left\| E_c(x; \theta_{model}) - E_p(t, s; \theta_{model}) \right\|_1 \quad (4)$$

3.2 Overall Model Architecture

Our model builds on FastSpeech2 [4], which is non-autoregressive TTS model, includes three parts: encoder, variance adaptor, decoder. In this acoustic model, with a arbitrary length text t_i^u, which paired with x_i^u in the dataset, through the phoneme embedding, the linguistic embedding sequence $\hat{t}_i^u \in \mathbb{R}^N$ is extracted with N is the length of hidden sequence. The output of encoder $h_i^u = E_p(\hat{t}_i^u, s_i^u; \theta_{model})$ where s_i^u is style vector extracted by Style extractor block. Then we combine this output with speaker embedding, then add with predicted energy, pitch and regularize the length by the predicted duration by the variance adopter. After adapted with prosody features, the output vector is used to reconstruct the mel-spectrogram with the style vector. In our work, we also employ SALN [17] as the building block of encoder and decoder to get a high adaptive model and address the challenge of *covariate shift* [18]. With this design, we train the model with objective is to minimize the total loss the model:

$$\mathcal{L} = \mathcal{L}_{acoustic} + \lambda_1 \mathcal{L}_{mi} + \lambda_2 \mathcal{L}_{content} \quad (5)$$

where, $\mathcal{L}_{rec} = \mathbb{E} \left\| \hat{x} - x \right\|_1$ is the L1 loss between reconstruction mel-spectrogram and reference mel-spectrogram. $\mathcal{L}_d, \mathcal{L}_e, \mathcal{L}_p$ are Mean Squared Error (MSE) between

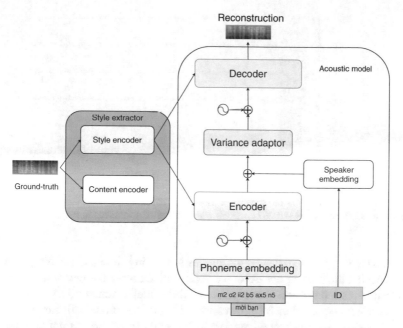

Fig. 2 Overal model architecture include two main components: (1) Style extractor which will try to directly capture the style from ground-truth mel-spectrogram. This module also use the extracted content of mel-spectrogram to **calibrate** the **style** vector (Calib-StyleSpeech). (2) Acoustic model extract the linguistic features from phonemes. And then use the the style vector extracted from Style extractor to get a pair of scalar value. Which will apply an affine transformation in each Feed-Forward Transformer block of the Encoder and Decoder of the acoustic model

ground-truth and predicted duration, and energy, and pitch, respectively. The λ_1 and λ_2 are used to control the influence of MI loss and content loss (Fig. 2).

4 Experiments

4.1 Imbalanced Dataset

First of all, we combine three dataset as mentioned: two female single-speaker dataset (1) OLLI-SPEECH-1.6 has 26034 utterance with south accent, (2) OLLI-SPEECH-NORTHERN has 20477 uterrances in north accent; and (3) OLLI-SPEECH-ASR multi-speaker dataset which has 1531 voices that still contain noise and resonant. After combined, we have a dataset with a total 336 h of 1533 voices over accent, age of Vietnamese. As depicted in Fig. 3, about 30% of speakers have less than 100 utterances (45–60 minutes) and most of them are southern voices. Besides, our assembled dataset also contains 2 female speakers with more than 20000 utterances.

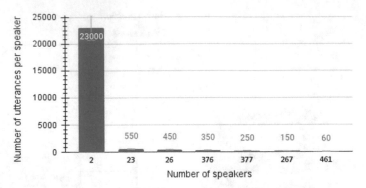

Fig. 3 Number of utterances per speaker of assembled dataset. Specially, there are 194 speakers with less than 50 utterances and 267 speakers with number of utterances in range [50,100)

This imbalanced problem causes difficulties in training time and the biggest consequence is that the bias on rich-resource voice will decrease the performance of TTS task on low-resource voice. However, our calibrate method shows its effectiveness in this dataset. As depicted in 4, by the calibrated style, our model still can well cluster the style vectors of speakers without the meta-learning method. It not only reduces the overfit of model on rich-resource voice but also keeps the good performance on naturalness and intelligence of synthesis voice.

4.2 Preprocessing and Training

We resample audio 48000 22050 Hz and down to a single channel on all dataset. With text data, we build a text-normalization tool to clean and ensure the content is correct with the audio. We extract the spectrogram with a FFT size of 1024, hop size of 256, and windows size of 1024 samples. In addition, we also use the Vietnamese grapheme-to-phoneme toolkit researched by Lam et al. [19] to convert Vietnamese text to a sequence of phonemes. Furthermore, we build a tool to preprocess input text for the purpose of cleaning up and handling ambiguity from text. We also use MFA toolkit [20] to train a new model for our dataset. And then, use it to generate the ground-truth duration. We also estimate the ground-truth pitch by DIO and Stonemask [21] algorithm with PyWorld[1] toolkit. The model parameters are set as follows: batch size 48, $\lambda_1 = \lambda_2 = 1$, Adam Optimizer with $\beta_1 = 0.9$, $\beta_2 = 0.98$, $\epsilon = 10^{-9}$. The model were trained on 2 GPUs GTX 1080 Ti with 3 days of total training time.

---

[1] https://github.com/JeremyCCHsu/Python-Wrapper-for-World-Vocoder

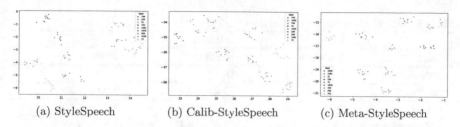

(a) StyleSpeech   (b) Calib-StyleSpeech   (c) Meta-StyleSpeech

Fig. 4 t-SNE visualization of the style vectors for seen speaker of **a** StyleSpeech, **b** Calib-StyleSpeech, and **c** Meta-StyleSpeech. Note that id-72 and id-1381 has only 28 and 41 utterances in dataset, respectively

4.3 Evaluations

Metric We evaluate our result in terms of naturalness by Mean Opinion Score (how natural the synthesized voice compares with real), and in terms of similarity by Similarity Mean Opinion Score (how similarity in timbre the synthesized compared with reference voice). We create two separate surveys for each MOS and SMOS test. There are 58 judges who listen to each sentence in random order in the MOS survey and 17 judges listen to pairs of synthesized and reference sentences also in random order in the SMOS survey. Both metrics are evaluated on a 1–5 scale and reported with 95% confidence intervals. The voice in the tests include: (1) **GT**, ground-truth records with no noise or resonant records, (2) **GT mel + vocoder**, the extracted mel from ground truth then synthesized by the our the same vocoder with the following samples, (3) **Adaspeech**, synthesized from Adaspeech model [8] (4) **StyleSpeech**, synthesized from StyleSpeech [17] (5) **Meta-StyleSpeech**, synthesized from Meta-StyleSpeech [17] (6) **Calib-StyleSpeech**, synthesized from our own model.

Naturalness and Intelligence According to the MOS results are shown in Table 1, we have 3 observations: (1) The Adaspeech model is sensitive to noise and resonant than others, we can see that our model can hold naturalness even the small of data in

Table 1 The results of MOS test

Model		Seen	Unseen–*Zero-shot*
GT	4.07 ± 0.13	–	–
GT mel + vocoder	3.67 ± 0.16	–	–
AdaSpeech	–	2.75 ± 0.15	–
StyleSpeech	–	3.25 ± 0.13	3.38 ± 0.14
Meta-StyleSpeech	–	$\mathbf{3.51 \pm 0.14}$	2.83 ± 0.16
Calib-StyleSpeech	–	3.25 ± 0.14	$\mathbf{3.49 \pm 0.14}$

The Multispeech column is the score of utterances which synthesized from **seen voice**. The Zero-shot is the MOS score of uterrances which synthesized from **unseen voice** by zero-shot approach, which mean that no re-train step required

Table 2 Similarity between synthesized and record of voice.

	Seen	Unseen	
Model	Northern	Southern	Northern
Adaspeech	2.94 ± 0.64	–	–
StyleSpeech	2.53 ± 0.77	3.94 ± 0.71	1.29 ± 0.30
Meta-StyleSpeech	**4.24 ± 0.56**	3.18 ± 0.58	1.24 ± 0.39
Calib-StyleSpeech	2.47 ± 0.72	**4.65 ± 0.31**	**1.65 ± 0.48**

training of voice. (2) The Meta-StyleSpeech is the best model on multispeaker TTS task. However, our model without meta-training still maintain the naturalness and intelligence as the StyleSpeech (the model that Meta-StyleSpeech use at Generator at meta training step). (3) Finally, and the most impressive is the performance on zero-shot voice cloning, our model outperform both StyleSpeech and Meta-StyleSpeech on our own dataset.

Similarity We compare the performance of our model with SOTAs to highlight the effectiveness of our proposed method on learning the style vector. However, Adaspeech synthesized low naturalness and intelligence voices on zero-shot and few-shot learning so we do not survey the similarity of Adaspeech synthesized voice. Even the noise and resonant in out dataset, experiment on our custom data show that the Calib-StyleSpeech can clone a voice from northern and southern accent better than others. Just by a small piece of speech, our model able to speak in a new voice without Meta learning approach.

The result in Table 2 show the effectiveness of calibrated method on learning the style of a new speaker, specially for zero-shot approach learning. Both northern and southern accent of Vietnamese are better learn by Calib-StyleSpeech. The poor SMOS of northern on zero-shot voice cloning cause by the low-resource problem of this accent on our own training dataset. Almost of dataset are southern accent, so the performance on learning a northern accent is more difficult than southern. However, our model still outperform the StyleSpeech and Meta-StyleSpeech. Calib-StyleSpeech can also keep a good performance on similarity with the seen accent of Northern.

Speaker embedding To better understand the effectiveness of calibrating method on improving the style information of style vector, we visualize the 10 style vector per speaker in total of 9 speakers which is extracted by three models as shown in Fig. 4. There is also a chart showing the imbalanced data of these speakers with different accents, genders. We can see that while StyleSpeech does not explicitly separate the vectors of the different speakers, our model creates clear clusters for each speaker. With sparse clusters, Calib-StyleSpeech makes performance better than StyleSpeech and competes with Meta-StyleSpeech

5 Conclusion

In this paper, we present a novel method using abstract content from the text to calibrate the style vector by MI function and use this effective style extraction on both multispeaker tasks and zero-shot voice cloning tasks. Calib-StyleSpeech can clone a voice with a piece of about 5 s of a new voice. According to the evaluation, our proposed method on disentangling the style and content embedding representation space shows effectiveness as expected. This method provides a solution to the speaker embedding problem, it also can be used in other tasks, such as speaker identification.

Acknowledgements We would like to thank OLLI Technology JSC for building three data sets OLLI-SPEECH-1.6, OLLI-SPEECH-NORTHERN, and OLLI-SPEECH-ASR. Specials thank for all computing resource and human support for this study.

References

1. Arik S, Chrzanowski M, Coates A, Diamos GF, Gibiansky A, Kang Y, Li X, Miller J, Ng AY, Raiman J, Sengupta S, Shoeybi M (2017) Deep voice: real-time neural text-to-speech. In: Precup D, Teh YW, (eds) Proceedings of the 34th international conference on machine learning, ICML 2017, Sydney, NSW, Australia, 6-11 August 2017, volume 70 of Proceedings of machine learning research, pp 195–204. PMLR
2. Wang Y, Skerry-Ryan RJ, Stanton D, Wu Y, Weiss RJ, Jaitly N, Yang Z, Xiao Y, Chen Z, Bengio S, Le QV, Agiomyrgiannakis Y, Clark R, Saurous RA (2017) Tacotron: towards end-to-end speech synthesis. In: Interspeech 2017, 18th annual conference of the international speech communication association, Stockholm, Sweden, August 20-24, 2017. ISCA, pp 4006–4010
3. Shen J, Pang R, Weiss RJ, Schuster M, Jaitly N, Yang Z, Chen Z, Zhang Y, Wang Y, Skerry-Ryan R, Saurous RA, Agiomvrgiannakis Y, Wu Y (2018) Natural tts synthesis by conditioning wavenet on mel spectrogram predictions. In: 2018 IEEE International Conference on Acoustics, Speech and Signal Processing (ICASSP), pp 4779–4783
4. Ren Y, Hu C, Tan X, Qin T, Zhao S, Zhao Z, Liu T-Y (2021) Fastspeech 2: fast and high-quality end-to-end text to speech. In: 9th International Conference on Learning Representations, ICLR 2021, Virtual Event, Austria, May 3-7, 2021. OpenReview.net
5. Wang Y, Stanton D, Zhang Y, Skerry-Ryan RJ, Battenberg E, Shor J, Xiao Y, Jia Y, Ren F, Saurous RA (2018) Style tokens: unsupervised style modeling, control and transfer in end-to-end speech synthesis. In: Dy JG, Krause A (eds) Proceedings of the 35th international conference on machine learning, ICML 2018, stockholmsmässan, Stockholm, Sweden, July 10-15, 2018, volume 80 of Proceedings of machine learning research. PMLR, pp 5167–5176
6. Ping W, Peng K, Gibiansky A, Arik SÖ, Kannan A, Narang S, Raiman J, Miller J (2018) Deep voice 3: scaling text-to-speech with convolutional sequence learning. In: 6th international conference on learning representations, ICLR 2018, Vancouver, BC, Canada, April 30 - May 3, 2018, Conference Track Proceedings. OpenReview.net,
7. Chen Y, Assael YM, Shillingford B, Budden D, Reed SE, Zen H, Wang Q, Cobo LC, Trask A, Laurie B, Gülçehre Ç, van den Oord A, Vinyals O, de Freitas N (2019) Sample efficient adaptive text-to-speech. In: 7th International Conference on Learning Representations, ICLR 2019, New Orleans, LA, USA, May 6-9, 2019. OpenReview.net
8. Chen M, Tan X, Li B, Liu Y, Qin T, Zhao S, Liu T-Y (2021) Adaspeech: adaptive text to speech for custom voice. In: 9th International Conference on Learning Representations, ICLR 2021, Virtual Event, Austria, May 3-7, 2021. OpenReview.net

9. Yuan S, Cheng P, Zhang R, Hao W, Gan Z, Carin L (2021) Improving zero-shot voice style transfer via disentangled representation learning. In: 9th International Conference on Learning Representations, ICLR 2021, Virtual Event, Austria, May 3-7, 2021. OpenReview.net

10. Liu S, Cao Y, Wang D, Wu X, Liu X, Meng H (2021) Any-to-many voice conversion with location-relative sequence-to-sequence modeling. IEEE ACM Trans Audio Speech Lang Process 29:1717–1728

11. Ren Y, Ruan Y, Tan X, Qin T, Zhao S, Zhao Z, Liu T-Y (2019) Fastspeech: Fast, robust and controllable text to speech. In: Wallach HM, Larochelle H, Beygelzimer A, d'Alché-Buc F, Fox EB, Garnett R (eds) Advances in neural information processing systems 32: annual conference on neural information processing systems 2019, NeurIPS 2019, December 8-14, 2019, Vancouver, BC, Canada, pp 3165–3174

12. Lancucki A (2021) Fastpitch: parallel text-to-speech with pitch prediction. In: IEEE International Conference on Acoustics, Speech and Signal Processing, ICASSP 2021, Toronto, ON, Canada, June 6-11, 2021. IEEE, pp 6588–6592

13. Beliaev S, Ginsburg B (2021) Talknet 2: non-autoregressive depth-wise separable convolutional model for speech synthesis with explicit pitch and duration prediction. CoRR, abs/2104.08189

14. Yan Y, Tan X, Li B, Qin T, Zhao S, Shen Y, Liu T-Y (2021) Adaspeech 2: adaptive text to speech with untranscribed data. In: IEEE International Conference on Acoustics, Speech and Signal Processing, ICASSP 2021, Toronto, ON, Canada, June 6-11, 2021. IEEE, pp 6613–6617

15. Neekhara P, Hussain S, Dubnov S, Koushanfar F, McAuley JJ (2021) Expressive neural voice cloning. CoRR, abs/2102.00151

16. Cheng P, Hao W, Dai S, Liu J, Gan Z, Carin L (202) CLUB: a contrastive log-ratio upper bound of mutual information. In: Proceedings of the 37th international conference on machine learning, ICML 2020, 13-18 July 2020, Virtual Event, volume 119 of Proceedings of machine learning research. PMLR, pp 1779–1788

17. Min D, Lee DB, Yang E, Hwang SJ (2021) Meta-stylespeech: multi-speaker adaptive text-to-speech generation. In: Meila M, Zhang T (eds) Proceedings of the 38th International Conference on Machine Learning, ICML 2021, 18-24 July 2021, Virtual Event, volume 139 of Proceedings of machine learning research. PMLR, pp 7748–7759

18. Ioffe S, Szegedy C (2015) Batch normalization: accelerating deep network training by reducing internal covariate shift. In: Bach FR, Blei DM (eds) Proceedings of the 32nd international conference on machine learning, ICML 2015, Lille, France, 6-11 July 2015, volume 37 of JMLR workshop and conference proceedings, pp 448–456. JMLR.org

19. Lam QT, Do DH, Vo TH, Nguyen DD (2019) Alternative vietnamese speech synthesis system with phoneme structure. In: 2019 19th International Symposium on Communications and Information Technologies (ISCIT), pp 64–69

20. Michael M, Michaela S, Sarah M, Michael W, Sonderegger M (2017) Trainable text-speech alignment using kaldi. In: INTERSPEECH, Montreal forced aligner

21. Morise M, Kawahara H, Nishiura T (2010) Rapid f0 estimation for high-snr speech based on fundamental component extraction. In: IEICE transactions on information and systems, vol 93. AAAI Press, pp 109–117

Providing Syntactic Awareness to Neural Machine Translation by Graph-Based Transformer

Binh Nguyen, Binh Le, Long Nguyen, Viet Pham, and Dien Dinh

Abstract Graph Neural Networks (GNNs) are a powerful tool to mine graph-structural information. To utilize GNNs in Neural Machine Translation (NMT) tasks, previous research applies a semantic parser to transform a sentence into a semantic graph, while this heuristic approach can be regarded as the data augmentation technique. Following this idea, this manuscript investigates the Dependency Parsing graph via GNNs to improve the current performance of the Transformer in NMT tasks. Briefly, we employ a Graph Transformer layer with graph-based input to probe graph-level information, then integrate the learned information into an encoder side. The experimental results on IWSLT15 English-Vietnamese datasets yield significant improvements over the scaled parameter Transformer by +0.46 and +0.47 BLEU, while the parameters are reduced by 53.2%. Further analyses reveal that the proposed is more robust in long and complex sentences than the Transformer, thus implying GNNs can capture non-trivial information in such groups to benefit the overall performance. Overall, the proposed method outperforms the Transformer, meanwhile cutting down the cost of training.

Keywords Graph neural networks · Neural machine translation · Transformer

B. Nguyen · B. Le · L. Nguyen (✉) · V. Pham · D. Dinh
Faculty of Information Technology, University of Science, Ho Chi Minh City, Vietnam
e-mail: nhblong@fit.hcmus.edu.vn

B. Nguyen
e-mail: nqbinh17@apcs.fitus.edu.vn

B. Le
e-mail: ltbinh@apcs.fitus.edu.vn

D. Dinh
e-mail: ddien@fit.hcmus.edu.vn

Vietnam National University, Ho Chi Minh City, Vietnam

1 Introduction

Graph Neural Networks (GNNs) [1–3] are a powerful tool to learn the inductive bias on graph data [4], while current datasets of translation tasks are absent of graphs. To employ GNNs into Neural Machine Translation (NMT), previous works [5, 6] propose a composition of (1) a semantic or syntactic parser to transform text-based data into graph-based data, (2) Message Passing Neural Networks (MPNNs) to learn the graph-based data, and (3) a baseline encoder-decoder NMT. Moreover, every semantic graph has a specific structure based on its theories and assumption, leading to the MPNNs are performed well in a parser can be underperformed in the others.

A theoretical expressive power of MPNNs is often evaluated by the Weisfeiler-Lehman (WL) test in terms of distinguishing non-isomorphic graphs [7]. Intuitively, the 3-WL test is strictly more powerful than the 2-WL test. However, MPNNs are not powerful enough [8], and still suffer from the over-smoothing problem [9], including combinatorial problems [10], computational heavily and memory hungry [11] to achieve 3-WL MPNNs. Even though, the applications of GNNs in Natural Language Processing (NLP) still gain a plethora of attention [12, 13] and potentially improve the current state-of-the-art performances [14, 15].

In this work, we exploit the linguistic properties of Dependency Parsing that provides relations among words and reveals complex phrase structure in terms of its phrasal constituents or lexical categories. Moreover, dependency parsing is dependency grammars that abstract the information from word composition variants, where it facilitates complex clausal relations in some languages such as English and Vietnamese. Intuitively, we hypothesize that GNNs will embed this knowledge into the latent feature space of word embeddings via training time, thus the NMT system will be acquired syntactic awareness to improve the translation qualities. To this end, we first preprocess English data into dependency parsing graph data via the parser, as a result, we augment English sentence data with graph data. At the encoder side, self-attention in the Transformer is applied to retrieve features from sequential data, and GNNs are supposed to exploit information from graph data. On the other hand, the decoder is kept as the original Transformer. Finally, we modify the Transformer's encoder to integrate the sentence-level embedding with the graph-level embedding, while our further analyses show that the proposed method can capture better information in long and complex sentences. We argue that the proposed method provides syntactic awareness to NMT systems, since GNNs learn the syntactic information underlying under the Dependency Parsing. The main contributions of this work are as follows:

- We investigate the linguistic properties of Dependency Parsing in Neural Machine Translation tasks.
- The proposed method outperforms the scaled parameters Transformer, while the parameters are reduced by 53.2%.
- We conduct further experiments comparing the proposed method and the standard Transformer revealing a comprehensive ability in translating long and complex sentences.

For the rest of the paper, the structure is as follows. Section 2 describes the background of Neural Machine Translation and Graph Notation. Section 3 provides the implementations of the proposed method in detail. Section 4 shows the experimental results and configurations. Section 5 is for the conclusion.

2 Background and Related Work

We provide brief NMT background and present the existing work related to graph neural networks in NMT systems.

2.1 Neural Machine Translation

NMT currently adopts an encoder-decoder framework [16] and the most reliable baseline NMT is empirically the Transformer. Overall, given a source language sentence $W = w_1, w_2, \ldots, w_n$ with n is numbers of tokens, the embedding of each token is $x_1 = Embedding(w_1)$. Let $Encoder$ and $Decoder$ are the stacks of L encoder and decoder layers, where the output of $Encoder^{(t)}$ is the input of $Encoder^{(t+1)}$. Let $\hat{Y}$ is the predicted sequence generating from $\hat{Y} = Decoder(Encoder(X))$. Finally, a corresponding target language sentence $Y = y_1, y_2, \ldots, y_m$ with m is numbers of tokens, we want the predicted sentence $\hat{Y}$ close to the truth sentence Y, so that the translation quality is better.

The performance of NMT relies on the retrieved information on the source language of an encoder side and the inference step of a decoder side. However, we argue that the expressive power of Encoder-Transformer can not fully capture the semantic and structure information of the source language, due to the self-attention module holding non-assumption on data structures. Thus, the most common way to advance the overall translation quality is to employ GNNs at the encoder such as [6, 17, 18], since they apply the heuristic data structure on the encoder side. On the other hand, the decoder is harder to modify and more rigid because the decoder at the interfere step is autoregressively generated, making it harder to apply data augmentation. One of the prevailing research trends is to exploit the graph information on the source language, to do this, they heuristically employ a semantic parser to transform the text into a semantic or syntactic graph. Where that graph is supposed to contain a phenomenon amount of linguistic properties based on that graph theoretics. Consequently, GNNs are employed to exploit this information to enrich the features of the encoder leading to the overall translation improvement.

2.2 Graph Notation

An Universal Conceptual Cognitive Annotation (UCCA) graph is represented as $G = (V, E)$, where V is the set of nodes and E is the set of edges. Let $v_i \in V$ to denote a node and $e_{ij} = (v_i, v_j) \in E$ to denote an edge directing from v_i to v_j. The neighboorhood of a node v is defined as $N(v) = \{u \in V | (v, u) \in E\}$. Meanwhile, an edge e_{ij} has a label $l_{ij} \in L$, where L is the pre-defined set of labels.

A dependency parse graph G is defined as $G = (V, E, L)$, where V is the set of all nodes (or words) in a sentence, E is the set of edges between nodes, and L. For each node $v_i \in V$ in graph G, a directed edge from node i to j is defined as $e_{ij} = (v_i, v_j) \in E$. Moreover, on each edge e_{ij} contains a label $l_{ij} \in L$.

3 Methodology

In this section, we provide the implementation of the proposed integrated graph structure (Fig. 1) in detail.

Fig. 1 The integrated graph structure in the transformer-based model

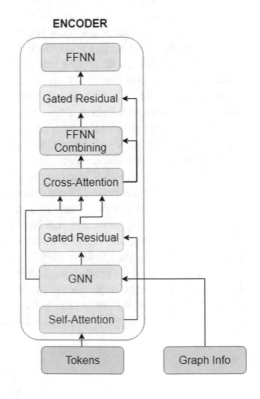

3.1   Graph Transformer

Graph Transformer imitates the multi-head attention of the Transformer where message computations transform the node embeddings into query, key, value.

$$q_i^{(t)} = W_q^{(t)} x_i^{(t)}$$
$$k_i^{(t)} = W_k^{(t)} x_i^{(t)} \qquad (1)$$
$$v_i^{(t)} = W_v^{(t)} x_i^{(t)}$$

here $W_q^{(t)}$, $W_k^{(t)}$ and $W_v^{(t)} \in R^{DxD}$ are trainable parameters, D is the dimensional size. The query, key, and value messages of node $x_i^{(t)}$ are respectively $q_i^{(t)}$, $k_i^{(t)}$, and $v_i^{(t)}$. For simplicity, we abandon the symbol of multi-head attention in the Eq. 1.

A cross-attention takes query $q_i^{(t)}$ of node i, key/value pair $k_j^{(t)}/v_j^{(t)}$ of its neighboor node j, and an edge label $l_{ij}^{(t)}$ to enhance node message as Eq. 2.

$$l_{ij}^{(t)} = W_l^{(t)} l_{ij}^{(t-1)}$$
$$\alpha_{ij}^{(t)} = \frac{score(q_i^{(t)}, k_j^{(t)} + l_{ij}^{(t)})}{\Sigma_u^{N(i)} score(q_i^{(t)}, k_u^{(t)} + l_{iu}^{(t)})} \qquad (2)$$

where $score(a, b) = exp(q^T k/\sqrt{d})$ is a scaled dot-product attention to compute attention score and $m_i^{(t)}$ is the final message of node i at t^{th} layer. The Graph Transformer applies sum aggregation over messages from neighbors as Eq. 3.

$$m_i^{(t)} = \Sigma_j^{N(i)} \alpha_{ij}^{(t)} (v_j^{(t)} + l_{ij}^{(t)})$$
$$h_i^{(t)} = Sum(\{m_j^{(t)}, j \in N(i)\}) \qquad (3)$$

To prevent the over-smoothing problem, the Graph Transformer applies a gated residual connection to compute node representation as Eq. 4.

$$s_i^{(t)} = W_s^{(t)} h_i^{(t)} + b_s^{(t)}$$
$$\beta_i^{(t)} = Sigmoid(W_g^{(t)} Concat(h_i^{(t)}; s_i^{(t)}; h_i^{(t)} - s_i^{(t)}) \qquad (4)$$
$$x_i^{(t+1)} = \beta_i^{(t)} s_i^{(t)} + (1 - \beta_i^{(t)}) h_i^{(t)})$$

where $W_s^{(t)} \in R^{DxD}$, $W_g^{(t)} \in R^{3DxD}$ are trainable parameters, and $Concat$ is a concatenate function.

3.2   The Transformer-Based Model

In this model (Fig. 1), we only modify the encoder side of the Transformer, whose decoder side is kept as original. First, the token embeddings are passed to the self-attention module to produce the sentence-level information, next, the graph embeddings are transferred to the GNNs to generate the graph-level information. The Gated Residual Connection is employed to blend the graph-level and sentence-level messages into one as Eq. 5.

$$out = \alpha * a + (1 - \alpha) * b$$
$$\alpha = Sigmoid(W_a a + W_b b) \tag{5}$$

where a and b are the two inputs from GNNs and self-attention modules. Moreover, we argue that the graph-level is non-trivial, thus the cross-attention module takes the graph-level messages as key/value pair and the blended messages as a query to produce higher-level information. Last, the Feedforward Neural Networks (FFNN) and FFNN Combining modules are similar, where the subtle difference is the FFNN Combining first concatenating two inputs vectors into one before applying the FFNN as Eq. 6.

$$x = \sigma(x W_1 + b_1) W_2 + b_2 \tag{6}$$

where W_1, W_2, b_1, and b_2 are trainable parameters, and σ is the Rectified Linear Unit (RELU) activation function.

4   Results and Analyses

In this section, we first provide details about the hyper-parameters configurations and the experimental results.

4.1   Datasets and Configurations

We use the International Workshop on Spoken Language Translation (IWSLT) 15 English-Vietnamese dataset[1] with 117k trained sentences and two test sets tst2013 and tst2015 (more information in Table 2). Table 1 shows the hyper-parameters for the standard and scaled Transformer, we note that the proposed method shares the same settings as the standard Transformer. N denotes the number of encoder and decoder layers. d_{model} and d_{ff} are the dimensional embeddings of the model and hidden sizes, where h is the number of attention heads in attention layers, and d_h is

---

[1] IWSLT datasets can be found in https://wit3.fbk.eu/2015-01.

Table 1 Hyper-parameters settings on English-Vietnamese dataset

Model	N	d_{model}	d_{ff}	h	d_h	P_{drop}	$P_{attn-drop}$	$P_{act-drop}$
Standard En-Vi	6	512	2048	8	64	0.3	0.0	0.0
Scaled En-Vi	6	1024	4096	16	64	0.3	0.2	0.2

Table 2 Statistics of the English to Vietnamese datasets

MT dataset	# tokens		#types		# sents	avg length		# docs
	en	vi	en	vi		en	vi	
Train	2435K	2867K	44,573	21,611	117,055	20.81	24.50	1192
Dev (tst2012)	27,988	34,298	3518	2170	1553	18.02	22.08	14
Test (tst2013)	26,729	33,683	3676	2332	1268	21.08	26.56	18
Test (tst2015)	20,850	26,235	3127	2059	1080	19.31	24.29	12

the dimension for each head. Lastly, P_{drop}, $P_{attn-drop}$, and $P_{act-drop}$ are respectively the percentages of dropout, attention dropout, and activation dropout.

We use case-sensitive BLEU (Bilingual Evaluation UnderStudy) scores [19] which uses an n-gram to evaluate the translation quality. We also use Meteor metric [20] that alleviates the disadvantage of BLEU by considering the synonyms. Additionally, the bootstrapping re-sampling method [21] is applied to do the significance tests to gain the most reliable comparison between our method and the Transformer. Finally, we use the dependency parser in Spacy[2] to preprocess data.

4.2 Ablation Study

Table 3 shows the experimental results on English to Vietnamese datasets, in which our baseline is described in Fig. 1 but without the GNNs and gated residual modules.

The standard Transformer gains 27.79 BLEU and 56.89 METEOR in the tst2013 test set, while the parameters of our baseline are 18.% higher but slightly improved by +0.06 BLEU and +0.22 METEOR. After employing the graph-level information, however, the proposed method significantly outperforms the standard and scaled Transformer in both evaluation metrics. Even more, the parameters of our model are 53.2% lower than that of the scaled Transformer while delivering better translation qualities, +0.46 and +0.47 BLEU in tst2013 and tst2015, respectively.

[2] https://spacy.io/models/en

Table 3 The ablation study is carried out to show the effectiveness of the proposed method

Model	#params (M)	tst2013		tst2015	
		BLEU	METEOR	BLEU	METEOR
Standard transformer	88.10	27.79	56.89	26.15	53.91
Scaled transformer	**252.96**	28.77	57.31	26.69	54.39
Our baseline	107.63	27.87	57.11	26.32	54.02
Our baseline + GNNs	118.31	**29.23**	**58.09**	**27.16**	**54.48**

The higher BLEU and METEOR scores are, the better translation qualities are

In conclusion, the Graph Transformer can provide comprehensive graph-level information that substantially improves the performance of NMT systems. The proposed method advances the current translation qualities and reduces the cost of training.

4.3 Performance Analyses on Complex Sentences

In this section, we want to answer where the proposed method improves and how it improves. First, we hypothesize that longer sentences have more information, and the higher tree depth contains a more complex structure. We introduce two groups based on sentence length and tree depth, then compute BLEU scores on each group to highlight differences between groups. Finally, we carry this method on IWSLT'17 English to Vietnamese tst2013.

Figure 2a indicates the results of the grouped-by-length method, in which the performance of the Graph Transformer is consistently higher than that of the standard

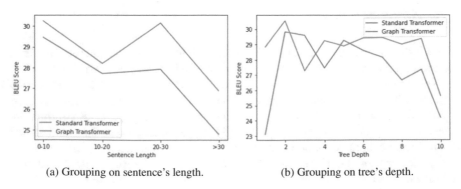

(a) Grouping on sentence's length.   (b) Grouping on tree's depth.

Fig. 2 Performance analyses on length and depth of sentences

Transformer. Especially for sentences longer than 20, the proposed method is greater than 2 BLEU scores, while the shorter one fluctuates around 0.5 BLEU.

Figure 2b illustrates the results of the grouped-by-depth method, where the performance of the Graph Transformer is diminished at group depths 3 and 5. To explain this phenomenon, we find that sentences with the same length do not imply the same depth, while the Transformer performs best in these groups. On the other hand, the proposed method remarkably exceeds the standard Transformer at a group depth smaller than 2 (+5.7 BLEU). When the tree depth becomes higher, the Graph Transformer is more consistently outperforms the standard Transformer. This implies that our method can extract more information at higher tree depths, leading to improve performances.

5 Conclusion

In this work, we propose the novel approach to provide syntactic awareness to NMT systems where syntactic information underlines in the Dependency Parsing. The proposed method outperforms the scaled parameters Transformer, while the parameters are reduced by 53.2%. Moreover, our method has a comprehensive ability in translating long and complex sentences over the standard Transformer.

In future work, we will evaluate our method on more language pair datasets (e.g. IWSLT En-Fr, En-Cs, En-De) and larger datasets (e.g. WMT En-De, En-Fr). We might also replace dependency trees by constituent trees to explore more syntactic relations between tree nodes.

Acknowledgements This research is funded by University of Science, VNU-HCM under grant number T2021-40.

References

1. Hamilton WL, Ying R, Leskovec J (2017) Inductive representation learning on large graphs. In: Proceedings of the 31st international conference on Neural Information Processing Systems, NIPS'17, (Red Hook, NY, USA), Curran Associates Inc., 2017, pp 1025–1035
2. Velickovic P, Cucurull G, Casanova A, Romero A, Liò P, Bengio Y (2018) Graph attention networks. In: 6th International Conference on Learning Representations, ICLR 2018, Vancouver, BC, Canada, April 30 - May 3, 2018, Conference Track Proceedings, OpenReview.net, 2018
3. Defferrard M, Bresson X, Vandergheynst P (2016) Convolutional neural networks on graphs with fast localized spectral filtering. In: Proceedings of the 30th international conference on neural information processing systems, NIPS'16, (Red Hook, NY, USA), Curran Associates Inc., pp 3844–3852
4. PW Battaglia JB, Hamrick VB, Sanchez-Gonzalez A, Zambaldi V, Malinowski M, Tacchetti A, Raposo D, Santoro A, Faulkner R, Gulcehre C, Song F, Ballard A, Gilmer J, Dahl G, Vaswani A, Allen K, Nash C, Langston V, Dyer C, Heess N, Wierstra D, Kohli P, Botvinick M, Vinyals O, Li Y, Pascanu R (2018) Relational inductive biases, deep learning, and graph networks

5. Yin Y, Meng F, Su J, Zhou C, Yang Z, Zhou J, Luo J (2020) A novel graph-based multi-modal fusion encoder for neural machine translation. In: Proceedings of the 58th annual meeting of the association for computational linguistics, (Online). Association for Computational Linguistics, pp 3025–3035

6. Song L, Wang A, Su J, Zhang Y, Xu K, Ge Y, Yu D (2020) Structural information preserving for graph-to-text generation. In Proceedings of the 58th annual meeting of the association for computational linguistics, (Online). Association for Computational Linguistics, pp 7987–7998

7. Morris C, Ritzert M, Fey M, Hamilton WL, Lenssen JE, Rattan G, Grohe M, Weisfeiler and leman go neural: higher-order graph neural networks. In: The Thirty-Third AAAI Conference on Artificial Intelligence, AAAI 2019, The Thirty-First Innovative Applications of Artificial Intelligence Conference, IAAI 2019, The Ninth AAAI Symposium on Educational Advances in Artificial Intelligence, EAAI 2019, Honolulu, Hawaii, USA, January 27 - February 1, 2019. AAAI Press, , pp 4602–4609

8. Xu K, Hu W, Leskovec J, Jegelka S (2019) How powerful are graph neural networks? In: 7th International Conference on Learning Representations, ICLR 2019, New Orleans, LA, USA, May 6-9, 2019, OpenReview.net

9. Chen D, Lin Y, Li W, Li P, Zhou J, Sun X (2019) Measuring and relieving the over-smoothing problem for graph neural networks from the topological view

10. Sato R, Yamada M, Kashima H (2019) Approximation ratios of graph neural networks for combinatorial problems. Curran Associates Inc., Red Hook, NY, USA

11. Balcilar M, Héroux P, Gaüzère B, Vasseur P, Adam S, Honeine P (2021) Breaking the limits of message passing graph neural networks. In: Proceedings of the 38th International Conference on Machine Learning (ICML)

12. Yang X, Feng S, Zhang Y, Wang D (2021) Multimodal sentiment detection based on multi-channel graph neural networks. In: Proceedings of the 59th Annual Meeting of the Association for Computational Linguistics and the 11th International Joint Conference on Natural Language Processing (Volume 1: Long Papers), (Online). Association for Computational Linguistics, pp 328–339

13. Schumann R, Riezler S (2021) Generating landmark navigation instructions from maps as a graph-to-text problem. In: Proceedings of the 59th Annual Meeting of the Association for Computational Linguistics and the 11th International Joint Conference on Natural Language Processing (Volume 1: Long Papers), (Online), Association for Computational Linguistics, pp 489–502

14. Shen W, Wu S, Yang Y, Quan X (2021) Directed acyclic graph network for conversational emotion recognition. In: Proceedings of the 59th annual meeting of the association for computational linguistics and the 11th international joint conference on natural language processing (Volume 1: Long Papers), (Online). Association for Computational Linguistics, pp 1551–1560

15. Yan H, Gui L, Pergola G, He Y (2021) Position bias mitigation: a knowledge-aware graph model for emotion cause extraction. In: Proceedings of the 59th annual meeting of the association for computational linguistics and the 11th international joint conference on natural language processing (Volume 1: Long Papers), (Online). Association for Computational Linguistics, pp 3364–3375

16. Cho K, van Merriënboer B, Bahdanau D, Bengio Y (2014) On the properties of neural machine translation: encoder–decoder approaches. In: Proceedings of SSST-8, Eighth workshop on syntax, semantics and structure in statistical translation, (Doha, Qatar), Association for Computational Linguistics, pp 103–111

17. Zhong L, Cao J, Sheng Q, Guo J, Wang Z (2020) Integrating semantic and structural information with graph convolutional network for controversy detection. In: Proceedings of the 58th annual meeting of the association for computational linguistics, (Online). Association for Computational Linguistics, pp 515–526

18. Zhao Y, Chen L, Chen Z, Cao R, Zhu S, Yu K (2020) Line graph enhanced AMR-to-text generation with mix-order graph attention networks. In: Proceedings of the 58th annual meeting of the association for computational linguistics, (Online). Association for Computational Linguistics, pp 732–741

19. Papineni K, Roukos S, Ward T, Zhu W-J (2002) Bleu: a method for automatic evaluation of machine translation. In: Proceedings of the 40th annual meeting of the association for computational linguistics, (Philadelphia, Pennsylvania, USA). Association for Computational Linguistics, pp 311–318
20. Banerjee S, Lavie A (2005) METEOR: an automatic metric for MT evaluation with improved correlation with human judgments. In: Proceedings of the ACL workshop on intrinsic and extrinsic evaluation measures for machine translation and/or summarization, (Ann Arbor, Michigan). Association for Computational Linguistics, pp 65–72
21. Koehn P (2004) Statistical significance tests for machine translation evaluation. In: Proceedings of the 2004 conference on empirical methods in natural language processing, (Barcelona, Spain). Association for Computational Linguistics, pp 388–395

Vietnamese Text Summarization Based on Neural Network Models

Khang Nhut Lam, Tuong Thanh Do, Nguyet-Hue Thi Pham, and Jugal Kalita

Abstract Text summarization produces a shortened or condensed version of input text highlighting its central ideas. Generating text summarization manually takes time and effort. This paper investigates several text summarization models based on neural networks, including extractive summarization, abstractive summarization, and abstractive summarization based on the re-writer approach and bottom-up approach. We perform experiments on the CTUNLPSum dataset in Vietnamese comprising 95,579 documents collected from commonly read Vietnamese online newspapers. The summarization model based on the bottom-up approach creates the best summaries. The F1-scores of ROUGE-1, ROUGE2, and ROUGE-L of the bottom-up approach are 0.598, 0.260, and 0.455, respectively.

Keywords Extractive summarization · Abstractive summarization · Re-writer · Bottom-up approach · LSTM · BiLSTM · BERT · Copy generator · AllenNLP · OpenNMT

1 Introduction

Hovy and Lin [1] define "a summary is a text that is produced out of one or more (possibly multimedia) texts, that contains (some of) the same information of the original text(s), and that is no longer than half of the original text(s)". The purpose of text summarization is to extract important information and to generate a summary such that the summary is shorter than the original and preserves the content of the text. Manually summarizing text is a difficult and time-consuming task when working with large amounts of information. This study will focus its discussion on automatically creating summaries for Vietnamese online articles.

K. N. Lam (✉) · T. T. Do · N.-H. T. Pham
Can Tho University, Can Tho, Vietnam
e-mail: lnkhang@ctu.edu.vn

J. Kalita
University of Colorado, Colorado Springs, USA
e-mail: jkalita@uccs.edu

© The Author(s), under exclusive license to Springer Nature Switzerland AG 2022
N. H. T. Dang et al. (eds.), *Artificial Intelligence in Data and Big Data Processing*,
Lecture Notes on Data Engineering and Communications Technologies 124,
https://doi.org/10.1007/978-3-030-97610-1_8

The two main approaches for automatically summarizing text [2] are extractive text summarization and abstractive text summarization. Most extractive summarization approaches perform 3 tasks: creating a representation of the original text (possibly to in terms of important keywords), scoring sentences based on the representation, and creating a summary by selecting sentences. Several approaches generated extractive text summaries in Vietnamese using different statistical-based approach [3], SVMs-based approach [4], graph-based approach [5–7], and genetic algorithm [8]. Abstractive summarization methods distill the main ideas of the original text and then generate a new shorter text, including new sentences that convey the main ideas in the original text. Extractive summarization methods are faster than abstractive summarization methods. In the summaries created using extractive methods, sentences are grammatical with correct spelling, but the connection between sentences is usually not good. In contrast, sentences in summaries created using the abstractive methods have good connections, but these approaches may produces ungrammatical sentences. Existing studies created abstractive Vietnamese text summaries using discourse rules, syntactic constraints, and word graph [9], LSTM with rich features [10], pointer generator networks using Bahdanau's attention mechanism [11].

In this paper, we focus on performing approaches based on neural networks for Vietnamese text summarization. First, we experiment with both the extractive and abstractive summarization models. Then, we study a re-writer approach that uses an abstractive summarization model to create summaries based on sentences chosen by an extractive summarization model. Finally, we discover an abstractive summarization model based on the bottom-up attention approach. The rest of this paper is organized as follows. In Sect. 2, we describe summarization models used for creating text summarization. Section 3 presents an evaluation method. Experiments are presented in Sect. 4. Section 5 concludes the paper.

2 Text Summarization Models

Our goal is to investigate the behavior of models to construct summaries for documents. Each document D consists of many sentences $sent_i$, each of which has many words w. We denote N_D to be the number of sentences in the document D.

2.1 GRU-RNN Model

An extractive summarizer using GRU-RNN model was introduced by Nallapati et al. [12]. The summarizer obtains that the summaries containing sentences from the documents, such that they maximize the ROUGE score [13]. This model is based on sequence classification. The first stage of the summarization process is to label sentences in a document D with 0 or 1. If a sentence $sent_i$ is labeled 0, this sentence has no or very little relation to the summary; in contrast, a sentence $sent_i$ is labeled

1 has relevance to the summary. A greedy approach and the ROUGE metric are used to label sentences. The collection of sentences labeled 1 is used for the training step to create summaries.

Sequentially processing sentences from the beginning to the end of the document D, a binary decision is made to add a particular sentence to the summary or not. A decision of 0 means a sentence is not included in the summary, and conversely. The GRU-RNN model consists of 2 layers. The first layer is a bidirectional RNN operating at word level in each sentence. An RNN runs in the forward direction from the first word to the last word, computes a hidden state representation at every word position based on the current embedding and the previous hidden state. Another RNN runs backward from the last word to the first word. The backward RNN and the forward RNN run simultaneously. The second layer is also a bidirectional RNN operating at sentence level in each document. The input to this RNN is the concatenation of hidden states of the bidirectional RNN at the word level. The top of this model is the classification layer to decide whether or not a sentence should be included in a summary. This model is used to construct summaries for Vietnamese documents by Lam et al. [14].

2.2 Sequence-to-Sequence Model with Attention Mechanism

A sequence-to-sequence (seq2seq) model consists of a neural network called encoder and another one called decoder. The encoder reads an input word sequence and encodes it into a single fixed-length vector called the context vector; the decoder reads that context vector and produces an output word sequence. The main issue with this model is the fixed-length context vector, especially for long input sequences. To address this issue, Luong et al. [15] and Bahdanau et al. [16] introduced attention mechanisms to help the model concentrate on the most relevant information at the right time in the input sequences.

Nallapati et al. [17] augmented a seq2seq model with Bahdanau's attention mechanism for text summarization. The sequence of words w_i in the document D are fed into the encoder to produce a sequence of encoder hidden states h_i. At each step t, the decoder receives the word embedding of the previous word and has decoder state s_t. The probability distribution of source words or the attention distribution of the output at each time step t, denoted a^t, is calculated by applying a softmax function as below,

$$e_i^t = v^T \tanh(W_h h_i + W_s s_t + b) \qquad (1)$$

$$a^t = softmax(e^t) \qquad (2)$$

where v, W_h, W_s, and b are learnable parameters. The context vector c_t is the sum of the product of probability distributions and encoder hidden states. The concatenation of the decoder state s_t and the context vector c_t is fed to a neural network model to produce the vocabulary distribution P_{vocab},

$$p_{vocab} = softmax(V'(V[s_t, c_t] + b) + b') \tag{3}$$

where V, V', b, and b' are learnable parameters. The seq2seq model is usually trained as a conditional language model. The training process maximizes the likelihood of each output word. At each step t, the decoder needs to decide which word to generate at the position t in a sentence. The problem is that the model does not know which word sequence should be generated to maximize the overall probability. To handle this issue, greedy search or beam search [18] is used to search through a list of output word candidates. In our experiment, we will use RNNs as encoder and decoder, and beam size of 5.

2.3 Sequence-to-Sequence Model with Pointer Generator

Constructing text summarization using the seq2seq model with attention mechanism faces two common issues: the summaries may be inaccurate because of out-of-vocabulary tokens or the summaries may include repeated words or phrases. To overcome these problems, See et al. [19] proposed a seq2seq model with a pointer network [20] to decide whether or not words should be generated from vocabulary or copied from the input word sequences. See et al. [19] also introduce the coverage model [21] to solve the repetition problem by using the coverage vector c^t to keep track of the attention history, to be used as an extra input to the attention mechanism, to penalize repeated attending at the same locations, See et al. [19] compute a coverage loss as below,

$$covloss_t = \sum_i \min(a_i^t, c_i^t). \tag{4}$$

The generation probability p_{gen} at time step t is computed using a sigmoid function.

$$p_{gen} = \sigma(w_{h*}^T h_t^* + w_s^T s_t + w_x^T x_t + b_{ptr}) \tag{5}$$

where w_{h*}, w_s, w_x, and b_{ptr} are learnable parameters. The copy distribution probability p_{copy}, which tells us when we should copy words from the input document, is obtained by the sum of attention distributions a^t. Finally, the combination of the copy distribution and vocabulary distribution is used to compute the probability of a word w. If w is OOV, then $p_{vocab}(w)$ is 0, and if the w is not in the source document, then the $p_{copy}(w)$ is 0.

$$p(w) = p_{gen} p_{vocab}(w) + (1 - p_{gen}) p_{copy}(w). \tag{6}$$

2.4 Bottom-Up Approach

A drawback of the neural network approaches for summarization is that they have difficulty in selecting content in the document. The bottom-up attention approach, which is a state-of-the-art model for image processing [22] that detects the bounding boxes of objects and applies attention on them, is applied to create the content selector of abstractive summarization [23]. The bottom-up approach for summarization eliminates the selection of redundant words and enhances copying of important words in the source documents by using attention masks. The bottom-up approach for constructing summaries uses (i) the Content-Selector model based on the AllenNLP[1] platform to select words or phrases in the documents that appear in summaries based on the probability distribution of words, and (ii) the OpenNMT[2] framework to construct abstractive summaries.

The Content-Selector consists of a tagging model and a predictor. Words in a document that are important can be copied if they are parts of the longest possible sub-sequences in the reference summary, as shown in Fig. 1.

The tagging model computes the probability of words p_{copy} using the following equation:

$$p_{copy} = \sigma(W_z \widetilde{h}_t + b) \tag{7}$$

where the $\widetilde{h}_t$ is the representation of the word w_t obtained by concatenating the word embedding and the contextual embedding into a vector and feeding it to a neural network model. These selection probabilities of words in the document are later used to modify the copy attention distribution of words $p(a_t)$ by OpenNMT. The predictor predicts the probability distribution of the words in any text based on the Tagging model and the processed source text. The OpenNMT framework supports sequence-to-sequence models used to generate abstractive summarization. In our work, we use the pointer generator model using Luong's attention mechanism [15] as

A document   A reference summary

Fig. 1 An example of important words in an input document and its reference summary

---

[1] https://allennlp.org.

[2] https://opennmt.net/.

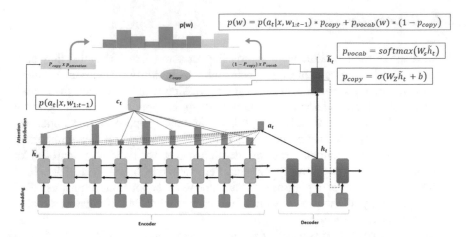

Fig. 2 The copy generator model using Luong's attention mechanism.

presented in Fig. 2. The Bottom-up approach for creating abstractive summarization combines the tags extracted from the Content-Selector and replaces attention probability $p(a_t|x, w_{1:t-1})$ by $p'(a_t|x, w_{1:t-1}, tag) = p(a_t|x, w_{1:t-1}) \times q(tag) \times 2$.

2.5 BERT Model

BERT [24] has been used to fine-tune a variety of NLP tasks, including text summarization. A token "CLS" is added at the beginning each sentence $sent_i$, and the vector for "CLS" represents $sent_i$. BERTSUM [25] is built on the output of BERT by applying L Transformer layers on the sentence representations. A decision whether to extract a sentence $sent_i$ for summary is made by computing the predicted score $\hat{Y}_i$:

$$\hat{Y}_i = \sigma(W_o h_i^L + b_o), \tag{8}$$

where the h_i^L is the vector of $sent_i$ from the top layer of the Transformer. Beside, the authors also introduced an abstractive summarization model using encoder-decoder framework, where the encoder is BERTSUM and the decoder is a 6-layer Transformer. We use the same parameters as suggested by the authors [25] (L=2, Adam optimizer with $\beta_1 = 0.9$, and $\beta_2 = 0.999$).

3 Evaluation Method

The two common approaches for evaluating text summarization are manual and automatic evaluation. Manual evaluation of summaries is performed by experts using their own criteria. Automatic evaluation of summaries uses algorithms to evaluate results automatically by comparing summaries created by the summarizer with reference summaries created by humans. Currently, the ROUGE (Recall-Oriented Understudy for Gisting Evaluation) [13] and BLEU (Bilingual Evaluation Understudy) [26] metrics are commonly used to evaluate summaries. These two methods are N-gram overlap metrics. The focus of the BLEU metric is N-gram precision, while this of the ROUGE metric is N-gram recall. In particular, BLEU measures how many of the N-gram in the system summaries appear in the reference summaries; whereas ROUGE measures how many N-grams in the reference summaries appear in the system summaries.

In this paper, ROUGE-N (e.g., ROUGE-1 and ROUGE-2) and ROUGE-L are used to evaluate experimental results. ROUGE-N computes N-gram overlaps between summaries created and reference summaries. The precision, recall, and F1-score values are calculated using the following formulas:

$$Precision = \frac{|\text{N-gram overlaps between system and reference summaries}|}{|\text{N-grams in the system summary}|} \quad (9)$$

$$Recall = \frac{|\text{N-gram overlaps between system and reference summaries}|}{|\text{N-grams in the reference summary}|} \quad (10)$$

$$F1\text{-}score = \frac{|Precision \times Recall|}{|Precision + Recall|} \times 2. \quad (11)$$

4 Experimental Results

We perform experiments on the CTUNLPSum dataset comprising 95,579 articles collected from commonly read Vietnamese newspapers such as vnexpress,[3] tuoitre,[4] dantri,[5] danviet,[6] laodong.[7] Some newspapers provide both Vietnamese edition and English edition as presented in Fig. 3, but we do not know whether they are always truly comparable. Therefore, we experiment on the Vietnamese edition only.

Pre-processing data is the first step to build text summaries. The NLTK[8] toolkit is used to segment paragraphs and sentences. Then, the Underthesea[9] toolkit is used

---

[3] https://vnexpress.net/.

[4] https://tuoitre.vn/.

[5] https://dantri.com.vn/.

[6] https://danviet.vn/.

[7] https://laodong.vn/.

[8] https://www.nltk.org/.

[9] https://github.com/undertheseanlp.

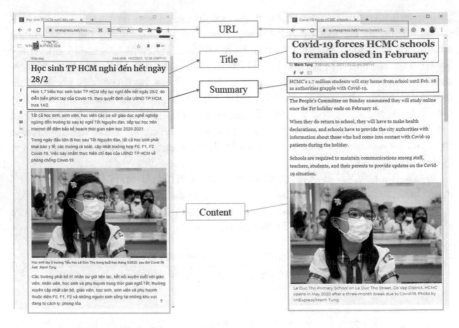

Fig. 3 An article in Vietnamese and its translation in English from the website vnexpress.net

to segment Vietnamese words. The final stage of the preprocessing step is to remove special characters such as

We perform experiment in the Google Colab environment with 12GB RAM with GPU. We evaluate models using the pyrouge[10] package. The datasets are divided into 80%, 10% and 10% for training, validation and testing, respectively.

In our experiments, the summarization models are used to create summaries that contain 3 sentences for each input document. First, we evaluate extractive summarization models and the abstractive summarization models, presented in Table 1. We found that BERT extracts the best sentences. We also perform experiments with abstractive models based on the Transformer and BERT models by putting the copy-generator model on the top of the decoder as proposed by Aksenov et al. [27]. The results show that the copy generator model helps BERT abstractive summarization model achieve the best F1-score again.

In addition, we develop an approach, called the re-writer, which combines the strength of extractive and abstractive summarization models. Given an document, the Content-Selector model is used to compute the distribution probabilities of words. Each sentence is assigned a score calculated by the average distribution probabilities of words in the sentences. Then, we simply extract five sentences with the highest scores. Next, these five extracted sentences are fed to OpenNMT at the inference step to create abstractive summaries with copy generator model. Table 2 shows that

---

[10] https://pypi.org/project/pyrouge/0.1.0/.

Table 1 ROUGE scores of extractive and abstractive summarization models

Model	ROUGE-1	ROUGE-2	ROUGE-L
Extractive summarization models			
RNN	0.384	0.200	0.327
GRU-RNN [14]	0.440	0.192	0.326
LSTM	0.384	0.198	0.327
BiLSTM	0.392	0.200	0.331
BERT	0.449	0.186	0.325
Abstractive summarization models			
RNN with attention	0.480	0.130	0.368
BERT abstractive	0.442	0.179	0.333
Abstractive summarization models with copy generator			
RNN with copy generator	0.548	0.221	0.428
Transformer with copy generator	0.536	0.179	0.406
BiLSTM with copy generator	0.543	0.228	0.433
BERT with copy generator	0.550	0.211	0.425

Table 2 ROUGE scores of abstractive summarizers using the re-writer approach

Model	ROUGE-1	ROUGE-2	ROUGE-L
RNN using the re-writer approach	0.548	0.202	0.420
LSTM using the re-writer approach	0.548	0.204	0.422
BiLSTM using the re-writer approach	0.553	0.212	0.427

F1-score of abstractive summarization model based on re-writer approach is higher, indicating improved the quality of summaries.

We also extract 5 sentences from given documents using the BERTSUM extractive summarization model, and feed them to the BERT abstractive summarization model to rewrite the summary. The F1-scores of ROUGE-1, ROUGE-2, and ROUGE-L of this method are 0.465, 0.210, and 0.399, respectively.

Finally, Table 3 shows F1-score of the abstractive summarization models using the bottom-up approach. The BiLSTM abstractive summarization model using the bottom-up approach outperforms other summarization models.

Our experiments show that BERT extracts the best sentences; the BiLSTM with the copy generator is not as good as the BERT with copy generator for constructing abstractive summaries, but the BiLSTM using the bottom-up approach gives the best

Table 3 ROUGE scores of abstractive summarizers using the bottom-up approach

Model	ROUGE-1	ROUGE-2	ROUGE-L
RNN using the bottom-up approach	0.528	0.207	0.412
LSTM using the bottom-up approach	0.533	0.208	0.414
BiLSTM using the bottom-up approach	0.598	0.260	0.455

results. We have not experimented with the BERT using the bottom-up approach. Besides, the experiments of Luu et al. [28] show that the combining of PhoBERT model [29], Multi-Layer Perceptron for selecting sentences and Maximal Marginal Relevance for eliminating repeated words helps generate Vietnamese text summarization with the highest F1-scores of 0.525, 0.247, and 0.378 for ROUGE-1, ROUGE-2, and ROUGE-L, respectively. In addition, the PhoBERT2PhoBERT model is the best model for abstractive summarization [30] with the F1-scores of 0,604, 0,291, and 0,394 for ROUGE-1, ROUGE-2, and ROUGE-L, respectively. These indicate that the combination of BERT or PhoBERT with the bottom-up approach may improve the quantity of summaries.

5   Conclusion

We experimented with text document summarization models based on neural networks, including extractive summarization models, abstractive summarization models, abstractive summarization models based on the re-writer approach, and abstractive summarization model based on the bottom-up approach. The extractive summarization models have better F1-score than abstractive model; however, the sentences in the extracted summaries are not well connected. The abstractive summarization models with copy generation increase the quality of the summaries by eliminating redundant words while handling well the OOV issue. The re-writer approach helps improve the F1-scores of models because it uses important sentences extracted from a document as input and simply rewrites these sentences. The bottom-up approach handles the content selection in the source document very well. We notice that the BERT and PhoBERT summarization models perform very well. For future work, we plan to develop a method that integrates the BERT and PhoBERT models, the copy generator model, the re-writer approach, and the bottom-up approach together to exploit the strengths of each model.

References

1. Hovy E, Lin CY (1999) Automated text summarization in SUMMARIST. Advances in automatic text summarization 14:81–94
2. Allahyari M, Pouriyeh S, Assefi M, Safaei S, Trippe ED, Gutierrez JB, Kochut K (2017) Text summarization techniques: a brief survey. arXiv preprint arXiv:1707.02268
3. Ha TL, Huynh QT, Luong CM (2005) A primary study on summarization of documents in Vietnamese. In: Proceedings of the first world congress of the international federation for systems research (ISFR'2005), 329–344 (2005)
4. Minh LN, Shimazu A, Xuan HP, Tu BH, Horiguchi S (2005) Sentence extraction with support vector machine ensemble. In: Proceedings of the first world congress of the international federation for systems research: the new roles of systems sciences for a knowledge-based society. JAIST Press, pp 69–73
5. Dinh TQ, Dung NQ (2012) Mot giai phap tom tat van ban tieng Viet tu dong. In: Proceedings of the 15th national conference
6. Hoang TAN, Nguyen HK, Tran QV (2020) An efficient Vietnamese text summarization approach based on graph model. In: Proceedings of the RIVF international conference on computing and communication technologies, research, innovation, and vision for the future (RIVF). IEEE, pp 1–6
7. Phuc D, Hung MX, Phung NT (2008) Using SOM based graph clustering for extracting main ideas from documents. In: Proceedings of the RIVF international conference on computing and communication technologies. IEEE, pp 209–214
8. An NN, Bac NQ, Hieu ND, Anh TN (2014) Determining the text feature coefficients by genetic algorithm for Vietnamese text summarization. J Military Sci Technol- JMST 32:36–46
9. Le HT, Le TM (2013) An approach to abstractive text summarization. In: 2013 International conference on Soft Computing and Pattern Recognition (SoCPaR). IEEE, pp 371–376
10. Quoc VN, Le Thanh H, Minh TL (2019) Abstractive text summarization using LSTMs with rich features. In: International conference of the pacific association for computational linguistics. Springer, pp 28–40
11. Anh DT, Trang NTT (2019) Abstractive text summarization using pointer-generator networks with pre-trained word embedding. In: Proceedings of the tenth international symposium on information and communication technology, pp 473–478
12. Nallapati R, Zhai F, Zhou B (2017) Summarunner: a recurrent neural network based sequence model for extractive summarization of documents. In: Proceedings of the AAAI conference on artificial intelligence, vol 31 (2017)
13. Lin CY (2004) ROUGE: a package for automatic evaluation of summaries. In: Text summarization branches out, pp 74–81
14. Lam KN, Phan KC, Tran NB (2019) Experiments on generating text summarization using extractive methods. In: Proceedings of the 12th national conference on Fundamental and Applied Information Technology Research (FAIR'12), pp 468–475
15. Luong MT, Pham H, Manning CD (2015) Effective approaches to attention-based neural machine translation. arXiv preprint arXiv:1508.04025
16. Bahdanau D, Cho K, Bengio Y (2014) Neural machine translation by jointly learning to align and translate. arXiv preprint arXiv:1409.0473
17. Nallapati R, Zhou B, Gulcehre C, Xiang B et al (2016) Abstractive text summarization using sequence-to-sequence RNNs and Beyond. arXiv preprint arXiv:1602.06023
18. Wiseman S, Rush AM (2016) Sequence-to-sequence learning as beam-search optimization. arXiv preprint arXiv:1606.02960
19. See A, Liu PJ, Manning CD (2017) Get to the point: summarization with pointer-generator networks. arXiv preprint arXiv:1704.04368
20. Vinyals O, Fortunato M, Jaitly N (2015) Pointer networks. arXiv preprint arXiv:1506.03134 (2015)
21. Tu Z, Lu Z, Liu Y, Liu X, Li H (2016) Modeling coverage for neural machine translation. arXiv preprint arXiv:1601.04811

22. Anderson P, He X, Buehler C, Teney D, Johnson M, Gould S, Zhang L (2018) Bottom-up and top-down attention for image captioning and visual question answering. In: Proceedings of the IEEE conference on computer vision and pattern recognition, pp 6077–6086
23. Gehrmann S, Deng Y, Rush AM (2018) Bottom-up abstractive summarization. In: Proceedings of the 2018 conference on empirical methods in natural language processing, pp 4098–4109
24. Devlin J, Chang MW, Lee K, Toutanova K (2018) BERT: pre-training of deep bidirectional Transformers for language understanding. arXiv preprint arXiv:1810.04805
25. Liu Y, Lapata M (2019) Text summarization with pretrained encoders. arXiv preprint arXiv:1908.08345
26. Papineni K, Roukos S, Ward T, Zhu WJ (2002) Bleu: a method for automatic evaluation of machine translation. In: Proceedings of the 40th annual meeting of the association for computational linguistics, pp 311–318
27. Aksenov D, Moreno-Schneider J, Bourgonje P, Schwarzenberg R, Hennig L, Rehm G (2020) Abstractive text summarization based on language model conditioning and locality modeling. arXiv preprint arXiv:2003.13027
28. Luu MT, Le TH, Hoang MT (2021) An effective deep learning approach for extractive text summarization. Ind J Comput Sci Eng (IJCSE)
29. Nguyen DQ, Nguyen AT (2020) PhoBERT: pre-trained language models for Vietnamese. arXiv preprint arXiv:2003.00744
30. Nguyen H, Phan L, Anibal J, Peltekian A, Tran H (2021) VieSum: how robust are transformer-based models on Vietnamese summarization? arXiv preprint arXiv:2110.04257

Adapting Cross-Lingual Model to Improve Vietnamese Dependency Parsing

Duc Do, Dien Dinh, An-Vinh Luong, and Thao Do

Abstract Dependency parsing is the task of analyzing the syntax of a sentence into a direct binary relational graph. Many languages have had the state-of-the-art model in this task and used it as a knowledge base to solve more complex problems. However, to achieve high accuracy in a dependency parsing model, it takes significant time and labor to build a large amount of annotated treebanks. For languages with little or no annotated treebanks, some approaches have been studied to induce a dependency parser from treebanks of high-resource languages to solve this problem. In this paper, we propose an approach to building a cross-lingual model to parse Vietnamese as a low-resource target language. The model uses English as a supportive high-resource source language to induce a Vietnamese parser. To remove the differences in syntaxes and lexicons of English and Vietnamese when training the model, the approach uses a filtering algorithm to choose English sentences having syntaxes as same as Vietnamese sentences based on Euclidean distance. The result shows that the proposed model significantly improves accuracy compared with models using only supervised mono-lingual treebanks.

Keywords Deep biaffine attention · Dependency parsing · Low-resource language · Cross lingual method · Transfer learning

D. Do (✉) · D. Dinh · A.-V. Luong
University of Science, Ho Chi Minh City, Vietnam
e-mail: ddien@fit.hcmus.edu.vn

Vietnam National University, Ho Chi Minh City, Vietnam
e-mail: dotrananhduc@gmail.com

A.-V. Luong
Saigon Technology University, Ho Chi Minh City, Vietnam

T. Do
University of Bath, Bath, UK
e-mail: tnd32@bath.ac.uk

97

1 Introduction

The process of building dependency parsing models is primarily based on languages having high resources of annotated syntactic trees, or *treebanks*, to train a high-accuracy parsing model. In some languages with large treebanks (e.g., English, Chinese), their model has achieved certain success in building a state-of-the-art dependency parser [13, 26]. However, it is not feasible and expensive for languages with few resources to achieve such success due to the significant time, effort, and cost required in building treebanks.

To solve this problem, many solutions have been proposed, notably, the work of [5, 11, 22] to use the cross-lingual method. These methods have similarities of utilizing treebanks from different high-resource languages, and extract the features to support building parsing models for low-resource languages. Some other tasks also use this method to train their model to overcome the resource shortage, such as sentiment analysis [10], and plagiarism detection [20].

Moreover, in the dependency parsing task, using the cross-lingual method has emerged as one of the recent prominent methods to solve the problem of building treebanks. Many researchers have further developed the technique into a separate branch of cross-lingual dependency parsing to support building a multitasking and multilingual parsing model for different languages [1, 2, 11]. Specifically, the model uses treebanks from high-resource languages to extract the linguistic features, synthesizes them into a unified knowledge base, and then applies them to the parser of low-resource languages. Corpus to support the process of building a cross-lingual model can be in dictionaries and bi-lingual for both source and target languages. In some cases, a bi-lingual corpus is not available, thus, the cross-lingual issue represents itself as a necessary problem and offers a new method to build a parsing model for low-resource languages.

In this paper, we present a cross-lingual dependency parsing model using English as the support language to resolve the low resource issue in Vietnamese. The model uses lexical features as the main feature to help limit the dependence on different sets of labels between two linguistic corpora. In addition, word embedding aids in eliminating the lexical difference between the two languages. Furthermore, Euclidean distance is used based on two coefficients to choose suitable syntactic English sentences to train the parsing model for Vietnamese. Finally, the model is trained on the Deep Biaffine Attention [4] to obtain improved performance than traditional algorithms. The result shows that the proposed model achieves a significantly increased accuracy of +0.17% in unlabeled attachment score (UAS) and bridges the gap to a fully supervised dependency parsing model. To sum up, major contributions in this paper can be listed as follow:

1. The Euclidean distance algorithm is presented to filter English sentences having syntaxes as same as Vietnamese sentences based on two coefficients of a graph.
2. The differences in annotated labels between two corpora is reduced by using cross-lingual word embeddings.

3. A flexible pipeline of training a cross-lingual dependency parsing model is proposed using Deep Biaffine Attention with supportive languages, which can effectively improve the performance of the low-resource language parser.

2 Proposed Model

In this section, the proposed model and its components are explained in details.
Specifically, the parser model is built by modifying work from [25, 27] to adapt
to the given problems. Based on the general graph-based algorithm for dependency
parsing [9], the detailed architecture of the cross-lingual dependency parsing model
is designed. The processing of training starts with filtering the suitable treebanks for
the parsing model; then these filtered treebanks are provided into the main algorithm
to induce a parser.

2.1 Treebank Filtration

To find suitable treebanks, based on Oya's work [19], a filtering algorithm to choose
English sentences having a similar syntax with Vietnamese is proposed. Oya [19]
measured the difference between the syntactic structure between the source sentence
and its translated sentence by the Euclidean distance between the two coefficients of
the graph representing a sentence. According to the author, this method, in addition
to calculating the difference between two pairs of translated languages, can also
calculate the difference of sentence pairs in two different languages.

Using translated bilingual corpus in Oya [19], the features of the author's method
is applied to the problem of cross-lingual parsing, which utilizes the same sentences
in the supportive language with the sentences in the boosted language.

Thus, a sentence whose syntax is represented as a graph has the degree centrality
coefficients and the closeness centrality coefficients expressed in the Cartesian coordinate. The difference between them can be calculated by Euclidean distance, with
the vertical axis representing the degree centrality coefficients and the horizontal
representing the closeness centrality coefficients. In addition, the two sentences are
syntactically similar when the Euclidean distance is zero.

Degree Centrality In a graph, the total number of edges associated with a node
(regardless of direction) is represented by the number of degrees, conventionally
$d(n_i)$, or the degree centrality of a node [24]. The higher the number of degrees,
the more significant degree centrality. Following that, the group degree centrality
coefficient measures the direct relationships of a group of nodes. The higher the
value, the greater the degree centrality of the group of nodes. The central position
of the graph, thus, can also be described. The equation used to calculate the group
centrality coefficient is as follow [6, 7, 24]:

$$C_D = \frac{\sum_{i=1}^{g}[C_D(n^*) - C_D(n_i)]}{\max \sum_{i=1}^{g}[C_D(n^*) - C_D(n_i)]} \tag{1}$$

where, $C_D(n_i)$ is the degree centrality coefficient of a node in the graph. $C_D(n^*)$ is the maximum degree centrality coefficient of a node in that graph.

The resulting centrality coefficient of a group is always in the range $[0 : 1]$. In describing the syntax of a sentence, the degree centrality coefficient indicates how central the words are to a particular word in that sentence. A coefficient of 1 means a node connected to all the remaining nodes in the graph if an adjacent node presents. In syntactic representation, it indicates that there exists a word that joins all the rest of the words in the sentence. When the coefficient result approaches 0, it signifies that the association of a node with the remaining nodes in the graph is decreasing. In a sentence, a word with a degree centrality coefficient close to zero shows how little the associative dependence of other words is on that word.

Closeness Centrality In a graph, the closeness centrality coefficient measures how close the node is to all other nodes in the set of nodes [24]. A word close to other words shows that the word has a lot of information and influence on the whole graph. To measure the metric in a graph, a method that uses geodesic path counting from one word to all other words is applied. The higher the coefficient is, the more likely a node is the closest node in that graph. According to the authors [24], the geodetic distance between two nodes in the conventional graph is $d(n_i, n_j)$. The total distance from the node n_i to the rest of the nodes in the graph is conventionalized by the equation $\sum_{i=1}^{g} d(n_i, n_j), \forall j \neq i$. Therefore, the equation of the closeness centrality coefficient of a node is expressed as the inverse of the sum of the distances of any node to the other node in the graph. The equation by authors [21] is defined as follows:

$$C_C(n_i) = \left[\sum_{i=1}^{g} d(n_i, n_j)\right]^{-1} \tag{2}$$

From Eq. 2, a higher coefficient result shows that a node can go to other nodes very easily and quickly. A low closeness centrality coefficient result indicates that a node has difficulty trying to connect with the rest of the nodes in the graph. The closeness centrality coefficient for a word is in the range $[0 : 1]$, representing the inverse average distance of that word to the rest of the others in the graph. The coefficient reaches its maximum value when a node connects to all other nodes in the graph; this means that node has the shortest total geodesic path to all other nodes.

Thus, the closeness centrality group equation is calculated as follows [6, 7, 24]:

$$C_C = \frac{\sum_{i=1}^{g}[C'_C(n^*) - C'_C(n_i)]}{[(g-1)(g-2)]/(2g-3)} \tag{3}$$

Euclidean Distance Each syntax of the sentence has only one group closeness centrality coefficient and one group degree centrality coefficient; thus, it can be represented on the Cartesian coordinate. Additionally, it is possible to use Euclidean distance to calculate the syntax similarity. The equation for calculating Euclidean distance is shown as follows, with $p = (p_1, \ldots, p_n)$ and $q = (q_1, \ldots, q_n)$ being two edges containing points in the coordinate [19]:

$$d(p.q) = d(q, p) = \sqrt{(q_1 - p_1)^2 + \cdots + (q_n - p_n)^2} = \sqrt{\sum_{i=1}^{n} (q_i - p_i)^2} \quad (4)$$

The Euclidean distance represents the similarity between the two syntaxes of two sentences, and a result of 0 represents two sentences with similar syntax. Conversely, a non-positive result greater than 0 shows that the syntax of the two sentences is different.

2.2 Deep Biaffine Attention

In this section, the approach to train the parsing model is explained in details. Specifically, PhoBERT [14] is selected as pre-training word embedding to eliminate the different vocabulary in the cross-lingual dependent parsing problem. A sentence containing n words is denoted as follow: $s = w_1 w_2 \ldots w_n$. Equation 5 shows $i^{th} vector$ for i^{th} word in the sentence. Figure 1 represents the detailed architecture of the cross-lingual dependency parsing model.

$$x_i = e(w_i) \quad (5)$$

In the second layer of the model, the BiLSTM layer is applied after the word embedding layer [8]. In this layer, the word vectors will be computed based on the context surrounding them, using equation $v_i = BiLSTM(x_{1:n}, i)$. The output of the BiLSTM layer is a vector representing the context of the word in the sentence.

Subsequently, the MLP layer adopted from work [8] uses the vector results from the BiLSTM layer as input. It is used as a filtering funnel for unimportant and irrelevant information from the results of the BiLSTM layer. The MLP layer, thus, could reduce the amount of information that needs to be processed to train the model [4].

To calculate the score of label between two words in a sentence, an equation defines the probability that label goes with a word m_i and word m_j as follow:

$$l_i^{arc-dep} = MLP^{arc-dep}(v_i)$$
$$l_j^{arc-head} = MLP^{arc-head}(v_j)$$
$$score^{label}(l_i) = L^{arc-head} U l_i^{arc-dep} + L^{arc-head} u \quad (6)$$

Similarly, to calculate the arc's score between two words, for each word m_i and dependent word m_j, the equation is as follows:

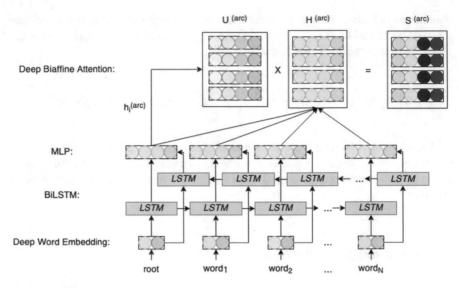

Fig. 1 The detailed architecture of the cross-lingual dependency parsing model

$$h_i^{arc-dep} = MLP^{arc-dep}(v_i)$$
$$h_j^{arc-head} = MLP^{arc-head}(v_j)$$
$$score^{arc}(h_i) = H^{arc-head} U h_i^{arc-dep} + H^{arc-head} u \qquad (7)$$

After calculating the score for each word pair in a sentence, the final graph will include weights on the arc, and the algorithms will be applied to find the maximum spanning tree in the graph, representing the dependent syntax of a sentence.

3 Experimental Setup

This experiment section introduces the Vietnamese and English treebanks and the strategies to train the proposed cross-lingual dependency parsing model.

Datasets: In the experiment, POS-predicted VnDT [16, 17] is used for Vietnamese treebanks. The treebanks are built based on automatically converting the constituent syntax to dependent syntax. According to the authors [16], the VnDT corpus includes 10,197 sentences and 218,749 words, with each sentence formatted according to the general formula CoNLL of the parsing problem [3, 18]. On the other hand, Penn Treebank [12] is used for English. Penn Treebanks are constituent syntax trees built for English, including 43,948 sentences and 1,046,829 words. In response to the dependency parsing problem, the Stanford Dependency converter version 3.3.0 is used to convert to the dependency parsing trees.

Table 1 Statistical table of the number of sentences corresponding to each length

Length	Vietnamese sentences	English sentences
Greater than 2	8965	43,722
From 5 to 20	4372	16,100
From 20 to 30	2453	14,726
From 30 to 40	1178	8566
From 40 to 50	479	2923
From 5 to 50	8514	42,451

During the training process, 8977 Vietnamese sentences for the training, 200 sentences for the development, and 1020 sentences for the testing are utilised. English treebanks are clustered according to the number of sentence lengths to filter out English sentences whose length is likely to affect the effectiveness of the model.

Additionally, the English corpus is divided into smaller parts consisting of tree-banks which have a length greater than 2, length 5–20, length 20–30, length 30–40, length 40–50, length 5–50, length 5–20 combined with length 20–30, length 5–20 combined with length 30–40, length 5–20 combined with length 40–50, length 20–30 combined with length 30–40, length 20–30 combined with 40–50, length 30–40 combined with length 40–50. Table 1 shows the statistics of the number of sentences corresponding to each length interval.

Parameters: Some parameters and configurations are adopted as in [4] to make a fair comparison. The training will stop after 10 consecutive loops without increasing the model's accuracy.

Evaluation metrics: The metrics to evaluate parsing models are LAS (Labeled Attachment Score) and UAS (Unlabeled Attachment Score). LAS is a point that evaluates the percentage of correctly annotating labels and dependency arcs to words. In comparison, UAS is a point that only evaluates the percentage of correctly annotating dependency arcs to words.

4   Results and Discussion

In this section, a summary of the findings is presented with further discussion about model improvement.

4.1   Results

In the training process, the model only uses English supporting sentences with a similar grammatical structure to Vietnamese with completely similar values, filtered

Table 2 Results of proposed model trained on different lengths

Models	UAS (%)	LAS (%)
Length >2	85.15	78.35
Length 5–20	85.17	78.29
Length 20–30	85.09	78.27
Length 30–40	85.19	78.27
Length 40–50	84.97	78.18
Length 5–50	85.21	78.40
Length 5–20 + 20–30	85.08	78.14
Length 5–20 + 30–40	84.92	78.15
Length 5–20 + 40–50	**85.39**	**78.66**
Length 20–30 + 30–40	85.15	78.46
Length 20–30 + 40–50	85.17	78.53
Length 30–40 + 40–50	85.15	78.32

Table 3 The results of the proposed model with the other models

Model	UAS (%)	LAS (%)
Our model	**85.39**	78.66
$PhoBERT_{base}$ [15]	85.22	**78.77**
$PhoBERT_{large}$ [15]	84.32	77.85
$DeepBiaffine$ [15]	81.19	74.99
$VnCoreNLP$ [23]	77.35	71.38

by the Euclidean distance. The results of the proposed model are shown in the Table 2 using the $PhoBERT_{base}$.

In addition, the proposed model is also presented to compare with other models [15, 23]. Table 3 shows that the proposed model achieves high results on the UAS score of +0.17%, lower than the $PhoBERT_{base}$ model is −0.11% at the LAS score. The result can be explained by the $PhoBERT_{base}$ model used extra features (e.g. POS tags) when training the model, thereby having the highest LAS score in the compared models.

4.2 Discussion

Figure 2 shows the distribution of Vietnamese and English sentence lengths on the Cartesian coordinate, with the vertical axis representing the closeness centrality coefficient and the horizontal axis representing the degree centrality coefficient. The blue represents Vietnamese sentences, and the red represents English sentences.

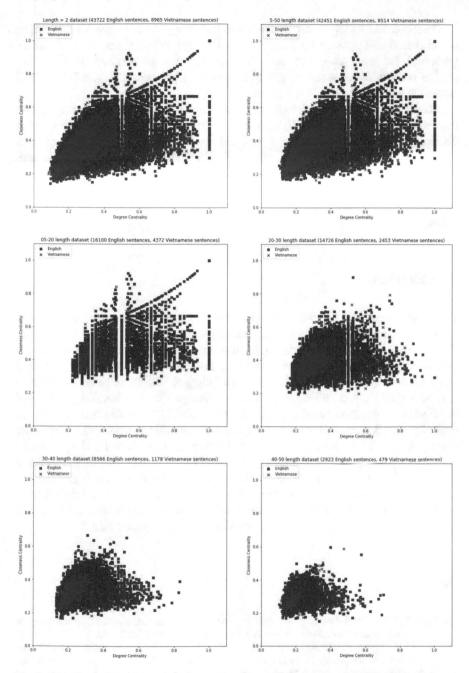

Fig. 2 Distribution of Vietnamese sentence length and English sentence length based on degree centrality and closeness centrality

Based on the results, most of the Vietnamese sentences in the VnDT corpus are clustered and covered by Penn Treebank's English sentences. Thus, it can be perceived that the syntaxes of the two languages, Vietnamese and English, have similar characteristics when compared by group degree centrality coefficients and group closeness centrality of the graph.

The proposed model, which uses English as supporting treebanks with length from 5 to 20 combined with 40 to 50, achieved the highest accuracy with 85.39% UAS. To explain the results, according to the authors [14], the $PhoBERT_{base}$ training corpus, which is about 20 GB text, can be mixed with English texts and thus, creates the cross-lingual word embedding for lexical vectors of English and Vietnamese. Therefore, the proposed model can be influenced by the supporting language and can increase the accuracy significantly, higher than that of the monolingual training model.

5 Conclusion

This paper presents a cross-lingual approach that uses high-resource language to induce a parser model for low-resource language. Besides, the approach uses lexical features as the main feature, helping to limit the model's dependence on different sets of labels between two linguistic corpora. Furthermore, the utilization of vector representation space from PhoBERT and multilingual BERT has aided in eliminating the lexical difference between the two languages. In addition, the paper used the Euclidean distance based on two closeness centrality coefficients and degree centrality coefficients to filter suitable syntactic sentences to train and improve the parsing model for Vietnamese. Finally, the model is trained on the Deep Biaffine Attention algorithm of the authors [4]. The result from the experiments shows that a new method can be explored to achieve higher accuracy with a larger high-resource treebank. The cross-lingual dependency parsing can be trained with treebanks from multiple languages. In future work, more methods and languages other than English can be further investigated to improve the accuracy of the parser model.

References

1. Agić Ž (2017) Cross-lingual parser selection for low-resource languages. In: Proceedings of the NoDaLiDa 2017 workshop on universal dependencies (UDW 2017), pp. 1–10
2. Ahmad W, Zhang Z, Ma X, Hovy E, Chang KW, Peng N (2019) On difficulties of cross-lingual transfer with order differences: a case study on dependency parsing. In: Proceedings of the 2019 conference of the North American chapter of the association for computational linguistics: human language technologies, vol 1 (Long and Short Papers), pp 2440–2452
3. Buchholz S, Marsi E (2006) Conll-x shared task on multilingual dependency parsing. In: Proceedings of the tenth conference on computational natural language learning (CoNLL-X), pp 149–164

4. Dozat T, Manning CD (2016) Deep biaffine attention for neural dependency parsing. arXiv preprint arXiv:1611.01734
5. Duong L, Cohn T, Bird S, Cook P (2015) Cross-lingual transfer for unsupervised dependency parsing without parallel data. In: Proceedings of the nineteenth conference on computational natural language learning, pp 113–122
6. Everett MG, Borgatti SP (1999) The centrality of groups and classes. J Math Soc 23(3):181–201
7. Freeman LC, Roeder D, Mulholland RR (1979) Centrality in social networks: Ii. experimental results. Soc Netw 2(2):119–141
8. Kiperwasser E, Goldberg Y (2016) Simple and accurate dependency parsing using bidirectional lstm feature representations. Trans Assoc Comput Linguist 4:313–327
9. Kübler S, McDonald R, Nivre J (2009) Dependency parsing. Synthesis Lect Human Lang Technol 1(1):1–127
10. Li S, Xue Y, Wang Z, Zhou G (2013) Active learning for cross-domain sentiment classification. In: Twenty-Third Int Joint Conf Artif Intell
11. Litschko R, Vulić I, Agić Ž, Glavaš G (2020) Towards instance-level parser selection for cross-lingual transfer of dependency parsers. In: Proceedings of the 28th International Conference on Computational Linguistics, pp 3886–3898
12. Marcus M, Santorini B, Marcinkiewicz MA (1993) Building a large annotated corpus of english: the penn treebank
13. Mrini K, Dernoncourt F, Tran Q, Bui T, Chang W, Nakashole N (2019) Rethinking self-attention: towards interpretability in neural parsing. arXiv preprint arXiv:1911.03875 (2019)
14. Nguyen DQ, Nguyen AT (2020) PhoBERT: pre-trained language models for Vietnamese. Findings of the association for computational linguistics: EMNLP 2020:1037–1042
15. Nguyen DQ, Nguyen AT (2020) Phobert: pre-trained language models for vietnamese. arXiv preprint arXiv:2003.00744
16. Nguyen DQ, Nguyen DQ, Pham SB, Nguyen PT, Nguyen ML (2014) From treebank conversion to automatic dependency parsing for vietnamese. In: Proceedings of 19th international conference on application of natural language to information systems, pp 196–207
17. Nguyen LT, Nguyen DQ (2021) PhoNLP: a joint multi-task learning model for Vietnamese part-of-speech tagging, named entity recognition and dependency parsing. In: Proceedings of the 2021 conference of the North American chapter of the association for computational linguistics: demonstrations
18. Nivre J, Hall J, Kübler S, McDonald R, Nilsson J, Riedel S, Yuret D (2007) The conll 2007 shared task on dependency parsing. In: Proceedings of the 2007 Joint Conference on Empirical Methods in Natural Language Processing and Computational Natural Language Learning (EMNLP-CoNLL), pp 915–932
19. Oya M (2020) Syntactic similarity of the sentences in a multi-lingual parallel corpus based on the euclidean distance of their dependency trees. In: Proceedings of the 34th pacific Asia conference on language, information and computation, pp 225–233
20. Potthast M, Barrón-Cedeno A, Stein B, Rosso P (2011) Cross-language plagiarism detection. Language Res Eval 45(1):45–62
21. Sabidussi G (1966) The centrality index of a graph. Psychometrika 31(4):581–603
22. Tran KM, Bisazza A (2019) Zero-shot dependency parsing with pre-trained multilingual sentence representations. In: Proceedings of the 2nd workshop on deep learning approaches for low-resource NLP (DeepLo 2019), pp 281–288
23. Vu T, Nguyen DQ, Nguyen DQ, Dras M, Johnson M (2018) VnCoreNLP: a vietnamese natural language processing toolkit. In: Proceedings of the 2018 conference of the North American chapter of the association for computational linguistics: demonstrations. Association for computational linguistics, New Orleans, Louisiana, pp 56–60. https://doi.org/10.18653/v1/N18-5012, https://aclanthology.org/N18-5012
24. Wasserman S, Faust K et al (1994) Social network analysis: methods and applications
25. Zhang Y, Li Z, Min Z (2020) Efficient second-order TreeCRF for neural dependency parsing. In: Proceedings of ACL, pp 3295–3305. https://www.aclweb.org/anthology/2020.acl-main.302

26. Zhang Y, Li Z, Zhang M (2020) Efficient second-order treeCRF for neural dependency parsing. arXiv preprint arXiv:2005.00975
27. Zhang Y, Zhou H, Li Z (2020) Fast and accurate neural CRF constituency parsing. In: Proceedings of IJCAI. pp 4046–4053. https://doi.org/10.24963/ijcai.2020/560

An Enhanced Ant Colony Algorithm for Vehicle Path Planning Optimization Problem

Thi-Kien Dao, Trinh-Dong Nguyen, Truong-Giang Ngo, Trong-The Nguyen, and Quoc-Tao Ngo

Abstract When confronted with a complicated issue in a particular combinatorial situation, the ant colony optimization algorithm (ACO) can easily slip into optimal local solutions, poor convergence speed, and other drawbacks. This study suggests an enhanced version for ACO (EACO) using information guide period ants search matrix with varied search times and pheromone volatilization factors to achieve algorithm balance between "exploration" and "exploitation. The EACO applies for logistics vehicle path planning optimization (VPP) by using the optimal solution of an "opts" method. The experimental results reveal that the proposed algorithm outperforms the ACO and the other algorithms in the literature for logistical distribution paths and delivery vehicle routes to satisfy customers' needs.

Keywords Optimization algorithm · Vehicle path planning · Enhanced ant colony algorithm · Logistics distribution · A pheromone

T.-K. Dao
Fujian Provincial Key Lab of Big Data Mining and Applications, Fujian University of Technology Fujian, Fujian, China

T.-D. Nguyen · T.-T. Nguyen
University of Information Technology, VNU-HCM, Ho Chi Minh City, Vietnam
e-mail: thent@uit.edu.vn

Vietnam National University, Ho Chi Minh City, Vietnam

T.-G. Ngo (✉)
Faculty of Computer Science and Engineering, Thuyloi University, Hanoi, Vietnam
e-mail: giangnt@tlu.edu.vn

T.-D. Nguyen
e-mail: dongnt@uit.edu.vn

Q.-T. Ngo
Institute of Information Technology, Academy of Science and Technology, Hanoi, Vietnam
e-mail: nqtao@ioit.ac.vn

© The Author(s), under exclusive license to Springer Nature Switzerland AG 2022
N. H. T. Dang et al. (eds.), *Artificial Intelligence in Data and Big Data Processing*,
Lecture Notes on Data Engineering and Communications Technologies 124,
https://doi.org/10.1007/978-3-030-97610-1_10

1 The Introduction

Dispatching delivery vehicles in the vehicle path planning optimization (VPP) problem is essential for modern logistics in social life distribution [1]. It is to load and unload, distribute goods, deliver goods and other links in the distribution center according to the needs of consumers, and finally deliver the goods to the customer [2]. In logistics distribution link, there have been many kinds of optimization decision-making, the distribution of goods in the vehicle route optimization problem is logistics company to speed up the delivery and high quality of service considering the main factors, how to quickly and efficiently to adjust delivery vehicle lines [3], to meet the needs of consumers, has become a logistics company and consumer attention important questions [4].

Much research on intelligent optimization algorithms [5, 6] is developed and applied widely in the engineering, industry, financial fields, and special transportation like VRP problems [3]. Intelligent optimization algorithms [7] are inspired by natural phenomena like the gravity algorithm-GSA [8], genetics optimization-GA [9], and swarm intelligent PSO [10, 11]. The ant colony optimizing algorithm (ACO) [12] is one of the powerful algorithms in intelligent optimization algorithms. The ACO is a bionic intelligent optimization algorithm inspired by the foraging behavior of real ants in nature, which uses ants to work together to find food as a model for optimization issues. It stimulates the behavioral characteristics and movement rules of ants that use pheromone as a traction guide during the movement process to control the ants to find the optimal path.

When faced with the complication problem in a special combinatorial complex issue like the VRP, the ACO algorithm is prone to locally optimal solutions, slow convergence, and other flaws. Variable search periods in pheromone volatilization, this work proposes an enhanced form of ACO (EACO). An a priori information guide period ants search matrix with various search times and pheromone volatilization factors is used to achieve algorithm balance between "exploration" and "exploitation."

The motivation is precise because the ACO has achieved satisfactory results after improving the integration of the local optimization capability of the primary ACO and the global optimization capability of the adaptive reaction optimization. The concepts between the "discovery" and "use" are used to achieve better balance changes into the algorithm as a distribution center. It can adjust the route of the delivery vehicles quickly and efficiently to meet the needs of consumers.

2 Vehicles Path Planning Mathematical Model

Delivery from a distribution center to several consumers, with the distribution center's location remaining constant, each customer's demand being specific, and the vehicle load not exceeding the load [3]. Assuming that the distribution center can meet the needs of all clients, it is necessary to design the distribution route of vehicles so that the total length of the route is minimized while completing all distribution responsibilities [13]. First, the total demand of customers on each distribution route shall not exceed the load of vehicles. Second, each customer's request is delivered by a delivery truck and can only be delivered once. Third, after completing each distribution task, the distribution vehicle must return to the distribution center.

Let A be a representation of the distribution center. There are m customers wanted goods to deliver from the distribution center. Let h and V be the number of vehicles and the quantity load of each vehicle. Let N_i be the demand of each customer ($i = 1,2,..., m$). Let $dics_{i,j}$ and d_{Aj} be the distance path between i-th and j-th customers ($i, j = 1, 2...m$), the distance from the customer in the delivery. Let Z be the total length of the distance traveled by the distribution vehicle. The following mathematical model of the distribution vehicle scheduling problem can be established.

The objective function

$$\min Z = \sum_{i=1}^{m}\sum_{j=1}^{m}\sum_{k=1}^{h} x_{ijk}d_{ij} + \sum_{j=1}^{m}\sum_{k=1}^{h} \omega_{Aj}d_{Aj} \tag{1}$$

The constraints are presented as follows.

$$\sum_{k=1}^{h} y_{ik} = 1, \quad \forall i \tag{2}$$

$$\sum_{i=1}^{m} N_i y_{ik} \leq V \tag{3}$$

Among them are given as follows.

$$x_{ijk} = \begin{cases} 1 & \text{Vehicle k goes from i−th to j−the customers} \\ 0 & \text{otherwise} \end{cases} \tag{4}$$

$$y_{ik} = \begin{cases} 1 & \text{Customer I's demand is delivered by vehicle k} \\ 0 & \text{other} \end{cases} \tag{5}$$

$$w_{Aj} = \begin{cases} 1 & \text{Direct from the center to customer j} \\ 0 & \text{otherwise} \end{cases} \tag{6}$$

That each vehicle starts from the determined distributing center. For dealing with the constraint condition, the Eqs. (2–5) are given details. Equation (2) ensures that each customer's demand is delivered by one vehicle; Eq. (3) indicates the vehicle's load capacity cannot exceed the load capacity. Equation (4) guarantees vehicle k ($k = 1, 2, ..., h$) through the customer to customer; Eq. (5) ensures that each customer has a car to complete the assignment; Eq. (6) indicates that each vehicle starts from the determined distributing center.

3 Enhancement of Ant Colony Algorithm

When dealing with the complexity problem, the ACO algorithm is prone to locally optimum solutions, resulting in sluggish convergence. Variable search times in pheromone volatilization are used in this section to show an upgraded form of ACO (EACO). To establish algorithm balance between searching technique exploration and exploitation is better, an a priori information guide period ants search matrix with varying search times and pheromone volatilization factors is utilized.

3.1 Original Ant Algorithm (ACO)

The ACO algorithm is one of the new evolutionary algorithms inspired by the ants foraging behavior of an intelligent algorithm. The ant from nest to food can always find the shortest path and logistics vehicle scheduling route. The optimization problem is the carrier of the goods delivered to the customer, for the shortest path for delivery of goods vehicles, based on ant colony counting method for logistics car bus line into line. Each ant starts to search in the ACO from any city and selects the next city with probability P at time t, calculated as follows.

$$
p_{ij}^k(t) = \begin{cases} \frac{\tau_{ij}^\alpha(t)\aleph_{ij}^\beta}{\sum_{j\in N_j^k} \tau_{ij}^\alpha(t)\aleph_{ij}^\beta} & j \in \text{permitted}_i \\ 0 & \text{otherwise} \end{cases}
\tag{7}
$$

where α and β represents the importance of pheromone, and the heuristic factor; N_j^k represents the unvisited city yet, $p_{ij}^k(t)$ represents the probability that the k ant chooses city j in city i at the time of t, and $\aleph_{ij} = 1/d_{ij}$ represents the heuristic factor on the path ij. τ_{ij} represents the size of the pheromone left on the path.

ij, and every time the city is visited, τ_{ij} is a way of updated pheromone. The expression for the pheromone is

$$
\tau_{ij}(t+1) = (1-\sigma)\tau_{ij}(t) + \sum_{k=1}^m \Delta\tau_{ij}^k
\tag{8}
$$

where, σ represents the volatilization factor of pheromone, $0 < \sigma < 1$, $\Delta\tau_{ij}^k = C$ as it is initial time currenet, and then $\Delta\tau_{ij}^k$ is amount of changing the pheromone as follows.

$$\Delta \tau_{ij}^k = \begin{cases} Q/L_k & j \in \text{allowed}_i \\ 0 & \text{otherwise} \end{cases} \tag{9}$$

where L_k and k represents the total path length of the ant and current visit after a trip, and Q is set to a constant value.

3.2 Enhanced Ant Colony Optimizing (EACO)

An enhanced ant colony optimizing algorithm is implemented through two factors, e.g., selection probability and volatile scheme. An a priori information guide period ants search matrix with varied search times by using variable search periods in pheromone volatilization. The pheromone volatilization factors are used to achieve algorithm balance between searching technique exploration and exploitation types.

Selection probability is an a priori information element that guides the ant in choosing the next traveling direction. Late in the search, due to the ant colony algorithm, the pheromone guide ants search accounted for an increasing share of the total, and the heuristic factor's function steadily weakened. Consider increasing a priori information in the search process of ants searching guidance, such as the inclusion of the same matrix as a priori information to guide the ant to the next city. The following is an example of a saving matrix.

$$U(x, y) = dics(x, 1) + dics(y, 1) - dics(x, y) \tag{10}$$

where $U(x,y)$ is an adjusting parameter for selection probability, the adjusted parameter of the savings matrix is added to the probability equation to avoid 0 in the selection probability. The ant cannot choose the next city and waste a lot of time. The selection probability is improved as follows.

$$p_{ij}^k(t) = \begin{cases} \dfrac{\tau_{ij}^\alpha(t)\eta_{ij}^\beta U_{ij}^\gamma + 1}{\sum_{j \in N_j^k} (\tau_{ij}^\alpha(t)\eta_{ij}^\beta U_{ij}^\gamma + 1)} & j \in \text{allowed}_i \\ 0 & \text{otherwise} \end{cases} \tag{11}$$

3.3 Improvement of Volatile Factors

Improved volatile factors are used to influence the amount of pheromone on the path, altering the global search capabilities. The amount of pheromone on the path is directly affected by the size of the volatile factor, influencing the ACO algorithm's global search ability and convergence rate. The pheromone quantity on the better path obtained in the previous search accelerates the convergence and gets the final better result. Therefore, σ is improved by using the piecewise function as follows.

$$\sigma = \begin{cases} a, & N < \frac{N_{max}}{3} \\ a - \left(N - \frac{N_{max}}{3}\right)/\frac{N_{max}}{3}(a - b), & \frac{N_{max}}{3} < N < \frac{2N_{max}}{3} \\ b, & N > \frac{2N_{max}}{3} \end{cases} \tag{12}$$

In the $a, b \in (0, 1)$, and $a > b$, N for the current iteration number, N_{max} for ant colony algorithm is the most significant number of iterations. In the traditional ant colony algorithm, the volatilization factor σ is constant (0,1). If the given σ is too large, the number of pheromones on the path will be reduced, resulting in the algorithm's convergence rate being reduced. If the given σ is small, the number of pheromones on the path will increase, making the algorithm converge sharply. Although the search time of the algorithm is saved, it may cause the algorithm to fall into local search. To make the ants between "exploration" and "exploitation" of balance, the ant colony algorithm in the search process can avoid the stagnation and has strong global search ability, considering not and iteration time use with volatile factor to adjust the size of the amount of pheromone, namely volatile factor σ changes with the number of iterations: at the beginning of the search, a given volatile factor to older, reduce the amount of pheromone path, is conducive to iterate over all path and get the optimal global solution; Then the volatilization factor is reduced appropriately so that the algorithm can be transformed from the global search to the optimal path search within a specific range, to avoid the sharp convergence of the algorithm. Finally, the volatilization factor is reduced to the minimum.

4   EACO for Vehicle Path Planning Optimization Problem

In order to test the feasibility of the EACO algorithm, the data in [14, 13] with scenarios, e.g., one center is set (0,0) serves N customers in grid map with coordinates to optimize the route planning. The 2-opt algorithm is a local search algorithm, which improves the initial solution by exchanging two edges [14]. Only the path order corresponding to the optimal solution of each iteration is exchanged and adjusted, and the optimal solution satisfying the condition is finally obtained. Its principle is shown in Fig. 1.

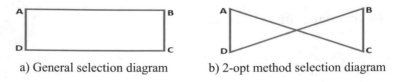

a) General selection diagram   b) 2-opt method selection diagram

Fig. 1 The exchanged and adjusted routes for optimal solution satisfying the condition

Figure 1 shows the exchanged and adjusted routes for optimal solution satisfying the condition.

In Fig. 1, traverse A, B, C and D 4 points and return to starting point A, whose order is $A \rightarrow B \rightarrow C \rightarrow D \rightarrow A$; If the 2-opt method is applied, the node order in it is exchanged to compare the path size. As shown in Fig. 1b, the node order $A \rightarrow C \rightarrow B \rightarrow D \rightarrow A$ is the path length in the order of switching B and C. The 2-opt method was used to adjust the order of the nodes optimized by the ant colony algorithm and meet the load requirements, and finally, the optimal global solution was obtained.

The EACO's primary steps are outlined here

The search steps are defined at the beginning of each distribution, and ants are placed on the distribution customer point, according to the preceding description of the improvement form. After the job begins, ants choose the next distribution customer point at random; the best option is discovered throughout each iteration. When the iteration is finished, the best solution from each iteration is combined to generate an ideal solution pool. 2-opt technique [13, 14], which is then fine-tuned to get the global best solution. The steps of the algorithm are as follows.

(1) Initialize ant colony algorithm parameters and pheromones in the path;
(2) Number of iterations $N = N + 1$;
(3) Put m ants on n cities;
(4) m ants select the next distribution customer point according to the probability function Eq. (11) and complete their own travels;
(5) Apply the 2-opt method to update the better solution;
(6) Update the pheromone according to Eq. (8), where the volatile factor is determined according to formula (12), and the taboo table is cleared;
(7) Determine if the iteration meets the condition, if not, go to step (3); if yes, go to step (8);
(8) Output result.

5 Experimental Results

In order to test the feasibility of the EACO, the data scenarios, customers M, and vehicles N, e.g., $10 \times 3, 40 \times 6$, are used to optimize the route planning. The logistics of 1 logistics center and, e.g., 10, 40 delivery points has a vehicle load of 1000 kg. Simulation is with Matlab language programming, running under Windows 10. The simulation parameters are set as follows: the number of ants is *50*, and the maximum iterations is set to *1000*; α, and β are set to *1*; $\gamma = 2$, $Q = 14$, P_{ij}^k and σ are shown in Eqs. (10) and (11); $a = 0.9$, $b = 0.1$ v_{min} is set to *0.2*; v (*iter* = *0*) is set to *0.9*; Q is set to *1501*, C is set to *20* respectively. The number of runs is set to *30* to the algorithms. The following are the stages taken by applying EACO for VRP planning.

- Create a model of the VRP planning of searching space, including the positions and forms of delivery points and the vehicle loads.
- The parsing approach consists of mapping search agents ants to the route planning model during optimization.
- Using the proposed EACO to find the best pathways for the point using the model. Update the position and pheromone of the ants to the best possible position. Calculate the fitness function values and the degree to which each ant's pheromone violates the constraint.
- Use the optimum path scheduling to guide the vehicles to the target spot.

The obtained results of the suggested approach's feasibility and effectiveness are compared with the other algorithms, e.g., A* [15], GA [16], and ACO [17] algorithms. A* algorithm has three parameters, e.g., f-the sum fitness, h-heuristic value, g- fitness cost value. The path in maps with hindrances considered for optimizing path planning in a 2D grid has started from a source point cell to a target cell.

Table 1 shows the comparison of the obtained results of the suggested approach with the A* algorithm [15], GA [16], and ACO [17] algorithms for the optimal path planning lengths. From Table 1, the optimal average path length of the A* algorithm is 68.339 km, the GA algorithm is 50.545 km, the ACO algorithm is 48.813 km, but the EACO algorithm is only 49.8134 km; the optimal path length of the EACO algorithm is 47.6844 km. It can be seen that the EACO finds shorter paths than the other algorithms.

The load of the transport vehicle is 9000 kg. The other parameters were selected in the same way as in the previous example. Each runs 10 times, and the optimal results are shown in Fig. 2 that shows the optimal path obtained using the EACO, with a total length of 47.85112 km,

Table 1 Comparison of the obtained results of the suggested approach with the A* algorithm, GA, and ACO algorithms

Serial number	A*	GA	ACO	EACO
1	70.447	50.143	50.319	47.305
2	69.118	51.094	49.370	48.202
3	70.137	50.722	50.098	47.851
4	71.618	50.795	51.156	47.920
5	70.445	50.722	50.318	47.851
6	71.359	50.916	50.971	48.034
7	70.315	49.526	50.225	46.723
8	68.972	50.911	49.265	48.029
9	66.992	49.673	47.851	46.862
10	67.986	50.953	48.561	48.069
Average path length	**68.339**	**50.545**	**48.813**	**47.684**

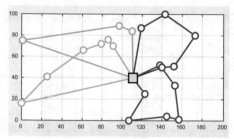 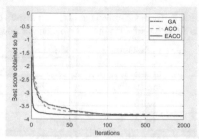

a) An example of planning routes of graphics for   b) Comparison of the convergence speed of
a selected instance.   the algorithms

Fig. 2 The obtained curve results from the EACO compared with the other methods for a selected instance of 20 points and 4 vehicles

$0 \rightarrow 16 \rightarrow 11 \rightarrow 10 \rightarrow 2 \rightarrow 0 \rightarrow 1 \rightarrow 19 \rightarrow 3 \rightarrow 4 \rightarrow 7 \rightarrow 0 \rightarrow 18 \rightarrow 6 \rightarrow 12 \rightarrow 0 \rightarrow 17 \rightarrow 14 \rightarrow 8 \rightarrow 0 \rightarrow 13 \rightarrow 5 \rightarrow 15 \rightarrow 9 \rightarrow 0$ (0 stands for distribution center).

Figures 2 and 3 display the obtained curve results from the EACO compared with the other methods for selected instances, e.g., r105c6, r202c6 with set 20 × 4 and 40 × 6 that means the selected set with the instances of 20 points and 4 vehicles and example of 40 points and 6 vehicles. It can be seen that the EACO produces the obtained curve results of coverage speed faster than the other algorithm in comparison.

Table 2 shows the comparison of the obtained results of the suggested approach with the A* algorithm, GA, and ACO for selected the VRP benchmark dataset [18]. It can be seen that the optimal path length of the suggested algorithm finds shorter paths than the other algorithm.

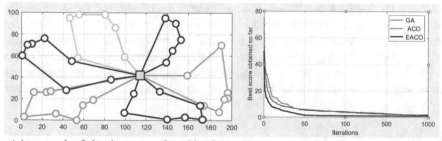

a) An example of planning routes of graphics for a   b) Comparison of the convergence speed of
selected instance.   the algorithms

Fig. 3 The obtained curve results from the EACO compared with the other methods for a selected instance of 40 points and 6 vehicles

Table 2 Comparison of the obtained results of the suggested EACO with the A* algorithm, GA, and ACO for selected the VRP benchmark dataset

#	Instance	A*	GA	ACO	EACO
		Avg	Avg	Avg	Avg
1	c101c6	1385.98	1333.18	1294.58	**1293.58**
2	c103c6	1211.44	1165.29	1132.67	1130.68
3	c206c6	1405.80	1352.25	1312.10	1312.08
4	c208c6	947.82	911.72	884.04	884.64
5	r104c6	1080.94	1039.76	1011.88	**1008.88**
6	r105c6	872.88	839.62	**814.89**	814.68
7	r202c6	929.91	894.49	867.96	867.92
8	r203c6	1182.72	1137.66	1103.89	1103.87
9	rc105c6	1538.94	1480.32	1436.55	**1436.35**
10	rc108c6	1419.92	1365.82	1326.20	**1325.25**

6   Conclusion

This study presented a new enhancement to the ants' colony optimization technique (EACO) for vehicle path planning problems (VPP). The enhanced performance prevents the algorithm from falling prematurely, based on applying the ant guidance search with the saving matrix distance and adjusting the amount of pheromone at different iteration periods with varying volatility factors. The dynamically adjusts the pheromone volatilization coefficient, pheromone concentration range, and distribution of the initial pheromone to balance exploration and exploitation of the algorithm. The superior solution is also optimized using the 2-opt approach. The experimental results reveal that the proposed algorithm outperforms the other algorithms handling the logistics distribution path optimization problem and rapidly and efficiently alters the delivery vehicle route to satisfy customers' needs. The findings of simulation experiments show that the EACO algorithm finds a shorter path than the other methods, allowing it to handle the VPP line optimization problem more effectively. In future work, we will extend implementation with a complicated benchmark and compare the results with the variety of improved versions of the ACO algorithm.

References

1. Lin C, Choy KL, Ho GTS, Chung SH, Lam HY (2014) Survey of green vehicle routing problem: past and future trends. Expert Syst Appl 41:1118–1138. https://doi.org/10.1016/j.eswa.2013.07.107
2. Daneshzand F (2011) The vehicle-routing problem. In: Logistics operations and management. https://doi.org/10.1016/B978-0-12-385202-1.00008-6
3. Utama DM, Dewi SK, Wahid A, Santoso I (2020) The vehicle routing problem for perishable

goods: a systematic review. Cogent Eng 7:1816148. https://doi.org/10.1080/23311916.2020.1816148

4. Nguyen T-T, Qiao Y, Pan J-S, Chu S-C, Chang K-C, Xue X, Dao T-K (2020) A hybridized parallel bats algorithm for combinatorial problem of traveling salesman. J Intell Fuzzy Syst 38:5811–5820. https://doi.org/10.3233/jifs-179668

5. Nguyen TT, Pan JS, Dao TK (2019) An improved flower pollination algorithm for optimizing layouts of nodes in wireless sensor network. IEEE Access 7:75985–75998. https://doi.org/10.1109/ACCESS.2019.2921721

6. Nguyen TT, Pan JS, Dao TK (2019) A compact bat algorithm for unequal clustering in wireless sensor networks. Appl Sci 9:1973. https://doi.org/10.3390/app9101973

7. Dao T-K, Pan T-S, Nguyen T-T, Chu S-C (2019) a compact articial bee colony optimization for topology control scheme in wireless sensor networks. J Inf Hiding Multimed Signal Process 06:297–310

8. Rashedi E, Nezamabadi-Pour H, Saryazdi S (2009) GSA: a gravitational search algorithm. Inf Sci (Ny) 179:2232–2248

9. Srinivas M, Patnaik LM (1994) Genetic algorithms: a survey. Computer (Long. Beach. Calif). 27:17–26. https://doi.org/10.1109/2.294849

10. Kennedy J, Eberhart RBT-IC on NN (1995) Particle swarm optimization. In: Proceedings of IEEE international conference on neural networks. IEEE, Perth, pp 1942–1948

11. Shi Y, Eberhart R (1998) A modified particle swarm optimizer. In: 1998 IEEE international conference on evolutionary computation proceedings. IEEE world congress on computational intelligence (Cat. No.98TH8360). pp 69–73

12. Dorigo M, Birattari M, Stutzle T (2006) Ant colony optimization. IEEE Comput Intell Mag 1:28–39

13. Pillay N, Qu R (2018) Vehicle routing problems. In: Natural computing series. https://doi.org/10.1007/978-3-319-96514-7_7

14. Psaraftis HN, Wen M, Kontovas CA (2016) Dynamic vehicle routing problems: three decades and counting. Networks 67:3–31

15. Zhang Y, Tang G, Chen L (2012) Improved A* algorithm for time-dependent vehicle routing problem. In: Proceedings of the 2012 international conference on computer application and system modeling. Atlantis Press, pp 1341–1344

16. Vaira G (2014) Genetic algorithm for vehicle routing problem

17. Yu B, Yang ZZ (2011) An ant colony optimization model: the period vehicle routing problem with time windows. Transp Res Part E Logist Transp Rev 47:166–181

18. Philip K, Prosser P, Shaw P (1998) Dynamic VRPs: a study of scenarios. Uni of Strathclyde Technical Report, pp 1–11. cs.strath.ac.uk/apes/ apereports.html

Generating Recommendations via Trust-Aware Recommendation System by the Topological Impact of Users in Social Trust Networks

Kamal Al-Barznji

Abstract There are a large number of current recommendation methods that have issues with cold starts and sparsity. In this study, these issues are addressed by proposing a novel trust-based recommendation method, and the proposed method uses trust information along with rating values to deal with "cold-start" users and items. Because in most real-world applications, only a few items are given feedback by the users. Therefore, we were faced with a sparse user-item matrix. Here, similar users are grouped using a random-walk-based method that calculates the influence of users in social networks. Then cluster seeds are identified among the most influential users. Assign unique labels to cluster seeds and use a novel label propagation method to spread labels to unassigned users. Finally, the combinations identified in the prediction process are used to predict missing ratings. To assess the efficiency of the proposed approach, several experiments were performed on the well-known and widely used real-world dataset called FilmTrust. The results are compared based on several known evaluation metrics, which are F1-Measure, Precision, Recall, Mean Absolute Error (MAE), and Root Mean Square Error (RMSE). The proposed method achieved the lowest values of MAE and RMSE and the highest values of F1, Precision, and Recall in comparison to the other recommended methods. Results showed that the proposed method is superior to the traditional and modern methods in terms of accuracy and efficiency in most cases. Therefore, it can be concluded that using trust information leads to more accurate rating predictions.

Keywords Recommender systems · Collaborative filtering · Cold-start · Sparsity · Trust statements · Propagation of labels · Clustering

K. Al-Barznji (✉)
Department of Computer Science, University of Raparin, Ranya, Kurdistan Region, Iraq
e-mail: kamal.barznji@uor.edu.krd

© The Author(s), under exclusive license to Springer Nature Switzerland AG 2022   121
N. H. T. Dang et al. (eds.), *Artificial Intelligence in Data and Big Data Processing*,
Lecture Notes on Data Engineering and Communications Technologies 124,
https://doi.org/10.1007/978-3-030-97610-1_11

1   Introduction

Due to the growth of online social networks and online e-commerce websites, faced with the exponential growth of digital information, the recommendation is a suggestion which can help to make appropriate decisions more quickly. Recommender systems (RS) are one of the most effective solutions for handling information overload by providing users with personalised recommendations based on their preferences. Real-world applications of recommendation systems include movie and music recommendation; web-service recommendation; and hotel recommendation. In the literature, a variety of recommendation methods have been proposed. These approaches are essentially grouped as content-based (CB) and collaborative filtering (CF). In CB, recommended items are generated by analysing user profiles or textual content. Instead of a target user's preferences, CF uses a group of users who have similar preferences. The most common method of recommendation is to utilise collaborative filtering (CF) approaches to improve a user-specific suggestion of an item based on ratings or use patterns without the need for extra information about both the item and the user [1].

In other words, one of the most important tasks in CF is to identify a group of similar users and use their ratings to predict future ratings. The primary CF methods use a similarity metric to identify a set of top users who are most similar to the target user in terms of their interests and behavior [2]. However, most of the available similarity measures consider enough ratings on common items to calculate similarity values. While in real-world applications, users give feedback on only a few items. This problem is known as "cold-start" as well. This problem occurs when there aren't enough resources to support the new user or item. Commonly used methods for avoiding this limitation include taking into account some side information (e.g., trust statements) with the ratings in similarity computations [3, 4].

Social networks provide trust statements of friendship connections as an effective source of information to be considered in the similarity computations in CF-based methods to improve the quality of their outputs. Trust-based recommenders use both explicit and implicit trust relationships in their rating prediction processes. Explicit trust is made directly by users, whereas implicit trust is estimated using explicit feedback from customers on items.

To group users into similar groups, most trust-based recommenders use trust statements and rating values to create the cluster. In [5] the authors employ a k-means clustering method to group the users. Moreover, the major disadvantage of k-means is its sensitivity to the initial seeds as the cluster centres. Due to random seed placement, clusters may fall into local optimum conditions. Therefore, in the traditional k-means, selecting the right initial seeds is crucial. To tackle this issue recently [6], the authors proposed a random walk method called LeaderRank to identify the most influential users as an initial set for the k-means clustering method. They then apply the random walk process to cluster the items. Their obtained results show that their method can choose better initial centres for the k-means clustering method in comparison with the random selection of the cluster centres.

This study proposes a recommendation system based on trust. The proposed method uses the random walk method on the graph of users to find the most influential users. These users are then used as a set of initial cluster seeds. Then a novel label propagation method is employed to group customers according to similarity. The clustering results are further used to predict missing rating values for the target user. More specifically, five steps are included in the proposed method's design. Rating values and trust statements are used in the first step to create the user graph. An algorithmic random walk is then used to rank each user based upon their importance. An initial group of high-ranking users is then selected to act as seeds for a label propagation process, and they are used as seeds in the next step. An entirely new label propagation mechanism is used in the fourth step to propagate the initial seeds' labels to their k-nearest neighbors. According to their similarities, a set of clusters is created. In the final step, each target user's rating gaps are predicted by looking at users in the same cluster as the target user. In comparison to traditional and modern methods, the proposed method has the following novelties:

1. The proposed method uses a random walk process based on the trust-based graph for choosing an initial set of cluster centres. Compared to [7], which uses an approximate method for revealing the densest subgraph, which is a time-consuming process, the proposed method uses the random walk method, which is faster than [7]. Therefore, the proposed method can be implemented in large-scale apps.
2. The proposed technique uses a novel label propagation mechanism for clustering. In comparison with the LeaderRank method [6], which uses the k-means for clustering users, the suggested method employs a novel label propagation method for grouping users, which groups the users into various sizes and various forms of clusters.
3. The LeaderRank [6] method clusters items by only using the rating scores. In contrast, the suggested solution relies on trust assertions as a rich extra source, combined with rating scores to address the sparsity issue.

FilmTrust, a well-known and widely used dataset, is used to evaluate the suggested technique's efficiency. The results reveal that the proposed method surpasses traditional and modern recommenders in most cases.

Continue reading to find out how the remaining part of the paper is structured. The related work is discussed in Sect. 2. The proposed method is presented and described in Sect. 3. Section 4 explains the experimental evaluation and discusses the outcomes and lastly, Sect. 5 closes the paper with a conclusion.

2 Related Works

Recently, there has been a lot of interest in clustering-based recommender systems. In [8], a clustering technique is used to group similar users. The authors of [9] proposed a pattern mining approach that groups similar users to the target user in a tree structure. In this structure, most similar users are placed on the first level near the target user. In [10], a fuzzy clustering, collaborative filtering method called IFCCF is proposed to group users based on their similarities. In similar research, the authors of [11] hybridised fuzzy c-means clustering with collaborating filtering to improve the quality of the recommendations. In some recent studies, evolutionary-based clustering methods have been used in recommendation engines. In [12], for example, the authors used k-means and an artificial bee colony (ABC-KM) to create effective movie recommendation systems. In similar research, the authors of [13] proposed a method called GWO-FCM, which uses the grey wolf optimization (GWO) method with fuzzy c-means to cluster similar users. To identify the initial centres of the method of k-means clustering and employ them in the rating prediction process, the Cuckoo search is used by the authors in [14]. In all these methods, evolutionary methods such as GA, ABC, Cuckoo, and GWO search methods are used to identify initial clusters and then fine-tune clusters by applying well-known methods of clustering such as FCM and k-means. Recently, in a different work [6], the authors applied a random walk-based method called leader rank to identify the initial centres of the k-means clustering approach. Although the methods of clustering-based are effective in alleviating the cold-start issue, most of them rely on the rating values. We know that in most real-world systems, users offer feedback on only a limited number of items. Thus, when faced with high spatial data, it results in deteriorating clustering-based methods. For instance, some new research employs some side information, such as trust statements, in the recommendation procedure. For example, the authors of [4] hybridised the trust-based regularisation terms in the SVD method objective function. The authors of [7] suggested a graph clustering approach for categorising users into multiple clusters. The authors presented a unique similarity measure that assesses user similarities based on trust assertions and feedback on items in the rating values form. A trust-based CF technique called TCFACO was proposed by the authors of [15], which uses the optimization of an ant colony to find a similar user list to the target user. The proposed technique uses the advantages of the novel clustering method in [16] to improve clustering quality based on the leader-rank idea in [6] and trust content to make a strong recommendation. In addition, implementing machine learning (ML) for the task of text classification entails converting a wide range of textual sources into structured information. The authors of [17] developed a prediction model for the issues using an Ensemble Classification Approach (ESA), and they believe that the combination of Machine Learning (ML), Natural Language Processing (NLP), and Ambient Intelligence (AI) offers up new research issues. The summarization of the main properties of related recommender methods is shown in Table 1.

3 Proposed Method

This section will go over the specifics of the technique that has been proposed. The study proposed a trust-aware recommender system. To solve the cold-start and

Table 1 Main properties of related works

Algorithm	Trust	Grouping users	Search	Year
TCFACO [15]	Yes	Ranking	ACO	2019
ABC-KM [12]	No	Clustering	ABC	2018
GWO-FCM [13]	No	Clustering	GWO	2018
LeaderRank [6]	No	Clustering	Random walk	2018
RTCF [3]	Yes	Graph clustering	Heuristic	2015
K-means Cuckoo [14]	No	Clustering	Cuckoo	2017
MLTRS [18]	No	Ranking	MA	2017
TrustMF [19]	Yes	–	MF	2016
DSTNMF [20]	No	–	MF	2017
TrustSVD [4]	Yes	–	MF	2015
UOCCF [21]	No	Ranking	Heuristic	2017
FPSO-CF [22]	No	List of users	PSO	2015
Proposed method	Yes	Clustering	Random walk	–

sparsity challenges, the proposed method employs trust relationships as well as rating values. In other words, friendship connections in the recommendation system and trust statements are utilised as supplementary sources of information. The proposed technique consists of five major phases. The first phase is to generate a graph of users based on the trust-based similarity measures. Then, in the second phase, a process of random walk is implemented on the graph in several iterations to calculate the influential user's values. The influential values of users are computed in such a way that they are similar to each other and that they are located in the dense regions of the graph. In the third phase, a set of the most influential users is used as cluster seeds. A new label propagation mechanism is used in the fourth phase to spread the labels of cluster seeds to their comparable friends. The labelled users also propagate the labels to their nearest neighbors. This procedure is continued until the cluster labels have been assigned to all users. The result of this phase is a set of clusters that hold those users who are similar to each other. Finally, in the fifth phase, these clusters are used to predict rating values for the target user's unseen items. The proposed technique is presented in Fig. 1 as an overview. The specifics of these processes are explained in the sections that correspond to them.

3.1 Generating a Graph of Users

The goal of this phase is to generate a user graph based on rating values and trust assertions. To achieve that goal, follow the idea proposed in [7] to generate the trust-based graph of users. Each user is a node in this graph, and the weights of the edges connecting two users' u and u' are calculated as follows:

Fig. 1 The proposed
method flow graph

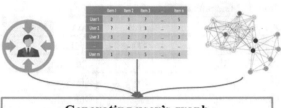

Generating user's graph	
• Compute trust-based similarity values	

Compute influence values
• Remove those edges where $w_{u,u'} < \lambda$
•Create binary graph
•Adding the ground node
•Apply LeaderRank Method

Identifying cluster seeds
•Identify top k influential users as cluster seeds

Label Propagation
•Propagate the labels of each seed to its k-nearest neighbors
•Process is repeated until all nodes are labeled

Rating prediction
•Define weight between two users
•Predict user missed rating item using similar users

$$
W_{u,u'} = \begin{cases}
\frac{2 \times sim(u,u') \times T_{u,u'}}{sim(u,u') + T_{u,u'}} & if\ sim(u,u') + T_{u,u'} \neq 0 \\
& and\ if\ sim(u,u') \times T_{u,u'} \neq 0 \\
T_{u,u'} & else\ if\ sim(u,u') = 0\ and\ T_{u,u'} \neq 0 \\
sim(u,u') & else\ if\ sim(u,u') \neq 0\ and\ T_{u,u'} = 0 \\
0 & else
\end{cases} \quad (1)
$$

where $T_{u,u'}$ is the explicit trust declaration between the users, which is calculated as
follows:

$$
T_{u,u'} = \frac{d_{\max} - d_{u,u'} + 1}{d_{\max}} \quad (2)
$$

where $d_{\max}$ is the maximum distance length between trustworthy network users and $d_{u,u'}$ is the number of hops or links between u and u'.

$$d_{\max} = L^R = \frac{\ln(n)}{\ln(k)} \tag{3}$$

where L^R is the network's average path length, n is the size of the trusted network, and k is the network's average degree. The interested authors are referred to [7] for more details.

Traditional similarity metrics cannot be utilised in sparse situations because they cannot efficiently utilise rating data. Thus, employ a similarity measure called CjacMD (Cosine Jaccard Mean Measure of Divergence) proposed in [6], which is suitable for sparse rating data. The following formula is used to determine the hybrid user similarity between two users:

$$sim(u, u') = sim(u, u')^c + sim(u, u')^j + sim(u, u')^m \tag{4}$$

where $sim(u, u')^c$, $sim(u, u')^j$ and $sim(u, u')^m$ are **Cosine, Jaccard and MMD** similarity values between users respectively and they are computed as follows:

$$sim(u, u')^c = \frac{\vec{r}.\vec{r}'}{\|\vec{r}\|.\|\vec{r}'\|} \tag{5}$$

$$sim(u, u')^j = \frac{|I_u \cap I_{u'}|}{|I_u \cup I_{u'}|} \tag{6}$$

$$sim(u, u')^m = \frac{1}{1 + \left(\frac{1}{r}\sum_{i=1}^{r}\left\{(\theta_u - \theta_{u'})^2 - \frac{1}{|I_u|} - \frac{1}{|I_{u'}|}\right\}\right)} \tag{7}$$

where $\vec{r}$ denotes the mean of rating values rated by u, I_u represents the set of items rated by u, and $|I_u|$ is the number of items rated by u. Here, θ_u represents a vector based on user u ratings, and r denotes the number of the co-rated item between u and u'.

3.2 Computing Influential Values

This section aims at computing the influential values of users by applying the LeaderRank algorithm [23]. To this end, first, those edges whose corresponding weights are lower than a threshold λ is eliminated from the graph. To apply the random walk method the graph needed to be converted to a weighted and undirected graph. To this end, all the positive weights are set to 1 and the other ones are set to zero. Finally, a

Fig. 2 Ground node with user's graph

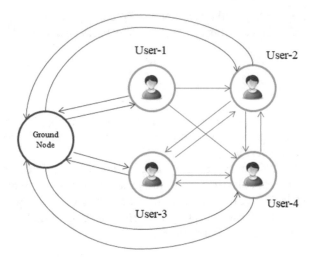

ground node is added to the graph in which the nodes are connected to the ground node. Figure 2 shows the users' graph with the ground node. In the next step, the random walk-based method called LeaderRank [23] is applied to this graph. Every node is assigned a score by the LeaderRank, with the score indicating its significance. Initially, scores are denoted as $r_g(0) = 0$ for the ground node (g) and $r_{u'}(0) = 1$ for all other nodes. Next to that, the score of each node is immediately updated in the following phases by directly performing the normal random walk procedure. As a result, if the score of node u is $r_u(t)$, at time phase t, the dynamics may be represented iteratively as follows:

$$r_u(t+1) = \sum_{u'}^{N+1} \frac{d_{u,u'}}{k_{u'}^{out}} r_{u'}(t) \tag{8}$$

where $k_{u'}^{out}$ is the out-degree of a node u' and $d_{u,u'}$ is an element of the equivalent (N + 1) dimensional adjacency matrix. If there is a directed connection from u' to u, then the value of $d_{u,u'}$ is one; else, it is zero. When all nodes' resources $r_u(t)$ converge to a unique stable state at the time t_c, The resource is subsequently spread equitably to all other nodes from the ground nodes. On node u, the ultimate resource distribution (leadership score R_u) is:

$$R_u = r_u(t_c) + \frac{r_g(t_c)}{U} \tag{9}$$

where R_u shows the influential value of user u.

3.3 Identifying Cluster Seeds

In this step, those users whose corresponding influential values R_u are higher than a threshold are identified as cluster seeds.

3.4 Propagation of Labels

Here, the goal of the stage is to group the users based on their similarities. To achieve this, they proposed a new label propagation algorithm. Using this algorithm, each cluster seed propagates its label to its k-nearest neighbors. Their neighbours also propagate their labels to their neighbors. This process is continued until all users are assigned cluster labels. Finally, each group of users is assigned the same cluster label. These identified clusters are then employed in the next step for prediction rating values of unseen items.

3.5 Rating Prediction

This step aims at predicting the rating values of unseen items by the target user. To this end, first, identify the cluster label of the target users. The label of the nearest cluster to the target user is assigned to him. All users of this cluster are used in the rating prediction process.

$$P_{u,i} = \bar{r}_u + \frac{\sum_{u' \in N(u)} W(u, u')(r_{u,i} - \bar{r}_{u'})}{\sum_{u' \in N(u)} W(u, u')} \tag{10}$$

where, $W_{u,u'}$ is the weight of the edge between two user u and u' and $r_{u,i}$ is the rating provided to the item by user u, $\bar{r}_u$ are the average ratings of user u and $\bar{r}_{u'}$ are the average ratings of the user u', and $N(u)$ is the neighbours of user u.

Algorithm 1 Provides the proposed method's pseudo-code

Input R: Rating Matrix, T: Trust Relationships

Output $\hat{R}$: Predicted unknown ratings

Begin

1.   W = *Compute* Trust-Based Similarity values using Eq. (2)
2. $G = $ *Create* Users Graph (W)

3. $G' = Add$ Ground Node (G)
4. $r_g(0) = 0; r_u(0) = 1; \forall u \in U$
5.   For $t = 1:t_c$
6. $r_u(t) = \sum\limits_{u'=1}^{N+1} \frac{d_{u',u}}{k_{u'}^{out}} r_u(t-1); \forall u \in U \cup g$
7.   End
8. $R_u = r_u(t_c) + \frac{r_g(t_c)}{|U|}; \forall u \in U$
9. $C = Identifying$ Cluster Centers (R)
10. $N(u) = Apply$ the proposed label propagation algorithm (C)
11. $\hat{R} = Predict$ Missing Ratings$(N(u)); \forall u \in U$ using Eq. (10)
12.   Return $\hat{R}$

End

4   Experimental Results

4.1   Experimental Settings

In this paper, many experiments are carried out to assess the efficiency of the proposed technique on the FilmTrust real-world dataset. The FilmTrust dataset is a social movie website where users can form friendships and share their movie ratings. The FilmTrust dataset's items are rated on a scale of 0.5 (minimum) to 4.0 (maximum). It has 35,497 ratings from 1508 users on 2071 films (items) with 1853 trust ratings [24]. We estimated the behaviour of all procedures on a machine that has the configuration of an i7 processor with 2.2 GHz and 4 GB of RAM. The results are compared based on well-known metrics, which are: F1-Measure, Precision, Recall, Mean Absolute Error (MAE), and Root Mean Squared Error (RMSE), as assessment measures which have been frequently utilised to compare and evaluate the effectiveness of recommendation systems, and a certain number of items are also recommended for a particular user [25]. These metrics are defined as follows:

$$MAE = \frac{1}{|S|} \sum_{i=1}^{|S|} |Pred_i - r_i| \qquad (11)$$

$$RMSE = \sqrt{\frac{1}{|S|} \sum_{i=1}^{|S|} (Pred_i - r_i)^2} \qquad (12)$$

4.2 Resuts

Figure 3a, b shows the results of MAE and RMSE for a varying number of clusters from 5 to 30 while the experiments are implemented on the FilmTrust dataset. The results show that the proposed technique obtained lower MAE and RMSE values in comparison to k-means and LeaderRank methods. Here, the proposed technique uses trust information along with rating values to deal with cold-start users and items. Because in most real-world applications only a few items are given feedback by the users. So we were faced with a too sparse user-item matrix.

Furthermore, the methods were also compared based on some other evaluation metrics, which are precision, recall, and F1-Measure, in which precision (P) is the fraction of recommended items that are related to the user, whereas recall (R) is the fraction of related items that are part of the set of recommended items as well, and F1-measure helps to simplify precision (P) and recall (R) into a single metric [26]. The following formulae are used to compute them:

$$Precision(P) = \frac{Correctly\ Recommended\ Items}{Total\ recommended\ Items} \tag{13}$$

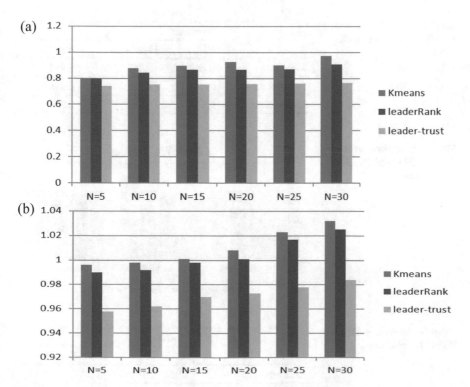

Fig. 3 Evaluation metrica results of (**a** MAE and **b** RMSE) on FilmTrust dataset

$$Recall(R) = \frac{Correctly\ Recommended\ Items}{Total\ Useful\ Recommended\ Items} \tag{14}$$

$$F1 - Measure = \frac{2 * P * R}{P + R} \tag{15}$$

The results are reported and shown in Table 2 and Fig. 4, which shows that the proposed method generates better results than the others.

From another point of view, the methods are performed on cold-start users. Users who rate fewer than five items are considered cold-start users. As a group of users with fewer than five ratings, this experiment aims to evaluate the algorithms' capabilities in two sparse domains. In this scenario, these users have limited access to information. As a result, the aim is to look at how extra sources of information, such as trust statements, may be utilised in conjunction with rating information to enhance rating prediction accuracy. The results of Table 3 and Fig. 5 show that the proposed method also generates better results because the proposed method achieved the lowest MAE and RMSE. So it can be concluded that using trust information results in a more accurate rating prediction.

Table 2 F1, Precision and Recall Comparison for all users

	FilmTrust		
	F1	Precision	Recall
K-means	0.668	0.548	0.855
LeaderRank	0.671	0.551	0.858
Leader-trust	0.684	0.567	0.863

Fig. 4 Evaluation metrics (F1, Precision and Recall) for all users

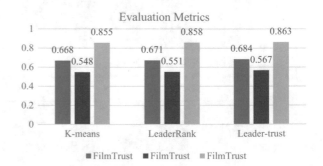

Table 3 Average MAE and RMSE for cold users

	FilmTrust	
	MAE	RMSE
K-means	1.068	1.32
LeaderRank	0.958	1.298
Leader-trust	0.791	0.979

Fig. 5 Average MAE and
RMSE for cold users

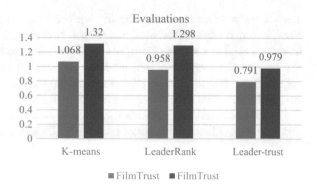

Fig. 6 Average execution
time in seconds

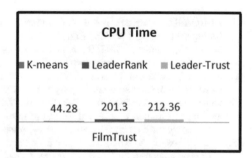

4.3 Execution Time

Many experiments were conducted to compare the execution times of the approaches
in this research, and the results are shown in Fig. 6. Results show that the suggested
method uses more processing resources than the others because of the need to
compute trust-based similarity weights in the graph of users, which requires additional CPU resources. Therefore, while the other results demonstrate that the
proposed technique achieves higher accuracy, it can be concluded that there is a
tight relationship between the accuracy of the methods and their computational
requirements.

5 Conclusion

The study proposed a trust-aware recommender system. In real-world applications,
the majority of users only provide feedback on a limited number of items. When it
comes to "cold-start" users and items, most rating-based recommendation systems
fail. Users' trust relationships and feedback are examined in this paper to address
this issue. An initial graph of user similarity is generated, and then the LeaderRank
method (which uses a random walk algorithm) is used to calculate the influence of

users. Cluster seeds are chosen by the top users with the highest influence values. Next, a unique label will be assigned to each cluster seed. A novel label propagation method is used to distribute the labels to all users. It is necessary to use the clustering results for the rating prediction process. To assess the efficiency of the proposed approach, several tests were performed on the well-known and widely used real-world dataset called FilmTrust. The results in Tables 2 and 3 show that LeaderRank surpasses K-means and Leader-trust performs better than LeaderRank. It is shown that the proposed method is better than the traditional and modern methods.

References

1. Al-Barznji K, Atanassov A (2017) A framework for cloud-based hybrid recommender system for big data mining. J Sci Eng Educ 2:58–65
2. Golbeck J, Hendler J (2006) Inferring binary trust relationships in web-based social networks. ACM Trans Internet Technol 6:497–529
3. Moradi P, Ahmadian S (2015) A reliability-based recommendation method to improve trust-aware recommender systems. Exp Syst Appl 42:7386–7398
4. Guo G, Zhang J, Yorke-Smith N (2015) TrustSVD: collaborative filtering with both the explicit and implicit influence of user trust and of item ratings. In: Proceedings of the twenty-ninth AAAI conference on artificial intelligence, Austin, Texas
5. Kant S, Ansari IA (2016) An improved K means clustering with Atkinson index to classify liver patient dataset. Int J Syst Assur Eng Manag 7:222–228
6. Kant S, Mahara T, Jain VK, Jain DK, Sangaiah AK (2018) LeaderRank based k-means clustering initialization method for collaborative filtering. Comput Electr Eng 69:598–609
7. Moradi P, Ahmadian S, Akhlaghian F (2015) An effective trust-based recommendation method using a novel graph clustering algorithm. Physica A 436:462–481
8. Xue G-R, Lin C, Yang Q, Xi W, Zeng H-J, Yu Y et al (2005) Scalable collaborative filtering using cluster-based smoothing. In: Presented at the proceedings of the 28th annual international ACM SIGIR conference on RDIR, Salvador, Brazil
9. Ramezani M, Moradi P, Akhlaghian F (2014) A pattern mining approach to enhance the accuracy of collaborative filtering in sparse data domains. Physica A 408:72–84
10. Birtolo C, Ronca D (2013) Advances in clustering collaborative filtering by means of fuzzy C-means and trust. Exp Syst Appl 40:6997–7009
11. Koohi H, Kiani K (2016) User based collaborative filtering using fuzzy C-means. Measurement 91:134–139
12. Katarya R (2018) Movie recommender system with metaheuristic artificial bee. Neural Comput Appl 30:1983–1990
13. Katarya R, Verma OP (2018) Recommender system with grey wolf optimizer and FCM. Neural Comput Appl 30:1679–1687
14. Katarya R, Verma OP (2017) An effective collaborative movie recommender system with cuckoo search. Egypt Inform J 18:105–112
15. Parvin H, Moradi P, Esmaeili S (2019) TCFACO: trust-aware collaborative filtering method based on ant colony optimization. Exp Syst Appl 118:152–168
16. Seyedi A, Lotfi A, Moradi P, Qader NN (2019) Dynamic graph-based label propagation for density peaks clustering. Exp Syst Appl 115:314–328
17. Kumar S, Nezhurina MI (2019) An ensemble classification approach for prediction of user's next location based on Twitter data. J Ambient Intell Humanized Comput 10:4503–4513
18. Wang S, Gong M, Li H, Yang J, Wu Y (2017) Memetic algorithm based location and topic aware recommender system. Knowl-Based Syst 131:125–134

19. Yang B, Lei Y, Liu J, Li W (2016) Social collaborative filtering by trust. IEEE Trans Pattern Anal Mach Intell 39:1633–1647
20. Li G, Zhang Z, Wang L, Chen Q, Pan J (2017) One-class collaborative filtering based on rating prediction and ranking prediction. Knowl-Based Syst 124:46–54
21. Li Y, Wang D, He H, Jiao L, Xue Y (2017) Mining intrinsic information by matrix factorization-based approaches for collaborative filtering in recommender systems. Neurocomputing 249:48–63
22. Wasid M, Kant V (2015) A particle swarm approach to collaborative filtering based recommender systems through fuzzy features. Procedia Comput Sci 54:440–448
23. Li Q, Zhou T, Lü L, Chen D (2014) Identifying influential spreaders by weighted LeaderRank. Physica A 404:47–55
24. Guo G, Zhang J, Yorke-Smith N (2013) A novel Bayesian similarity measure for recommender systems. In: Proceedings of the 23rd IJCAI, pp 2619–2625
25. Al-Barznji K, Atanassov A (2018) Comparison of memory based filtering techniques for generating recommendations on large data. Проблемы машиностроения и автоматизации 1:44–50
26. Al-Barznji K, Atanassov A (2017) Collaborative filtering techniques for generating recommendations on big data. In: International conference automatics and informatics, CAN, Sofia, Bulgaria

Dimensional Reduction Layer: The Simple Way to Build Lightweight Models

Trang Phung⬤, Thi Hong Thu Ma⬤, and Van Truong Nguyen⬤

Abstract Convolutional neural network (CNN) has been making great achievements in computer vision and artificial intelligence. To achieve state-of-the-art performance, the CNN models are designed to become deeper and deeper, using more feature maps to extract many features from the input image such as ResNet-152 with 60.3M parameters and 11 GFLOPs. Therefore, running these models in real-time is a big issue, especially, running real-time on PC and/or embedded devices without using GPU. In this work, we propose a simplest layer named "Dimensional Reduction Layer" (DR layer) that reduces the number of feature maps and makes models lighter and less computational. The experiments on various network backbones including basic model, mobile model, and res model show that the DR layer reduces significantly the performance of these models when we concatenate more feature maps together. Besides, we found that the DR layer works well on models using residual learning blocks such as ResNet.

Keywords Dimensional reduction layer · Convolution neural network · Deep learning

1 Introduction

Machine learning, especially deep learning, has achieved great accomplishments in many areas recently. Recurrent Neural Networks (RNN) and Long Short Term Memory (LSTM) Networks with the idea that they might be able to connect previous information to the present task have been applied to solve many problems in speech recognition and natural language processing (NLP) effectively. Along with

T. Phung
Thai Nguyen University, Thai Nguyen, Vietnam

T. H. Thu Ma (✉)
Tan Trao University, Tuyen Quang, Vietnam
e-mail: thutq7@gmail.com

V. T. Nguyen
Thai Nguyen University of Education, Thai Nguyen, Vietnam

© The Author(s), under exclusive license to Springer Nature Switzerland AG 2022   137
N. H. T. Dang et al. (eds.), *Artificial Intelligence in Data and Big Data Processing*,
Lecture Notes on Data Engineering and Communications Technologies 124,
https://doi.org/10.1007/978-3-030-97610-1_12

the development of NLP, image processing and computer vision also have break-throughs. Models built based on Convolutional Neural Network (CNN) achieve high performance. For example, Alex et al. [5] built the network called AlexNet that won the object identification contest (ImageNet) with 57% top-1 accuracy and 80.3% top-5 accuracy in 2012. In some following years, many variant models based on CNN, such as ZFNet [12] in 2013, GoogleNet [9] in 2014, VGGNet [8] in 2014, ResNet [1] in 2015, won in the contest. Apart from image classification, CNN is usually applied to many image problems such as multi-object detection, image captioning, image segmentation.

With the skip connections method in [1], ResNet has been able to avoid vanishing gradient problem, without sacrificing network performance. That helps deep layers at least not worse than shallow layers. Moreover, with this architecture, the upper layers get more information directly from the lower layers so it will adjust the weight more effectively. After ResNet, a series of variations of this method was introduced. Experiments show that these architectures can be trained with CNN models with a depth of up to thousands of layers. ResNet has quickly become the most popular architecture in computer vision. The models with thousands of layers mean that the computational cost in the networks is extremely large. For example, AlexNet is proposed in 2012 uses 56M parameters and 0.73 GFLOPs. In 2015, Resnet-152 uses 60.3M parameters and 11 GFLOPs. With huge numbers of parameters and computational costs, these models are very difficult to predict in real-time, especially when they are deployed on embedded devices with low memory and without GPU support.

Recently, several lightweight models are published with workable results such as MobileNet [2], MobileNet V2 [7], ShuffleNet [13], ShuffleNet V2 [6] and so on. Most of the above methods are focus on reducing complex calculations in Convolution operation. Therefore, these networks can easily run real-time on PCs, mobiles, or embedded devices without using GPU. In this paper, we present a novel approach to reduce parameters and FLOPs of CNN models. Our key contributions are:

(1) We propose a new layer named "Dimensional Reduction" that helps to reduce the number of channels in each input of convolution operation. Thus, Dimension Layer has the effect of reducing the parameters and FLOPs.
(2) We conducted experiments with three types of networks including basic CNN, Depthwise CNN, and Residual CNN. We see that Dimensional Reduction Layer is suitable for the architecture of the ResNet with approximate performance in terms of accuracy and the speed is improved significantly that compare with the original network.

The remainder of the paper is organized as follows. Section 2 provides a review of the related work. The novel approach is proposed in Sect. 3. The experimental results, comparisons and component analysis are presented in Sect. 4 and the final the conclusions are given in Sect. 5.

2   Related Work

AlexNet was suggested in 2012 by Krizhesky et al. [5]. In that year, AlexNet won in the ImageNet contest with a top-5 error rate 11% lower than the second place. AlexNet is built upon the architecture of the LeNet network but instead of using the tanh function, the author proposed the ReLU function for faster speed and faster converging. Following the success of AlexNet, ZFNet [12] won in 2013 with the top-5 validation error rate is 16.5%. ZFNet is improved from AlexNet by using 7×7 filter size and stride $= 2$ in the first convolution layer instead of 11×11 filter size and stride $= 4$ in the AlexNet. In addition, the author proposed the Deconvnet technique for the visualization of layers. This technique helps us to understand clearly about activation processing and learning features from the input image. The year 2014 is marked with deep learning models achieving error rates below 10%, like GoogleNet, VGGNet. VGGNet is presented by Simonyan and Zisserman [8], VGGNet uses 3×3 filters instead of larger sizes, which reduces the number of parameters used in the model. However, VGGNet is still considered a very big model with 138M parameters and 144M parameters for the VGG-16 and the VGG-19, respectively. Unlike VGGNet, GoogleNet (Inception v1) uses 1×1 filters that are used as a dimension reduction module to reduce the computation bottleneck. In addition, the Inception Module applied in GoogleNet helps the model to learn many different features from different kernel sizes of convolution layers. As network architectures become deeper and deeper, it is difficult to train because of the vanishing gradient.

ResNet [1] is is one of the very deep networks with up to 152 layers and is model won the 2015 ImageNet competition. To solve the vanishing gradients problem, ResNet used Skip Connection (or Shortcut Connection). This helps to architecture at least no less than the shallow architecture. Moreover, with this architecture, the upper layers get more direct information from the lower layers so it will adjust the weight more effectively. ResNcXt [11] is an updated version of ResNet using Grouped Convolution in AlexNet. With an equal number of parameters and FLOPs (e.g. For example, ResNet-50 and ResNeXt-50 both have 25.5M parameter and 4.2GFLOPs) but ResNeXt gives the top 1 error rate about 1.7 % lower than ResNet. Another updated version of ResNet is proposed in 2017 called the Residual Attention Network [10]. With add the Attention Module and reduced the number of iterations in the residual unit, The Attention-56 network resulted in a top1 error rate of 0.4% and 0.26% for top 5 error compared to ResNet-152. Moreover, the Attention-56 network is only built with 52% traffic number and 56% FLOP compared to ResNet-152. Different from ResNet, for each class whose input was added from the previous one earlier, but in Densenet, each layer obtains additional inputs from all preceding layers. Received by each class feature maps from all previous layers, the network may be thinner and more compact, i.e. the number of channels may be less. Because, it has higher calculation efficiency and memory efficiency. SENet [3] with use Squeeze-and-Excitation (SE) Block in network architecture, SENET won the ImageNet competition in 2017 with a top 5 error of 3.79%.

It can be seen that each year, the models are designed to be more complex, deeper and more parameters aim to learn more features from input images. This led to models becoming bulky. It is difficult to deploy on embedded devices or run in real-time. In 2017, Howard et al. [2] proposed a model named MobileNet. Instead of using normal covolution, MobileNet model used Depthwise and Pointwise convolution which 8 to 9 times less computation than normal convolutions with kernel size is 3×3 and a small reduction in accuracy. An updated version of MobileNet is new called MobileNet v2 proposed by Sandler et al. [7] in 2018. With the use of linear bottlenecks and inverted residuals, MobileNet v2 gave better results than MobileNet on various problems like image classification, object detection and image segmentation. ShuffleNet [13] is a light weight model which takes its ideas from swapping locations channels and use 1×1 group convolution to limit the number of calculations. ShuffleNet has given better results and numbers. The amount of FLOPs is much lower than that of MobileNet. ShuffleNet v2 proposed channel split technique, with engineering. This time the convolution layer only works on half of the feature maps and is joined the next layers. This is as a kind of feature reuse which help ShuffleNet v2 better than ShuffleNet.

Far apart from the previous above methods, in this paper, we introduce a new layer named Dimensional Reduction to reduce computational cost in CNN models. Our approach not only achieves promising performance but also easily deploys on PC and embedded devices in real-time without GPU support.

3 Proposed Approach

In this section, we propose our novel approach to reduce the numbers of parameters and FLOPS for CNN models. First, we introduce DR layer, and then we present how to apply DR layer in three main network architectures including basic architecture, MobileNet architecture, and ResNet architecture.

3.1 Dimensional Reduction Layer

Let denote the output of a Conv block (including convolutional layer, batch normalization layer and activation layer) are is a tensor array of size $B \times H \times W \times D$ where B, H, W, D are batch size, height, width and dimension respectively. Remember that the batch normalization layer and activation layer don't change the input and output size. Dimensional reduction layer is applied after each Conv block aims to reduce the dimension D of feature maps. An example of the dimensional reduction layer is illustrated in Fig. 1.

In Fig. 1, number of dimensions reduce to $\times 2$. A dimensional reduction layer's operation is described as follows: First, we split feature maps into several groups where each group has m feature matrices. For example, $m = \{2, 4, 8, 16, \ldots\}$ (in

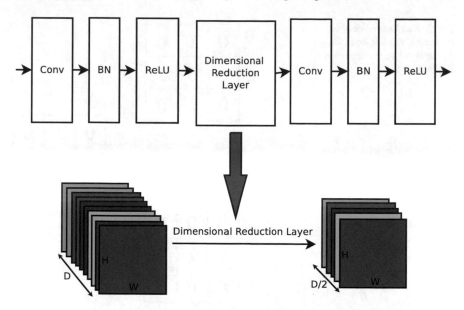

Fig. 1 An example of the Dimensional Reduction Layer with $m = 2$. In which Conv, BN, ReLU are Convolution, BatchNormalization and ReLU activation layers, respectively. The H, W, D are height, width and channel dimension, respectively

Fig. 1, we choose m is 2). We then applied one of the following operations (including {max, add, average}) for these groups. The purpose of this process is to merge m feature matrices in each group into one feature matrix. Figure 2 shows examples of the max operation in a group of two feature matrices. For max operation, all feature matrices are merged into one feature matrix where each element in the output matrix is calculated by the max of each corresponding element in the input feature matrices. The average and add operations are the same idea as the max operation.

Let denote the input of a Conv block is $H \times W \times D$ where a convolution layer with D' filters and the kernel size as $k \times k$. We get the Conv block's output is $H \times W \times D'$ of shape (we assume that the output feature map has the same spatial dimensions as the input). The number of parameters is calculated in the following:

$$k \times k \times D \times D' \tag{1}$$

Standard convolutions have the computational cost as:

$$k \times k \times D \times D' \times H \times W \tag{2}$$

However, if the input dimension is reduced by m through the dimensional reduction layer, the number of parameters to use in the Conv layer is calculated in the Eq. 3. Note that the dimensional reduction layer does not use any additional parameters.

Fig. 2 An example of max operator in the DR layer with the input is two feature maps 3 × 3 of size

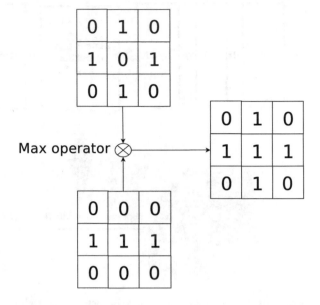

$$\frac{k \times k \times D \times D'}{m} \tag{3}$$

After applying the dimensional reduction layer, the computational cost of Conv block is reduced by m times as in Eq. 4.

$$\frac{k \times k \times D \times D' \times H \times W}{m} \tag{4}$$

We group the m feature matrices into a group thus we have D/m groups. For each group, we need to repeat $H \times W \times m$ to apply one operation. Therefore, the computational cost of the dimensional reduction layer would be:

$$\frac{D}{m} \times H \times W \times m = D \times H \times W \tag{5}$$

Comparison the sum of Eqs. 4 and 5 to Eq. 2, we have:

$$\frac{(\frac{k \times k \times D \times D' \times H \times W}{m}) + (D \times H \times W)}{k \times k \times D \times D' \times H \times W} = \frac{1}{k \times k \times D'} + \frac{1}{m} \tag{6}$$

As can be seen in Eq. 6, we demonstrate that using a dimensional reduction layer reduces the computational cost more than the normal Conv block. For example, the convolution layer with a kernel size of 3 × 3. and the number of filters is 128, in case $m = 2$, the computational cost is reduced approximately 0.5 times compared to the model without using the DR layer.

However, there exists a question for the dimensional reduction layer is "Does Merging these feature matrices through the dimensional reduction layer affect the model's ability and performance during training?". We see visually that when it comes to reducing the number of feature maps, it is very likely that the model's performance will be reduced. Because reducing the number of feature maps make it harder for the next Conv layer(s) to "learn". Simultaneously, with several lightweight models such as MobileNet and ShuffleNet, reducing computation costs in lightweight models reduces the model's performance relative to deep models. The lightweight models can easily overcome deep models in computation cost, but they cannot outperform deep models in terms of performance. Details of the dimensional reduction layer's performance are presented in Sect. 4.

3.2 Network Architecture and Training

To evaluate the effectiveness of the dimensional reduction layer, we adopt three different models, including the basic model (using normal convolution), the mobile model (using depthwise convolution) and the res model (using residual convolution). For more convenience in presenting the model architecture, we group BatchNormalization (BN) and ReLU with three types of convolution layers into blocks named Conv_block, DW_block and Res_block as shown in the Fig. 3.

The three models are described in Table 1. In which, conv_block(n), DW_block(n) and res_block(n) mean the convolution layer in each block has the number of filters

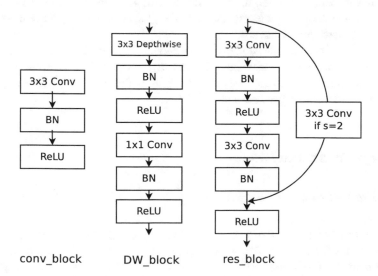

Fig. 3 Three type blocks of each model. In which Conv_block is described in the left, DW_block is illustrated in the centre and Res_block is shown is the right

Table 1 Models architecture of three models. In which, FC is the the fully connected layer and Global AVE is the global average pooling layer

Basic Model	Mobile Model	Res Model
Conv_block(32)	Conv_block(32)	Conv_block(32)
MaxPool(2,2)	DW_block(32) ×3	Res_block(32) ×3
Conv_block(64)	DW_block (64,s=2)	Res_block(64, s=2)
MaxPool(2,2)	DW_block (64) ×2	Res_block(64) ×2
Conv_block(128)	DW_block (128,s=2)	Res_block(128, s=2)
MaxPool(2,2)	DW_block (128) ×2	Res_block(128) ×2
Global AVE	Global AVE	Global AVE
FC	FC	FC

= n. All convolution layer have a default stride equal to 1. In case stride=2 is denoted by s=2. MaxPool(2,2) is the MaxPooling layer that has pool size = 2 and stride = 2.

The first model is called the basic model, in this model, the layers are designed by alternating Conv_block and maxpooling, and the last layers are Global average pooling and the fully connected (FC) layer with the activation function softmax to predict the classification probability.

Mobile Model is built on the depthwise and pointwise convolution [2]. These two convolutions combine with BN, and ReLU layers are grouped into DW_block. Each DW_block receives the corresponding number of filters of 32, 64, 128, and each DW_block is repeated three times. The first block of each repeat has stride=2.

Similar to Mobile Model, the Res Model uses Res_block with the number of filters 32, 64 and 128 respectively and each Res_block is repeated three times. The first block of each repeat has stride=2.

For Basic Model, Mobile Model and Res Model that uses the dimensional reduction layer, we apply a dimensional reduction layer before the convolution layer in each block except conv_block (32) at the beginning of each model. All other parameters for the models are preserved as in Table 1.

4 Experiment and Result

4.1 Dataset and Implementation

Dataset: To evaluate the Reduced Dimension Layer efficiency on the above three models, we perform on the CIFAR-100 dataset. The CIFAR-100 dataset consists of 60,000 images with 32 × 32 RGB resolution in 100 classes where 6000 images per class. The CIFAR-100 is built by Krizhevsky et al. in 2009 [4]. In this dataset, there are 50,000 training images and 10,000 test images.

Implementation Details: All models with and without using DR Layer are trained with the same optimization function, Adam optimizer. The initial learning rate is set to 0.001 and reduced by a factor of $10\times$ manually when the validation loss did not reduce. Training images are split into multiple parts with a batch size of 32 images. Data augmentation was applied with training images includes: horizontal flip, shift horizontally, shift vertically and rotation images. All networks are trained on the same PC with E5-2678 v3 dual CPUs, 2070S GPU, 64GB of RAM.

Evaluation Metrics: For evaluation, we measure the following metrics:

- The top-1 accuracy: is a standard performance measure for multi-class classification in action recognition. This measure is simply calculated as the ratio between the number of correctly predicted scores per the total number of points in the test set.
- The number of parameters or model size: is the total of parameters that are used in the model. This measure affects the model size. Typically, the more layers in the model, the larger the number of parameters.
- The complexity or computational cost or number of float-point operations (FLOPs): is the measure of the number of multiply-adds in the model. It is an indirect metric and an approximation [6]. Typically, a model requires computation at millions of FLOPs (MFLOPs) or billions of FLOPs (GLOPs).

4.2 Result and Analysis

The Top-1 accuracy of the three models, including Basic, Mobile, and Res networks with and without using the DR layer is shown in Table 2. We can see that with the DR layer the performance of models is reduced. However, in different models, the decrease in accuracy is also different. As in the Basic model, the performance is reduced by up to 6.18% with $m = 2$ and up to 23.54% with $m = 8$. With the Res model, the accuracy is reduced by only 0.89% with $m = 2$ and 9.13% with $m = 8$. The experiment shows that the DR layer works well for models that use shortcut connections such as ResNet. As can be seen in Table 2, the max and add operator's performance are better than the average operator's performance in the same model architecture.

Table 3 shows the number of parameters used and the FLOP of each model. Specifically, the number of parameters and FLOPs of the basic network without DR layers are 106K and 10.35 MFLOPs, respectively. Opposite, in the case, using DR layers with $m = 2$, the number of parameters is 54K (reduce 50.87%) and the computational cost is 5.62 MFLOPs (reduce 42.7%). In the case, using DR layers with $m = 4$, the number of parameters is 28K (reduce 73.69%) and the computational cost is 3.25 MFLOPs (reduce 68.6%). In the case, using DR layers with $m = 8$, the number of parameters is 15K (reduce 85.97%) and the computational cost is 2.07 MFLOPs (reduce 80.0%).

Table 2 Top-1 accuracy of three models with and without using the DR layer. All models are performed on the CIFAR-100 dataset. w/o DR layer denotes that these models without DR layer

	Basic model (%)	Mobile model (%)	Res model (%)
w/o DR layer	**54.02**	**62.17**	**67.71**
Add, m=2	44.99	**56.77**	**66.82**
Max, m=2	**47.84**	55.35	66.48
Average, m=2	45.81	53.80	65.45
Add, m=4	39.82	**49.11**	61.01
Max, m=4	**39.94**	44.72	61.43
Average, m=4	37.31	46.89	**62.14**
Add, m=8	28.48	**36.65**	**58.58**
Max, m=8	**30.48**	35.81	55.81
Average, m=8	25.19	37.89	56.22

Table 3 Comparison models using the DR layer to other networks without the DR layer in terms of the number of parameter and FLOP

		w/o DR layer	$m = 2$	$m = 4$	$m = 8$
Basic model	# Params	106,820	54,340	28,100	14,980
	MFLOPs	10.35	5.62	3.25	2.07
Mobile model	# Params	78,340	48,180	33,100	25,560
	MFLOPs	10.74	6.5	4.37	3.31
Res model	# Params	1,100,132	565,604	298,340	164,708
	MFLOPs	162.79	82.58	42.47	22.41

The number of parameters and FLOPs of the mobile network without DR layers are 78K and 10.74 MFLOPs, respectively. Opposite, in the case, using DR layers with $m = 2$, the number of parameters is 48K (reduce 38.49%) and the computational cost is 6.5 MFLOPs (reduce 39.47%). In the case, using DR layers with $m = 4$, the number of parameters is 33K (reduce 57.74%) and the computational cost is 4.37 MFLOPs (reduce 59.31%). In the case, using DR layers with $m = 8$, the number of parameters is 25K (reduce 67.37%) and the computational cost is 3.31 MFLOPs (reduce 69.18%).

The number of parameters and FLOPs of the res network without DR layers are 1100K and 162.79 MFLOPs, respectively. Opposite, in the case, using DR layers with $m = 2$, the number of parameters is 48K (reduce 48.58%) and the computational cost is 82.58 MFLOPs (reduce 49.27%). In the case, using DR layers with $m = 4$, the number of parameters is 298K (reduce 72.88%) and the computational cost is 42.47 MFLOPs (reduce 73.91%). In the case, using DR layers with $m = 8$, the number of parameters is 164K (reduce 85.02%) and the computational cost is 22.41 MFLOPs (reduce 86.23%). Besides, we found see that when m increases, the number of parameter parameters and the FLOP of the models decreases significantly. All the experiment results of all models using the DR layer are illustrated in the Fig. 4.

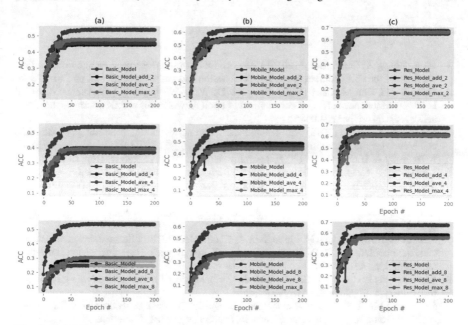

Fig. 4 Individual validation accuracy for each model versus epoch. The **a** (left column) shows basic models performance that includes $m = 2$, $m = 4$ and $m = 8$ in order from top to bottom. **b** (the center column) illustrates mobile models performance that includes $m = 2$, $m = 4$ and $m = 8$ in order from top to bottom. **c** (the right column) illustrates res models performance that includes $m = 2$, $m = 4$ and $m = 8$ in order from top to bottom

5 Conclusion

This paper presents a new method called a dimensional reduction layer that reduces the number of parameters and FLOP in CNN models. The dimensional reduction layer works by grouping feature matrices in groups together into a single feature matrix. We also propose three computations of feature matrices: max, average, and add. The dimensional reduction layer experiment on three models with three different architectural types shows that the more feature matrices we lump together, the less the model's accuracy. Experiment results show that the DR layer works well for models that use the shortcut connection, such as ResNet.

In the future, we will improve the dimensional reduction layer for better results on many different architectures of the CNN network. Besides, we will also test a reduced dimension on some other problems in computer vision.

Acknowledgements This research is funded by Thai Nguyen University, Tan Trao University and Thai Nguyen University of Education.

References

1. He K, Zhang X, Ren S, Sun J (2016) Deep residual learning for image recognition. In: Proceedings of the IEEE conference on computer vision and pattern recognition, pp 770–778
2. Howard AG, Zhu M, Chen B, Kalenichenko D, Wang W, Weyand T, Andreetto M, Adam H (2017) Mobilenets: efficient convolutional neural networks for mobile vision applications. arXiv preprint arXiv:1704.04861
3. Hu J, Shen L, Sun G (2018) Squeeze-and-excitation networks. In: Proceedings of the IEEE conference on computer vision and pattern recognition, pp 7132–7141
4. Krizhevsky A, Hinton G et al (2009) Learning multiple layers of features from tiny images
5. Krizhevsky A, Sutskever I, Hinton GE (2012) Imagenet classification with deep convolutional neural networks. In: Advances in neural information processing systems, pp 1097–1105
6. Ma N, Zhang X, Zheng HT, Sun J (2018) Shufflenet v2: practical guidelines for efficient cnn architecture design. In: Proceedings of the European conference on computer vision (ECCV), pp 116–131
7. Sandler M, Howard A, Zhu M, Zhmoginov A, Chen LC (2018) Mobilenetv2: inverted residuals and linear bottlenecks. In: Proceedings of the IEEE conference on computer vision and pattern recognition, pp 4510–4520
8. Simonyan K, Zisserman A (2014) Very deep convolutional networks for large-scale image recognition. arXiv preprint arXiv:1409.1556
9. Szegedy C, Liu W, Jia Y, Sermanet P, Reed S, Anguelov D, Erhan D, Vanhoucke V, Rabinovich A (2015) Going deeper with convolutions. In: Proceedings of the IEEE conference on computer vision and pattern recognition, pp 1–9
10. Wang F, Jiang M, Qian C, Yang S, Li C, Zhang H, Wang X, Tang X (2017) Residual attention network for image classification. In: Proceedings of the IEEE conference on computer vision and pattern recognition, pp 3156–3164
11. Xie S, Girshick R, Dollár P, Tu Z, He K (2017) Aggregated residual transformations for deep neural networks. In: Proceedings of the IEEE conference on computer vision and pattern recognition, pp 1492–1500
12. Zeiler MD, Fergus R (2014) Visualizing and understanding convolutional networks. In: European conference on computer vision. Springer, pp 818–833
13. Zhang X, Zhou X, Lin M, Sun J (2018) Shufflenet: an extremely efficient convolutional neural network for mobile devices. In: Proceedings of the IEEE conference on computer vision and pattern recognition, pp 6848–6856

A Power Generating Distribution Planning Using Swarm Moth-Flame Optimization

Trong-The Nguyen, Van-Chieu Do, and Thi-Kien Dao

Abstract Power generating distribution planning (PDP) is a critical component of the power system's financial performance. An effective PDP model has been constructed by carefully considering various system operation restrictions typical of multi-constrained nonlinear optimization problems. This work suggests solving the PDP model based on a new swarm-based technique called moth flame optimization (MFO) to achieve energy-saving and cost consumption reduction. In the experimental section, the IEEE-bus benchmark test systems are used to verify the performance of the proposed scheme system. The results show that the proposed scheme can solve the power system PLP problem with good robustness and significant economic benefits.

Keywords Power distribution planning · Economic dispatch · Swarm moth flame optimization · Optimization algorithm

1 Introduction

Power generating distribution planning (PDP) is a critical component of the power system's financial performance [1]. Because PDP determines each unit's output in a given time under the given load demand [2], in the power system, each generating unit can share the load demand to meet the actual constraints and seek to minimize the total operation cost of the whole system [3]. The PDP is considered a fundamental problem in the operation of modern power systems and plays an essential role in improving the operation economy of power systems [4]. The economic dispatch model of a power system has only the constraints of unit capacity limitations and power balance

T.-T. Nguyen · T.-K. Dao (✉)
School of Computer Science and Mathematics, Fujian University of Technology, Fuzhou 350118, China

T.-T. Nguyen
e-mail: vnthe@hpu.edu.vn

T.-T. Nguyen · V.-C. Do
Haiphong University of Management and Technology, Haiphong, Vietnam

© The Author(s), under exclusive license to Springer Nature Switzerland AG 2022
N. H. T. Dang et al. (eds.), *Artificial Intelligence in Data and Big Data Processing*,
Lecture Notes on Data Engineering and Communications Technologies 124,
https://doi.org/10.1007/978-3-030-97610-1_13

but also considers many practical nonlinear conditions existing in the power system operation process, e.g., the unit forbidden area, valve point effect, and climbing rate limit [5]. Therefore, the power system's PDP problem is essentially a non-smooth, highly nonlinear multi constraint optimization problem [6]. The traditional classical mathematical algorithm, e.g., Lagrange relaxation method, linear programming (LP) for the complex problem, would have faced challenging time complicated for the computation. The development of swarm computing methods is one of the most effective ways [7, 8] can deal with the complex issues of nonlinear like the PDP problem [9, 10].

Moth flame optimization (MFO) is a new swarm intelligence algorithm in recent years [11]. Its idea comes from the simulation of the lateral flight mechanism behind the moth's behavior. MFO algorithm has advantages such as model simple and easy to implement. The FMO has been widely used in the engineering field, e.g., analysis of pumping test data, which provides an effective method for accurately estimating confined aquifer parameters, optimizing the Muskingum model's parameters with higher simulation accuracy, etc. [12].

This paper introduces solving the PDP model of the power system with the MFO algorithm to overcome the premature and local convergence problems and to improve the global optimization ability. In addition, we also suggest a method combining the balancing unit with the penalty function to deal with power balance equality constraint and unbalanced power allocation problem to improve the algorithm's convergence speed and calculation accuracy.

2 Power Generating Distribution Planning Model

As a nonlinear optimization problem of the PDP model [5], we considered it multiple constraints under the premise of ensuring the safe and stable operation of the system. The output of the unit of system operation is determined by the decision to meet the output plan and minimize the generation cost of the system. Two components in modeling for a system power system are objective function and dealing with its constraints detailed as follows.

Objective function

The unit operation cost would determine the economic distribution of load among operating units is generally expressed as a function of unit output. If the fuel cost function of the power system is regarded as the summation of a series of quadratic polynomials, the PDP's fitness function is presented as follows.

$$\min \sum_{i=1}^{n'} F_i(P_i) = \sum_{i=1}^{n'} \left(\alpha_i P_i^2 + \beta_i P_i + \gamma_i \right) \tag{1}$$

where, $F_i(P_i)$ is the function of fuel/coal cost in i-th unit; α_i, β_i and γ_i are the cost function's coefficients; P_i and n' are the i-th unit's output and the total number of units with their operations. In the practical power system, the PDP objective function considering "valve point effect" can be expressed as

$$F_{cost} = \min \sum_{i=1}^{n'} F_i(P_i) = \sum_{i=1}^{n'} \left(\alpha_i P_i^2 + \beta_i P_i + \gamma_i + e_i \left| \sin[f_i(P_i^{\min} - P_i)] \right| \right) \quad (2)$$

where e_i and f_i are the coefficients reflecting the "valve point effect" of the i-th unit; $P_i^{\min}$ is the i-th unit's lower limit output.

Constraints

In order to solve the above problems, a relatively comprehensive PDP model closer to the actual operation will be established by comprehensively considering many practical constraints in power system operation, including power balance, generation capacity, restricted area, and climbing rate limit conditions.

(a)   The power balance constraint is

$$\sum_{i=1}^{n'} F_i = P_{load} + P_{loss} \quad (3)$$

where: P_{load} and P_{loss} are the power system's total load and the network loss, which can be approximately presented as a function of unit output by coefficient matrix $\mathbf{B}$

$$P_{loss} = \sum_{i=1}^{n'} \sum_{j=1}^{n'} P_i B_{ij} P_j + \sum_{j=1}^{n'} B_{0i} P_i + B_{00} \quad (4)$$

where, B_{ij}, B_{00}, B_{0i} are the known transmission loss parameters; P_i and P_j are the output of the i and j units respectively.

(b)   The generation capacity constraint is

$$P_i^{\min} \leq P_i \leq P_i^{\max} \quad (5)$$

where: $P_i^{\max}$ is the i-th thermal power unit's upper limit of the output.

(c)   The restricted area is

$$\begin{cases} P_i^{\min} \leq P_i \leq P_{i,1}^{L}, \\ P_{i,j-1}^{U} \leq P_i \leq P_{i,1}^{L}, \text{ with } j = 1, 2, \ldots, n_i, i = 1, 2, \ldots, n' \\ P_{i,n_i}^{U} \leq P_i \leq P_i^{\max}, \end{cases} \quad (6)$$

where: $P_{i,1}^L$ is the lower limit of the first forbidden area of the i-th unit; $P_{i,j-1}^U$ is the upper limit of the $j-1$ forbidden area of the i-th unit; n_i is the number of the forbidden areas of the i-th unit; P_{i,n_i}^U is the i-th unit's upper limit of the last forbidden area.

(d) Climbing rate limit.

When the output increases, it is

$$P_i - P_i^0 \le U_{Ri} \tag{7}$$

When the output is reduced, it is

$$P_i - P_i^0 \le D_{Ri} \tag{8}$$

where P_i^0 is the output of the previous stage of unit i; U_{Ri} and D_{Ri} are the unit (i) 's maximum upward climbing rate MW/time period and the maximum downward climbing rate MW/period.

3 PDP Optimization Based on Moth-Frame Algorithm

3.1 Principle of Moth Frame Algorithm

MFO inspired from the moth and fire are two crucial components considered the candidate optimization solution of the problem, moth flies in the decision space, and the fire is the best position that the moth has found up to now [11]. Therefore, the fire is applied to the "wind vane" for moths to search in the searching space. Each moth-frame is considered an agent searching to search around a fire and updates its position when it finds a better solution. The searching moth and fire agents would be presented with a matrix M and F respectively as follows [11].

$$M = \begin{pmatrix} m_{11} & m_{12} & \dots & m_{1d} \\ m_{21} & m_{22} & \dots & m_{2d} \\ \vdots & \vdots & \vdots & \vdots \\ m_{n1} & m_{n2} & \dots & m_{nd} \end{pmatrix}, \quad F = \begin{pmatrix} F_{11} & F_{12} & \dots & F_{1d} \\ F_{21} & F_{22} & \dots & F_{2d} \\ \vdots & \vdots & \vdots & \vdots \\ F_{n1} & F_{n2} & \dots & F_{nd} \end{pmatrix} \tag{9}$$

where, n is the number of moths and d is the dimension of the variable. The vectorstoring all moth and fire fitness can be expressed as

$$O_M = [O_{M1}, O_{M2}, \dots, O_{Mn}]^T, \quad O_F = [O_{F1}, O_{F2}, \dots, O_{Fn}]^T \tag{10}$$

where: O_M is the fitness matrix of the moth; O_{M1} is the fitness of the first moth; O_{Mn} is the fitness of the nth moth; O_F is the adaptability matrix of the fire; O_{F1} is the fitness of the first fire; O_{Fn} is the fitness of the nth fire.

Each moth updates its position around its corresponding frame fire:

$$M_i = S(M_i, F_j) \tag{11}$$

where M_i is the position of the ith moth; F_j is the position of the jth fire; S is the logarithmic spiral function. The logarithmic spiral function is

$$S(M_i, F_j) = D_{ij} e^{bt} \cos(2\pi t) + F_j \tag{12}$$

where b is the constant coefficient for logarithmic spiral function with distance D_{ij}; t is a random number [-1 to -2] in the iteration loops. The smaller t is, the closer the search position is to the fire. By taking different values of T, the other position relations between moth and fire can be obtained. The update mechanism of MFO algorithm with abscissa t is a random number in the range of r~1, and the ordinate is the position of the moth. The logarithmic spiral update formula searches new positions around the fire by moth agent and effectively utilizes historical optimization solution as local development ability and exploring the unknown area as global search ability to realize the resolution of the optimization problem. The exploring and exploiting the search ability of the algorithm, the number of agent's fires are used to update the position changing dynamically under iteration times. The number of agents' fires under the current iteration is stated in the following formula.

$$N_{\text{flame}} = N - L\frac{N-1}{T} \tag{13}$$

where: N_{flame} and N is the number of fires under the current iteration number and population the maximum number; T and L the maximum and the current numbers of iterations; if N = n, it is the same as the number of moths or the number of the maximum fire updating the position.

3.2 Solution of PDP Model Using Moth Fire Algorithm

Agent moth code variables

The optimal variable of the PDP problem is the active power output of each unit, and the dimension of the problem is equal to the total number of units. Combined with the characteristics of the MFO algorithm and PDP model, each moth represents a candidate solution of the power system load economic dispatch problem, and the position of each moth is a vector composed of the output of each unit.

$$P = \begin{pmatrix} p_1^1 & \cdots & p_i^1 & \cdots & p_{n'}^1 \\ \vdots & \vdots & \vdots & \vdots & \vdots \\ p_1^i & \cdots & p_i^i & \cdots & p_{n'}^i \\ \vdots & \vdots & \vdots & \vdots & \vdots \\ p_1^n & \cdots & p_i^n & \cdots & p_{n'}^n \end{pmatrix} \tag{14}$$

where: P is the output moth matrix of the generator set, and the row vector of the matrix P represents the specific position of each moth.

Only the modified position should be in the area, meeting the conditions for the output capacity constraint and the forbidden area constraint. Due to the limitation of unit capacity and climbing rate, the output capacity of the first unit has upper and lower limits. The lower bound of output is expressed as follows.

$$l_i = \max\left(P_i^{\min}, P_i^0 - D_{Ri}\right) \tag{15}$$

The upper bound of output is

$$u_i = \max\left(P_i^{\min}, P_i^0 - U_{Ri}\right) \tag{16}$$

In updating the position, the moth may cross the feasible boundary formed. When one or more units fail to meet the output capacity constraint, the maximum range of the modified constraint is mapped symmetrically to the interior of the boundary according to the number of units violating the boundary constraint. Then the corrected unit output is randomly determined between the limit and the mapping positions within the boundary.

$$P_i = \begin{cases} l_i + R(l_i - P_i), & P_i < l_i \\ u_i + R(u_i - P_i), & P_i > u_i \\ P_i & others \end{cases} \tag{17}$$

where R is a random number generated by uniform distribution between 0 and 1.

The existence of the forbidden zone makes the decision-making space no longer continuous, and there is an interruption area that initializes the output of a unit randomly between the upper and lower boundaries. The power balance constraint is an equality constraint. The penalty function method is used alone to deal with the conditions. Most of them are infeasible solutions because of the large amount of time spent in the early stage of the algorithm.

New power balance constraint

Some steps consider network transmission loss: Set the number of moths n, the maximum cycle iteration number Q and the acceptable unbalanced power value P_0, and the cycle count $C_0 = 1$. T is the number of units that may have a balancing unit, the initial value is equal to the total number of units n', A random sequence x generated by the ordered sequence $\{1, 2, ..., g\}$ is generated. Generate a balancing unit sequence with $(n' - T + 1)$ element from random sequence X. Calculate the network transmission loss P_{loss}, and then approximately determine the output of the balancing unit

$$t_{emp} = P_{load} + P_{loss} - \left(\sum_{i=1}^{n'} P_i - P_I \right) \tag{18}$$

where P_I is the power used to balance the unit. If the output of the balancing unit is within the upper and lower limits of the unit output, check whether it falls in the forbidden area; if it falls in the forbidden area, the output of the balancing unit is set as the nearest boundary value of the prohibited area; if not, the value of temp is directly set as the output value of the balancing unit; If the output of the balancing unit exceeds the unit output, the output of the balancing unit is set as the output value of the balancing unit If t is greater than 0, return to step 3 to re-enter the internal circulation; otherwise, it means that the cycle has completed all units and no balancing unit has been found.

Recalculate the network transmission loss P_{loss}, and use Eq. (19) to calculate the unbalanced power ΔP. if $C_0 \leq Q$ and $\Delta P > P_0$, $C_1 = C_0 + 1$ and return to step 2 to re-enter the external loop iteration, otherwise, the constraint processing is finished.

$$\Delta P = P_{load} + P_{loss} - \sum_{i=1}^{n'} P_i \tag{19}$$

Figure 1 shows a flowchart of the moth flame optimization for the power system PLP problem.

Algorithm implementation process

The specific process of applying moth to fire algorithm to solve the PDP model of power system is as follows.

Step 1: the moth position is initialized, and all moths are randomly initialized between the upper and lower bounds of the unit output capacity constraint shown in Eqs. (15) and (16).

$$P_i = l_i + R(u_i - l_i) \tag{20}$$

Step 2: unit operation constraint processing, check the position of all moths. If the output of a moth falls in the forbidden area, re-initialize the output of the unit whose particle falls in the banned area at random until the constraint in the forbidden area is met. If the output of some units of a particle exceeds the upper and lower limits, the unit output shall be reassigned according to Eq. (16).

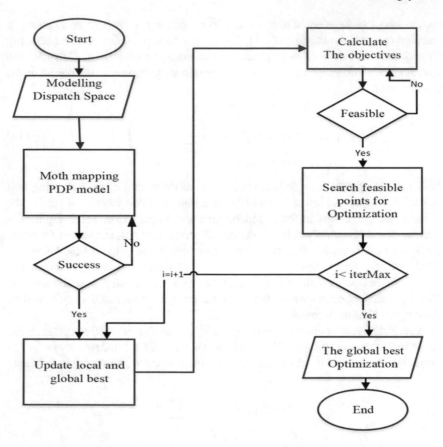

Fig. 1 A flowchart of the moth flame optimization for the power system PLP problem

Step 3: power balance constraint treatment. The power balance constraint is treated by the balance unit method. The new moth population can be obtained by combining step 2 and step 3.

Step 4: fitness assessment, using the cost function shown in Eqs. (1) or (2) to evaluate the fitness of all moths. Suppose the balance unit method still does not meet the equality constraints. In that case, the penalty function method shown in Eq. (21) is used to punish the unbalanced power, forcing the moth to fly away from the infeasible region and explore the feasible area. Among them is the total fuel cost (total fitness) considering penalty cost, is the fuel cost function of unit I, and is the penalty factor.

$$F_{cost} = \sum_{i=1}^{n'} F_i(P_i) + k\left(\sum_{i=1}^{n'} P_i - P_{load} - P_{loss}\right) \tag{21}$$

Step 5: update the location and fitness of the fire. If it is in the first generation, the initial population of the moth is directly used as the initial population of the fire. In the iteration process, the first m individuals with the best fitness were selected from the fire population and the updated moth population as the new fire population, and the fire fitness was updated.

Step 6: moth position update, moth use Eq. (12) to update its position with reference to the corresponding fire, and the number of fires decreases dynamically according to Eq. (13).

Step 7: determine the optimal launch and termination conditions. If the maximum number of iterations is reached, the iteration will be terminated. The optimal scheduling scheme (the best fire location) and the generation cost (the best fire fitness) of the output economic dispatch problem; otherwise, the iterative process is repeated until the termination condition is satisfied.

4 Simulation Results

In order to verify the reliability and effectiveness of the proposed method, IEEE 6 and 15 units systems [13] are used for testing the proposed performance. The system of 6-units is an IEEE 30 bus system, and the total load demand is 1263 MW. In the model calculation, the climbing rate limit, network transmission loss, and forbidden zone constraints are considered most. The total load demand of 15 unit system is 2630 MW, and the actual constraints such as forbidden area constraint, climbing rate limit, and network transmission loss are taken into account. However, others some other units, e.g., 2, 5, and 6, have three prohibited areas, and unit 12 has two. The decision interval of the system is a nonconvex decision space with 210 convex intervals. The existence of the transmission point effect is considered in the model.

Figure 2 shows the graph of the daily power generating distribution for the test system of 6 units.

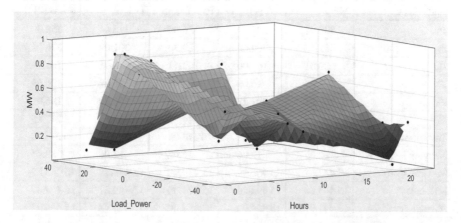

Fig. 2 The graph of the daily power generating distribution for the test system of 6 units

The parameters of the MFO algorithm are set as follows: the number of moth population is 40; when applied to three different examples, the maximum iteration times are 2000 [11]. In Eq. (12), B is a constant 1, t is a random number in the range [-1 to -2]. The penalty factor is 500. The maximum number of iterations is 50, and the allowable value of unbalanced power deviation is 0.01 MW.

The obtained results of the proposed scheme are compared with the other previous methods, e.g., GA [4], PSO [5], DE [6], and gray GWO [9]. The number of search agents individual is uniformly set to 40 in all algorithms during the simulation, and the maximum number of iterations is 2000. Each instance is run separately multiple times to ensure the tests' effectiveness, comparability, and robustness. For example, the parameters of setting for 15 machines test system, e.g., system prohibited generating units, the total load demand, and the coefficients B: [B_{ij}, B_{oi}, B_{oo}] power factors.

Tables 1 and 2 show the comparison of the suggested MFO with the GA, PSO, DE and GWO [9] for test systems of the IEEE 30 buses with 6 and 15 units, respectively; where the P_1, P_2, ..., P_n are the generator output power of each branch, n is number of units of the solution. The values in the Tables are the average value, minimum value, maximum value, maximum deviation (the difference between the maximum generation cost and the minimum generation cost) and standard deviation of the generation cost of the 6, and 15 generators systems. It can be seen from Table 1 that for a 6-machine system, the best result is $15,433.07/h, the worst result is $15,453.08/h, the average result is $15,443.07/h, and the standard deviation is 0.0010. The difference between the worst and best results is only 0.01$/h, indicating that MFO has good robustness.

It can be seen from Table 2 that for a 15 machine system, the minimum, maximum, and average cost of the 50 test results of the MFO algorithm are 32,697.15 $/h, 33,398.04 $/h, and 32,727.95 $/h, respectively, and the standard deviation is 0.0293.

Table 1 Comparison of results of different algorithms (for IEEE 6 unit system)

Units output	MFO	GA [4]	PSO [5]	DE [6]	GWO [9]
P_1	339.91	306.47	356.65	286.44	353.08
P_2	331.18	364.00	329.30	301.02	326.01
P_3	88.31	104.00	102.41	92.96	101.39
P_4	104.00	16.00	97.51	55.28	96.53
P_5	291.21	207.22	198.90	350.88	196.91
P_6	268.36	368.00	244.59	268.50	242.15
Total power/MW	**1275.45**	1276.01	1276.03	1276.95	1276.40
Network loss/MW	**12.45**	12.96	13.02	12.96	12.44
Min cost ($/h)	**15 433.07**	15 450.00	15 459.00	15 449.77	15 443.82
Max cost ($/h)	15 443.08	15 492.00	15 524.00	15 449.87	16 449.87
Average cost($/h)	**15 443.07**	15 454.00	15 469.00	15 449.78	15 446.95
Std. deviation($/h)	0.0010	0.0011	0.0012	0.0013	0.582 2
Max deviation($/h)	0.21	42.00	5.00	2.11	0.98

Table 2 Comparison of results of different algorithms (for IEEE 15-unit test system)

Units output	MFO	GA [4]	PSO [5]	DE [6]	GWO [9]
P_1	424.88	383.09	445.81	358.05	441.36
P_2	413.97	455.00	411.63	376.28	407.51
P_3	110.38	130.00	128.02	116.20	126.74
P_4	130.00	20.00	121.88	69.10	120.67
P_5	364.01	259.03	248.62	438.59	246.13
P_6	335.45	460.00	305.74	335.62	302.68
P_7	338.10	465.00	202.33	376.97	200.31
P_8	171.00	60.00	220.81	144.09	218.60
P_9	122.87	148.02	152.46	73.77	150.93
P_{10}	89.84	42.97	154.58	142.02	153.03
P_{11}	53.69	73.96	72.45	61.42	71.73
P_{12}	26.19	45.37	52.16	46.55	51.64
P_{13}	31.40	25.00	74.27	65.57	73.53
P_{14}	29.22	37.00	41.06	36.35	40.65
P_{15}	29.69	52.99	44.52	29.66	44.07
Total power/MW	**2660.08**	2662.40	2668.40	2662.29	2660.36
Network loss/MW	**30.08**	32.43	38.28	32.28	30.36
Min cost ($/h)	**32,697.15**	32,858.00	33,063.54	32 751.39	32,732.95
Max cost ($/h)	**33,398.04**	33,331.00	33,337.00	32,945.00	32,756.01
Average cost ($/h)	**32,727.95**	33,039.00	33,228.00	32,756.01	32,735.06
Std. deviation ($/h)	**0.0293**	0.09	0.81	0.05	0.36
Max deviation ($//h)	**0.89**	473.00	273.46	193.61	23.06

Compared with other algorithms, MFO still shows efficient and stable optimization ability, and the maximum deviation is only 0.89 $/h, which indicates that the MFO algorithm has good robustness and stability.

Figure 3 shows the comparison obtained result curves of the suggested FMO with the other algorithms. e.g., DE, PSO, and GWO for the test systems of 15 units. The observed figure shows that the MFO optimization method has better quality performance in convergence speed and time consumption than PSO and GWO methods. In general, we can say that the MFO can solve the PDP problem in the power system with good robustness and significant economic benefits.

5 Conclusion

This paper suggested a new solution to solving the power generating distribution planning (PDP) problem based on a new swarm-based technique called moth flame

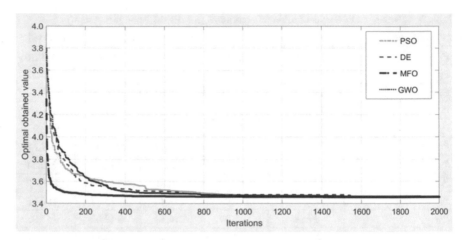

Fig. 3 The comparison obtained result curves of the suggested algorithm with the other algorithms for the test system of 15 units

optimization (MFO) to reduce power loss and cost consumption. PDP model is a critical component of the power system's financial performance, considering various system operation restrictions that are typical multi-constrained nonlinear optimization problems. The experimental results of the suggested method were compared to the findings of other algorithms, the IEEE 30 buses with 6-unit and 15-unit systems. Compared results show that the proposed MFO produced more optimal and stable derivatives that can effectively handle the PDP problem, resulting in significant cost gains and save fuels.

References

1. Tsai CF, Dao TK, Pan TS, Nguyen TT, Chang JF (2016) Parallel bat algorithm applied to the economic load dispatch problem. J Internet Technol 17:761–769. https://doi.org/10.6138/JIT. 2016.17.4.20141014c
2. Nguyen T-T, Wang M-J, Pan J-S, Dao T, Ngo T-G (2020) A load economic dispatch based on ion motion optimization algorithm BT—advances in intelligent information hiding and multimedia signal processing. Presented at the (2020)
3. Dao TK, Pan TS, Nguyen TT, Chu SC (2015) Evolved bat algorithm for solving the economic load dispatch problem. In: Advances in intelligent systems and computing, pp 109–119. https://doi.org/10.1007/978-3-319-12286-1_12
4. Warsono, King DJ, Özveren CS, Bradley DA (2007) Economic load dispatch optimization of renewable energy in power system using genetic algorithm. In: 2007 IEEE Lausanne POWERTECH, Proceedings (2007). https://doi.org/10.1109/PCT.2007.4538655
5. Safari A, Shayeghi H (2011) Iteration particle swarm optimization procedure for economic load dispatch with generator constraints. Exp Syst Appl 38:6043–6048
6. Noman N, Iba H (2008) Differential evolution for economic load dispatch problems. Electr Power Syst Res 78:1322–1331

7. Nguyen T-T, Wang H-J, Dao T-K, Pan J-S, Liu J-H, Weng S-W (2020) An improved slime Mold algorithm and its application for optimal operation of cascade hydropower stations. IEEE Access 8:1. https://doi.org/10.1109/ACCESS.2020.3045975

8. Nguyen T-T, Pan J-S, Chu S-C, Dao T-K, Do V-C (2019) Improved performance of wireless sensor network based on fuzzy logic for clustering scheme BT—advances in smart vehicular technology, transportation, communication and applications. Presented at the (2019)

9. Pradhan M, Roy PK, Pal T (2016) Grey wolf optimization applied to economic load dispatch problems. Int J Electr Power Energ Syst 83:325–334

10. Nguyen T-T, Pan J-S, Dao T-K (2019) A compact bat algorithm for unequal clustering in wireless sensor networks (2019).https://doi.org/10.3390/app9101973

11. Mirjalili S (2015) Moth-flame optimization algorithm: a novel nature-inspired heuristic paradigm. Knowl-Based Syst 89:228–249

12. Mehne SHH, Mirjalili S (2020) Moth-flame optimization algorithm: theory, literature review, and application in optimal nonlinear feedback control design. In: Nature-inspired optimizers. Springer, pp 143–166

13. Chakrabarti S, Kyriakides E, Eliades DG (2008) Placement of synchronized measurements for power system observability. IEEE Trans Power Deliv 24:12–19

Deep Convolutional Embedded Fuzzy Clustering with Wasserstein Loss

Tianzhen Chen and Wei Sun

Abstract Deep clustering combines embedding and clustering to obtain the optimal nonlinear embedding space, which is more effective in real scenes than traditional clustering methods. However, for a class of deep embedding clustering represented by DCEC, the loss function is mostly KL divergence, and KL divergence will fail in some cases. To avoid KL failure, this paper proposes to use Wasserstein distance as the loss function of the clustering layer, and introduce the fuzzer into the clustering layer, and then develop a deep convolution embedded fuzzy clustering method with Wasserstein loss (DCEFC). Through comprehensive experiments on three popular image datasets with the comparison methods, the experimental results show that our proposed method is superior to the comparison methods, and the clustering results are effective.

Keywords Unsupervised learning · Pattern recognition · Deep clustering · Wasserstein distance

1 Introduction

With the provision of computer computing power and the advent of the information age, the data that people obtain is more and more diverse and complex. Traditional methods are sensitive to the dimension of data, will be affected by the dimension disaster, and cannot find the hidden information in the data well in terms of data performance [1]. And because of its powerful data reconstruction ability and data performance ability, deep learning has been deeply researched and applied by a wide range of scholars once it was proposed [2–5].

T. Chen (✉) · W. Sun
Dalian Neusoft University of Information, Dalian, China
e-mail: chentianzhen@neuedu.com

Neusoft Institute of Modern Industry, Meizhouwan Vocational Technology College, Putian, China

W. Sun
e-mail: sunwei@neusoft.edu.cn

© The Author(s), under exclusive license to Springer Nature Switzerland AG 2022   163
N. H. T. Dang et al. (eds.), *Artificial Intelligence in Data and Big Data Processing*,
Lecture Notes on Data Engineering and Communications Technologies 124,
https://doi.org/10.1007/978-3-030-97610-1_14

Unsupervised deep learning is the organic integration of traditional methods ideas and deep learning, which has attracted the attention of scholars at home and abroad [6, 7]. The reason is that unsupervised deep learning avoids tedious data annotation and other preliminary work, which greatly saves the cost of constructing data. At the same time, it can also express the implicit information in the data, which helps to improve the accuracy of grouping data into different clusters [8].

The typical representatives of unsupervised deep learning are: autoencoder, which is a symmetric type of neural network. Its main design purpose is to encode and compress the input data into a meaningful representation, and then decode it and reconstruct it with the original input data that is as similar as possible [9]. Autoencoders have many variants, such as adversarial autoencoders (AAE) [10], variational autoencoders (VAE) [11], Convolutional Autoencoder (CAE) [12], and so on. Another type of unsupervised deep learning is the deep clustering model, such as Deep Embedding Clustering (DEC) [13], which is composed of a self-encoder and a clustering layer. Adversarial learning for robust deep clustering (ALRDC) [14], which defines adversarial samples in the embedding space of clustering networks. Contrastive clustering (CC) [15]: it is comparative learning between instance level and clustering level. There are other excellent deep clustering models. Then DEC has proved that deep clustering can be better than the traditional shallow clustering methods in different tasks, which provides a new research direction for deep learning and clustering [13]. Deep Convolutional Embedding Clustering (DCEC) is a typical example, which uses CAE to replace part of the original AE results [16]. However, it is the deformation of DEC, IDEC, DCEC, and other DECs, KL divergence is used as the loss function of clustering, and KL will fail in the case of no overlap or very little overlap in the distributed support set [17].

To solve the above problems, we propose to use Wasserstein distance to replace KL divergence as the loss function of the clustering layer and introduce the concept of the fuzzer into the distribution, and then propose a deep convolution embedding fuzzy clustering with Wasserstein loss.

2   Related Research

2.1   Deep Convolutional Embedded Clustering

Deep Convolutional Embedded Clustering (DCEC) is a deep clustering model, which conducts unsupervised learning based on deep learning and autoencoder, extracts the clustering characteristics of data, and then divides the data set samples into several clusters so that the samples with high similarity are in the same cluster, and the samples with low similarity are in different clusters [16]. The convolution autoencoder of deep convolution embedded clustering is divided into two parts: encoder h and decoder g. The loss function they use is MSE (as shown in Formula 1). The task of the encoder is to convert the input data into a low-dimensional representation, and the task of the decoder is to reconstruct the input data from the low-dimensional representation.

In our method, the embedding point $Z^{C \times N} = \{z_j\}_{j=1}^N$, $z_j \in \mathbb{R}^C$ is obtained by reducing the dimension of data $X^{N \times W} = \{x_j\}_{j=1}^N$, $x_j \in \mathbb{R}^W$ through a convolution autoencoder. Like DEC, only the encoder part is retained.

Then, the K-means clustering method is used to initialize the weight $V = \{v_i\}_{i=1}^C$ of the clustering layer, and then the $Q_j = \{q_{ij}\}_{i=1}^C$ distribution and $P_j = \{p_{ij}\}_{i=1}^C$ distribution are obtained according to Formulas (2) and (3), respectively.

Finally, the loss function (as shown in Formula 4) is used to fine-tune the clustering results, and the model optimization and clustering are completed under the premise of ensuring that the embedding space is not distorted to the maximum extent.

$$L = \frac{1}{2N} \sum_{j=1}^N \left\| x_j - g_{\theta'}\left(h_\theta(x_j)\right) \right\|^2 \tag{1}$$

$$q_{ij} = \frac{\left(1 + \|v_i - z_j\|^2 / \alpha\right)^{-\frac{\alpha+1}{2}}}{\sum_{k=1}^C \left(1 + \|v_k - z_j\|^2 / \alpha\right)^{-\frac{\alpha+1}{2}}}, \sum_{i=1}^C q_{ij} = 1, \forall j \tag{2}$$

$$p_{ij} = \frac{q_{ij}^2 / \sum_j q_{ij}}{\sum_{k=1}^C \left(q_{kj}^2 / \sum_j q_{kj}\right)}, \sum_{i=1}^C p_{ij} = 1, \forall j \tag{3}$$

$$L = \mathrm{KL}(P \| Q) = \sum_{j=1}^N \sum_{i=1}^C p_{ij} \log \frac{p_{ij}}{q_{ij}} \tag{4}$$

2.2 Fuzzer

Fuzzer is a classic method in the field of automation control. Its principles are fuzzy mathematics and fuzzy logic. The paper [18] introduces the concept of membership function, breaking the classic mathematics "non The limitation of 0 is 1", using real numbers between [0, 1] to describe the intermediate state.

Fuzzer is the presentation of the essence of the fuzzy theory, which is widely used. One of the most successful applications is Fuzzy C-means Clustering (FCM) [19]. Fuzzy sets make FCM have natural, non-probabilistic characteristics, and can obtain more flexible clustering results than K-means.

2.3   Wasserstein Distance

Wasserstein distance, also called Earth Mover's Distance, bulldozer distance, abbreviated as EMD, is used to indicate the degree of similarity between two distributions. In other words, Wasserstein distance measures the amount of movement required to "move" the data from the original distribution q to the target distribution p the minimum value of the average distance (similar to the minimum value wga of the work required to move a pile of soil from one shape to another). The paper [17] has been proved that the superiority of Wasserstein distance over KL divergence is that even if the support sets of two distributions do not overlap or overlap very little, the Wasserstein distance can still reflect their distance, and can tell us how they are different, that is, how to transform from one distribution to another. Wasserstein distance is the minimum cost under optimal path planning.

3   Proposed Method

Because embedded clustering such as DEC and DCEC uses KL divergence as the loss function of the model, but KL divergence does not take into account the relationship between p and q, we use Wasserstein distance to replace KL divergence as the loss function of the model. At the same time, inspired by the fuzzy theory in FCM, we introduce the concept of fuzzer into Formula (2) and successfully apply it to DCEC. Finally, a deep convolution embedded fuzzy clustering with Wasserstein loss (DCEFC) is proposed by us.

Our method is mainly divided into two stages, namely the pre-training stage and the clustering stage. The network structure used in the pre-training stage is a three-layer convolution plus a symmetric mirror network structure with a layer of the network layer. The purpose is to process the data set in batches and then build the convolution automatic encoder required for pre-training and performing pre-training. Secondly, the parameters θ (the weight W and the bias B) of the encoder and the hidden feature Z_t of the last pre-training encoder are retained. Then in the clustering stage, the network structure of the convolutional autoencoder in the pre-training is retained in the next stage, and the network parameters of this part are inherited. The cluster center is represented by the weight of the hidden feature Z_t of the last pre-training encoder, and the clustering layer is stacked behind the pre-training encoder to form the clustering model. Finally, the hidden feature Z_t is clustered by the K-means clustering method to obtain the cluster prediction labels $\{s_j\}_{j=1}^{N}$ and cluster centers $\{v_i\}_{i=1}^{C}$ which are used to initialize the weight of the cluster layer.

To improve the clustering effect and avoid KL divergence failure, we use Formula (5) to calculate the fuzzy distribution q_{ij} of embedding points and Formula (3) to calculate the auxiliary distribution p_{ij}, and Formula (7) to minimize the Wasserstein distance according to the model results, so that the target distribution is as close as possible to the clustering output distribution. The update frequency T is set. When the auxiliary distribution p_{ij} is updated, the label assigned to the embedding points x_j by the current model will also be updated. The label is obtained by Formula (8). If the label $\{s_j \in S\}_{j=1}^N$ difference rate (percentage) between the two consecutive updates of the auxiliary distribution is less than the threshold δ, the training is stopped. This can further improve the final clustering results after each update training.

$$q_{ij} = \frac{\left(m + \|v_i - z_j\|^2\right)^{\frac{-2}{m-1}}}{\sum_{k=1}^C \left(m + \|v_k - z_j\|^2\right)^{\frac{-2}{m-1}}}, \sum_{i=1}^C q_{ij} = 1, \forall j \qquad (5)$$

Then Wasserstein loss can be defined as:

$$W(P, Q) = \inf_{\gamma \prod(P,Q)} \mathbb{E}_{(p,q)\ \gamma}\left[\|p - q\|\right] \qquad (6)$$

There is γ represents a distribution in all possible joint distribution $\prod(P, Q)$, namely $\gamma \in \prod(P, Q)$. Then, for each possible joint distribution γ, a sample p and q can be obtained from sampling (p, q) γ, and the distance $\|p - q\|$ of the pair of samples can be calculated. Therefore, the expected value $\mathbb{E}_{(p,q)\ \gamma}\left[\|p - q\|\right]$ of the sample pair of distances under the joint distribution can be calculated.

We assume that the joint distribution γ satisfies the condition $\gamma \prod(p, q)$, which means:

$$\int \gamma(p, q)dq = P(p) \text{ and } \int \gamma(p, q)dp = Q(q) \qquad (7)$$

That is to say, γ is a joint distribution, its edge distribution is the target distribution P and the original distribution Q.

The tags of embedding points are obtained as follows:

$$s_j = \arg\max_j q_{ij} \qquad (8)$$

which $\{q_{ij}\}_{i,j=1}^{C \times N}$ is calculated by the Formula (2).

To better introduce our DCEFC, we give the schematic diagram of our method as shown in Fig. 1, and give the method description of our method as shown below.

Our DCEFC structure is shown in Fig. 1. $\{x_j\}_{j=1}^N$ represents the input data in Fig. 1. The encoding layer and the decoding layer have three layers of convolution, and the data are first input into the encoder part. After passing through three layers of convolution layer with 2 steps, they are flattened into a one-dimensional feature vector. After the output features of the encoder are received by the decoder and the clustering layer, the clustering results are continuously fine-tuned based on the loss function until the convergence criterion is met. The final result indicates the output data $\{\hat{x}_j\}_{j=1}^N$ after the encoder extracts the features and is restored by the decoder.

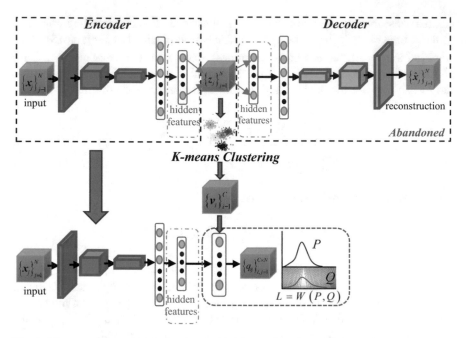

Fig. 1 Diagram of deep convolutional embedded fuzzy clustering with Wasserstein loss

$\{q_{ij}\}_{i,j=1}^{C \times N}$ denotes the optimal clustering result $\{x_j\}_{j=1}^{N}$ generated by the clustering layer.

Our method is described as follows:

DCEFC: Deep Convolutional Embedded Fuzzy Clustering with Wasserstein Loss

INPUT: Dataset: $X^{N \times W} = \{x_1, \cdots, x_j, \cdots, x_N\}$, $x_j \in \mathbb{R}^W$; Final number of clusters: C; Batch size: N_b; Number of pre-training iterations: *PreEpoches*; Target distribution update frequency: T; Stop threshold: δ; The maximum number of iterations: *MaxIter*.

OUTPUT: Clustering centers: V; Clustering pseudo-labels: $\hat{Y}$.

1   Divide samples X into N / N_b batches, Each batch of training samples is represented by X_t, where $t \in \{1, 2, \cdots, N / N_b\}$;

2   **for** $epoch \in \{1, 2, \cdots, PreEpoches\}$ **do**

3   **for** $batch\_number \in \{1, \cdots, t, \cdots, N / N_b\}$ **do**

4   Encoder: $Z_t = h_\theta(X_t)$;

5   Decoder: $\hat{X}_t = g_{\theta'}(Z_t)$;

6   The MSE loss is calculated and optimized parameters θ, θ';

7   The initialization weight and bias θ of Encoder are pre-trained, and the hidden feature Z_t obtained by Encoder is initialized by K-means for cluster center V and pseudo label $\hat{Y}$.

8   **for** $iter \in \{1, 2, \cdots, MaxIter\}$ **do**

9   **if** $iter \% T == 0$ **then**

10   Computing all embedding points $\{z_j = h_\theta(x_j)\}_{j=1}^N$ with Encoder;

11   Use the formula (5) update fuzzy membership $\{q_{ij} \in Q\}_{i,j=1}^{C \times N}$;

12   According to formula (3), Q and $\{z_j\}_{j=1}^N$ are used to update the auxiliary target distribution $\{p_{ij} \in P\}_{i,j=1}^{C \times N}$;

13   Save the last pseudo-label: $\hat{Y}_{old} = \hat{Y}$;

14   Update pseudo-label $\{\hat{y}_j \in \hat{Y}\}_{j=1}^N$ by formula (8);

15   Computing $Sum(\hat{Y}_{old} \neq \hat{Y}) = \sum_{j=1}^N 1(\hat{y}_{old_j} \neq \hat{y}_j), \hat{y}_{old_j} \in \hat{Y}_{old}, \hat{y}_j \in \hat{Y}$;

16   **if** $Sum(\hat{Y}_{old} \neq \hat{Y}) / N < \delta$ **then**

17   Stop training;

18   Disrupting sample X into N / N_b batches $\{X_1, \cdots, X_t, \cdots, X_{N/N_b}\}$;

19   **for** $batch\_number \in \{1, \cdots, t, \cdots, N / N_b\}$ **do**

20   Calculating partial embedding point $\{z_j = h_\theta(x_j)\}_{j=1}^{N_b}$ with Encoder;

21   Update fuzzy membership $\{q_{ij} \in Q_t\}_{i,j=1}^{C \times N_b}$ by (5);

22   Segmenting $\{p_{ij} \in P\}_{i=1, j=batch\_number*N_b}^{C \times N_b}$;

23   Calculation Wasserstein distance according to formula (7) optimization parameters θ and cluster center V;

Table 1 Details of each dataset

Datasets	Samples	Features	Clusters
Fashion-MNIST	70,000	784	10
MNIST	70,000	784	10
USPS	9298	356	10

4 Time Complexity Analysis

To conveniently approximate the time complexity with asymptotic symbols, only the calculation of the output of the automatic encoder network, including the reconstruction of input data X, is considered. Without considering the activation function and loss term, when L is large enough, the time complexity of the automatic encoder is $O\left(Lt_p\left(N^2W + NW^2\right)\right)$, where t_p is the number of pre-training times. The time complexity of K-means is $O(t_k NWC)$, where t_k is the number of iterations. Let t_f be the number of iterations of the clustering layer, then the time complexity of fine-tuning in this stage is $O\left(\frac{L}{2}t_f\left(N^2W + NW^2\right)\right)$. So the total time complexity is $O\left(t_k NWC + L\left(t_p + \frac{t_f}{2}\right)\left(N^2W + NW^2\right)\right)$.

Generally, $t_k \leq t_f \leq t_p \leq C \leq L \leq W \ll N$, so the time complexity of our DCEFC is $O\left(N^2\right)$.

5 Experimental Results and Analysis

5.1 Datasets

To test the effectiveness of our method, we used three common standard image datasets in our experiment, and the properties of each dataset are shown in Table 1.

5.2 Evaluation Method

ACC, NMI, and ARI are selected as the evaluation metrics of this experiment, and their respective calculation methods are shown in Formulas (9–11).

$$ACC = \frac{1}{N} \sum_{j=1}^{N} \left(y_j = g(\hat{y}_j)\right) \tag{9}$$

where $g(\cdot)$ is the mapping function to determine the best match between the unsupervised prediction label $\hat{y}$, and the original real label y.

$$NMI = \frac{2 \cdot I(y, \hat{y})}{H(y) + H(\hat{y})} \quad (10)$$

$H(\cdot)$ is the entropy of each variable, and $I(\cdot)$ is the mutual information between the predicted label $\hat{y}$ and the real label y.

$$ARI = \frac{RI - E[RI]}{\max(RI) - E[RI]} \quad (11)$$

where $E[RI]$ is the expectation of RI, $RI = \frac{TP+TN}{TP+FP+TN+FN}$, where TP is the number of true positives, TN is the number of true negatives, FP is the number of false positives, and FN is the number of false negatives.

Generally, the above metrics are usually used in various clustering papers [13, 16, 20–22]. Although each evaluation method has its advantages and disadvantages, combining them is enough to prove the effectiveness of the clustering methods. It is worth noting that the range of ACC and NMI is within [0, 1], while the range of ARI is [−1, 1], and the larger the value is, the better the clustering performance is.

5.3 Introduction of Comparison Methods

- **K-means** [23]: a partition-based hard clustering method, each data only belongs to one cluster.
- **DEC** [13]: a deep clustering model based on Student's-t distribution, which abandons the decoder. The comparison experiment uses the DEC code published by the authors.
- **IDEC** [22]: it is an improved version of DEC, retaining local structures.
- **DCEC** [16]: it is composed of CAE, local preserving structure, and clustering layer.
- **GrDNFCS** [21]: it uses the deep fuzzy clustering method proposed by the original data reconstruction based on autoencoder, considering the separation between clusters and affinity regularization based on pseudo-labels.
- **DECCA** [20]: it uses the Frobenius norm as the penalty term, and combines it with the deep embedding clustering model proposed by the shrinkage autoencoder.

The parameters set by these methods are the same as those set in the reference except that m is set as the default value of 2.5. We carried out five random executions of all experiments and took the average as the experimental results. The experimental results of each method are shown in Table 2.

Table 2 Experimental results of each method on Fashion-MNIST, MNIST, and USPS

Dataset	Metrics	K-means	DEC	IDEC	DCEC*	GrDNFCS*	DECCA*	DCEFC
Fashion-MNIST	ACC	0.5123	0.5791	0.5772	0.6332†	0.6351	0.6099	*0.6370*
	NMI	0.5178	0.6275	0.6029	0.6636†	0.6609	*0.6698*	*0.6698*
	ARI	0.3643	0.4558	0.4481	–	0.5028	–	*0.5142*
MNIST	ACC	0.5324	0.8847	0.8851	0.8897	0.9145	*0.9637*	*0.9589*
	NMI	0.4997	0.8525	0.8637	0.8849	0.9074	0.9074	*0.9152*
	ARI	0.3652	0.8243	0.8382	–	0.8626	–	*0.9119*
USPS	ACC	0.6681	0.7277	0.7541	0.7900	0.7652	0.7731	*0.8037*
	NMI	0.6265	0.7368	0.7362	0.8257	0.7761	0.8053	*0.8329*
	ARI	0.5463	0.6639	0.6796	–	0.6903	–	*0.7593*

* Represents the data of the method in the original article. Numbers for other methods produced by us are marked with a †

5.4 Experimental Results and Analysis

From Table 2, we can see that our DCEFC is better than the comparison methods in NMI and ARI, and is close to other methods in ACC.

Our DCEFC does experiments on the selection of m on the MNIST data. It can be seen from Fig. 2. that when m is between 1.5 and 3.5, the NMI value of our method is greater than 0.9, especially when $m = 2.1$, the NMI can reach 0.93829. To pursue the best result, we choose 2.1 as the default value of m in the experiment.

We conducted a comparative experiment between DCEFC and DCEC on the MNIST dataset. m = 2.1 in DCEFC and $\alpha = 1.0$ in DCEC were consistent with other parameters, so the experimental results of the three indicators are shown in Fig. 3. It can be seen from Fig. 3 that our method can effectively improve the accuracy of the method and reduce the number of iterations, which is attributed to the effect of the fuzzer.

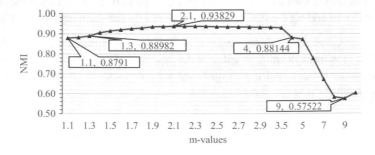

Fig. 2 DCEFC selects different m values on the MNIST dataset to obtain different NMI

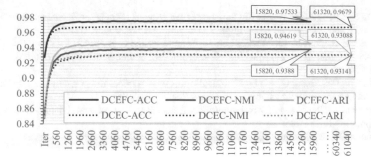

Fig. 3 The experimental results of DCEFC and DCEC on the MNIST dataset

6 Conclusions

In this paper, the convolution automatic encoder (CAE) is used to learn to preserve the good features of the local structure, and by introducing the concept of a fuzzer into the clustering layer, and Wasserstein distance is used as the loss function of the two distributions, so we propose a deep convolution embedded fuzzy clustering method with Wasserstein loss (DCEFC). In the experiment of this paper, the effectiveness of DCEFC in the image clustering task is proved by experience, and it is verified that the fuzzer and Wasserstein loss is essential for the deep clustering of the image. The future work is devoted to optimizing the model structure, making full use of the valuable information provided by the abandoned decoder in the pre-training stage, and trying to solve the accumulated error in the iterative optimization process.

Acknowledgements This work was funded by the Liaoning Province Key Laboratory for the Application Research of Big Data.

References

1. Jain A, Singh T, Sharma SK, Prajapati V (2021) Implementing security in iot ecosystem using 5g network slicing and pattern matched intrusion detection system: a simulation study. Interdisc J Inf Knowl Manage 16
2. Chen W, Yang L, Zha B, Zhang M, Chen Y (2020) Deep learning reservoir porosity prediction based on multilayer long short-term memory network. Geophysics 85(4):WA13–WA25
3. Georgiou T, Liu Y, Chen W, Lew M (2020) A survey of traditional and deep learning-based feature descriptors for high dimensional data in computer vision. Int J Multimedia Inf Retrieval 9(3):135–170
4. Huang SC, Pareek A, Seyyedi S, Banerjee I, Lungren MP (2020) Fusion of medical imaging and electronic health records using deep learning: a systematic review and implementation guidelines. NPJ Digital Med 3(1):1–9
5. Wang G, Ye JC, De Man B (2020) Deep learning for tomographic image reconstruction. Nature Mach Intell 2(12):737–748

6. Deng N, Liu J (2021) Where did you take those photos? tourists' preference clustering based on facial and background recognition. J Destination Mark Manag 21:100632
7. Nikparvar B, Thill JC (2021) Machine learning of spatial data. ISPRS Int J Geo Inf 10(9):600
8. Dogan A, Birant D (2021) Machine learning and data mining in manufacturing. Exp Syst Appl 166:114060
9. Li X, Chen W, Zhang Q, Wu L (2020) Building auto-encoder intrusion detection system based on random forest feature selection. Comput Secur 95:101851
10. Makhzani A, Shlens J, Jaitly N, Goodfellow I, Frey B (2015) Adversarial autoencoders. arXiv preprint arXiv:1511.05644
11. Kingma DP, Welling M (2019) An introduction to variational autoencoders. arXiv preprint arXiv:1906.02691
12. Cheng Z, Sun H, Takeuchi M, Katto J (2018) Deep convolutional autoencoder-based lossy image compression. In: 2018 Picture coding symposium (PCS). IEEE, pp 253–257
13. Xie J, Girshick R, Farhadi A (2016) Unsupervised deep embedding for clustering analysis. In: International conference on machine learning. PMLR, pp 478–487
14. Yang X, Deng C, Wei K, Yan J, Liu W (2020) Adversarial learning for robust deep clustering. Adv Neural Inf Process Syst 33:9098–9108
15. Li Y, Hu P, Liu Z, Peng D, Zhou JT, Peng X (2021) Contrastive clustering. In: 2021 AAAI conference on artificial intelligence (AAAI)
16. Guo X, Liu X, Zhu E, Yin J (2017) Deep clustering with convolutional autoencoders. In: International conference on neural information processing. Springer, pp 373–382
17. Arjovsky M, Chintala S, Bottou L (2017) Wasserstein generative adversarial networks. In: International conference on machine learning. PMLR, pp 214–223
18. Dzitac I, Filip FG, Manolescu MJ (2017) Fuzzy logic is not fuzzy: World-renowned computer scientist lotfi a. zadeh. Int J Comput Commun Control 12(6):748–789
19. Askari S (2021) Fuzzy c-means clustering algorithm for data with unequal cluster sizes and contaminated with noise and outliers: Review and development. Exp Syst Appl 165:113856
20. Diallo B, Hu J, Li T, Khan GA, Liang X, Zhao Y (2021) Deep embedding clustering based on contractive autoencoder. Neurocomputing 433:96–107
21. Feng Q, Chen L, Chen CP, Guo L (2020) Deep fuzzy clustering—a representation learning approach. IEEE Trans Fuzzy Syst 28(7):1420–1433
22. Guo X, Gao L, Liu X, Yin J (2017) Improved deep embedded clustering with local structure preservation. In: Ijcai, pp 1753–1759
23. MacQueen J, et al (1967) Some methods for classification and analysis of multivariate observations. In: Proceedings of the fifth Berkeley symposium on mathematical statistics and probability, vol 1. Oakland, CA, USA, pp 281–297

Predicting the Future Actions of People in the Real World to Improve Health Management

Thu Nguyen, Ngoc-Mai Bui, Thu-Thuy Ta, and Tu-Anh Nguyen-Hoang

Abstract Busy life makes people do not have time to exercise, stay up late, often eat fast food. Therefore, predicting and recommending what actions should do at what times of day are necessary and also help people improve their health without functional foods or advising of doctors. However, human actions are extremely complex and affected by many subjective and objective factors. The previous proposed methods gradually improve the accuracy of action prediction, but not yet focus on the health-affecting activities and the processed time. Therefore, it is still difficult to apply the predicted results in the reality, such as health app. Therefore, we focus our research on how to solve this problem. Based on user action data, identify attributes related to daily activities that affect health. From there, apply data processing techniques and machine learning methods in the field of data science to solve this problem. In this paper, we propose a PAS-KNNTP (Periodic Action Sequences—K Nearest Neighbor for Time Prediction) method based on personalized key features to predict the next action and when it occurs. To evaluate the effectiveness of the proposed method, we experiment on real-world user action data for the period of 17 months. Experimental results show that the method still ensures the accuracy of predicting future actions, improving the accuracy of predicting the time of happening action up to 20% and reducing the prediction time from 20 min to 15 s. Experimental results prove that our method is capable to apply the reality.

Keywords Action sequences · Machine learning · Data science

T. Nguyen (✉) · N.-M. Bui · T.-T. Ta · T.-A. Nguyen-Hoang
University of Information Technology, Ho Chi Minh City, Vietnam
e-mail: 17520731@gm.uit.edu.vn

T.-T. Ta
e-mail: thuyta@uit.edu.vn

T.-A. Nguyen-Hoang
e-mail: anhnht@uit.edu.vn

T. Nguyen · N-M. Bui · T.-T. Ta · T.-A. Nguyen-Hoang
Vietnam National University, Ho Chi Minh City, Vietnam
e-mail: thunta@uit.edu.vn

© The Author(s), under exclusive license to Springer Nature Switzerland AG 2022
N. H. T. Dang et al. (eds.), *Artificial Intelligence in Data and Big Data Processing*,
Lecture Notes on Data Engineering and Communications Technologies 124,
https://doi.org/10.1007/978-3-030-97610-1_15

1 Introduction

The world is developing more and more modern. People increasingly have high requirements for quality of life and health protection. However, modern life makes people too busy with their work and have little time to take care of their health. Besides, the monitoring health of patients also helps doctors make early diagnoses to save the patient's life [1]. We need the intelligent healthcare monitoring systems to solve this problem.

In the Industry 4.0, the IoT (Internet of Things) is one of the technologies that is expected to bring many benefits to people. The IoT technology is applied in many fields and attracts many investment sources. One of the important applications is building the smart healthcare system in the smart city [2]. The smart healthcare systems are developed on the IoT environment in the smart city. The IoT devices will monitor health and collect data, such as sensors, smartphones, etc. Data science, artificial intelligence and machine learning are applied to process the collected data and make recommendations in an intelligent healthcare system [3]. Therefore, the problem of predicting what actions will happen at some point in the future plays an important role in the intelligent healthcare system. Thanks to these results, we can build the virtual assistants to consult health and assist doctors in diagnosing diseases [4]. Specifically, for personal smart health monitoring devices, the accurately predicting what actions will happen in the future increases the commercial value of these devices. Through the user's activities that log on these devices, the health monitoring system predicts the appropriate activity to occur. This system help to make personalized recommendations for each user. In particular, these functions can provide a service to users, they can reach their target more easily. In addition, when users follow the recommended activities correctly, they can prevent negative external health effects and promote healthy living. As a result, through these regular reminders, it will be easier for users to achieve their personal health goals. For example, prompting users to measure their weight and exercise every morning can make them consciously lose weight or exercise. On the other hand, predicting actions and times those actions will happen in the future acts as a virtual assistant for the user. The virtual assistant will be tasked with recommending and supporting the user to properly perform healthy activities. A patient's daily activity history also helps doctors easily monitor health and make early diagnosis to save patient's lives.

However, the previous studies on the problem of predicting people's future actions often focused on the activities on social networks [5], or limit the group of users to observe [6] , or the movement process of users in the AR field [7]. For example, interaction on Facebook, predicting student learning trends, predicting how people move between locations. Therefore, we focus on the predicting actions that affect human health and reducing processing time so that we can improve the applicability of smart health management systems in the practice. We propose the PAS-KNNTP method that still ensures the accuracy of predicting human actions in the future, improving the accuracy of predicting the time happening action to 20% and reducing the processing time from 20 min down to 15 s.

We present the contents related to this method as follows. We present the related previous studies in part 2. The theoretical background and procession of the PAS-KNNTP method are presented in Part 3. Section 4 presents how we experiment to evaluate the effectiveness of the method. Part 5 is the conclusion of the paper about the proposed method.

2 Related Work

It is generally agreed today that there are a lot of research work about predicting actions in the future. The models that are frequently used are Markov, LSTM, RNN, or Weibull distribution. In addition, due to the property of actions which is extremely complicated, those work combine the models with several algorithms and probability distribution. Specifically, this work is related to the three fields below.

2.1 Predicting Action

The issue of predicting the next actions has been experimented in many work such as predicting the next SQL query [8], human behavior in size-variant repeated games [9], action progress in videos [10], etc. From these examples, we can realize that many projects used Markov and Recurrent Neural Network (RNN) to process the discrete-time sequence actions. However, because of the discreteness of actions, the Markov model has a low accuracy rate when predicting actions affected by long-term cycles [11]. In addition to, the RNN model has the function of assuming discrete time, and it fits into dataset about user's history in the past. However, the disadvantage of RNN is limited ability to memorize in long-term.

In general, there are many work researching the issue of predicting the next action in the future. Nevertheless, in the field of predicting daily actions such as eating, drinking, sleeping, resting,—there are only some work. Furthermore, these work did not show when the predicted actions will happen. Therefore, our topic solves the above problems that process discrete-time, property long-term action, and predict additionally the time. Especially, our model improved the accuracy rate of Markov and the capability of remembering data with the RNN.

2.2 Repeated Action

People always have a habit of repeating actions which happened in the past. For example, if a person walks on a road, there is a high probability that he or she will repeat that road in the near future. More generally, the human behaviors often change over time, but they are interdependent and occur in a certain cycle. Over the last decade, the work that have studied the architecture of repeat action often focus on the user's product consumption behaviors such as buying some items again [18], music listening [13], or repeated web search queries [14], ...These work researched the nature of repeating old actions and performing new actions. Then, they have modeled the actions of people.

Besides, the work of repeating action architecture are also significant surveys to give a comprehensive and general overview. Thanks to these researches, we can conclude that human actions in the real world have the nature of repeating old actions as a habit. In addition, the time when this action is repeated will vary and is affected by the nature of each specific action.

On another hand, most of the activities in this related work are unaffected by different short- and long-term cycles and generally change over time. Therefore, it is contrary to the goal of building a predictive model of the topic. However, these work are the premise to develop the action modeling process based on the repetitive nature of users' habitual actions.

2.3   Temporal Point Processes

For the study of the transient point process, the topics often revolve around the problem of predicting the time of subsequent change actions or even the growth of the virus [15] through the Poisson model and Hawkes process. They model the influence of users in the social networks [16] and then, analyze the evolution of information and network structure, document clustering, time users return to perform old actions, ...

The Poisson model is a regression model. This model uses the discrete probability distribution method to train the model. Unlike conventional probabilistic models, Poisson does not rely on the probability of events occurring or the number of occurrences of events. The Poisson calculates the average number of times that action occurs in a given time instead. Besides, the Hawkes process is known as a self-excited and point process. Therefore, its impending events will have dependencies on previous events.

The problem of predicting the sequence of human actions related to each other has broadened the scope of this work and used exponential distribution and Weibull kernels to model the tendency of people to change actions over time. Thereby, it proves that this property is so important when predicting user actions and when actions will occur. These are the reason that the research works on the provisional scoring process are so important for the topic. This is also a premise to decide to choose a model suitable for the requirement of the problem and complex dataset.

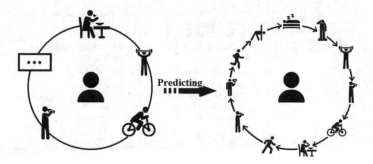

Fig. 1 Predicting daily actions and the time these actions happen

3 The Proposed Method

Firstly, we will detail the problem of predicting action and when that action will take place in the future. Secondly, the scope of the problem will also be clearly defined. Finally, we will introduce the PAS-KNNTP method, the theory platform and how it works.

3.1 Problem Statement

Predicting human actions is a problem that many researchers are interested in and applied in many fields, such as recommendation systems, e-commerce, healthcare systems, etc. However, previous work are often concerned with predicting user actions on the Internet. Therefore, in this paper, we focus on predicting daily actions that affect human health. At the same time, we are also interested in when the action will occur. This is visually represented in Fig. 1.

The results of the problem can be applied to smart healthcare systems. This system acts as a virtual assistant to help users manage their health, set performance goals and help doctors make accurate diagnoses early. The problem is specifically defined with input and output values as follows:

– **Input** The information of the user, such as parameters, actions, time action lasting, time when action occurs. The actions of user are categorized into 10 labels. These labels are drink, sleep, heart rate, running, weight, food, walking, biking, workout, and stretching actions. The time of the action is divided into 4 time frames: 0–6h, 6–12h, 12–18h, 18–24h.
– **Output** Actions and when that actions will happen in the future.

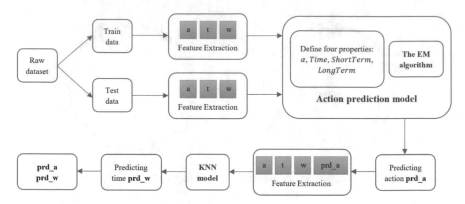

Fig. 2 The PAS-KNNTP method

3.2 Our Framework

Our goal is to predict the people's future action and the time of this happening action. To predict when an action will occur, we need to predict the action in advance. Therefore, our method needs to solve the problem in sequence: predicting the action that will happen and predicting when the action will occur. The way the PAS-KNNTP method works proposed by us is presented in detail in Fig. 2. Our model consists of 2 main phases as follows:

- **State 1: Data processing** First, we integrate data collected from health monitoring devices. Then, select data related to daily actions that affect health. Finally, transform the data to analyze and synthesize the necessary features for the model's training and prediction process.
- **State 2: Training and prediction** The training and prediction process is divided into two parts: predicting next action and predicting the time of happening action. We use information about the past actions to predict the future actions. These are action (a), time (t), and window of time (w). After we have the action result will happen in the future (prd_a), we use this result together with the dataset to predict when the action will occur (prd_w).

3.3 Predicting Next Action

Model definition By experimentally observed, user actions have the following properties: personalization, time-varying propensities, the dependence of actions in the short-term cycle, and influence of the long-term cycle. Therefore, the model is similar to user actions using a multivariate temporal point process with time-varying intensity [4].

$$\lambda_u(t, a) = \alpha_{ua} + Time_u(t, a) + ShortTerm_u(t, a) + LongTerm_u(t, a) \quad (1)$$

Inside, $\lambda_u(t, a)$ is the sum of properties personalized action preferences α_{ua}, time-varying propensities $Time_u(t, a)$, the dependence of actions in the short-term cycle $ShortTerm_u(t, a)$, and influence of the long-term cycle $LongTerm_u(t, a)$.

Specifically:

- $\alpha_{ua} \geq 0$ represents the personalized action, so α_{ua} can be considered a constant.
- $Time_u(t, a)$ represents the time-varying tendency of the action, the ongoing events can be affected by the actions that have happened before. This property is modeled by a mixed Gaussian model with the following formula:

$$Time_u(t, a) = \sum_{z \in Z} \frac{\beta_{az}}{\sqrt{2\pi\sigma^2_{az}}} \exp(-\frac{(l_t - \mu_{az})^2}{2\sigma^2_{az}}) \quad (2)$$

where $z \in Z$ represents the variable of the mixture Gaussian model and the mixture number can be decided through the cross-validation method; μ_{az}, σ_{az} of each action a and the mixture variable z are the mean and standard deviation of the Gaussian distribution. The most important value of the model is $\beta_{az} \geq 0$ which determines the value of the model. Finally, l_t is latent that represents the hours of the day.
- $ShortTerm_u(t, a)$ represents the interdependence of actions in short-term cycles. To model the action a occurring at the time t as being influenced by the action a' occurring at previous time t' with $t' < t$, we use the Hawkes process combined with the self-exciting and Exponential decay function. $ShortTerm_u(t, a)$ is represented by the following formula:

$$ShortTerm_u(t, a) = \sum_{(t',a' \in H_{ut})} \theta_{a'a}\omega_{a'a}exp(-\omega_{a'a}\Delta_{t't}) \quad (3)$$

where $H_{ut} = \{(t', a')| (t', a') \in H_u, t' < t\}$ is the sequence of actions that occurred before time t in the history u; $\Delta_{t't} = t - t'$ is the time difference between t and t'; $\omega_{a'a} \geq 0$ is the parameter that determines the influence of action trigger a' on the action a. This is also the shape of the exponential distribution. Lastly, $\theta_{a'a} \geq 0$ is also a parameter that determines the effect of action trigger a' on the action a, which is the size of the exponential distribution (scaling).
- $LongTerm_u(t, a)$ represents the action property that is affected by the long-term cycle. With this property, we use the Weibull distribution to model actions. Action a occurring is affected by action a'. Action a is occurring at the time t. Action a' is occurring at the previous time t'. $(t' < t)$ is defined as follows:

$$LongTerm_u(t, a) = \sum_{t', a' \in H_{ut}^a} \phi_{c_{t'}a} \gamma_{c_{t'}a} k_{c_{t'}a} \Delta_{t't}^{k_{c_{t'}a}-1} exp\left(\gamma_{c_{t'}a} \Delta_{t't}^{k_{c_{t'}a}-1}\right) \quad (4)$$

where $H_{ut} = \{(t', a')| (t', a') \in H_u, t' < t\}$ is the sequence of actions that occurred before time t in the history u; $\Delta_{t't} = t - t'$ is the time difference between t and t'; $c_{t'} \in C$ represents the time frames where the action occurs during the day (there are 4 time frames: 0–6h, 6–12h, 12–18h, 18–24h). Besides, $\phi_{c_{t'}a} \geq 0$, $\gamma_{c_{t'}a} \geq 0$ are the triggers of action a, deciding the influence of action a' on action a at $c_{t'}$ and $k_{c_{t'}a} \geq 0$ is the shape of the Weibull distribution.

Training model To make the model more efficient, we use Maximum likelihood estimation to further determine the model's parameters [4]. The sum of these parameters is called $\Psi = \{\alpha, \beta, \mu, \sigma, \Theta, \Omega, \Phi, \Gamma, K\}$. With action sequence in the past $\mathcal{H} = \{H_u\}_{u \in U}$ and taking place in the time from 0 to T, the log-likelihood function can be represented as follows:

$$\mathcal{L}(\Psi|\mathcal{H}) = \sum_{u \in U} \sum_{n=1}^{N_u} log\lambda_u(t_{un}, a_{un}) - \sum_{u \in U} \int_0^T \sum_{a \in A} \lambda_u(t, a)dt \quad (5)$$

Next, we use an algorithm to find the maximum value of log-likelihood based on the Expectation Maximization (EM) algorithm [4]. First, in step E-step, use 3 variables p, q, r as triggers for the action's properties introduced at [4]. These triggers will estimate the probability of action, this action will happen for the $l - th$ time at time t.

$$p_{z,un}^k = \frac{1}{R_{un}} \frac{\beta_{a_{un}z}^k}{\sqrt{2\pi(\sigma_{a_{un}z}^k)^2}} exp(-\frac{(l_{t_{un}} - \mu_{a_{un}z}^k)^2}{2(\sigma_{a_{un}z}^k)^2}) \quad (6)$$

$$q_{um,un}^k = \frac{1}{R_{un}} \theta_{a_{um}a_{un}}^k \omega_{a_{um}a_{un}}^k exp(-\omega_{a_{um}a_{un}}^k \Delta_{t_{um}t_{un}}) \quad (7)$$

$$r_{ul,un}^k = \left\{\frac{1}{R_{un}} \phi_{c_{ul}a_{un}}^k \gamma_{c_{ul}a_{un}}^k k_{c_{ul}a_{un}}^k \times \Delta_{t_{ul}t_{un}}^{k_{c_{ul}a_{un}}-1} exp(\gamma_{c_{ul}a_{un}}^k \Delta_{t_{ul}t_{un}}^{k_{c_{ul}a_{un}}})\right\} \quad (8)$$

Continuously, in step M-step, will take the results calculated in step E-step to estimate the values of the model's parameters and update them after each calculation. More specifically, through the Jensen inequality to calculate the lower bound of the log-likelihood function. Then, the result of this following function is used to take the derivative of the parameters and set them equal to 0.

$$\alpha_{ua}^{k+1} = \frac{\sum_{n=1}^{N_u} I(a_{un} = a) p_{0,un}^k}{T} \quad (9)$$

$$\beta_{az}^{k+1} = \frac{2T}{|U|T} \times \frac{\sum_{u \in U} \sum_{n=1}^{N_u} I(a_{un} = a) p_{z,un}^k}{erf\left(\frac{\mu_{az}^k}{\sqrt{2}\sigma_{az}^k}\right) + erf\left(\frac{T - \mu_{az}^k}{\sqrt{2}\sigma_{az}^k}\right)} \tag{10}$$

$$\theta_{a'a}^{k+1} = \frac{\sum_{u \in U} \sum_{n=1}^{N_u} \sum_{m=1}^{n-1} I(a_{um} = a', \ a_{un} = a) p_{um,un}^k}{\sum_{u \in U} \sum_{n=1}^{N_u} I(a_{un} = a')(1 - \exp(-\omega_{a'a}^k(T - t_{un})))} \tag{11}$$

$$\phi_{ca}^{k+1} = \frac{\sum_{u \in U} \sum_{n=1}^{N_u} \sum_{l=1}^{n-1} I(a_{ul} = a', \ a_{un} = a, \ c_{ul} = c) r_{ul,un}^k}{\sum_{u \in U} \sum_{n=1}^{N_u} I(a_{un} = a', c_{un} = c)(1 - \exp(-\gamma_{ca}^k(T - t_{un})^{r_{ca}^k}))} \tag{12}$$

$k_{ca}^{k+1}, \mu_{az}^{k+1}, \sigma_{az}^{k+1}$ are being estimated by the maximum value of the lower bound through gradient-based optimization along with Newton's method. Therefore, $\omega_{a'a}^{k+1}, \gamma_{ca}^{k+1}$ is calculated through the formula of [4].

The result that the Action prediction model return will be an action that will happen in the future. Therefore, this result will be used to solve the problem of predicting the time when the action will occur in the future.

3.4 Predicting the Time of Happening Action

Combining the available dataset and the action prediction results predicted by the Action prediction model, we start conducting predicting the time. Now, the data has 4 features: a, t, w, and prd_a. Among many multi-layer classification methods, the KNN model is a promising way to enable efficient visualization mining in low-dimensional space [17]. Therefore, we choose this classification method to predict the time (prd_w) that the action will occur in 4 time frames: 0–6h, 6–12h, 12–18h, 18–24h.

KNN model determines the class for sample (E) by the following steps:

1. Calculate the distance between E with all the samples in the training set. The distance between points can be calculated by Euclidean.
2. Choose K samples closest to E in the training set.
3. Assign E to the class with the most samples among those K neighboring samples (or E will be equal to the mean of K samples).

Nevertheless, with the KNN model, there is no exact method to estimate the exact K value. Because, many studies have shown that there is no specific K value that is suitable for all types of data, which is different from sizes and properties [18]. If we choose a small value of K, this will affect predicting results. If we choose a high value of K, this process has cost a lot. Furthermore, many studies have also demonstrated that, when choosing a small value of K, there will be flexibility and low bias but the high variance and many neighbors will have smoother decision boundaries, that is, the lower error but higher bias. As a result, we have determined this value by selecting an odd number if the number of labels is even and vice versa. In addition, by modeling different values of K and testing their performance by Grid Search, we have chosen a suitable K value when predicting the time of action to be k = 115 for the number class = 4.

4 Experiments

This section evaluates our capable of prediction approach by the experimental studies. We will introduce specifically our datasets and then, we will show the evaluation methods and experimental settings. Finally, conducting experiments, we show experimental results.

4.1 Dataset

The topic experiment on a real-world user action dataset consisting of 234,665 actions taken by 295 people in 17 months [4]. The dataset consists of 5 main components as follows: **a** (user action id), **t** (the duration of action a, calculated from the time of experimental observation to the time when the action is recorded and has the unit of minutes), **w** (time when action a occurs), **prd_a** (user's next action id after action a), **prd_w** (time when action prd_a occurs).

4.2 Evaluation Methods

We used 3 measures to compare the effective between our model with baseline method.

- **Accuracy Score** This measure will calculate the ratio of the correct prediction results to the existing results of the dataset.
- **Macro Average Recall** Macro Average Recall is the average of Recalls by class, and Recall is defined as the rate of the number of true positives among those that are actually positives [19].
- **Process cost** In addition to prediction accuracy, we also need to consider implementation costs. Therefore, we measure and compare the execution time of each experimental approach.

4.3 Experimental Settings

We evaluate the effectiveness of the method through two phases: predicting action (prd_a) and predicting the time the action will occur in the future (prd_w). For each stage, we set up 4 methods for evaluation. We use Gaussian Naive Bayes (Gaussian NB) method, Logistic Regression method, Support Vector Classifier (SVC) method, and PAS-KNNTP method to experiment.

4.4 Experiment Results

We conduct experiments and evaluate the effectiveness of the proposed method in two aspects: predictive efficiency and computational cost. The process of predicting future action and when this action will take place are both fully appreciated. Then, we evaluate the effect of applying the KNN method in predicting the time that action occurs. Here are the specific stats (Fig. 3).

After the experimental settings, our model has successfully predicted the user's actions in the real world while still ensuring the properties of the action (personalization, time-varying propensities, the dependence of actions in the short-term cycle, and influence of the long-term cycle). In addition, the application of the KNN method to predict the time the action will occur in the future is very effective. Because the KNN model is a promising way to enable efficient visualization mining in low-dimensional space (Table 1).

Experimental results show that the PAS-KNNTP method still ensures the accuracy of predicting future actions, improving the accuracy of predicting the time of happening action up to 20% and reducing the prediction time from 20 min to 15 s. Experimental results prove that our method is capable to apply the reality.

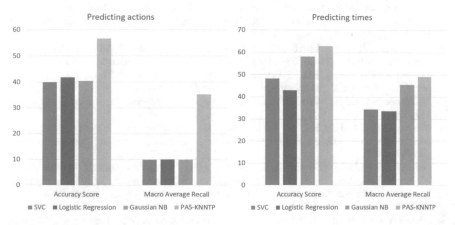

Fig. 3 Predicting actions and times

Table 1 The execution time of predicting times

Methods	SVC	Logistic regression	Gaussian NB	PAS-KNNTP
Time (s)	1200	8	1	15

5 Conclusions

In this paper, we focus on the problem of predicting actions that affect human health and when this action will take place in the future. We propose the PAS-KNNTP method that still ensures the accuracy of predicting human actions in the future, improving the accuracy of predicting the time happening action to 20% and reducing the processing time from 20 min down to 15 s. Our method has solved challenges in predicting actions and time. Especially, the result of predicting still meeting the properties of actions in the real world. Therefore, our model can contribute to predicting actions and when they will occur in accordance with health and safety standards. From that, it can apply to modeling human actions and building a recommendation system for user healthcare applications and devices.

Acknowledgements This research is funded by University of Information Technology, Ho Chi Minh City, Vietnam (VNU-HCMC) under grant number D1-2021-06.

References

1. Kishor A, Chakraborty C (2021) Artificial intelligence and internet of things based healthcare 4.0 monitoring system. In: Wireless personal communications, pp 1–17
2. Poongodi M, Sharma A, Hamdi M, Maode M, Chilamkurti N (2021) Smart healthcare in smart cities: wireless patient monitoring system using IoT. J Supercomputing 1–26
3. Islam MM, Rahaman A, Islam MR (2020) Development of smart healthcare monitoring system in IoT environment. SN Comput Sci 1:1–11
4. Kurashima T, Althoff T, Leskovec J (2018) Modeling interdependent and periodic real-world action sequences. In: Proceedings of the 2018 world wide web conference, pp 803–812
5. Chowdhury FA, Liu Y, Saha K, Vincent N, Neves L, Shah N, Bos MW (2021) CEAM: the effectiveness of cyclic and ephemeral attention models of user behavior on social platforms. In: Proceedings of the international AAAI conference on web and social media, vol 15, pp 117–128
6. Yao M, Zhao S, Sahebi S, Feyzi Behnagh R (2021) Stimuli-sensitive Hawkes processes for personalized student procrastination modeling. In: Proceedings of the web conference 2021, pp 1562–1573
7. Gjoreski M, Janko V, Slapničar G, Mlakar M, Reščič N, Bizjak J, Drobnič V, Marinko M, Mlakar N, Luštrek M, Gams M (2020) Classical and deep learning methods for recognizing human activities and modes of transportation with smartphone sensors. Information Fusion 62:47–62
8. Meduri VV, Chowdhury K, Sarwat M (2021) Evaluation of machine learning algorithms in predicting the next SQL query from the future. ACM Trans Database Syst (TODS) 46(1):1–46

9. Vazifedan A, Izadi M (2021) Predicting human behavior in size-variant repeated games through deep convolutional neural networks. In: Progress in artificial intelligence, pp 1–14

10. Becattini F, Uricchio T, Seidenari L, Ballan L, Bimbo AD (2020) Am i done? Predicting action progress in videos. ACM Trans Multimedia Comput Commun Appl (TOMM) 16(4):1-24

11. Du N, Dai H, Trivedi R, Upadhyay U, Gomez-Rodriguez M, Song L (2016) Recurrent marked temporal point processes: embedding event history to vector. In: Proceedings of the 22nd ACM SIGKDD international conference on knowledge discovery and data mining, pp 1555–1564

12. Zhang H, Dong J (2020) Prediction of repeat customers on E-commerce platform based on blockchain. In: Wireless communications and mobile computing

13. Kapoor K, Subbian K, Srivastava J, Schrater P (2015) Just in time recommendations: modeling the dynamics of boredom in activity streams. In Proceedings of the eighth ACM international conference on web search and data mining, pp 233–242

14. Teevan J, Adar E, Jones R, Potts M (2006) History repeats itself: repeat queries in Yahoo's logs. In: Proceedings of the 29th annual international ACM SIGIR conference on research and development in information retrieval, pp 703–704

15. Chiang WH, Liu X,Mohler G (2021) Hawkes process modeling of COVID-19 with mobility leading indicators and spatial covariates. Int J Forecasting

16. Tanaka Y, Kurashima T, Fujiwara Y, Iwata T, Sawada H (2016) Inferring latent triggers of purchases with consideration of social effects and media advertisements. In: Proceedings of the ninth ACM international conference on web search and data mining, pp 543–552

17. Zhu H, Zhu M, Feng Y, Cai D, Hu Y, Wu S, Wu X, Chen W (2021) Visualizing large-scale high-dimensional data via hierarchical embedding of KNN graphs. Vis Informatics 5(2):51–59

18. Zhang Z (2016) Introduction to machine learning: k-nearest neighbors. Ann Translational Med 4(11)

19. Sokolova M, Lapalme G (2009) A systematic analysis of performance measures for classification tasks. Information Process Manage 45(4):427–437

An Enhanced Moth Flame Algorithm for Peak Load Distribution Optimization

Thi-Kien Dao, Thi-Xuan-Huong Nguyen, Trong-The Nguyen, and Minh-Thu Dao

Abstract Peak load distribution optimization (PLD) is a typical multi-constrained nonlinear optimization problem considered an essential vital part of the power system to achieve energy-saving and consumption reduction. This study introduces an enhanced version of the moth flame optimization algorithm (EMFO) for the PLD problem. The actual operation constraints of the power system of the PLD are modeled for the objective function of optimization. In the experimental section, the IEEE-bus benchmark system is used as the case study to test the performance of the proposed scheme. The results show that the proposed scheme can solve the power system PLD problem with feasibility and significant economic benefits.

Keywords Peak load distribution optimization · Regulation load distribution power · Peak shaving cost · Enhanced moth flame algorithm

1 Introduction

Peak load distribution optimization (PLD) is one of the essential load dispatch problems to allocate power generations system optimally [1]. A power generation system includes multiple power generation units to achieve minimum power generation cost under the system constraints [2]. The optimization problem is to find a decision that

T.-K. Dao
Fujian Provincial Key Lab of Big Data Mining and Applications, Fujian University of Technology, Fujian, China

T.-X.-H. Nguyen · T.-T. Nguyen (✉)
University of Management and Technology Haiphong, Haiphong, Vietnam
e-mail: vnthe@hpu.edu.vn

T.-X.-H. Nguyen
e-mail: huong_ntxh@hpu.edu.vn

M.-T. Dao
Faculty of Information Technology, VNU-University of Engineering and Technology, Hanoi, Vietnam
e-mail: thudm@vnu.edu.vn

© The Author(s), under exclusive license to Springer Nature Switzerland AG 2022
N. H. T. Dang et al. (eds.), *Artificial Intelligence in Data and Big Data Processing*,
Lecture Notes on Data Engineering and Communications Technologies 124,
https://doi.org/10.1007/978-3-030-97610-1_16

makes one or more relationship indicators reach the maximum (or minimum) under the restriction of a series of objective or subjective conditions [3]. Increasing the clean energy installed capacity share is an unavoidable prerequisite for the power system's healthy and long-term development [4]. Thermal power, however, continues to account for a significant component of the country's power supply structure due to its long history of evolution [5]. As the penetration rate of clean energy in the grid rises, the area available for thermal power generation will inevitably shrink, and thermal power units will gradually transition to a peak-shaving role [6].

The power generation cost has attracted much attention from scholars as a review of the literature. Current research works primarily focus on calculating the economic benefits of thermal power peak shaving [7]. The goal function of maximizing clean energy consumption, development of peak shaving auxiliary markets, power peaking load distribution, and power peaking load distribution are complex nonlinear problems [8].

The traditional optimization methods, e.g., linear programming, gradian climbed hilling, square optimum, would be faced the computation complex times when dealing with the complicated problem like the nonlinear PLD [9]. The metaheuristic algorithms [10], e.g., genetic algorithms (GA) [11], swarm algorithms (PSO) [12], evolution algorithms (DE), improved bat algorithms (IBA) [3, 13] effectively solve the typical PLD problems.

The moth flame optimizer (FMO) [14] is a newly proposed metaheuristic algorithm inspired by the moth behavior in flying spiral trajectory for lighting flame. Still, it has disadvantages, such as optimal local solution, slow convergence rate, etc., making the problem's solution unsatisfactory. Optimizing large-scale power systems such as the PLD is still falling into the local optimum if the algorithm lacks diversity agents.

This study proposes an enhanced version for the moth flame optimization algorithm (EMFO) based on chaotic sequence and quadratic interpolation to improve its diversity agent. It means the algorithm's exploration and development capabilities could be effectively balanced and enhanced—the algorithm's convergence speed. As the analysis statement, a solution to the peak-load cost estimation of power load distribution is modeled to minimize the cost based on a novel EFMO. The suggested resolution is implemented to provide a particular reference for the optimized operation of thermal power under a high-proportion clean energy grid.

2 Peak Load Distribution Model

The receiving-end power grid unit acts as a peak-shaving unit, compressing its own power generation space to guarantee clean energy consumption [6]. The peak shaving process compared with the original power generation plan [15]. This part of the on-grid electricity revenue loss is the unit opportunity cost. For peak shaving thermal power unit ith, its opportunity cost calculates as the following formula.

$$R_i = \sum_{t=1}^{T} \left(P_{i,t}^{sc} - P_{i,t}^{ac} \right) \times \Delta t \times \rho^{BG} \tag{1}$$

In the formula R is the opportunity cost of thermal power unit i; T is the scheduling period; $P_{i,t}^{sc}$ is the planned power generation; $P_{i,t}^{ac}$ is the power generation after peak shaving; Δt is the generation time of thermal power during this period; ρ^{BG} is the on-grid price of thermal power. The output reduction of peak-shaving thermal power units has lost part of the on-grid power. Due to the power output status change, the production cost of power generation has also changed correspondingly. The second-order function is modeled as follows.

$$C_{gi} = a P^2 + b P + c \tag{2}$$

where C_{gi} is the power generation cost of the production power unit I; P is the unit's output, and a, b, and c are the unit coefficients. The change in the production cost of peak-shaving thermal power plants can be expressed as follows.

$$\Delta C_{i,t} = C_{gi} \times P_{i,t}^{sc} - C_{gi} \times P_{i,t}^{ac} \tag{3}$$

In formula, $\Delta C_{i,t}$ is the change in power generation cost during t. When it is a positive number, it indicates that the cost of power generation has decreased, otherwise, it suggests that the cost of power generation has increased; $C_{gi} \times P_{i,t}^{sc}$ is the planned power generation cost, and $C_{gi} \times P_{i,t}^{ac}$ is the actual power generation cost. Based on the above processing, the total cost of peak shaving of thermal power units is shown in the following formula.

$$R_{ac,i} = R_i - \sum_{t=1}^{T} \Delta C_{i,t} \tag{4}$$

In the formula, $R_{ac,i}$ is the total cost of peak shaving of the thermal power unit. A model of minimizing peaking cost as the objective function of the power peaking load distribution will be presented in Sect. 4.

3   Enhanced Moth Flame Optimization

The moth flame optimization algorithm (MFO) [14] is a new metaheuristic algorithm that has advantages as understand and implement quickly and few parameters. Still, it has disadvantages, e.g., fulling local optimum and slow convergence when dealing with a complex problem. The enhanced moth flame optimization (EMFO) algorithm is proposed in this section using chaotic initialization and Gaussian mutation to improve the algorithm's ability to jump out of the local optimum. The brief presentation is listed as follows. First, the moth population is initialized using cubic chaotic mapping to make the moths more evenly distributed in the search space. Second, Gaussian mutation is used to perturb a small number of poor individuals population. Third, the Archimedes curve is used to broaden the search range and improve the ability to explore the previously unexplored territory.

3.1 Moth Flame Optimization

Let Mo be a matrix of the spatial position of the moth in the MFO algorithm [14] with n is a moth population size and d is a dimension of the space search problem. The position matrix in space is similar to the space matrix of moths, represented as follows.

$$Mo(n,d) = \begin{bmatrix} m_{o11} & m_{o12} & \cdots & m_{o1d} \\ m_{o21} & m_{o22} & \cdots & m_{o2d} \\ \vdots & \vdots & \vdots & \vdots \\ m_{on1} & m_{on2} & \cdots & m_{ond} \end{bmatrix} \tag{5}$$

Let OMo be the objective function value as of other moths' individual space stored by the matrix Mo as follows.

$$OMo = [om_{o1}, om_{o2}, \ldots, om_{on}]^T \tag{6}$$

Let Fr be the flame core of the algorithm with their objective function values as follows.

$$Fr(n,d) = \begin{bmatrix} f_{r11} & f_{r12} & \cdots & f_{r1d} \\ f_{r21} & f_{r22} & \cdots & f_{r2d} \\ \vdots & \vdots & \vdots & \vdots \\ f_{rn1} & f_{rn2} & \cdots & f_{rnd} \end{bmatrix} \tag{7}$$

The objective function value is calculated for the flame's position as follows.

$$OFr = [of_{r1} of_{r2} \ldots of_{rn}]^T \tag{8}$$

The flame reduction principle process is implemented as given in the following formula.

$$flame_r = Round\left(N - t \times \frac{N-1}{T}\right) \tag{9}$$

where i and T are the current number of iterations and the maximum number of iterations, respectively; N is the maximum number of flames.

$$D_i = |Fr_j - Mo_i| \tag{10}$$

where D_i is the distance of the flame to moth that are flying moth move towards the flame position. The movement path is given as a logarithmic spiral curve, and the curve is taken as the primary update mechanism of the moth. The logarithmic spiral curve of the algorithm is defined as follows.

$$S(Mo_i, Fr_j) = D_i e^{bt} \cos(2\pi t) + Fr_j \tag{11}$$

where $S(Mo_i, Fo_j)$ is the updated position of the moth; D is the distance between the ith moth and the jth flame; is a constant related to the shape of the spiral; b is a random number, the value interval is [-1,1]; $e^{bt} \cdot \cos(2\pi t)$ is the right a number spiral curve expression.

3.2 An Enhancement of FMO

Gaussian mutation mechanism is used to determine the flame space position, leading to the effective utilization of the current global optimal moth. With the updated formulas of the solution in the algorithm, the space vector position of each moth is directly related to the position of the closest flame space. The selected individual's worst fitness value is applied to mutate to the moth population scale; it represents the variance convenient for controlling and narrowing the update range and appropriately increasing the population diversity of the algorithm. The improved moth generation formula is described as follows.

$$\gamma_i = 4\gamma_{i-1}n^3 - 3\gamma_{i-1}n \tag{12}$$

where γ_i is a random generator as cubic chaotic; $-1 \leq \gamma \leq 1, n = 0, 1, \ldots, N$. The initializing the moth population using a cubic chaotic map is the map to the solution space as the following formula.

$$x_{id} = L_d + (1 + \gamma_{id}) \times \frac{U_d - L_d}{2} \tag{13}$$

where U and L are the upper and lower bound of the search space; d is the dimensional coordinate of the generated ith moth.

$$M_{new}^t = M_j^{*t} + Gaussian(\mu, \sigma^2) \tag{14}$$

where $Gaussion(\mu, \sigma^2)$ is generation of the population; n is the moth population size; σ is the proportion of variation; μ is the mean value; According to experience of the variance, the value of σ is set to 1/6 in the algorithm.

4 The EFMO for Load Distribution Optimization

The optimization problem of the PLD is modeled as its objective function with the total amount of incoming electricity from the receiving end power system grid. It is determined by the total amount of incoming electricity or additional local clean energy issuance at each period. The peak load share based on the peaking cost of each power unit is bear over time, minimizing the thermal power peaking load distribution [15].

$$obj = \min\left(\sum_{i=1}^{N} R_{ac,i}\right) \tag{15}$$

where obj is the objective function; N is the total number of peak-shaving thermal power units, and other parameters are the same as above. Subjective to the objective function is dealing with its constraints.

Power balance constraint is given as follows.

$$\sum_{i=1}^{N} \Delta P_{i,t} = \Delta P_t (t = 1, 2, \ldots, T) \tag{16}$$

In the formula: $\Delta P_{i,t}$ is the peak-shaving load allocated to the ith thermal power unit at time t; ΔP_t is the total peak-shaving demand of the power grid at time t.

The active output of the power units constraint is expressed as follows.

$$P_{i,min} \leq P_{i,t} \leq P_{i,max} \tag{17}$$

where $P_{i,min}$ and $P_{i,max}$ are the minimum and maximum technical output of the thermal power unit, respectively. Generating units climbing ability constraints are presented as follows.

$$\begin{cases} P_{i,t} - P_{i,t-1} \leq A_{u,i} \\ P_{i,t-1} - P_{i,t} \leq A_{d,i} \end{cases} (i = 1, 2, \ldots, N) \tag{18}$$

where $A_{u,i}$ and $A_{d,i}$ are the maximum climbing ability and the maximum downward climbing ability of the thermal power unit, respectively.

The minimum startup and shutdown constraints of the unit is expressed as below.

$$\begin{cases} \left(Y_{i,t-1}^{on} - U_{i,min}^{on}\right)\left(s_{i,t-1} - s_{i,t}\right) \geq 0 \\ \left(Y_{i,t-1}^{off} - U_{i,min}^{off}\right)\left(s_{i,t} - s_{i,t-1}\right) \geq 0 \end{cases} \tag{19}$$

where $Y_{i,t-1}^{on}$ and $Y_{i,t-1}^{off}$ are respectively when the thermal power unit i actually starts and stops time; $U_{i,min}^{on}$ and $U_{i,min}^{off}$ are the minimum start and stop time of thermal power unit i; $s_{i,t}$ are the state variables of thermal power unit, 1 means running, 0 means stop. Algorithm 1 shows a pseudo-code of the EFMO for the PLD as the steps of the process optimization.

Algorithm 1. EFMO's pseudo-code for the PLD

1: Initialization parameters: population size N, dimension D, maximum number of iterations *MaxT*;
2: Mapping modeling optimization space to moth solutions; Randomly initialize the moth position *Mo* in the search space;
3: **while** $(1 <= MaxT)$
4: Calculate the objective function value *OMo* of each moth, Eqs.(5),(6);
5: Calculate the number of flames, Eqs(7),(8)
6: If the current iteration number $l = 1$, update the flame population according to $OFr = sort\,(OMo), Fr = sort\,(Mo)$;
7: Otherwise, update the flame population according to $OFr = sort\,(OMol - 1, OMl), Fr = sort(Ml - 1, Ml)$;
8: Record the first flame as the best individual;
9: **for** $i = 1$ to N do
10:   Update the position of the moth;
11:   Determine whether the individual position of the moth exceeds the upper and lower limits of the search space;
12:   If it exceeds the boundary, re-initialize the position in the search space;
13: **end for**
14: **end while**
15: Output the optimal solution.

Figure 1 displays the solution process as a flowchart of the gusseted EFMO for the OLD problem.

5 Experimental Results

The selection standard is adopted the load power in typical calculation daily. All power plants in the power grid are turned on, and the load rate is relatively high: Table 1 displays the power generation cost and coefficient parameters of a power grid system. The amount of renewable energy generation that the power grid can absorb is directly proportional to the power system's peak shaving capacity.

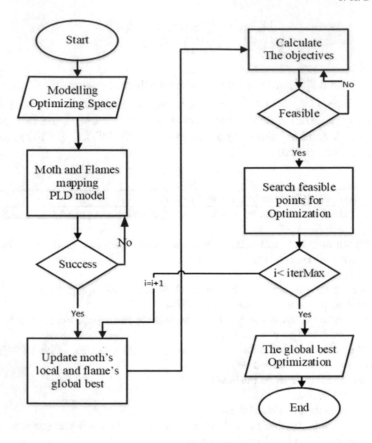

Fig. 1 A flowchart of the gusseted EFMO for the PLD problem

Table 1 Power generation cost coefficient parameters of the thermal power unit for peak load regulation	Plants installed capacity/MW	Power generation cost factors		
		a	*b*	*c*
	135	0.024 1	236.3	17 066
	150	0.024 0	236.02	18 082
	300	0.022 4	233	27 405
	350	0.022 0	232.3	31 621
	360	0.021 9	232.11	32 298
	467	0.020 8	230.12	39 541
	600	0.019 1	228	50 000
	660	0.018 9	226.53	52 606
	1000	0.015 0	220	75 000

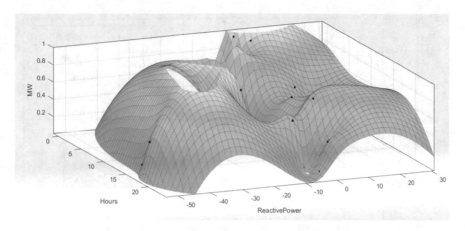

Fig. 2 A obtained peak-shaving load distribution curve from the EFMO

The peak-shaving demand curve is set to ensure the electricity grid's safe and steady operation. An IEEE benchmark power grid system [16] with fifteen plants is used to test the performance of the EFMO in order to verify its dependability and effectiveness. The unit test system also considers slow-changing rate, ascending, and output upper and lower limitations, and total load demand. The suggested EFMO's results are compared to the MFO [14], PSO [12], and IBA [3]. The number of search agents is uniformly set to 30 in all algorithms during the simulation, and the maximum number of iterations is 1000.

Figure 2a illustrates the visual graph of the introduced scheme for the peak-shaving load distribution of a power grid system.

Figure 3 shows the comparison of the suggested method with the other algorithms, e.g., the FMO [14], PSO [12], and IBA [3] algorithms for the LPD problem. It can be seen that the EFMO produces the convergence fastest and increases the quality of performance for the optimization problem.

Table 2 depicts a comparison of obtained results of the suggested FMO with the FMO, PSO, and IBA algorithms for the grid system of fifteen plant units. The compared results show that the total cost of system peak shaving of the EFMO is better than the FMO, PSO, and IBA. It means that the proposed EFMO has a more vital ability to avoid local optimal solutions. The EFMO can provide a specific reference for optimizing the power operation under the background of high clean energy penetration and calculating the power peak-shaving compensation costs.

6   Conclusion

This paper proposed an enhanced moth flame optimization algorithm (EMFO) to avoid the original one's falling local optimum for the peak load distribution (PLD)

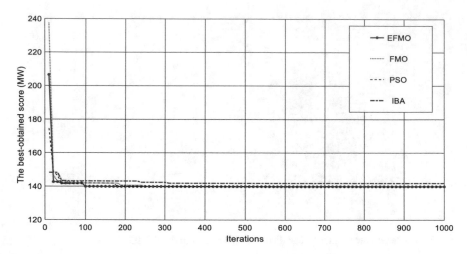

Fig. 3 Comparison of the suggested scheme with the other algorithms, e.g., the FMO [14], PSO [12], and IBA [3] algorithms for the LPD problem in terms of convergence curve

Table 2 Comparison of obtained results of the suggested EFMO with the FMO, PSO, and IBA algorithms for the grid system of fifteen plant units

Plant units	EFMO	FMO	PSO	IBA
P_1	297.42	268.17	312.07	250.63
P_2	289.78	318.50	288.14	263.40
P_3	77.27	91.00	89.61	81.34
P_4	91.00	14.00	85.32	48.37
P_5	254.81	181.32	174.03	307.02
P_6	234.81	322.00	214.02	234.94
P_7	236.67	325.50	141.63	263.88
P_8	119.70	42.00	154.56	100.87
P_9	86.01	103.61	106.72	51.64
P_{10}	62.89	30.08	108.21	99.42
P_{11}	37.58	51.77	50.72	42.99
P_{12}	18.33	31.76	36.51	32.59
P_{13}	21.98	17.50	51.99	45.90
P_{14}	20.45	25.90	28.74	25.45
P_{15}	20.78	37.09	31.16	20.76
Total power output (MW.)	**1869.48**	1860.20	1873.44	1869.19
Total generation cost ($/h.)	**23,500.37**	23,645.18	23,712.08	23,697.81
Power loss (MW.)	**27.10**	27.60	32.46	28.21
Deviation	0.02	0.03	0.02	0.02
Total CPU times (sec.)	0.33	**0.32**	0.85	1.14

problem. Based on the peak load cost theory, the objective function is constructed for the power peak load distribution model. The actual operation constraints of the PLD power system are dealing with using the penalty function in optimization. In the experimental section, the case study of the selected IEEE-bus benchmark system is used to test the performance of the proposed EFMO system. The suggested scheme results are compared with the previous schemes in the literature show that the proposed EFMO can solve the power system PLD problem with significant economic benefits. In future work, we will extend implementation with a complicated benchmark CEC 2019 function test suite to prove the proposed EFS performance and compare the results with various swarm algorithms.

References

1. Rahmani R, Othman MF, Yusof R, Khalid M (2012) Solving economic dispatch problem using particle swarm optimization by an evolutionary technique for initializing particles. J Theor Appl Inf Technol 46:526–536
2. Tsai CF, Dao TK, Pan TS, Nguyen TT, Chang JF (2016) Parallel bat algorithm applied to the economic load dispatch problem. J Internet Technol 17:761–769. https://doi.org/10.6138/JIT.2016.17.4.20141014c
3. Dao TK, Pan TS, Nguyen TT, Chu SC (2015) Evolved bat algorithm for solving the economic load dispatch problem. In: Advances in intelligent systems and computing. pp 109–119. https://doi.org/10.1007/978-3-319-12286-1_12
4. Nguyen TT, Wang MJ, Pan JS, Dao T, Ngo TG (2020) A load economic dispatch based on ion motion optimization algorithm. In: Smart innovation, systems and technologies, pp 115–125. https://doi.org/10.1007/978-981-13-9710-3_12
5. Wang C-H, Nguyen T-T, Pan J-S, Dao T-K (2017) An optimization approach for potential power generator outputs based on parallelized firefly algorithm. https://doi.org/10.1007/978-3-319-50212-0_36
6. Wang Z, Wang S (2013) Grid power peak shaving and valley filling using vehicle-to-grid systems. IEEE Trans power Deliv 28:1822–1829
7. Reihani E, Motalleb M, Ghorbani R, Saoud LS (2016) Load peak shaving and power smoothing of a distribution grid with high renewable energy penetration. Renew Energy 86:1372–1379
8. Pan T-S, Dao T-K, Nguyen T-T, Chu S-C (2014) Optimal base station locations in heterogeneous wireless sensor network based on hybrid particle swarm optimization with bat algorithm. J Comput 25
9. Nguyen TT, Pan JS, Dao TK (2019) An Improved flower pollination algorithm for optimizing layouts of nodes in wireless sensor network. IEEE Access 7:75985–75998. https://doi.org/10.1109/ACCESS.2019.2921721
10. Nguyen T-T, Wang H-J, Dao T-K, Pan J-S, Liu J-H, Weng S-W (2020) An improved slime mold algorithm and its application for optimal operation of cascade hydropower stations. IEEE Access 8:1. https://doi.org/10.1109/ACCESS.2020.3045975
11. Nguyen T-T, Shieh C-S, Horng M-F, Dao T-K (2014) A genetic algorithm with self-configuration chromosome for the optimization of wireless sensor networks. In: 12th international conference on advances in mobile computing and multimedia, MoMM 2014. https://doi.org/10.1145/2684103.2684132
12. Barisal AK, Mishra S (2018) Improved PSO based automatic generation control of multi-source nonlinear power systems interconnected by AC/DC links. Cogent Eng 5:1422228
13. Dao TK, Pan TS, Nguyen TT, Pan JS (2018) Parallel bat algorithm for optimizing makespan in job shop scheduling problems. J Intell Manuf 29:451–462. https://doi.org/10.1007/s10845-015-1121-x

14. Mirjalili S (2015) Moth-flame optimization algorithm: a novel nature-inspired heuristic paradigm. Knowledge-based Syst 89:228–249
15. Feng Z, Niu W, Wang W, Zhou J, Cheng C (2019) A mixed integer linear programming model for unit commitment of thermal plants with peak shaving operation aspect in regional power grid lack of flexible hydropower energy. Energy 175:618–629
16. Moeini A, Kamwa I, Brunelle P, Sybille G (2015) Open data IEEE test systems implemented in sim power systems for education and research in power grid dynamics and control. In: 2015 50th international universities power engineering conference (UPEC). IEEE, pp 1–6

A New Benchmark Data Set for Chemical Laboratory Apparatus Detection

Ze-sheng Ding, Shuo-yi Ran, Zhi-ze Wu, Zhi-huang He, Qian-qian Chen, Yun-sheng Wei, Xiao-feng Wang, and Le Zou

Abstract In chemistry laboratory, the safety of the experimenter is essential. The chemical laboratory apparatus is an important part of the chemical laboratory, and apparatus safety cannot be ignored. An auxiliary device that takes advantage of artificial intelligence and computer vision for the automatic detection of chemical laboratory apparatus can adapt to the changing realities to improve personnel safety. The most important task is to accurately identify and locate these apparatuses from the image. In this work, we captured 2246 images (data augmentation to 13,476) in the real chemical laboratory environment, containing 21 types of chemical laboratory apparatus, and produced an open Chemical Laboratory Apparatus Dataset (CLAD) by annotating the images. To implement feature extraction, classification and localization in one algorithm and automate the whole, we conducted experiments on this dataset using the deep learning models Faster R-CNN (3 backbones), Single Shot MultiBox Detector (SSD), YOLOv3-SPP and YOLOv5, to fully analyze the performance of this dataset. Experiments demonstrate Faster R-CNN (resnet50-fpn) performs the best in terms of average precision (99.0%), YOLOv5 performs the best in terms of average recall (94.5%), and in terms of inference speed YOLOv5 can process the most images per second (62) on our device, and it has satisfied real-time applications.

Keywords Chemical laboratory apparatus · Deep learning · Automatic detection · Object detection · Faster R-CNN

Z. Ding · Z. He · Q. Chen · X. Wang · L. Zou (✉)
School of Artificial Intelligence and Big Data, Hefei University, Hefei 230601, Anhui, China
e-mail: zoule1983@163.com

S. Ran · Z. Wu
School of Artificial Intelligence and Big Data, Institute of Applied Optimization, Hefei University, Hefei 230601, Anhui, China

Y. Wei
School of Energy Materials and Chemical Engineering, Hefei University, Hefei 230601, Anhui, China

© The Author(s), under exclusive license to Springer Nature Switzerland AG 2022   201
N. H. T. Dang et al. (eds.), *Artificial Intelligence in Data and Big Data Processing*,
Lecture Notes on Data Engineering and Communications Technologies 124,
https://doi.org/10.1007/978-3-030-97610-1_17

1 Introduction

The safety of chemical laboratories has always been a critical problem for most scientific research institutions and universities [1]. In the past few years, accidents such as explosions, fires, poisonings, burns, cuts and electrocution in chemical laboratories have been commonplace. However, safety is still neglected. Chemical laboratory apparatus is an important part of the chemical laboratory. It is essential to the completion of relevant chemical experiments by selecting properly and using rationally chemical laboratory apparatus. For this goal, we conceive a device for automatic detection of chemical laboratory apparatus, which can automatically identify and precisely locate chemical laboratory apparatus and assist experimenters in conducting dangerous chemical experiments, taking advantage of the fact that artificial intelligence can adapt to changing real-world environments and flexibly adjust its position according to the object state.

Object detection is one of the important fundamental research areas in computer vision [2]. The deep convolutional neural network (CNN) models have excellent feature extraction and representation capabilities, and deep learning-based object detection algorithms are making a splash in areas such as fire and smoke detection, garbage classification, and defect detection. Thereby, in this manuscript, we draw conceptually on these field applications, use several advanced models of object detection as a robust chemical laboratory apparatus detector. In this paper, the main contributions include three aspects:

First, we collected 2246 images from a real chemical laboratory environment, which contains 21 types of chemical laboratory apparatus. To improve the processing efficiency, the image resolution was compressed to 512×512 pixels; to enrich the dataset and prevent overfitting problems in training as well as to improve the generalization ability of the model, we use data augmentation to extend the chemical laboratory apparatus images to 13,476.

Second, we conducted experiments using the two-stage Faster R-CNN [3] object detection model on a chemical laboratory apparatus dataset and used VGG [4], ResNet50-fpn, and MobileNetV2 as their backbone networks for feature extraction, respectively. We also used the equally good one-stage object detection models YOLOv3-SPP [5], SSD [6], YOLOv5 for experiments on CLAD dataset. We compared the performance of these models on this dataset and analyzed the results of these methods. The resulted in a baseline that can be used for future studies in chemical laboratories.

Third, experiment results show that when the Faster R-CNN uses resnet50-fpn as the backbone for feature extraction at two stages, it achieves a maximum mAP of 99.0% and a recall of 91%; at one stage: the mAP of YOLOV5 is 97.8%, the average recall is better at 94.5%, more images per second are processed.

The remainder of this paper is organized as follows. In Sect. 2, we review the applications of object detection in existing domains. Section 3 describes the design and content when building CLAD. Section 4 evaluates the performance of several state-of-the-art target detection models on CLAD. Finally, conclusions and future work are drawn in Sect. 5.

2 Related Work

At present, deep learning techniques have achieved good results in face recognition, character recognition, and image classification, and even surpassed the level of human recognition.

With the widespread popularity of intelligent surveillance devices in recent years, deep learning-based object detection technology has been widely used in other fields, such as waste classification in resource recycling scenarios to turn waste into treasure. In the field of fire and smoke detection as well as industrial exhaust gas monitoring using the widely available video surveillance devices in the society for data collection, smoke detection, and fire identification using image recognition, thus achieving fast and wide range of fire detection. In the field of defect detection, such as glass surface defect detection, printing surface defect detection, chemical fiber surface defect detection, weld defect detection, using deep learning methods to replace the traditional human eye visual inspection can improve productivity, ensure product quality and ensure the safety of the working environment of inspectors. In these areas of industry-specific target detection applications, computer vision technology plays a vital role in these object detection applications.

The main task of object detection is to locate and identify all the objects that appear in an image. Before the emergence of deep learning network models, object detection was mainly studied by manually extracting the features needed for object detection, which is time-consuming and laborious. With deep learning network model's outstanding performance, CNN-based detectors have quickly become the hottest research directions in target detection, which are mainly divided into two categories: two-stage detection and one-stage detection. The two-stage object detection method first generates the region of interest (ROI) by filtering, and then does further category score prediction and bounding box regression on the filtered ROI. The classical algorithms have R-CNN [7], Fast-RCNN [8], Faster-RCNN [3], Mask R-CNN [9], etc. The two-stage structure is more accurate, but the two-stage requires separate classification and regression for each proposal; the speed is compromised. The one-stage object detection algorithms discard the ROI generation part of the two-stage object detection algorithm and do all the work of object detection directly in an end-to-end network models, mainly YOLO [10], SSD [6], and RetinaNet [11], etc. The one-stage structure is faster but far less accurate than the two-stage structure. Based on transfer learning, Wu et al. [12] proposed a benchmark data set for aircraft type recognition from remote sensing images. In a study related to the detection of chemical laboratory apparatus, Rostianingsih et al. [13] constructed a COCO dataset on chemical apparatus, described in detail the process of manually annotating images using an annotation tool, and synthesized the dataset by combining different backgrounds.

In this paper, we create a Chemical Laboratory Apparatus Dataset, CLAD, which contains 13,476 apparatus images collected in real chemical laboratories, and it is composed of 21 types of chemical apparatus. Experiments were conducted in advanced object detection models. For the two-stage detection method, the Faster R-CNN backbone uses three pre-trained models, VGG, MobileNetV2 and Resnet50-fpn. For the one-stage detection method, we use YOLOv3-SPP, SSD, YOLOv5, etc. We designed different training parameters and scales to evaluate their performance on CLAD.

3 Chemical Laboratory Apparatus Dataset: CLAD

To implement the detect of chemical laboratory apparatus, we use deep convolutional neural networks, so we need to collect a large amount of data and manually annotate it to make a dataset eventually. The quality and quantity of the collected data directly determine how good the prediction model is, so a high-quality dataset can help the model understand the data like a human. We produced such a dataset and provided baseline results for detecting chemical laboratory apparatus.

The chemical laboratory apparatus dataset was taken in a real chemical laboratory environment. We took photos from the Chemical Engineering and Process Laboratory, Chemical Experiment Center, and Inorganic Chemistry Laboratory at Energy Materials and Chemical Engineering school of Hefei University. The experimental apparatus is mainly divided into glass instruments and metal instruments, of which glass apparatus have a greater safety risk, like beakers, flasks, graduated cylinders, etc. Experimental instruments of glass have higher safety requirements, so our data set is mainly glass experimental apparatus. Some examples of real chemical laboratory apparatus images are shown in Fig. 1.

The CLAD, an image dataset of high-quality chemical laboratory apparatus was constructed in this manuscript. In general, CLAD includes 21 different types of chemical apparatus images, mainly glass apparatus. CLAD consists of the following 21 types of chemical apparatus: alcohol lamp, beaker, boiling flask, boiling flask-3-neck, conical flask, crucible, evaporating dish, filter flask, funnel, glass rod, graduated cylinder, medicine dropper, medicine spoon, mortar, plastic wash bottle, rubber suction bulb, separatory funnel, test tube, tube brush, volumetric flask, wide-mouth bottle All images of laboratory instruments were carefully labeled by me and four students in the chemical field. Each image contains one or more laboratory apparatus images, and the number of apparatus on each image is variable and usually multiple. Each sample of laboratory apparatus in the CLAD dataset is shown in Fig. 2.

To enrich the CLAD dataset and prevent the overfitting problem of training, better extract the features of chemical laboratory apparatus and improve the generalization ability of the deep learning-based objection detection model, the chemical laboratory apparatus dataset is expanded by using data augmentation techniques. The original images were rotated clockwise by 90°, 180° and 270°, flipped horizontally and flipped vertically. After data augmentation, there are 13,476 images in the dataset, and the

Fig. 1 Illustration of real chemical laboratory apparatus images

images are annotated in the standard format of the PASCAL VOC dataset. The number of sample images varies depending on the experimental apparatus type (see Table 1) and ranges from 324 to 852.

4 Experiments and Results

In all experiments, we use a computer with Windows 10 (64-bit) operating system, Intel Core i7-7700 K processor, 16 GB RAM and an NVIDIA GeForce GTX 1060 graphics card. The algorithms are implemented and executed based on the Pytorch open-source framework, equipped with CUDA parallel computing architecture and cuDNN deep neural network acceleration library.

We randomly selected 12,132 images from 13,476 images as the training set, 1344 images as the validation set, with no overlap between the training, validation. The performance of the three feature extractors on our CLAD was compared in

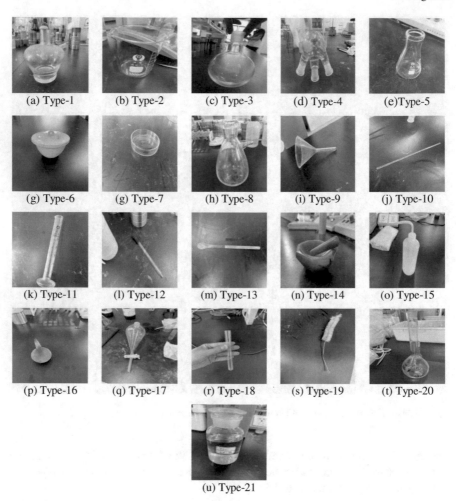

Fig. 2 Samples of the 21 laboratory apparatus types from CLAD

Table 1 Different laboratory apparatus and the number of images in each class of CLAD

Types	#Images	Types	#Images	Types	#Images
Alcohol lamp	552	Boiling flask	774	Beaker	738
Conical flask	606	Medicine spoon	324	Funnel	582
Plastic wash bottle	420	Evaporating dish	414	Glass rod	660
Graduated cylinder	636	Medicine dropper	750	Crucible	600
Rubber suction bulb	390	Filter flask	462	Mortar	648
Boiling flask-3-neck	618	Separatory funnel	852	Test tube	570
Wide mouth bottle	420	Volumetric flask	492	Tube brush	798

Table 2 Hyperparameter settings of the models

Model	Backbone	Batch size	Batch	Weight decay	Lr
Faster R-CNN	VGG16	4	3033	0.0005	0.005
	MobileNetV2	4	3033	0.0005	0.005
	Resnet50-fpn	4	3033	0.0005	0.005
SSD	Resnet50	4	3033	0.0005	0.005
YOLOv3-spp	DarkNet53	4	3033	0.0005	0.010
YOLOv5	CSPDarkNet53	4	3033	0.0005	0.010

two-stage Faster R-CNN; the performance on CLAD was also examined in one-stage SSD, YOLOv3-spp, YOLOv5. The configuration parameters of all methods are shown in Table 2.

We first trained VGG16, ResNet50-fpn and MobileNetV2 on two-stage Faster R-CNN and compared them using the training data. After data enhancement for each algorithm, we evaluated by training again. During the training process, we specified the point with the highest mAP value of the evaluation data as the endpoint.

Precision and recall are mathematically illustrated in Eqs. (1) and (2)

$$\text{Precision} = \frac{TP}{TP + FP} \tag{1}$$

$$\text{Recall} = \frac{TP}{TP + FN} \tag{2}$$

where TP = number of true positive; FP = number of false positive; and FN = number of false negative. The specific division is shown in Table 3.

The algorithm performance evaluation results are shown in Table 4.

In all one-stage Faster R-CNN experiments, we use average precision (mAP) and average recall (AR) to measure and calculate FPS. the highest mAP of 0.990 is achieved by applying the data augmentation algorithm, and making ResNet50-fpn as a backbone. Comparing the data augmentation used to the three feature extractors, the results all show a much better performance than before data augmentation.

We also used the state-of-the-art one-stage models YOLOv3-spp, YOLOv5 and SSD models to compare with the best Faster R-CNN in two-stage using ResNet50-fpn, and the results are shown in Table 5. The mAP of the four methods in the table, Faster R-CNN, reaches the highest 0.990, because it is a two-stage method, it can only detect nine images per second in FPS. In contrast, the mAP of SSD and YOLOv5 as one-stage is not as high as that of Faster R-CNN, but they also reach 0.944 and

Table 3 Confusion matrix

	Positives	Negatives
True	True positives	True negatives
False	False positives	False negatives

Table 4 Comparison of VGG, ResNet50-fpn and MobileNetV2 applied to faster R-CNN

Network	mAP	Recall	FPS
VGG16	0.885	0.676	6
ResNet50-fpn	0.928	0.810	8
MobileNetV2	0.896	0.690	20
VGG16 with Aug	0.975	0.823	7
ResNet50-fpn with Aug	*0.990*	*0.910*	9
MobileNetV2 with Aug	0.974	0.833	*22*

Performance improves with data augmentation for each network. (best marked with *Italic*)

Table 5 Comparison of Faster R-CNN, SSD, YOLOv3-spp and YOLOv5 on the CLAD dataset (best marked with *italic*)

Model	mAP	Recall	FPS
Faster R-CNN	*0.990*	0.833	9
SSD	0.944	0.788	41
YOLOv3-spp	0.859	0.737	22
YOLOv5	0.978	*0.945*	*62*

0.978 respectively. The mAP of SSD and YOLOv5 as one-stage is not as high as that of Faster R-CNN, but also reaches 0.944 and 0.978, respectively, which is an excellent result.

Figure 3 shows the process of mAP and recall changes on the validation set for the four methods during the training process. Faster R-CNN can already provide high mAP and high recall in the first few rounds of training, while slowly increasing almost constantly afterward. The performance of YOLOv3 is poorer both in mAP and Recall. Whereas YOLOv5 starts with both low, but as the number of rounds increases both are increasing rapidly, and eventually mAP gets good results and Recall exceeds the other methods.

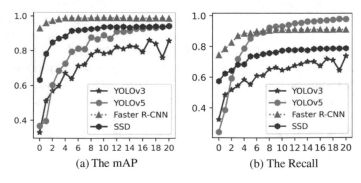

(a) The mAP   (b) The Recall

Fig. 3 The mAP and the Recall on the CLAD validation set over the training epochs

5 Conclusions

In this paper, we developed and produced a new benchmark data set for apparatus type recognition and location of chemical laboratory apparatus images, named CLAD, consisting of 13,476 images of chemical apparatus taken in real chemistry laboratories and associated hand-labeled label files.

CLAD can be used to evaluate the performance of object detection algorithms in real laboratory apparatus images. Then, we evaluate a representative set of object detection methods in one-stage: YOLOv3-spp, SSD, YOLOv5 and two-stage: Faster R-CNN and its variants. Experiment results conducted on our CLAD dataset, Faster R-CNN using resnet50-fpn as the backbone for feature extraction has better Map, its mAP can reach 0.990, but its speed is not satisfactory (9 FPS). Compared to Faster R-CNN, YOLOv5 mAP is only 0.012 lower than Faster R-CNN (0.978), but has a better Recall (0.945 vs. 0.833) and the test speed is significantly faster (62 FPS), which seems satisfactory so far.

In future, we will use CLAD to develop better methods for chemical apparatus identification. Our dataset is currently limited in variety and quantity which may reduce the accuracy of our trained models and make them more susceptible to overfitting problems; we will also consider expanding the number of apparatus based on CLAD, as well as collecting more types of chemical apparatus to further enrich the dataset.

Acknowledgements This work was supported in part by the grant of the National Natural Science Foundation of China, No. 61806068, in part by the grant of Anhui Provincial Natural Science Foundation, Nos. 1908085MF184, 1908085QF285, in part by the Key Research Plan of Anhui Province, No. 201904d07020002.

References

1. Walters AUC, Lawrence W, Jalsa NK (2017) Chemical laboratory safety awareness, attitudes and practices of tertiary students. Saf Sci 96:161–171
2. Liu L, Ouyang W, Wang X (2020) Deep learning for generic object detection: a survey. Int J Comput Vision 128(2):261–318
3. Ren S, He K, Girshick R (2015) Faster R-cnn: towards real-time object detection with region proposal networks. Adv Neural Inf Process Syst 28:91–99
4. Simonyan KZ (2014) A very deep convolutional networks for large-scale image recognition. ArXiv Preprint ArXiv:1409.1556
5. Huang Z, Wang J, Fu X (2020) DC-SPP-YOLO: dense connection and spatial pyramid pooling based YOLO for object detection. Inf Sci 522:241–258
6. Liu W, Anguelov D, Erhan D, Szegedy C, Reed S, Fu CY, Berg AC (2016) Ssd: single shot Multibox detector. In: European conference on computer vision. Springer, Cham, pp 21–37
7. Girshick R, Donahue J, Darrell T, Malik J (2014) Rich feature hierarchies for accurate object detection and semantic segmentation. In: Proceedings of the IEEE conference on computer vision and pattern recognition, pp 580–587
8. Girshick R (2015) Fast R-cnn. In: Proceedings of the IEEE international conference on computer vision, pp 1440–1448

9. He K, Gkioxari G, Dollár P, Girshick R (2017) Mask R-cnn. In: Proceedings of the IEEE international conference on computer vision, pp 2961–2969

10. Redmon J, Divvala S, Girshick R (2016) You only look once: unified, real-time object detection. In: Proceedings of the IEEE conference on computer vision and pattern recognition, pp 779–788

11. Lin TY, Goyal P, Girshick R (2017) Focal loss for dense object detection. In: Proceedings of the IEEE international conference on computer vision, pp 2980–2988

12. Wu ZZ, Wan SH, Wang XF, Tan M, Zou L, Li XL, Chen Y (2020) A benchmark data set for aircraft type recognition from remote sensing images. Appl Soft Comput 89:106132

13. Rostianingsih S, Setiawan A, Halim CI (2020) COCO (creating common object in context) dataset for chemistry apparatus. Procedia Comput Sci 171:2445–2452

Power Quality Evaluation Based on the Fusion of Improved Entropy and Analytic Hierarchy Process

Haifeng Zhang, Dongfei Cui, Guanqun Zhuang, Jun Leng, Naiyu Liu, and Xiangdong Meng

Abstract With the increasing access of Distributed Generation (DG), power quality evaluation has become a research hotspot. However, The traditional entropy method has the problem of unreasonable weight distribution in power quality evaluation. In order to solve the above problems, this paper proposes an improved entropy weight method to solve the unreasonable problem that entropy value is close to 1 and the entropy weight changes drastically; Secondly, the fusion of analytic hierarchy process and improved entropy method is introduced for weight calculation; Finally, the simulation experiment shows that compared with the traditional power quality evaluation method, the method in this paper is more suitable for the situation where the power quality index fluctuates greatly after the introduction of DG.

Keywords Power quality · Improved entropy weight method · Analytic hierarchy process · Weight calculation

H. Zhang · G. Zhuang · J. Leng · N. Liu · X. Meng
Electric Power Research Institute, State Grid Jilin Electric Power Co., Ltd, Changchun, China
e-mail: zhanghf1107@163.com

G. Zhuang
e-mail: zhuangguanqun2021@163.com

J. Leng
e-mail: lengjun202111@163.com

N. Liu
e-mail: liuny1107@163.com

X. Meng
e-mail: xiangdongmeng@163.com

D. Cui (✉)
Northeast Electric Power University, Jilin, China
e-mail: nihongyin@163.com

1 Introduction

With the continuous growth of electricity, traditional electric energy is developing in the direction of high efficiency, intelligence, and sustainability. Under this background, DG is introduced into the power energy market, such as photovoltaic energy, wind energy, etc. However, the high penetration rate of DG access to the distribution network may lead to a series of problems such as increased short-circuit current, reduced power supply reliability, etc.

In 2008, the International Power Grid Conference (IPGC2008) was held in Paris, and concluded that the traditional passive distribution network model has not been able to deal with the above problems well. İn this conference, active distribution network technology was proposed for the first time. The so-called active distribution network can control the distribution of power flow through the network topology, and realize the supporting role of DG [1].

The power quality level is affected by many factors, and its operational status is usually described by multiple indicators. A single independent indicator cannot reflect the level of power quality systematically. Therefore, the core content of power quality evaluation research is to merge a multi-index problem into a single index. In the active distribution network, the introduction of DG has the characteristics of complex load changes and strong volatility, which will cause fluctuations in the power flow and bring difficulties to controlling the distribution network. In addition, due to using a large number of power equipment such as grid-connected inverter devices and solid-state switching devices, the power quality and reliability problems in active distribution networks have become increasingly complex and prominent [2–5].

There are many methods for power quality evaluation, including fuzzy mathematics[6–8], artificial neural networks and deep learning methods[9, 10], user demand methods [11], extension cloud Theory [12], analytic hierarchy process [13, 14], entropy weight methods [15], combination methods [16, 17], probability statistics and vector algebra method [18, 19]. Although these methods are effective in power quality assessment, there are still some areas to be improved. The methods based on fuzzy mathematics, user demand methods, and the analytic hierarchy process belong to expert weighting methods. The evaluation results are greatly affected by subjective factors. Experts may have large differences in the discrimination of various indicators. The judgment matrix needs to be tested for consistency. Based on entropy weight method, extension cloud theory, probability statistics, and vector algebra method generally do not need expert subjective weighting, only the actual fluctuation of electric energy index is considered. Reflect on the actual power quality situation. In the probability and statistics method, the selection of the benchmark value is inconsistent, which will lead to a large gap in the results. The methods based on the artificial neural network method, deep learning method, etc. need to collect a large number of samples to evaluate the network model. At the same time, the relevant feature extraction will also be combined with some methods above. For training, if the number of samples is not enough, the final evaluation will also have a large error.

After researching the existing power quality evaluation methods and their scenarios, this paper proposes an improvement strategy for the entropy weight method, which can avoid some of the shortcomings of the traditional entropy weight method. In the power quality evaluation, this paper fully considers the complex factors that affect power quality, combines the improved entropy weight method with the analytic hierarchy process to distribute the weights of the importance of different indicators in DG. Combined with the experimental verification, it is possible to judge the pros and cons of the power quality.

2 Traditional Entropy Method

According to the basic principles of information theory, the amount of information can be measured by information entropy. Entropy has been widely used in engineering technology, social economy, and other fields. Information entropy can judge the degree of dispersion of some given indexes. For example, the smaller of information entropy, the greater of indicator dispersion. Therefore, the information entropy can be used to calculate the indicator weight coefficient, which can provide a theoretical basis for the comprehensive evaluation of multiple indicators.

In power quality evaluation, the entropy weight method is used to construct a model for the comprehensive evaluation of power quality. The basic calculation steps are shown in Fig. 1:

2.1 Indicator Data Matrix and Standardization

There are n available energy indicators to form the original matrix X, which can be represented by the following matrix:

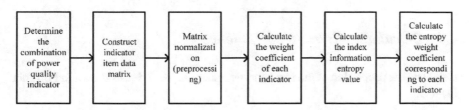

Fig. 1 The basic steps of the traditional entropy method for power quality evaluation

$$X = \begin{pmatrix} X_{11} & \cdots & X_{1m} \\ \vdots & \vdots & \vdots \\ X_{n1} & \cdots & X_{nm} \end{pmatrix}_{n \times m} \tag{1}$$

In (1), X_{ij} is the value of the j-th index of the i-th scheme. For a certain x_j, the greater of X_{ij}, the greater role of the index in the comprehensive evaluation. If all the indicator values are equal, the indicator does not play a role in a comprehensive evaluation.

2.2 Data Non-Negative Processing

After standardizing the indicators, if there are negative numbers in X, the data needs to be non-negative. In addition, to avoid the meaningless situation of the logarithm when calculating the entropy value, it is necessary to perform the matrix translation operation. The specific operation is divided into two cases:

For large indicators, the translation operation is computed according to the following formula:

$$X'_{ij} = \frac{X_{ij} - \min(X_{1j}, \cdots, X_{nj})}{\max(X_{1j}, \cdots, X_{nj}) - \min(X_{1j}, \cdots, X_{nj})} + 1,$$
$$i = 1, 2, \cdots, n; \quad j = 1, 2, \cdots, m \tag{2}$$

For small indicators, the translation operation is computed according to the following formula:

$$X'_{ij} = \frac{\max(X_{1j}, \cdots, X_{nj}) - X_{ij}}{\max(X_{1j}, \cdots, X_{nj}) - \min(X_{1j}, \cdots, X_{nj})} + 1,$$
$$i = 1, 2, \cdots, n; \quad j = 1, 2, \cdots, m \tag{3}$$

2.3 Calculating Entropy Coefficient

Calculate the proportion of the first scheme under the j-th index as shown in Formula (4)–(7):

$$P_{ij} = X_{ij} / \sum_{i=1}^{n} X_{ij} \quad (j = 1, 2, \cdots m) \tag{4}$$

$$e_j = -(\sum_{i=1}^{n} P_{ij} \log(P_{ij}))/ \log(m) \tag{5}$$

$$g_j = 1 - e_j \tag{6}$$

$$W_j = g_j / \sum_{j=1}^{m} g_j \ , j = 1, 2 \cdots m \tag{7}$$

3   Improve the Entropy Method

From the calculation of (6) and (7), it can be seen that when the calculated value of entropy e_j is closer to 1, the difference coefficient g_j tends to 0, and W_j is too sensitive. In [20], the author improved the calculation of the entropy weight coefficient to the above problems, and changed (7) into (8):

$$W_j = (\sum_{j=1}^{m} e_j + 1 - 2e_j)/ \sum_{j=1}^{m}(\sum_{j=1}^{m} e_j + 1 - 2e_j) \ , j = 1, 2 \cdots m \tag{8}$$

After modified in (8), it can indeed alleviate the situation when the entropy value tends to 1, but it cannot give a more reasonable entropy value for other situations. In [21], the author gives an example of (8) calculation unreasonable, when the entropy value vector is equal to (0.99, 0.98, 0.1), the entropy weight corresponding to the entropy value 0.1 should be much larger than 0.99 and entropy weight corresponding to 0.98 entropy value. However, the entropy weight vector obtained by the improved method given by (8) is (0.215, 0.219, 0.566), which does not reflect the weight between entropy values well. For this reason, the literature [21] modified (7) and changed it into (9):

$$W_j = (1 - e_j + \varepsilon)/ \sum_{j=1}^{m}(1 - e_j + \varepsilon) \ , j = 1, 2 \cdots m \tag{9}$$

ε is a very small value. When the value is $\varepsilon = 0$, it degenerates into the traditional entropy weight method, which is finally determined in the literature [21] after comprehensive consideration. At this time, the entropy vector is equal to (0.99, 0.98, 0.1), the final entropy weight is (0.089, 0.098, 0.813). Compared to the literature [20], setting a fixed value can only stretch the overall entropy weight coefficient calculation result, and does not solve the problem fundamentally.

According to the above analysis, it can be seen that when e_j tends to 1, W_j is too sensitive, and when e_j tends to 0, W_j is more reasonable. When e_j tends to 1, the same restriction is added to the situation where e_j tends to 0, which will inevitably affect e_j. Therefore, in order to solve this problem, this paper adds a certain limit to e_j like [21]. To obtain a more reasonable value distribution of entropy weight, the specific calculation formula is as follows:

$$w_j^{(1)} = (1 - e_j + e_j\varepsilon)/\sum_{j=1}^{m}(1 - e_j + e_j\varepsilon) \ , j = 1, 2 \cdots m \qquad (10)$$

4 Evaluation Process

The improved entropy weight method in this paper determines the index weight based on the difference of the index. This is an objective weighting method that avoids the deviation caused by human factors. Because the importance of indicators is ignored, sometimes the determined indicator weights are far from the expected results. In order to better reflect the importance of the indicators themselves, this paper uses the improved entropy weight method while introducing the tomographic analysis method to make subjective expert judgments on various indicators. In the end, the fusion method in this paper takes into account the opinions of experts and the actual fluctuation of power indicators, which can realize the division of the importance of various power quality indicators.

4.1 Analytic Hierarchy Process

Analytic Hierarchy Process can solve complex nonlinear problems through qualitative and quantitative analysis of specific problems. It is a subjective evaluation method that is currently used more frequently. This method can distinguish the importance of different indicators relative to the evaluation purpose, and give the order of each indicator. In addition, the mathematical processing of this method is more rigorous and development, many new ideas are constantly infiltrated. The steps of using the analytic hierarchy process to obtain the index weight coefficients are shown in Fig. 2.

The structure of the hierarchical structure model is to establish the hierarchical structure among the indicators according to a certain subordination relationship, and it is often analyzed according to Table 1.

Under normal circumstances, the more complex the research problem, the more levels established. For the 9-scale method, more information is required, and it is difficult to grasp the importance of indicators. The 3-scale method requires less information, and it is easy to judge the importance of the indicators, but there is a loss of judgment information, cumulative dominance, and consistency [22]. Therefore, this paper refers to Feng et al. [21], using the 5-scale method to divide the index

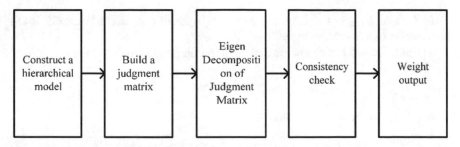

Fig. 2 Calculating index weight coefficient

Table 1 Description of 1–9 scales in analytic hierarchy process

Different scale	Description
1	Two elements are equally important
3	The former element is slightly more important than the latter
5	The former element is obviously more important than the latter
7	The former element is more important than the latter
9	The former element is extremely important than the latter
2,4,6,8	The middle value of the above two adjacent judgments
1/3	The former element is slightly less important than the latter
1/5	The former element is obviously less important than the latter
1/7	The former element is stronger less important than the latter
1/9	The former element is extremely unimportant than the latter
1/2,1/4,1/6,1/8	The middle value of the above two adjacent judgments

hierarchy to form a judgment matrix. Firstly, confirm the importance of each index based on expert opinions, and then construct a judgment matrix A based on the importance of each index.

For judgment matrix A, calculate the maximum eigenvalue and eigenvector using Formula (11):

$$A\varpi = \lambda_{\max}\varpi \tag{11}$$

In (11), $\lambda_{\max}$ is the largest eigenvalue of matrix A, and ϖ is the eigenvector corresponding to $[w_1^{(2)}, w_2^{(2)}, ..., w_m^{(2)}]$.

The consistency test can be carried out according to Formula (12)

$$\theta = (\lambda_{\max} - m)/(m - 1) \quad m \geq 1 \tag{12}$$

In (13), m is the order of matrix A.

4.2   Algorithm Fusion

Two different weights of the same index can be obtained after two power quality evaluation methods are calculated. The fusion adopts a simple linear combination form. The specific formula is as follows and α, β is a linear coefficient.

$$w_j = \alpha w_j^{(1)} + \beta w_j^{(2)} \tag{13}$$

5   Experimental Results

5.1   Improved Entropy Method Experiment

To complete the algorithm experiment, this paper uses the simulated entropy data in the literature [21] for comparison, and the test results using different methods are shown in Table 2.

Through algorithm experiments, when e_j tends to 1, the entropy weight coefficient calculated by the algorithm is very close, which is better than the traditional method and the method in Feng et al. [21]. For the case where e_j tends to 0, the algorithm can also be very well.

Table 2 Comparison between different entropy methods

No	Entropy	Traditional entropy weight method	The method in Feng et al. [21]	Our method
1	(0.999,0.998,0.997)	(0.167,0.333,0.5)	(0.330,0.333,0.337)	(0.243,0.247,0.249)
2	(0.3,0.2,0.1)	(0.292,0.333,0.375)	(0.296,0.333,0.371)	(0.121,0.130,0.143)
3	(0.003,0.002,0.001)	(0.33,0.333,0.337)	(0.333,0.333,0.334)	(0.242,0.247,0.248)
4	(0.9,0.5,0.1)	(0.067,0.333,0.6)	(0.111,0.333,0.555)	(0.104,0.163,0.143)
5	(0.99,0.98,0.1)	(0.011,0.022,0.968)	(0.089,0.098,0.813)	(0.211,0.226,0.143)

5.2 Electric Energy Evaluation Experiment

In order to complete the algorithm experiment, this paper selects five indicators, including voltage flicker, voltage fluctuation, harmonic voltage, and Three electrical unbalance. In order to obtain the above data, two collection points are set up in the two substations of a power supply company, which are recorded as collection point 1 and collection point 2. Each access point collects data for 100 consecutive days of indicators, and the percentage of statistics is shown in Fig. 3:

The power quality evaluation is carried out for the above two access points, and the evaluation results are shown in Table 3:

For collection point 1, since the fluctuation of the electric energy index is not large, the final evaluation result is not much different. For collection point 2, because the voltage fluctuation index changes drastically, the weight coefficient should be

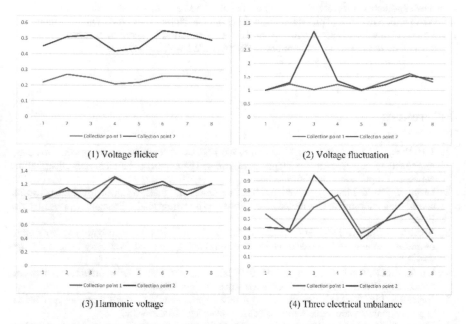

(1) Voltage flicker

(2) Voltage fluctuation

(3) Harmonic voltage

(4) Three electrical unbalance

Fig. 3 Samples of collected data for different indicators in 8 periods

Table 3 Power quality evaluation results of different access points

Collection point	Analytic hierarchy process	Entropy method	Our method
Collection point 1	1.65	1.97	1.75
Collection point 2	2.17	2.58	3.66

increased, but the separate analytic hierarchy process and entropy weight method does not conform to the objective situation.

6 Conclusion

This paper proposes an improved entropy weight method, which focuses on solving the unreasonable problem of the entropy value and the entropy weight changes drastically. the effectiveness of this method is proved by experiments. However, the sampled power data needs to be standardized and dimensionless using different preprocessing methods, which can bring different results. The next research work is the impact analysis of different pre-processing methods.

Acknowledgements Project supported by the Science and Technology Development Plan of State Grid Corporation of China [Title: Research and System Development of Power Quality Monitoring and Power Supply Reliability Evaluation of Active Distribution Network of Jilin Electric Power Research Institute Co. Ltd. Grant Number:JLDKYGSWWFW202106012]

References

1. Ji H, Wang C, Li P, Zhao J, Song G, Ding F, Wu J (2018) A centralized-based method to determine the local voltage control strategies of distributed generator operation in active distribution networks. Appl Energy 228:2024–2036
2. Qiang S, Jun Y, Lianlong W (2015) Reactive power optimization in distribution networks considering intermittent distributed generations and uncertain load. J Shanghai Univer Electric Power 135–139
3. Junjie L, Jiaming D, Shigong J, Yunfei W, Peng L, Xulu F (2021) A distribution network optimization planning method based on differential reliability demand. Power Syst Clean Energy 74–82
4. Lidi Z, Xun D, Wenhao Z, Mingtao H (2019) Islanding partition of active distribution network considering demand response. Electric Measure Instrum 75:63–68
5. Ouyang S, Liu ZW, Li Q, Shi YL (2013) A new improved entropy method and its application in power quality evaluation. Autom Electric Power Syst 164:156–159
6. Yiwang X, Haozhong C, Haiqun W et al (2009) Comprehensive assessment of power quality based on improved AHP and probability statistics. Power Syst Protect Control 48–52:71
7. Pengzhi S, Liu W (2014) The comprehensive assessment of power quality based on attribute recognition theory and AHP algorithm. In: Advanced materials research, vol 1044, pp 3530
8. Kai D, Wei LI, Yuchuan HU, Yuchuan HU, Pan HU, Yimin Q (2019) Power quality comprehensive evaluation for low-voltage DC power distribution system. In: 2019 IEEE 3rd information technology, networking, electronic and automation control conference (ITNEC), pp 15–17
9. Chen-Chen W, Ji M, Jia-Nan Z, Wen-Hui Z (2021) Power quality prediction and warning based on BP neural network optimized by genetic algorithm. In: Electrotechnics electric(Jiangsu Electrical Apparatus), pp 18–22
10. Sahani M, Dash PK (2021) FPGA-based deep convolutional neural network of process adaptive VMD data with online sequential RVFLN for power quality events recognition. In IEEE transactions on power electronics, vol 36, no 4, pp 4006–4015

11. Yuqing H, Jianchun P, Lilin M, Huan Y, Ming W (2010) Comprehensive evaluation of power quality considering customer demands. Autom Electric Power Syst 48–52
12. Ruqi L, Haoyi S (2012) A synthetic power quality evaluation model based on extension cloud theory. Autom Electric Power Syst 36:66–70
13. Meitei NM, Tamang D, Gao S (2021) A new harmonic analysis reporting technique to improve power quality in distribution system network applications. In: 2021 International conference on advances in electrical, computing, communication and sustainable technologies (ICAECT), pp 19–20
14. Yangwu S, Xiaotao P, Shiitongqin, et al (2012) A grey comprehensive evaluation method of power based on Optimal combination view. Autom Electronic Power Syst 67–73
15. Dou J, Ma H, Yang J, Zhang Y, Guo R (2021) An improved power quality evaluation for LED lamp based on G1-entropy method. İn IEEE Access, vol 9, pp 111171–111180
16. Rui W, Jie F, Ke Z et al (2007) Fuzzy sysynthetic evaluation of power quality based on entropy and AHP. Elector Measurem Instrum 44:21–25
17. Nan L, Zhengyou H (2009) Power quality comprehensive evaluation combining subjective weight with hobjective weight. Power Syst Technol 33:55–61
18. Yong W, Xinglei Y, Hongbing W, et al (2017) Comprehensive assessment of power quality of wind farms based on probability distribution. In: Power capacitors and reactive power compensation, pp 171–176
19. Wen G, Huixia W, Cuilan Y (2009) Comprehensive assessment of power quality based on probability and vector algebra. Shanxi Electric Power 65–67
20. Zhanan Z, Xingguo C (2014) Using entropy theory to determine pumped storage capacity. J Electric Machinery Control 34–39
21. Feng L, Bo S, Xuan W, Wenbao L, Haifeng J, Delong Z (2019) Power quality assessment for rural rooftop photovoltaic access system based on analytic hierarchy process and entropy weight method. Trans Chinese Soc Agric Eng 159–166
22. Zeshui X (1998) New analytic hierarchy process (AHP). Syst Eng Theo Practice. 75–78(1998)

Data and Big Data Processing and Analytics

Fairness Budget Distribution for Influence Maximization in Online Social Networks

Bich-Ngan T. Nguyen, Phuong N. H. Pham, Loi H. Tran, Canh V. Pham, and Václav Snášel

Abstract Influence maximization (IM) is an important problem in social influence, viral marketing, and economics. This paper studies a fairness constraint in the Influence Maximization problem, a general IM version that aims to find a k-size seed set distributed in target communities. Each has certain upper and lower bounds so that the influence spread is maximal. However, solving this problem faces two main challenges: it is an NP-hard problem, and the seed sets with strong influence may not satisfy the fairness constraint. To address this problem, we propose the Fairness Budget Influence Maximization algorithm. This algorithm combines an improved greedy strategy with generating sampling and a stop-and-stare technique. We conducted experiments on real social networks. The result shows that our proposed algorithm equalizes or outperforms the state-of-the-art algorithm in terms of the objective value, the running time, the memory usage, and especially the target communities coverage ratio of the seed set. It is almost greater than the compared algorithm and guarantees fairness constraint.

Keywords Fairness submodular maximization · Influence maximization · Approximation algorithm · Sampling · Viral marketing

B.-N. T. Nguyen · P. N. H. Pham · L. H. Tran
Faculty of Information Technology, Ho Chi Minh city University of Food Industry,
Ho Chi Minh, Vietnam
e-mail: nganntb@hufi.edu.vn; thi.bich.ngan.nguyen.st@vsb.cz

P. N. H. Pham
e-mail: phuongpnh@hufi.edu.vn; nguyen.huy.phuong.pham.st@vsb.cz

C. V. Pham (✉)
ORLab, Faculty of Computer Science, Phenikaa University, Hanoi 12116, Vietnam
e-mail: canh.phamvan@phenikaa-uni.edu.vn

B.-N. T. Nguyen · P. N. H. Pham · V. Snášel
Faculty of Electrical Engineering and Computer Science, VŠB-Technical University of Ostrava,
Ostrava, Czech Republic
e-mail: vaclav.snasel@vsb.cz

© The Author(s), under exclusive license to Springer Nature Switzerland AG 2022
N. H. T. Dang et al. (eds.), *Artificial Intelligence in Data and Big Data Processing*,
Lecture Notes on Data Engineering and Communications Technologies 124,
https://doi.org/10.1007/978-3-030-97610-1_19

225

1 Introduction

In the growing internet era, information diffusion through online social networks (OSNs) is a fascinating research theme that attracts many researchers because of its economic merits. OSNs have capable of influencing sharing behavior, spreading content and messages from one person to another quickly [11]. Therefore, organizations and brands have utilized OSNs to promote products, spread information or opinions, campaign for elections, etc. It is also known as viral marketing [12]. This process is to be successful. It needs to seed content with highly influential individuals in social networks to spread the influence to the largest number of people. Finding the set of such individuals is known as the *Influence Maximization* (IM) problem. The IM was first proposed by Kempe et al. [14] almost two decades ago.

IM is a well-known problem, attracting more and more attention of researchers over the years [4, 6, 21, 25]. There are some methods for IM evaluating the influence spread, such as the discrete optimization problem formulation [14]. But they were showed that it is NP-hard, using a greedy algorithm that adds the members with maximal marginal gain [4]. Another new approach—Reverse Influence Sampling (RIS) that was proposed by Borgs et al. [6], which captures the influence landscape of G through generating a set R of random Reverse Reachable sets [21].

However, this problem has many variations and constraints that must face in practice. Specifically, in a social network with many communities, to increase advertising effectiveness and competition, brands want to spread their products throughout the communities they target with minimal budget and maximally effective. This issue has some recent publications [1, 9, 16, 31], often called *Fairness influence maximization* (FIM). It asks to find a set of seeds distributed in each community to ensure coverage propagation in the target communities. To the best of our knowledge, existing literature do not consider both maximum and minimum budget constraints for each community to ensure equitable distribution.

Motivated by this phenomenon, we study the FIM by setting upper and lower bounds for choosing seeds in communities to guarantee fairness budget constraint. Specifically, our contributions are as follows. *Assume that $G = (V, E)$ is a social network under a diffusion model, k is a total budget, $C' = \{C_1, C_2, \ldots, C_K\}$ is a set of target communities. Each community C_i has a budget pair of lower bounded k_l^i and upper bounded k_u^i, $k_l^i \leq k_u^i$. The problem demands to find a seed set S, $|S| \leq k$ satisfying $k_l^i \leq |S \cap C_i| \leq k_u^i$ so that the number of influenced users by S is maximal.*

- We propose a Fairness Budget Influence Maximization (FBIM) algorithm to solve the above problem. This method inherits the DSSA algorithm of Nguyen et al. [21], one of the state-of-the-art methods, and combines fairness constrain. Our algorithm has a complexity less than or equal to the complexity of DSSA. In particular, our method resolves the fairness constraint while DSSA does not.
- We further investigate the performance of our algorithm by conducting some experiments on real social networks. The results indicate that FBIM outputs a result seed set S whose coverage ratio over communities is greater than the result of DSSA even our results cover almost 100% of the target communities. This result depends

on the right parameter setting for the dataset. A high coverage ratio means the number of selected seeds covering the target communities. Thus, it ensures that the affected individuals are the ones we want to affect. Besides, FBIM performs similarly to DSSA in terms of runtime, memory usage, and the expected number of influenced nodes.

Organization. We organize the next parts of this paper as follows. In Sect. 2, we discuss previous related works. The definitions and description of our problem are presented in Sect. 3. Our proposed algorithms and theoretical analysis are detailed in Sect. 4. We show and discuss experiment results in Sect. 5, and in Sect. 6, we conclude the paper.

2 Related Work

According to previous studies [2, 14, 29], the IM problem is solved by a greedy strategy, the result obtains a $(1 - 1/e)$-approximation. Although the greedy strategy is very effective for the IM, it is still complicated to compute the influence function $f(S)$. Because it is #P-hard [4]. Existing methods of the IM can be categorized into main groups based on the way that computes the influence function as follows:

(1) *The simulation-based methods* such as CELF [18], CELF++[5, 7], which use Monte-Carlo sampling to compute the influence function. They combined with the heuristic approaches based on the greedy algorithm to obtain highly scalable algorithms for influence maximization. These methods aim to generate an $(1 - 1/e)$ approximate solution.

(2) *The proxy-based methods* such as SimPath [8], Degree, PageRank [10, 19, 23]. The main idea of these approach is to devise proxy models to approximate influence function $f(S)$ for conquering the #P-hard. The obtained approximate solution is $(1 - 1/e - \epsilon)$ for any $\epsilon > 0$.

(3) *The sketched-based methods*, such as TIM, TIM+ [28], IMM [27]. The authors used a novel RIS sampling method that was introduced by Borgs *et al.* in [6]. These methods aim to generate an $(1 - 1/e - \epsilon)$ approximate solution with minimal numbers of RIS samples. However, the shortcoming of these methods is the number of sampling generated can be quite large, and their lower bound is unknown.

Subsequently, Nguyen et al. [21] introduced two novel sampling algorithms SSA and D-SSA, which aim towards achieving minimum number of RIS samples and guarantee to achieve $(1 - 1/e - \epsilon)$-approximations . Nevertheless, Huang et al. showed SSA/D-SSA contains some errors in [13]. They also present a revised version of SSA, referred to as SSA-Fix. Meanwhile, Nguyen et al. [20] also give a importantly modified version of D-SSA, which named D-SSA-Fix, that provides $(1 - 1/e - \epsilon)$-approximations. In other directions, IM problem has investigated under several constraints such as, budget constraint [22], time constraint [3, 24, 26], competitive constraint [24], etc.

Unfortunately, almost all the above approaches focus on finding the most influential nodes to maximize the number of influenced nodes. They ignore to ensure fairness in the distribution of selecting elements over partitions of the dataset. Recent studies have taken an interest in and incorporated the fairness factor into the IM problem. A current line of research focuses on fairness in methods developing for human data, with a particular emphasis on classification problems. For example, it applies fairness in machine learning to develop several methods for measuring and defining fairness [1], to pre-process biased data [9], to range from identifying statistical constraints on the output that based on the particularities of input dataset [15], or to determine the causal relationships of features in data that lead to bias [16]. Measurement approaches have also been developed to understand bias in large amounts of data [31]. To the best of our knowledge, the above studies' outcomes do not satisfy the fairness constraint on the upper and lower bounds of the partitions in the data.

3 Preliminaries

In this section, we first present the Linear Threshold (LT) model [14], a well-known information diffusion model for investigating social influence problems [14, 22, 27, 28]. Based on that, we then formally define the Fairness Budget Influence Maximization under LT model.

Definition 1 *(LT model). Given a directed graph $G = (V, E, w)$, V is the set of nodes and $|V| = n$, E is the set of edges and $|E| = m$. At the start, the seed set S nodes are active, while all remaining nodes are inactive. The process spreads according to the following rule. A node $v \in V$ is influenced by each neighbor u with a probability $p(u, v)$, $p(u, v) \in w$ such that $\sum_{u \, neighbor \, of \, v} p(u, v) \leq 1$. Each node v is assigned a threshold Λ_v, which uniformly random from the interval $[0; 1]$, this represents the weighted fraction of v's neighbors that must become active in order for v to turn active. At step t, all nodes that were active in step t-1 keep active, and the process activates any node v for which the total probabilities of its active neighbors is at least Λ_v:*

$$\sum_{u \, neighbor \, of \, v} p(u, v) \geq \Lambda_v \qquad (1)$$

Based on maximizing a submodular function $f(\cdot)$ under the fairness and cardinality constraints problem, the calculation of $f(\cdot)$ for the Fairness Budget Influence Maximization is in this way.

Denoting $\mathcal{R}$ as the set of random RR sets. In the study of Nguyen et al. [21], finding a seed set S and influence spread $f(S)$ is based on computing the covering most of the RR sets of S. Because, generating a set $\mathcal{R}$ of multiple random RR sets, influential nodes will possibly appear frequently in the RR sets. We find the nodes that appear in the most RR sets to add the seed set S. Therefore, a seed set S cover a RR set R_i if $S \cap R_i \neq \emptyset$. For simplicity, we denote the coverage of S on $\mathcal{R}$ as $\mathsf{Cov}_{\mathcal{R}}(S)$, and its formula is

$$\text{Cov}_\mathcal{R}(S) = \sum_{R_i \in \mathcal{R}} min\{1, |S \cap R_i|\} \tag{2}$$

Besides, the influence spread $f(S)$ on a random RR set R_i is proportional to the probability that S intersects with R_i. According to recent work [6], we can calculate influence spread function $f(\cdot)$ according to the following formula

$$f(S) = n\mathbb{E}[\text{Cov}_\mathcal{R}(S)] \cdot n/|\mathcal{R}| \tag{3}$$

Based on the above mentioned, we formally define the FBIM problem as follows.

Definition 2 *(Fairness Budget Influence Maximization—*FBIM*). Given a social network $G = (V, E, w)$, with the set of nodes V and the set of edges E, $|V| = n$, $|E| = m$ under the LT model. Given a set of groups $C' = \{C_1, C_2, \ldots, C_K\}$ where $C_i \subseteq V$, and $C_i \bigcap_{i \neq j} C_j = \emptyset$. Each group C_i has a lower bounded and upper bounded budgets k_l^i and k_u^i, $k_l \leq k_u$. Given a total budget k, the problem find seed set S, $|S| \leq k$ satisfying $k_l^i \leq |S \cap C_i| \leq k_u^i$ so that the influence spread function $f(S)$ is maximal.*

In selecting process for each node s to add in the seed set S, we need to check the feasibility of adding s to S so that S is an acceptable solution, called an *extendable set S*. Mathematically, the definition of an extendable set is as follows.

Definition 3 *(An extendable set S). A set $S \subseteq V$ is extendable iff*

$$|S \cap C_i| \leq k_u^i, \quad i = 1, \ldots, K \sum_{i=1}^{K} max(|S \cap C_i|, k_l^i) \leq k \tag{4}$$

4 Proposed Algorithms

In this section, we introduce our proposed algorithms, which combine an improved greedy strategy with generating sampling and the stop-and-stare technique that inherit the method of Nguyen et al. [21] for the FBIM problem. At the high level, the main idea of the algorithm is that *(1) we select k nodes, which appear the most in communities, to add the seed set S so that the coverage on the set of RR sets of S is maximum, and (2) we compute the influence spread of the set S by the stop-and-stare technique. If the result is still not good, we repeat the search for S on the set of RR sets, in which, the number of elements has doubled.*

4.1 Algorithm Description

In Algorithm 2, it receives a graph G, communities set C' in G, budget k, and parameters ϵ, δ. It first generates a set $\mathcal{R}$ that has Γ elements, where Γ is computed as formula in Line 1. Next, it finds the seed set S and compute influence spread $f(S)$ base on $\mathcal{R}$ by Algorithm 1. Generate a set of RR sets (called $\mathcal{R}'$) for the second time to evaluate the influence spread of set S by the stop-and-stare technique. If the result matches the condition in Line 13 then stop the algorithm, otherwise, repeat the process of finding $\langle S, f(S) \rangle$ with $\mathcal{R}$ doubled until $|\mathcal{R}|$ reach the threshold as Line 17. The detail of algorithms is fully presented in Algorithm 1 and Algorithm2.

Algorithm 1: Fairness-Max-Coverage procedure

Input: Graph $G, \mathcal{C}', k$, RR sets ($\mathcal{R}$).
Output: An seed set S that satisfy fairness constraint, $|S| \leq k$ and its estimated influence
$\quad\quad f(S)$.
1. $S \leftarrow \emptyset$
2. **for** $i = 1$ **to** k **do**
3. $\quad$ $\mathbb{U} \leftarrow \{s \in (C_{j_{C_j \in \mathcal{C}'}} \setminus S) | S + s$ is extendable $\}$
4. $\quad$ $v \leftarrow \arg\max_{\{v \in \mathbb{U}\}}(\mathsf{Cov}_{\mathcal{R}}(S \cup v) - \mathsf{Cov}_{\mathcal{R}}(S))$
5. $\quad$ $S \leftarrow v$
6. **return** $\langle S, n\mathbb{E}[\mathsf{Cov}_{\mathcal{R}}(S)].n/|\mathcal{R}| \rangle$.

Algorithm 2: Fairness Budget Influence Maximization - FBIM

Input: Graph $G, \mathcal{C}', k$, and $0 \leq \epsilon, \delta \leq 1$
Output: A seed set $S, |S| \leq k$.
1. $\Gamma \leftarrow \sqrt{8}(1+\epsilon)^2 \ln \frac{2}{\delta} \frac{1}{\epsilon^2}$
2. $\mathcal{R} \leftarrow$ Generate Γ random RR sets by RIS
3. $\langle S, f(S) \rangle \leftarrow$ **Fairness-Max-Coverage**$(G, \mathcal{C}', k, \mathcal{R})$
4. **repeat**
5. $\quad$ $\mathcal{R}' \leftarrow$ Generate $|\mathcal{R}|$ random RR sets by RIS
6. $\quad$ $f'(S) \leftarrow \mathsf{Cov}_{\mathcal{R}'}(S).n/|\mathcal{R}'|$
7. $\quad$ $\epsilon_1 \leftarrow f(S)/f'(S) - 1$
8. $\quad$ **if** $\epsilon_1 \leq \epsilon$ **then**
9. $\quad\quad$ $\epsilon_2 \leftarrow \frac{\epsilon - \epsilon_1}{2(1+\epsilon_1)}$
10. $\quad\quad$ $\epsilon_3 \leftarrow \frac{\epsilon - \epsilon_1}{2(1-1/e)}$
11. $\quad\quad$ $\delta_1 \leftarrow 1/e^{\frac{\mathsf{Cov}_{\mathcal{R}}(S).\epsilon_3^2}{2c(1+\epsilon_1)(1+\epsilon_2)}}$
12. $\quad\quad$ $\delta_2 \leftarrow 1/e^{\frac{(\mathsf{Cov}_{\mathcal{R}'}(S)-1).\epsilon_2^2}{2c(1+\epsilon_2)}}$
13. $\quad\quad$ **if** $\delta_1 + \delta_2 \leq \delta$ **then**
14. $\quad\quad\quad$ **return** S.
15. $\quad$ $\mathcal{R} \leftarrow \mathcal{R} \cup \mathcal{R}'$
16. $\quad$ $\langle S, f(S) \rangle \leftarrow$ **Fairness-Max-Coverage**$(G, \mathcal{C}', k, \mathcal{R})$
17. **until** $|\mathcal{R}| \geq (8 + 2\epsilon)n.\frac{\ln\frac{2}{\delta} + \ln\binom{n}{k}}{\epsilon^2}$;
18. **return** S.

4.2 Theoretical Analysis

Because we solve the FBIM problem based on the DSSA method, we do not perform the proofs given in [21]. We only analyze the complexity of the algorithms, stated in the following Theorem 2 and Theorem 1.

Theorem 1 *Algorithm 1 is an improved greedy algorithm, has $O(k.T)$ complexity with $T = \left| \bigcup_{C_i \in C'} (C_i) \right|$ and $T \leq n$.*

Proof Algorithm 1 loops k times to choose seeds for the size-k seed set S. Each loop scans at most all elements of C_i ($\forall C_i \in C'$) to choose an element that has the maximal coverage on the $\mathcal{R}$. Therefore, the algorithm takes $O(k.T)$ complexity. $\square$

Theorem 2 *Algorithm 2 has $O\left(k.T.log\left((8+2\epsilon)n.\frac{ln\frac{2}{\delta}+ln\binom{n}{k}}{\epsilon^2}\right)\right)$ complexity.*

Proof Algorithm 2 loops generating the $\mathcal{R}$ set of random RR sets with randomly selected source elements from C_i and finding $\langle S, f(S) \rangle$ bases on $\mathcal{R}$ through Algorithm 1. Algorithm 2 has two conditions to stop the loop at Lines 13 and 17. Because in each loop, $|\mathcal{R}|$ doubled and call Algorithm 1 to find a new S, in the worst case, this algorithm stops when it meets the condition line 17, meaning the maximum number of iterations of the algorithm is $log\left((8+2\epsilon)n.\frac{ln\frac{2}{\delta}+ln\binom{n}{k}}{\epsilon^2}\right)$. Thus, this algorithm takes $O\left(k.T.log\left((8+2\epsilon)n.\frac{ln\frac{2}{\delta}+ln\binom{n}{k}}{\epsilon^2}\right)\right)$

5 Experiment

5.1 Experiment Setting

Datasets. For a comprehensive experiment, we choose two datasets, including the Epinions social network (**Epinions**)[17] and the Orkut social network and ground-truth communities (**Orkut**)[30], which apply commonly in finding the seed set with Influence Maximization. The description of used datasets is presented in Table 1.
Environment. We conducted the experiments on a Linux machine with 2 Intel(R) Xeon E5-2630 v4 @ 2.20 GHz CPUs and 64GB of DDR4 @2400 MHz RAM. Our implementation is written in C/C++ language and compiled with g++ 11.
Parameter Setting We conducted experiments with two sets of parameter setting.

Table 1 Statistics of datasets.

Dataset	#Nodes	#Edges	#Com	Avg. degree	Type
Epinions	131828	841372	6359	13.4	Directed
Orkut	3072441	117185083	4745	78	Undirected

1. We set $|C'| = K$ with $K \in [50, 1000]$ (it), $k \in [1000; 20000]$, the pair $(k_l^i; k_u^i)$
   of C_i are (0.3; 0.5), called FBIM-1 and (0.1; 0.9), called FBIM-2. These values
   of K are quite small compared to the number of communities currently in the
   dataset.
2. We used the same k as the first setting but we did not set a specific K, the upper
   bound was fixed at 0.5 while the lower bound would be 0.1, 0.01, and 0.001 for
   FBIM-3, FBIM-4, and FBIM-5 respectively. We will explain these changes in
   the experiment results section.

All experiments are under the Linear Threshold model [14], which is a well-known
diffusion model that commonly used for IM problem. We set parameters $\epsilon = 0.1$,
$\delta = 1/n$ as the default setting.

5.2 Algorithms Compared

To our knowledge, there is currently no algorithm in resolving the fairness influence
maximization under constraints with a budget and lower/upper bounds for groups.
Therefore we experimented with two algorithms FBIM and DSSA over different
sets, which have input parameters to analyze and evaluate the effectiveness of these
algorithms. As mentioned above, both algorithms obtain S and almost $|S| \simeq k$ such
that $f(S)$ reaches the maximum. But the FBIM algorithm finds S satisfying the
fairness constraint, while DSSA does not. Thus, we compare the factors such as
influence $f(S)$, running time, memory usage, and coverage of S across the target
communities. The coverage of S across the target communities is the number of
communities, which have the number of elements selected into S to satisfy lower
and upper bounds constraints.

5.3 Experiment Results

We discuss the results of experiments to clarify the strengths and weaknesses of the
algorithms in this section. The results are clearly showed in Figs. 1 and 2.

For the first parameter setting, we want to show how FBIM perform in a general
setting, which include a large range $[k_l^i, k_u^i]$ and a small K. The results in Fig. 1 shows
that FBIM almost runs faster than DSSA, even in some cases, FBIM is 2 to 4 times
faster than DSSA. However, the influence $f(S)$ of FBIM is less than $f(S)$ of DSSA
because of the small value of K. When K is small, we have fewer communities to
cover. It leads to faster computation while still guaranteeing the fairness constrain.
Nevertheless, this is a double-edged sword. Because the K communities are randomly chosen, we only have a smaller range to choose the seeds from, although
we take less time to ensure fairness. Thus, it does not guarantee that the algorithm
chooses nodes with the strongest influence according to the greedy strategy. More-

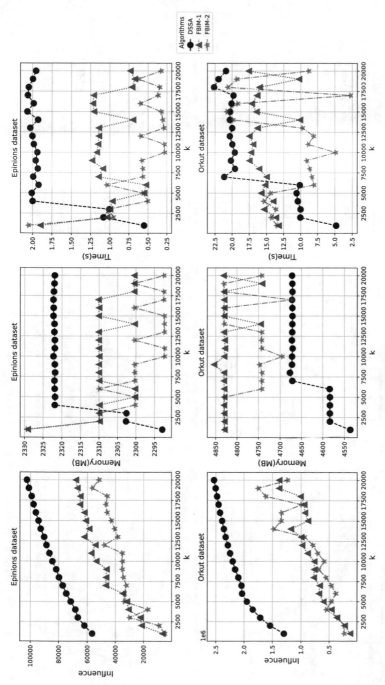

Fig. 1 Influence, memory usage and running time of DSSA, FBIM-1 and FBIM-2

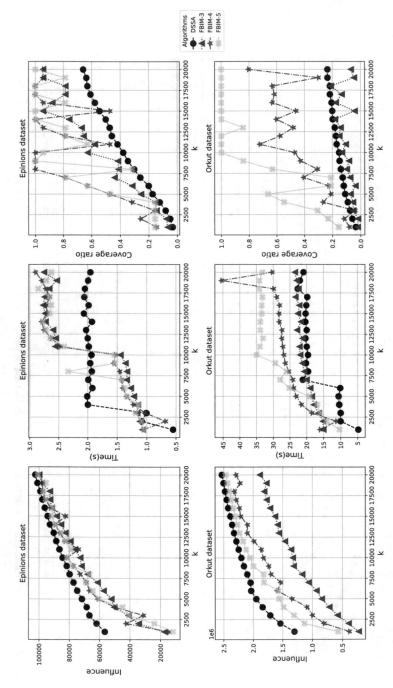

Fig. 2 Influence, running time and communities coverage ratio of DSSA, FBIM-3, FBIM-4 and FBIM-5

over, FBIM also takes more memory usage, which is mostly when the dataset contains large communities (as Orkut). It is understandable because FBIM must take an additional step to store and process communities' information. In summary, in this case, although FBIM obtains S with an influence spread $f(S)$ 2 to 3 times smaller than $f(S)$ of DSSA, the running time of FBIM is faster, and one important thing is it solves fairness constraint.

For the second parameter setting, we want to show the value of K to select a k-size seed set S, which can cover most of the target communities. Therefore, we do not set a specific value for K. In this case, Fig. 2 shows that the lower bound impacted the results of FBIM strongly. We can see in Fig. 2, the influence of FBIM-4 and FBIM-5 increase significantly compared to FBIM-3 (FBIM-3 has the same lower bound setting as FBIM-2), even in some cases, they were equal to the influence of DSSA. The communities coverage ratio also changed drastically in FBIM-4 and FBIM-5. At times, it even reached 1. On the other hand, the set S covers the desired communities, so the influenced individuals are the target ones we want to influence. In summary, we can control the quality of the seed set of FBIM by holding the lower bound k_l^i and k. But the drawback is the long-running time. The cause is to find S, DSSA only needs to see the first k seeds that satisfy its requirements, while FBIM must find k seeds such that (1) satisfy the exact requirements as DSSA's and (2) satisfy the requirements so that S is extendable. The more communities for S to cover, the more extensive the list of nodes that needs computing to ensure fairness.

6 Conclusion

This paper proposes the FBIM algorithm for the IM problem under fairness constraints. We compare our algorithm to the state-of-the-art algorithm (DSSA) by conducting experiments. The results indicate that our algorithms are highly scalable and outperform the adaptations. In the future work, we plan to improve the FBIM algorithm to obtain a shorter running time and a more significant influence spread in more extensive and more diverse datasets.

Acknowledgements This work was supported by Ho Chi Minh City University of Food Industry (HUFI), Vietnam. This work was partially supported by National research project under Grant No. D TD L.CN.46/20-C

References

1. Barocas S, Hardt M, Narayanan A (2019) Fairness and machine learning. fairmlbook.org. http://www.fairmlbook.org
2. Chekuri C, Vondrák J, Zenklusen R (2010) Dependent randomized rounding via exchange properties of combinatorial structures. In: FOCS 2010, USA. IEEE Computer Society, pp 575–584

3. Chen W, Lu W, Zhang N (2012) Time-critical influence maximization in social networks with time-delayed diffusion process. In: AAAI 2012
4. Chen W, Wang C, Wang Y (2010) Scalable influence maximization for prevalent viral marketing in large-scale social networks. In: ACM SIGKDD 2010. USA, pp 1029–1038
5. Chen W, Wang Y, Yang S (2009) Efficient influence maximization in social networks. In: SIGKDD 2019. KDD '09, New York, pp 199–208
6. Borgs C, Brautbar M, J.T.C.B.L (2014) Maximizing social influence in nearly optimal time. In: SODA 2014. SIAM, Portland, Oregon, USA, pp 946–957
7. Goyal A, Lu W, Lakshmanan LV (2011) Celf++: Optimizing the greedy algorithm for influence maximization in social networks. In: WWW 2011. New York, NY, pp 47–48
8. Goyal A, Lu W, Lakshmanan LV (2011) Simpath: an efficient algorithm for influence maximization under the linear threshold model. In: 2011 IEEE 11th international conference on data mining, pp 211–220
9. Hardt M, Price E, Srebro N (2016) Equality of opportunity in supervised learning. In: Advances in neural information processing systems, vol 29, Spain, pp 3315–3323
10. He X, Kempe D (2014) Stability of influence maximization. In: SIGKDD 2014. KDD '14, Association for Computing Machinery, New York, NY, pp 1256–1265
11. Heidemann J, Klier M, Probst F (2012) Online social networks: a survey of a global phenomenon. Comput Netw 56(18):3866–3878 (the WEB we live in)
12. Ho JY, Dempsey M (2010) Viral marketing: motivations to forward online content. J Bus Res 63(9):1000–1006 (Advances in Internet Consumer Behavior Marketing Strategy)
13. Huang K, Wang S, Bevilacqua G, Xiao X, Lakshmanan LVS (2017) Revisiting the stop-and-stare algorithms for influence maximization. Proc VLDB Endow 10(9):913–924
14. Kempe D, Kleinberg JM, Tardos É (2003) Maximizing the spread of influence through a social network. In: SIGKDD 2003, pp 137–146
15. Kleinberg J, Mullainathan S, Raghavan M (2017) Inherent trade-offs in the fair determination of risk scores. In: ITCS 2017. LIPIcs, vol 67. Schloss Dagstuhl–Leibniz-Zentrum fuer Informatik, pp 43:1–43:23
16. Kusner M, Loftus J, Russell C, Silva R (2017) Counterfactual fairness. In: NIPS'17. Curran Associates Inc., pp 4069–4079
17. Leskovec J, Huttenlocher D, Kleinberg J (2010) Signed networks in social media. In: SIGCHI 2010. CHI '10, Association for Computing Machinery, New York, NY, pp 1361–1370
18. Leskovec J, Krause A, Guestrin C, Faloutsos C, VanBriesen J, Glance N (2007) Cost-effective outbreak detection in networks. In: SIGKDD 2007. New York, NY, pp 420–429
19. Liu Q, Xiang B, Chen E, Xiong H, Tang F, Yu JX (2014) Influence maximization over large-scale social networks: a bounded linear approach. In: CIKM 2014. CIKM '14, New York, NY, pp 171–180
20. Nguyen HT, Dinh TN, Thai MT, Sen A, Li WW, Thai MT (2018) Revisiting of revisiting the stop-and-stare algorithms for influence maximization. In: Computational data and social networks. Springer International Publishing, Cham, pp 273–285
21. Nguyen HT, Thai MT, Dinh TN (2016) Stop-and-stare: optimal sampling algorithms for viral marketing in billion-scale networks. In: SIGMOD 2016. ACM, pp 695–710
22. Nguyen HT, Thai MT, Dinh TN (2017) A billion-scale approximation algorithm for maximizing benefit in viral marketing. IEEE/ACM Trans Netw 25(4):2419–2429
23. Page L, Brin S, Motwani R, Winograd T (1999) The pagerank citation ranking: Bringing order to the web. Technical Report 1999-66, Stanford InfoLab, sIDL-WP-1999-0120
24. Pham CV, Duong HV, Bui BQ, Thai MT (2018) Budgeted competitive influence maximization on online social networks. In: CSoNet 2018. LNCS, vol 11280. Springer, Heidelberg, pp 13–24
25. Pham CV, Pham DV, Bui BQ, Nguyen AV (2021) Minimum budget for misinformation detection in online social networks with provable guarantees. In: Optimization letters
26. Pham CV, Thai MT, Ha DK, Ngo DQ, Hoang HX (2016) Time-critical viral marketing strategy with the competition on online social networks. In: (CSoNet 2016). LNCS, vol 9795. Springer, Heidelberg, pp 111–122

27. Tang Y, Shi Y, Xiao X (2015) Influence maximization in near-linear time: a martingale approach. In: Sellis TK, Davidson SB, Ives ZG (eds) SIGMOD 2015. ACM, pp 1539–1554
28. Tang Y, Xiao X, Shi Y (2014) Influence maximization: near-optimal time complexity meets practical efficiency. In: SIGMOD 2014. ACM, pp 75–86
29. Udwani R (2018) Multi-objective maximization of monotone submodular functions with cardinality constraint. In: NIPS'18. Curran Associates Inc., pp 9513–9524
30. Yang J, Leskovec J (2012) Defining and evaluating network communities based on ground-truth. In: 12th IEEE International Conference on Data Mining, ICDM 2012. IEEE Computer Society, pp 745–754
31. Yang K, Stoyanovich J (2017) Measuring fairness in ranked outputs. In: SSDBM '17. Association for Computing Machinery

Deep Learning-Based Pneumonia Detection Using Big Data Technology

Anh-Cang Phan, Ho-Dat Tran, Thanh-Ngoan Trieu, and Thuong-Cang Phan

Abstract The number of people suffering from lung-related diseases accounts for a high percentage. Patients with lung diseases are of all ages and genders, especially children and the elderly with low resistance. To accurately diagnose lung diseases, doctors often diagnosis based on chest X-ray images. Chest X-ray is one of the most widely used techniques in medicine, helping doctors to assess and detect lung lesions. However, it is difficult to accurately and quickly diagnose lung lesions with a large number of patients taking the time of doctors in reading test results. Therefore, in this paper, we propose a novel method to accurately detect lung lesions by constructing deep learning networks in Spark parallel and distributed computing environments. Experimental results show that the proposed method achieves an accuracy of 96% with the training time reduced by 57% compared to training in a stand-alone computing environment. This supports doctors in quickly making a preliminary diagnosis of lung lesions for timely treatment.

Keywords Deep learning · Pneumonia detection · Lung lesions · VGG-16 · Resnet-50

A.-C. Phan (✉) · H.-D. Tran
Vinh Long University of Technology Education, Vinh Long, Vietnam
e-mail: cangpa@vlute.edu.vn

H.-D. Tran
e-mail: datth@vlute.edu.vn

T.-N. Trieu
Université de Bretagne Occidentale, Brest, France
e-mail: ngoan.trieuthanh@etudiant.univ-brest.fr

T.-N. Trieu · T.-C. Phan
Can Tho University, Campus II, 3/2 street, Ninh Kieu district, Can Tho, Vietnam
e-mail: ptcang@cit.ctu.edu.vn

© The Author(s), under exclusive license to Springer Nature Switzerland AG 2022   239
N. H. T. Dang et al. (eds.), *Artificial Intelligence in Data and Big Data Processing*,
Lecture Notes on Data Engineering and Communications Technologies 124,
https://doi.org/10.1007/978-3-030-97610-1_20

1 Introduction

The lung is the most important respiratory organ in the human body. This organ is contact with the external environment thus it is susceptible to many diseases such as pneumonia, bronchitis, tuberculosis, or lung cancer. When lung damage occurs, the lung cannot supply enough oxygen to the body, leading to the patient's whole body damage, even death. According to current statistics, there are an estimated 4–5 million people in the United States having pneumonia each year with about 55,000 deaths [7]. In the US, pneumonia, along with influenza, is the leading infectious cause of death. Pneumonia causes a number of symptoms such as chest tightness, shortness of breath, and fatigue that have long-term effects on the patient's health. Therefore, timely and accurate diagnosis helps patients quickly receive treatment, avoiding adverse effects on their lives. One of the ways to detect this disease is a diagnosis based on chest X-ray images. However, the number of people infected is increasing, while the capacity of hospitals is not commensurate. Covid-19 is a respiratory disease, one that especially reaches into the human respiratory tract, which includes the lung. In September 2021, Covid-19 has became a global pandemic, with an estimated number of cases in the US surpassing 40 million people [6]. The application of technology on lung damage detection will solve two problems, reducing the doctor's workload and increasing the capacity of medical examination and treatment.

Recently, there are a lot of related studies dealing with the problem of lung injury diagnosis. Chen et al. [2] used the UNet++ on CT images to diagnose pneumonia with an accuracy of 95%. The study saved about 65% time of doctors in reading test results. Elshennawy and Ibrahim [3] used deep learning techniques in the detection and classification of pneumonia. The authors experimented on 4 models of ResNet152V2, MobileNetV2, CNN, and LSTM with an accuracy of 91%. Gabruseva et al. [4] proposed a method to identify the opacities of the lung based on RetinaNet with Se-ResNext101 network pre-trained on the ImageNet dataset. The proposed method achieved high results in the trials of the Radiological Society of North America. Zeiser et al. [10] proposed a method using CNN to identify pneumonia on digital chest X-ray images. The research results showed that the VGG16 architecture achieved outstanding performance with an accuracy of 85.11% on the dataset of 5184 images. Abbas et al. [1] extracted features using CNN network and used PCA algorithm to reduce the dimension of X-ray images. The study achieved an accuracy of 93.1% on a dataset of 196 images. In this study, we propose to build deep learning models in big data environment developed on advanced architectures such as ResNet-50 & VGG-16 to diagnose lung lesions on X-ray images. This will help the clinical examination of doctors be simple, fast, and achieve high accuracy. The rest of the paper is structured as follows: an overview of the theory is presented in Sect. 2, the proposed method is introduced in Sects. 3 and 4 provides the experimental results based on the proposed models. Finally, we draw the conclusion in Sect. 5.

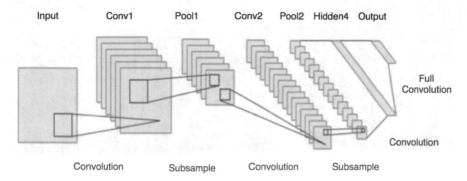

Fig. 1 An illustration of CNN network architecture

2 Background

2.1 Convolutional Neural Networks

The architecture of convolutional neural networks (CNN) is widely used in the field of computer vision. It is considered as one of the solutions to optimize the network training process, which is proposed to reduce weights to increase computational speed and reduce training time. The basic architecture of a CNN network includes convolution layer, nonlinear layer, pooling layer, and fully connected layer. These layers are linked together in a certain order. Normally, the input data will be propagated through the convolutional layers and nonlinear layers, then the computed values are propagated through the pooling layer. The triple convolution layer, nonlinear layer, and pooling layer are repeated many times in the network. The result is then passed through the fully connected layer for classification (Fig. 1).

VGG-16 network architecture One of the easiest ways to improve network models accuracy is to increase the networks' depth. The VGG-16 architecture (Fig. 2) is deeper than AlexNet, consisting of 13 two-dimension convolutional layers (instead of 5 compared to AlexNet) and 3 fully connected layers [8]. VGG-16 uses only small size filters (3×3) instead of multiple filter sizes as AlexNet. For example, using 2 filters of size 3×3 on a feature map with a depth of 3, we would need n_filters $\times$ kernel_size $\times$ kernel_size $\times$ n_channels = $2 \times 3 \times 3 \times 3 = 54$ parameters. If using a filter of size 5×5, we will need $5 \times 5 \times 3 = 75$ parameters. Small size filters will help to reduce the number of parameters, which brings more computational efficiency.

Resnet-50 network architecture ResNet is the most commonly used architecture at the moment. ResNet is also the earliest architecture to adopt batch normalization. Although it is a very deep network with a number of layers up to 152, the size of ResNet50 is only about 26 million parameters thanks to applying some special techniques. ResNet architecture with fewer parameters but effective result in the winner of the 2015 ImageNet competition [5]. ResNet has convolution blocks that

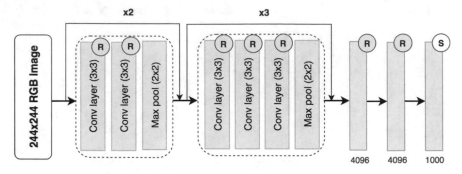

Fig. 2 VGG-16 network architecture

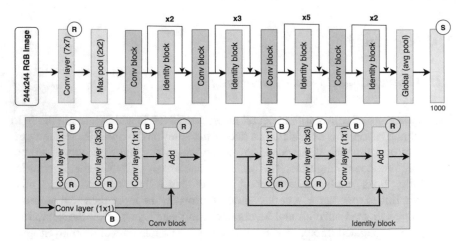

Fig. 3 Resnet-50 network architecture

use a 3×3 filter similar to that of InceptionNet (Fig. 3). The convolution block consists of two convolution branches where one branch applies a 1×1 convolution before adding it directly to the other branch. The identity block does not apply the 1×1 convolution, but directly adds the value of that branch to the other branch.

2.2 Distributed and Parallel Processing

Apache Spark is a scalable parallel and distributed data processing engine that is faster than many other data processing frameworks. It includes several libraries to build applications for machine learning (MLlib), stream processing (Spark Streaming), and graph processing (GraphX). Spark applications run as a collection of independent processes on a computer cluster, coordinated by the SparkContext object in the driver program [9]. SparkContext can connect to some kind of cluster manager such

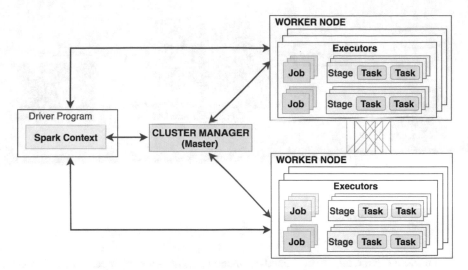

Fig. 4 Parallel and distributed computing model on spark

as Mesos and YARN to allocate resources across applications. The executors on the worker nodes are the processes that run computation and store data for applications. SparkContext will send the tasks to the executors to run (Fig. 4).

Data parallelization is a popular technique used to speed up training on a dataset that is divided into mini-batches. The distributed training model consists of a master node and worker nodes. Each worker keeps an identical copy of the network parameters running forward pass and backward pass. At the end of back propagation, each worker sends the calculated parameters to a parameter server at the master node. The parameter server aggregates parameters and updates the network parameters. The updated parameters are then sent to each worker and the process is repeated for a new mini-batch [8]. In Spark, the way it works is quite simple. The Keras is initialized in the Spark driver and sends mini-batches of the dataset to the workers. Each worker trains the model on its own piece of data and sends the parameter values back to the master. The master node computes and updates the parameters and sends them to the workers. This process is repeated until the optimal parameter values are found so that the training model has high accuracy. In this study, we install and use the BigDL library on the Spark environment to train the network models.

2.3 Evaluation Metrics

To evaluate the loss in model training, we use the cross-entropy function to calculate the difference between the two probability distributions of the prediction with the label. The loss function is presented in the Eq. 1, where y is the truth label, $(\hat{y}_i$ the predicted label, and M is the number of classes.

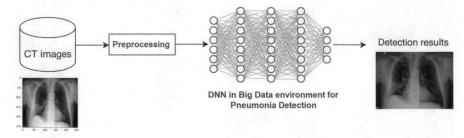

Fig. 5 General model of the proposed method for pneumonia detection

$$H = -\sum_{i}^{M} y_i log(\hat{y}_i) \tag{1}$$

The network models are evaluated by measuring AP (Average precision) and mAP (mean Average Precision). These metrics can be calculated by Precision and Recall (Eq. 2). The Precision is calculated as the total number of correct predictions (TP) divided by the total number of positive predictions (TP + FP). Recall is calculated on the total number of correct predictions (TP) divided by the total number of correct cases (TP + FN). The accuracy of predicted cases is giving high precision. High recall means that the rate of missing actual positive cases is low. $F1$ is the weighted average of precision and recall.

$$P = \frac{TP}{TP + FP}; \quad R = \frac{TP}{TP + FN}; \quad F1 = \frac{2 * P * R}{P + R} \tag{2}$$

3 Proposed Method

In this study, we propose a pneumonia detection system in a big data environment for fast and accurate diagnosis. The general model of the proposed method for pneumonia detection is shown in Fig. 5. We design and perfect deep learning networks in Spark parallel and distributed computing environments.

Figure 6 describes the implementation phases of the proposed method for pneumonia detection. It includes three phases of the preprocessing, training, and testing.

- Phase 1—Preprocessing: We propose to build two DNN network models developed based on VGG-16 & ResNet-50 architectures. The goal at this phase is to build the network models and fine-tune the parameters so that the models achieve high accuracy. The results of this phase are the training models in the local environment with optimal parameter values. These parameters are the initial weights for the training models on the big data environment in the next phase.
- Phase 2—Training model in Spark: Taking advantage of transfer learning, we use the pre-trained parameters in phase 1 as initialization parameters for training the

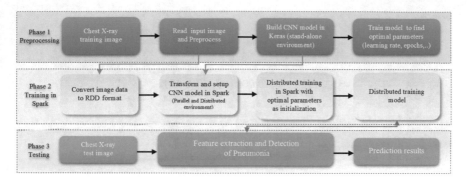

Fig. 6 Implementation phases of the proposed method for pneumonia detection using DNN in Spark parallel and distributed environment

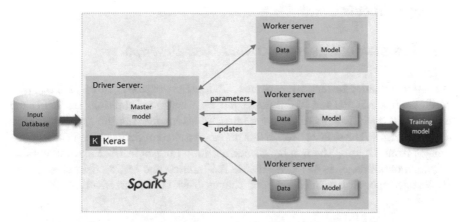

Fig. 7 Proposed model to train DNNs in Spark parallel and distributed environment

proposed DNN models in Spark. Then, we re-train these proposed DNN models for our case. The results of this phase are the optimal parameters of the distributed training models. We perform training models in a parallel and distributed computing environment as shown in Fig. 7. Each worker will receive a copy of the training model and its parameters. The dataset is distributed to workers through partitioning by splitting into mini-batches through Cluster manager. The parameters of each model are fine-tuned to achieve optimal parameters. Thus, each model has different training parameters and is trained on different batches. These parameters are then sent to the Master to calculate the average and update the parameter values. The Master contains global parameters responsible for distributing parts of the model to the workers and acts as a coordinator between the workers.
- Phase 3—The test data are put into the trained model in phase 2 for lung lesion detection to give the diagnosis results as shown in Fig. 6.

Executors

											Task Time	
Executor ID ▲	Address	Status	RDD Blocks	Storage Memory	Disk Used	Cores	Active Tasks	Failed Tasks	Complete Tasks	Total Tasks	(GC Time)	Input
0	192.168.100.189:43615	Active	0	3.2 MiB / 912.3 MiB	3.2 MiB	1	1	0	1	2	18 s (0.2 s)	0.0 B
driver	master:46719	Active	0	3.2 MiB / 5.2 GiB	0.0 B	0	0	0	0	0	0.0 ms (0.0 ms)	0.0 B
1	192.168.100.44:38131	Active	0	3.2 MiB / 912.3 MiB	3.2 MiB	1	1	0	1	2	21 s (0.5 s)	0.0 B
2	192.168.100.26:44163	Active	0	3.2 MiB / 912.3 MiB	3.2 MiB	1	1	0	0	1	0.0 ms (0.0 ms)	0.0 B

Fig. 8 Details of computing nodes for executing tasks

4 Experiments

4.1 Installation Environment and Data Description

We install a Spark Cluster on 4 machines (Ubuntu Server 18.04), in which the master configuration is 5CPU and 16 GB RAM and 3 workers configurations are 8CPU and 8 GB RAM each. Details of the computing nodes involved in the experiments are shown in Fig. 8. The dataset used in this paper consisted of 5856 labeled chest X-ray images from Kaggle. The number of normal lung images is 1583 and the number of pneumonia images is 4273. We divide the training and testing dataset in a ratio 80:20, corresponding to the number of images 4684/1172 respectively.

4.2 Models and Parameters

We design and perfect two deep learning models in parallel and a distributed processing environment developed on VGG-16 & ResNet-50 architectures. Taking advantage of transfer learning, we use the pre-trained parameters determined in phase 1 to re-train on Spark Clusters with the number of computing nodes of 1, 2, and 3. The training parameters for both models are shown in Table 1.

4.3 Training Results

To evaluate the models, we rely on the loss measure and the training time. Based on the results of Figs. 9 and 10, the proposed models have the high accuracy of over 95%. In which, model 2 has the loss value of 0.35, which is lower than model 1 with the loss value of 0.5. Model 2 has higher accuracy of 96% than model 1 with accuracy of 95%.

Table 1 Training parameters in the two models

Model	Method	N. class	Epoch	Batch size	Optimizer	Learning rate	Input	Activation
1	Proposed DNN in Spark developed on VGG-16	2	10	32	Adam	0.0001	256,256,3	Sigmoid
2	Proposed DNN in Spark developed on Resnet-50	2	10	32	Adam	0.0001	256,256,3	Sigmoid

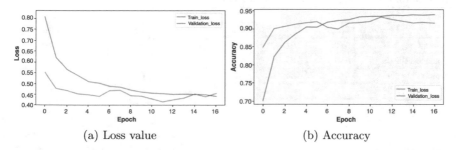

(a) Loss value   (b) Accuracy

Fig. 9 Loss and accuracy for model 1

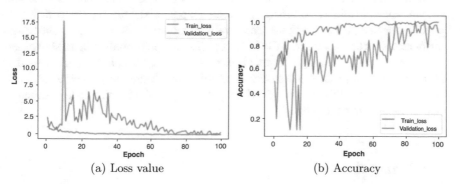

(a) Loss value   (b) Accuracy

Fig. 10 Loss and accuracy for model 2

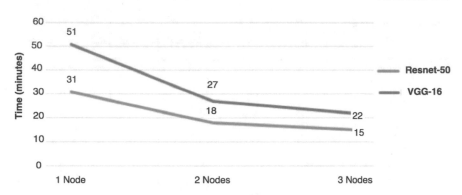

Fig. 11 Training time of the two models

Table 2 Measures of the models on 3 computing nodes

	Model 1				Model 2			
	Precision	Recall	F1	No. images	Precision	Recall	F1	No. images
Normal	0.91	0.90	0.91	335	0.93	0.95	0.94	335
Pneumonia	0.96	0.97	0.96	837	0.98	0.97	0.97	837
Average	0.94	0.93	0.93	1172	0.95	0.96	0.96	1172
Accuracy	**0.95**			1172	**0.96**			1172

We compare the training time on both models in both local and distributed environments. The results are shown in Fig. 11. Although both models have almost the same number of training parameters, Model 2 applies identity blocks with 1×1 convolutions resulting in 1.5 times faster training time than model 1 using the usual convolution blocks. In addition, the local and distributed training results show that model 2 gives better training time on all three cases using 1, 2, and 3 nodes. As the number of computing nodes increases, the training time is significantly reduced. This is one of the outstanding advantages of deep learning model training in a distributed parallel processing environment.

4.4 Testing Results

We use the confusion matrix to determine the precision, recall, and F1-score measures as shown in Table 2. Model 2 has higher accuracy than model 1.

We compare the accuracy of the proposed method with the preceding studies. The comparison results are shown in Table 3. It is difficult to accurately compare the studies due to different datasets and experimental environments. Therefore, we only compare the accuracy of the preceding studies with our proposed method. Based on the results of Table 3.

Table 3 Comparison with the related studies

Authors	Methods	Accuracy (%)
Chen et al. [2]	UNet++	95.24
Elshennawy and Ibrahim [3]	ResNet152V2, MobileNetV2, CNN, LSTM	91
Gabruseva et al. [4]	RetinaNet SSD with Se-ResNext101	87.5
Zeiser et al. [10]	VGG16	85.11
Abbas et al. [1]	PCA	93.1
Proposed method	Model 2	**96**

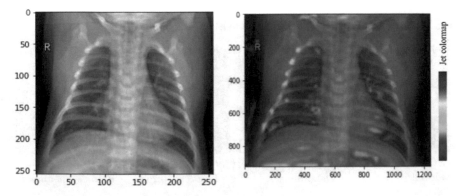

Fig. 12 Normal lung image detection results—Model 2

The results of identifying normal lung images and pneumonia images using model 2 are presented in Figs. 12 and 13. The regions of the lung lesion are detected in red color (Fig. 13). Lung lesion detection is a very important task to evaluate the accuracy of the models for classification.

Figure 14 shows the classification results of lung lesions. The area of lung lesion shows in different colors compared to the normal lung image. Figure 14a, d show the predicted normal lung and pneumonia images, respectively, which coincide with the truth labels. Figure 14b, c show the results of the incorrect predictions. Model 2 achieves high accuracy in detecting lung lesions. From the results, the doctor can easily identify and diagnose the patient's level of pneumonia for appropriate treatment.

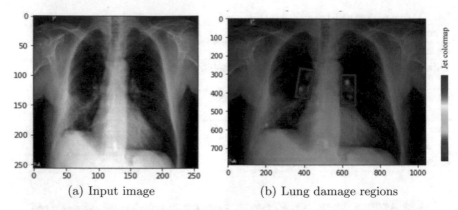

(a) Input image   (b) Lung damage regions

Fig. 13 Pneumonia image detection results—Model 2

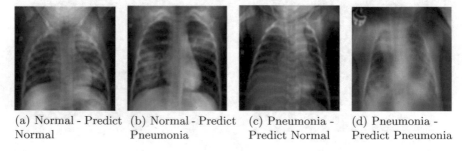

(a) Normal - Predict   (b) Normal - Predict   (c) Pneumonia -   (d) Pneumonia -
Normal   Pneumonia   Predict Normal   Predict Pneumonia

Fig. 14 Pneumonia classification—Model 2

5 Conclusion

In this paper, we propose a method to detect pneumonia using deep learning techniques in big data environment. We build and implement two deep learning networks developed on VGG-16 and ResNet-50 architectures in Spark parallel and distributed processing environment. Experimental results show high accuracy to identify lung lesions on X-ray images, assisting doctors in timely diagnosis and treatment. Comparison of the training and detection models in the traditional and Spark environment shows that the deep learning neural networks in the parallel and distributed processing environment significantly improve the training time compared to the stand-alone environment. We will make an extension of our further research with other deep learning networks and experiment on large datasets to automatically classify the injured lung regions. Besides, the increasing number of computing nodes is expected to improve the response time. Furthermore, this study can be extended to other lesion diagnosis systems on big data-driven medical images that are suitable for real-time detection systems.

References

1. Abbas A, Abdelsamea MM, Gaber MM (2021) Classification of covid-19 in chest x-ray images using detrac deep convolutional neural network. Appl Intelligence 51(2):854–864
2. Chen J, Wu L, Zhang J, Zhang L, Gong D, Zhao Y, Chen Q, Huang S, Yang M, Yang X et al (2020) Deep learning-based model for detecting 2019 novel coronavirus pneumonia on high-resolution computed tomography. Sci Rep 10(1):1–11
3. Elshennawy NM, Ibrahim DM (2020) Deep-pneumonia framework using deep learning models based on chest x-ray images. Diagnostics 10(9):649
4. Gabruseva T, Poplavskiy D, Kalinin A (2020) Deep learning for automatic pneumonia detection. In: Proceedings of the IEEE/CVF conference on computer vision and pattern recognition workshops, pp 350–351
5. He K, Zhang X, Ren S, Sun J (2016) Deep residual learning for image recognition. In: Proceedings of the IEEE conference on computer vision and pattern recognition, pp 770–778
6. Ortiz JL, Bacon J, Hayes C. U.S. covid-19 updates. https://eu.usatoday.com/story/news/health/2021/09/07/us-cases-40-million-unemployment-labor-day-travel-updates/5748549001/
7. Sethi S (2020) Overview of pneumonia. https://www.msdmanuals.com/professional/pulmonary-disorders/pneumonia/overview-of-pneumonia
8. Simonyan K, Zisserman A (2014) Very deep convolutional networks for large-scale image recognition. arXiv preprint arXiv:1409.1556
9. Spark A. Cluster mode overview. https://spark.apache.org/docs/latest/cluster-overview.html
10. Zeiser FA, da Costa CA, Ramos GDO, Bohn H, Santos I, Righi RDR (2021) Evaluation of convolutional neural networks for covid-19 classification on chest x-rays. arXiv preprint arXiv:2109.02415

A New Stratified Block Model to Process Large-Scale Data for a Small Cluster

Thanh Trinh, Long Pham Duc, Chinh Trung Tran, Thuong Ta Duy, and Tamer Z. Emara

Abstract Recently, big data analytics has been a hot topic in different fields for many researchers. Several big clusters based on Spark and Hadoop are developed to handle big data files. In this paper, we study big data analytics through the problem of classification. We propose a new stratified block (SB) model for a small cluster that can execute a big data file. In this model, a stratified sampling method is used to split a big file into a specific number k of small files (k data blocks). For each block, we build a classifier model (decision tree) to predict a result of a testing file. To save the memory of the cluster, we only keep the predicted result of each block in the memory; then, the next block is loaded to generate a new tree model. Hence, k blocks yield k predicted results of the testing file. Finally, we aggregate the k results into the final predicted result of the testing file. Three big data files are used to evaluate the performance of SB in terms of computing time and accuracy metrics.

Keywords Bigdata · Spark · Classification

T. Trinh (✉) · L. P. Duc · C. T. Tran · T. T. Duy
Faculty of Computer Science, Phenikaa University, Yen Nghia, Ha Dong, Hanoi 12116, Vietnam
e-mail: thanh.trinh@phenikaa-uni.edu.vn

L. P. Duc
e-mail: long.pd19010018@st.phenikaa-uni.edu.vn

C. T. Tran
e-mail: chinh.tt19010005@st.phenikaa-uni.edu.vn

T. T. Duy
e-mail: thuong.td19010032@st.phenikaa-uni.edu.vn

T. Trinh
Phenikaa Research and Technology Institute (PRATI), A&A Green Phoenix Group JSC, No.167 Hoang Ngan, Trung Hoa, Cau Giay, Hanoi 11313, Vietnam

T. Z. Emara
Faculty of Computer Science and Artificial Intelligence, Pharos University in Alexandria, Alexandria, Egypt
e-mail: tamer.emara@pua.edu.eg

© The Author(s), under exclusive license to Springer Nature Switzerland AG 2022
N. H. T. Dang et al. (eds.), *Artificial Intelligence in Data and Big Data Processing*,
Lecture Notes on Data Engineering and Communications Technologies 124,
https://doi.org/10.1007/978-3-030-97610-1_21

1   Introduction

Big data analytics gains a lot of interest from many researchers and big companies. Big data can be generated from social network data with millions of comments per hour. Many research problems are defined from big data analytics, such as the problems of computing time and accuracy. Spark [16, 17] and Hadoop [12] frameworks are famous cluster-based applications for parallel processing. The frameworks are very effective in handling big data files by dividing the big files into small files (data blocks) and distributing these blocks into all nodes of the cluster for parallel processing. Several distributed clusters [4, 5] are developed based on Spark and Hadoop to execute classification tasks for big data files.

However, their works are conducted on a big cluster with many nodes and huge memory. Therefore, given a small cluster, how to execute big files is a challenging problem. Another problem of dividing big files is how many small files are effective for a specific cluster. These problems yield computing time and accuracy issues of small clusters.

In this paper, we propose a new stratified block model (SB) to handle big data files with a small cluster. We set up a cluster based on Spark, and SB model is executed in the cluster. The statistics of large-scale data are very important when splitting them into data blocks. Hence, in SB, the stratified sampling method is used to generate a chain of data blocks. These blocks are distributed into the cluster. Each data block is converted into a RDD (Resilient Distributed Datasets) block in Spark environment. We build a model of a decision tree based on one RDD block. Then, the tree model is used to predict a testing file. SB model only stores a predicted result of the RDD block in order to save the memory of each worker node. And, SB model removes the RDD block and the tree model before loading a new data block to process. Finally, we have an aggregation of predicted results for the whole chain of data blocks. A majority vote is used to achieve the final predicted result of the testing file.

We aim to execute big data files by a small cluster. In experiments, three big files are used to evaluate the computing time and accuracy of SB. Furthermore, different strategies to split big files into k data blocks are used. From experiment results, we can conclude that if the size of each block is in a range of 50 and 90% of the memory of each worker node, SB model is effective in terms of computing time metric. However, if k is large (such as 100 blocks), SB spends too much time yielding better accuracy.

Our contributions are summarized as follows:

- We propose a new stratified block (SB) mode to execute large-scale data.
- The size of each block should be in a range of 50–90% of each worker node's memory.
- SB model is used as an ensemble method.

The remainder of this paper is organized as follows. In Sect. 2, we present the preliminaries that are used in this study. Section 3 gives the details of the stratified block model. Section 4 shows the empirical study. Section 5 gives some related works. And Sect. 6 draws some conclusions.

2 Preliminaries

2.1 Spark

Apache Spark has emerged as one of the strongest unified engines for large-scale data processing. Spark supports several programming languages, such as Python and Scala, to handle big data. Spark can be configured with/without Hadoop to be a cluster with different worker nodes. Figure 1 describes a standalone architecture of Spark. That includes five main entities as follows:

1. **Workers**. A cluster has at least one worker node (a computer). Workers supply their resources to a spark application, such as CPU, RAM, and storage.
2. **Executors**. A spark application is submitted to a cluster, and it will be run on each worker node as a distributed process. At each worker node, Spark will create a java virtual machine (JVM) process called an executor. The executor executes the spark application in multiple threads concurrently, and it can cache data in memory or disk. When the application is terminated, all executors also terminate.
3. **Tasks**. A task that is the smallest unit of work is created by Spark for per data partition. Each task is sent to an executor on a worker node and executed by a thread in the executor. One or more tasks can be concurrently run in one executor. That means more partition is, more task is executed in parallel.

Fig. 1 A standalone architecture of spark

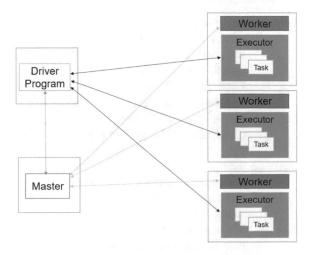

4. **Driver Programs**. A driver program is an application that supplies the data processing code as a job to be executed by Spark on worker nodes.
5. **Cluster Managers**. A Spark cluster needs to have cluster managers to control all worker nodes within the cluster. Managers deal with computing resources and provide low-level scheduling of cluster resource applications that run on the same worker nodes.

2.2 Decision tree

Decision tree is one famous example for supervised learning. Decision tree model is very easy to implement for classification tasks. Tree model is often used as a base classifier of ensemble learning methods, such as random forest, bagging, or boosting. Decision tree can be constructed by different algorithms, for example, ID3 [8], CART [2], C4.5 and C5.0 [10]. In general, the basic idea of these algorithms is described in the following details:

Step 1   Choose the best feature to split the remaining cases, and the feature is selected as a node.
Step 2   Repeat Step 1 recursively for each child node.
Step 3   The process of building a tree model stops when the stop conditions meet. For instance, the target classes of all remaining cases are the same.

3 Stratified Block Model

In this section, we present a stratified block (SB) model to execute big files in small clusters. Figure 2 illustrates SB which can be described as follows:

Srep 1:   Given a big training file $D = \{x_i, y_i\}$ with N samples (millions) where x_i represents a vector of m features and $y_i \in \{c_1, c_2, ..., c_l\}$ is a class label, we use stratified sampling to split D into k data blocks $B = \{B_1, B_2, ..., B_k\}$, and $B_i \cap B_j = \emptyset$. Then these blocks B are distributed into a cluster.
Step 2:   Each block B_j will be converted into a RDD (Resilient Distributed Datasets) block, then the RDD block is used to build a tree model.
Step 3:   The testing file T is also converted into a RDD. The tree model that is built in Step 2 is used to predict T. From that, SB has achieved a predicted result R_j.
Step 4:   The process stops when all blocks B are used to build tree models and generate predicted results. Otherwise, SB model only stores result R_j in memory of the cluster and removes RDD and the tree model out of the memory before going back to Step 2.
Step 5:   We have a set of results R_j. We use the majority vote to obtain the final result R_{final}, and R_{final} is used to give the prediction of testing file T.

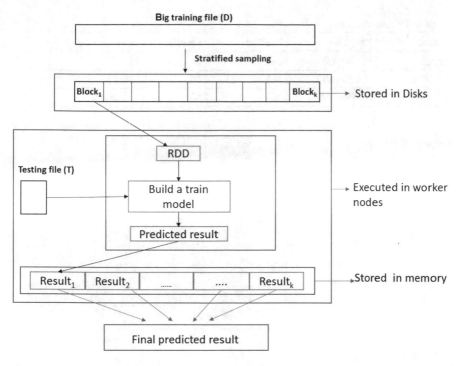

Fig. 2 Stratified block model

Obviously, SB is considered as an ensemble method which uses decision tree as a base classifier.

4 Empirical Study

4.1 *Cluster*

We set up a standalone Spark cluster with two nodes: one master node and one worker node. Each node has 4 cores with 2.9 GHz, 8 GB RAM, and 100 GB (SSD). The cluster runs inside Ubuntu system. Algorithms are implemented in Python and executed in the cluster.

Table 1 Dataset statistics

Dataset	Samples	Features	Classes	Size on disk (GB)
Higgs	11 000 000	28	2	≃7.3
DS10m	10 000 000	100	20	≃17.6
DS20m	20 000 000	100	50	≃35.2

Fig. 3 The difference in computing time executed in the same computer

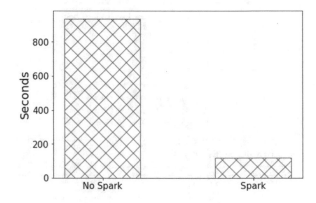

4.2 Data

Table 1 describes the three generated big data files used in experiments. Higgs was generated by Baldi et al. [1] and stored in UCI.[1] We used Gaussian to generate the other two big data files, DS10m and DS20m. DS10m is not difficult to classify. But DS20m is a challenge to classify.

In experiments, we used stratified sampling method to take 500,000 samples of each big file as a testing file (T), and the rest of the big file was considered a big training file (D).

4.3 Experimental Results

We start this empirical study by Higgs file to determine the number of blocks for the three big files.

Computing time. Figure 3 presents the difference in computing time when a tree model predicts Higgs with Spark and without Spark in the same computer. Obviously, the tree model with Spark runs 9 times faster than it without Spark.

To evaluate the performance of SB model, we carry out five different strategies to split the training file of Higgs into small data blocks. Strategies are represented by

---

[1] https://archive.ics.uci.edu/ml/datasets/HIGGS

Table 2 The number of blocks, size of each block, and percentage of block'size and memory

Higgs			DS10m			DS20m		
Blocks (k)	Size (GB)	% of Memory (%)	Blocks (k)	Size (GB)	% of Memory (%)	Blocks (k)	Size (GB)	% of Memory (%)
1	5	83.3	1	16.6	276	1	34.1	568
2	2.5	41.6	3	5.53	92.2	7	4.9	81.7
5	1	16.6	5	3.32	55.3	10	3.41	56.8
10	0.5	1.66	10	1.66	27.7	20	1.7	28.3
100	0.05	0.83	100	0.166	2.77	100	0.34	5.7

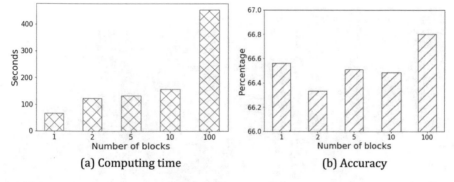

(a) Computing time   (b) Accuracy

Fig. 4 Computing time and accuracy of Higgs executed in the cluster

k number of blocks. In other words, a strategy $k = 1$ means that we have only one block, which is the whole training file. And another strategy $k = 10$ indicates that the training file is split into 10 blocks; the size of each block is 0.5 GB (seen in Table 2).

Figure 4 shows the computing time of five splitting strategies for Higgs, which are executed in the cluster. It is observed that the whole training file is executed in the cluster is fastest. However, strategy $k = 100$ yields the worst computing time. It can be explained as follows: (1) The size of the training file (5 GB) is less than the size of memory used for each worker node (6 GB) in the cluster; (2) In SB model, k blocks will load in a queue. After achieving the result R_i, SB model only stores R_i to save memory. (3) When SB model obtains all k results $R = \{R_1,..,R_k\}$, SB model aggregates these results to yield the final predicted result of the testing file. And, Fig. 5 shows the whole parallel process of Higgs in the cluster. We can observe that with the strategy $k = 100$, SB model takes a lot of time to load and build a new tree model and yield a result for each block, seen in Fig. 5e. Therefore, that are why the computing time with many blocks is slower than it with few blocks.

We observe that the computing time of the strategies ($k = 1$, $k = 2$, and $k = 5$) is very fast compared with those of the other two strategies for Higgs. And the size

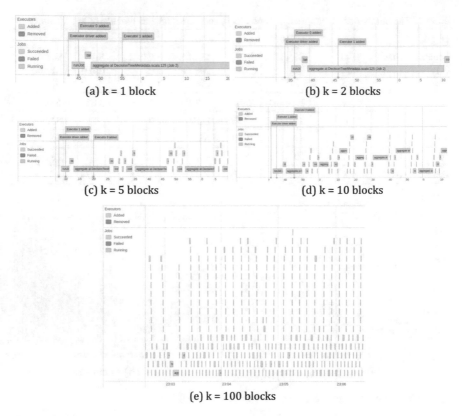

Fig. 5 The processing of Higgs with different strategies (number of blocks) in the cluster

of each block is from around 50–90% of the memory of each worker node, seen in Fig. 4a and Table 2.

Hence, we determine to split the training files of DS10m and DS20m by five different strategies. And the size of each block is less than the size of memory. Table 2 presents the number of blocks, the size of one block, and the ratio of block's size and memory of worker node with different splitting strategies for the three big files.

As we expected, the size of each block is from 50 to 90% of the memory (seen in Table 2), SB model processes DS10m and DS20m very fast. The computing time results are illustrated in Fig. 6a and Fig. 7a.

Accuracy. Figure 4b, Fig. 6b and Fig. 7b show the accuracy results of each strategy for three big data files respectively. When k is increased, the computing time also is slower. However, the accuracy is improved. SB model works based on a chain of blocks and yields a chain of results. Hence, SB models can be considered an ensemble method. Splitting big files into many blocks can enhance the accuracy of a classifier, such as decision tree model.

However, we can observe that the accuracy of Higgs (Fig. 4b) and DS20m (Fig. 7b) is not much improved, even we split the two files into 100 blocks. This is the problem of data and the effectiveness of decision tree model. For this kind of data, we should split them into a specific number of blocks with a condition, i.e., the size of each

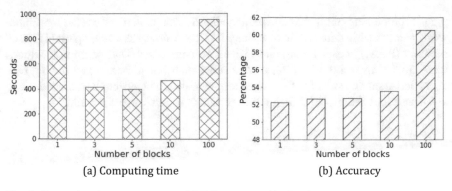

(a) Computing time   (b) Accuracy

Fig. 6 Computing time and accuracy of **DS10m** executed in the cluster

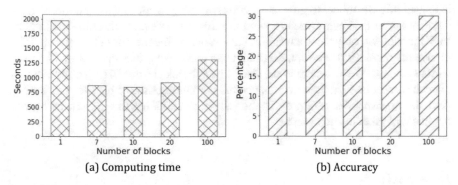

(a) Computing time   (b) Accuracy

Fig. 7 Computing time and accuracy of **DS20m** executed in the cluster

block is from 50 to 90% of memory in order to enhance the performance of SB model in terms of computing time.

5 Related Work

The volume of data has been increasing by every minute. For example, social data that are generated from online social networks provide a lot of research problems for scientists and companies [13, 15]. And, bio-information needs to have some techniques of feature selection to address the problem of high dimensional data, i.e., thousands of features compared with few samples [14]. Largescale data analysis has been addressing for the last few decades [3, 6, 11]. They proposed a new architectural paradigm called a cluster with several separate computers; and the computers connected through a local network to share their resources, such as CPU, memory, and disks. The cluster executed large-scale data in parallel processing within its computers. Shvachko et al. [12] proposed a new term of distributed files for the parallel processing of big data files called HDFS. The main work of Hadoop framework processes big data files in disks, and it is quite slow. Recently, Spark [17] used memory (RAM) to execute big files instead of using disks. Data sampling methods were studied in works [7, 9], they proposed a distributed data model for big data analysis. Another works [4] developed a new distributed system based on Hadoop and Spark for running big files, but this system is too expensive to build.

6 Conclusions

In this paper, we propose a stratified block (SB) model that can process big data files in a small cluster. First, we split a big training file into many blocks. Then, each block is loaded into SB model to generate a corresponding result of a testing file. Finally, the list of results is aggregated to yield the final predicted result of the testing file. In SB model, we only store a result of each block in the memory of cluster in order to save the memory. Therefore, SB model can execute big files under a small cluster with two nodes. We also suggest the size of each block is from 50 to 90% of memory of each worker node in order to enhance the computing time of big files processing. And, to improve the accuracy of SB model, we can split big files into many blocks, but this will take time to compute.

References

1. Baldi P, Sadowski P, Whiteson D (2014) Searching for exotic particles in high energy physics with deep learning. Nature Commun 5(1):4308
2. Breiman L, Friedman J, Olshen R, Stone C (1984) Classification and regression trees. Pacific Grove
3. DeWitt DJ, Gerber RH, Graefe G, Heytens ML, Kumar KB, Muralikrishna M (1986) GAMMA—a high performance dataflow database machine. In: Proceedings of the 12th international conference on very large data bases, VLDB '86. Morgan Kaufmann Publishers Inc. San Francisco, CA, USA, pp 228–237
4. Emara TZ, Huang JZ (2019) A distributed data management system to support large-scale data analysis. J Syst Softw 148:105–115
5. Emara TZ, Huang JZ (2020) Distributed data strategies to support large-scale data analysis across geo-distributed data centers. IEEE Access 8:178526–178538
6. Howard JH, Kazar ML, Menees SG, Nichols DA, Satyanarayanan M, Sidebotham RN, West MJ (1988) Scale and performance in a distributed file system. ACM Trans Comput Syst 6(1):51–81
7. Mahmud MS, Huang JZ, Salloum S, Emara TZ, Sadatdiynov K (2020) A survey of data partitioning and sampling methods to support big data analysis. Big Data Min Analytics, 3(2):85–101
8. Quinlan JR (1986) Induction of decision trees. Mach Learn 1(1):81–106
9. Salloum S, Huang JZ, He Y (2019) Random sample partition: a distributed data model for big data analysis. IEEE Trans Ind Inform 15(11):5846–5854
10. Salzberg SL (1994) C4.5: Programs for machine learning by J. Ross Quinlan. Morgan Kaufmann Publishers, Inc., 1993. Mach Learn 16(3):235–240
11. Shemer J, Neches P (1984) The genesis of a database computer. Computer 17(11):42–56

12. Shvachko K, Kuang H, Radia S, Chansler R (2010) The hadoop distributed file system. In: 2010 IEEE 26th symposium on mass storage systems and technologies (MSST). IEEE, May 2010, pp 1–10
13. Trinh T, Wu D, Huang JZ, Azhar M (2020) Activeness and loyalty analysis in event-based social networks. Entropy 22(1):119
14. Trinh T, Wu D, Salloum S, Nguyen T, Huang JZ (2016) A frequency-based gene selection method with random forests for gene data analysis. In: 2016 IEEE RIVF international conference on computing and communication technologies: research, innovation, and vision for the future, RIVF 2016—proceedings, pp 193–198
15. Trinh T, Wu D, Wang R, Huang JZ (2020) An effective content-based event recommendation model. Multimedia Tools Appl
16. Zaharia M, Chowdhury M, Franklin MJ, Shenker S, Stoica I (2010) Spark: cluster computing with working sets. In: Proceedings of the 2nd USENIX conference on hot topics in cloud computing, HotCloud'10, 2010. USENIX Association, USA, p 10
17. Zaharia M, Xin RS, Wendell P, Das T, Armbrust M, Dave A, Meng X, Rosen J, Venkataraman S, Franklin MJ, Ghodsi A, Gonzalez J, Shenker S, Stoica I (2016) Apache Spark. Commun ACM 59(11):56–65

Towards Meta-Learning Based Data Analytics to Better Assist the Domain Experts in Industry 4.0

Moncef Garouani, Adeel Ahmad, Mourad Bouneffa, Mohamed Hamlich, Gregory Bourguin, and Arnaud Lewandowski

Abstract Industrial Data Analytics (IDA) provide methods and tools to cope with the vast amounts of data. The big industrial data is generated continuously during the execution of manufacturing processes. Hence, the predictive maintenance is among the most critical activities of the manufacturing processes concerned by the IDA. We believe that the maintenance activity can be managed by using Machine Learning (ML) methodologies, especially the data analytics solutions based on meta-learning. The challenge is then to facilitate the application of ML by the industry 4.0 actors, who are supposedly not AI specialists. The automated machine learning (AutoML) seems to be the area dealing with this challenge. In this paper, we primarily discuss the challenges of assisting industry 4.0 actors to implement ML algorithms in the context of predictive maintenance. Later on, we present a novel AutoML based framework for the industry 4.0 actors and researchers, who presumably have limited expertise in ML domain. It aims to enable them to generate ML-based data analytics solutions and deploy these solutions in manufacturing workflows. Specifically, the framework implements the approaches based on meta-learning and uses the web semantic concepts in this regard. In the context of Industry 4.0 such approaches lead to the implementation of the smart factory concepts. It

M. Garouani · A. Ahmad · M. Bouneffa · G. Bourguin · A. Lewandowski
University of Littoral Côte d'Opale, UR 4491, LISIC, Laboratoire d'Informatique Signal et Image de la Côte d'Opale, 62100 Calais, France
e-mail: adeel.ahmad@univ-littoral.fr

M. Bouneffa
e-mail: mourad.bouneffa@univ-littoral.fr

G. Bourguin
e-mail: gregory.bourguin@univ-littoral.fr

A. Lewandowski
e-mail: arnaud.lewandowski@univ-littoral.fr

M. Garouani · M. Hamlich
CCPS Laboratory, ENSAM, University of Hassan II, Casablanca, Morocco

M. Garouani (✉)
Study and Research Center for Engineering and Management, HESTIM, Casablanca, Morocco
e-mail: moncef.garouani@etu.univ-littoral.fr

© The Author(s), under exclusive license to Springer Nature Switzerland AG 2022   265
N. H. T. Dang et al. (eds.), *Artificial Intelligence in Data and Big Data Processing*,
Lecture Notes on Data Engineering and Communications Technologies 124,
https://doi.org/10.1007/978-3-030-97610-1_22

makes the factory processes more proactive on the basis of predictive knowledge extracted from the various manufacturing devices, sensors, and business processes in real-time.

Keywords Machine learning · Data analytics · AutoML · Meta-learning · Ontologies · Industry 4.0

1 Introduction

The recent trends in the automation, data exchange, and interoperability in manufacturing technologies are intended towards the concrete implementation of the *smart factory* which is among the core concepts of the Industry 4.0 [1]. It is generally achieved by combining the use of hardware and software technologies including CPS (Cyber Physical Systems) and frameworks, which effectively deal with the knowledge and data management [2]. The AI, particularly ML, supports the knowledge extraction process using algorithms, methods, and tools to build models from diverse data representations. These data representations emerge from the different activities of a company and its environment. The careful literature study reveals that the ML has demonstrated its usefulness and benefits in many fields [1]. Its successful applications in the context of manufacturing require a lot of effort from human experts since there is not one fit all algorithm to perform well on all possible problems [3].

The industrial actors, engineers and researchers, yet lack the ML expertise required to handle the massive industrial heterogeneous data sources even though being familiar with manufacturing data [2]. The industrial researchers may collaborate with data science experts to confront such a lack. This collaboration may involve the actors of various expertise, for instance the domain experts can be the production planners, engineers in assembly lines, or the ones who devise strategies to pursue the goals [2]. However, the existing data analytical approaches do not sufficiently enable these domain experts to specify the problem formulation and devising strategies to achieve the reflected goals [2, 4]. The collaboration between domain experts and data scientists risks to be enacted in a continuous to and fro inputs among them without yielding useful results [4].

Thereby, the automated assistance for the required human expertise can allow the stakeholders of smart factories to rapidly build, validate, and deploy ML solutions. It may also improve their quality of service, productivity, and more importantly, reduce the need of the interventions from ML human experts. Motivated by this goal, the AutoML [3] has emerged across smart factories as a new research field that aims to automatically select, compose, and parameterize machine learning models which are able to achieve an optimal performance on a given task.

Despite the advances achieved in the field of AutoML, few works have been conducted to apply these techniques in the manufacturing field and hence, the industrial needs are yet to be fulfilled [2]. In this paper, we present the architecture of a novel AutoML-based framework that assists domain experts in selecting and configuring

ML-based data analytics solutions. It mainly attempts to respond the question that which type of algorithm fits well on the problem at hand and how related parameters should be set to get most satisfactory results. The framework is designed in order to implement two approaches. The first one is based on meta-learning [5] concepts. It means that the system learns from its past experiences to improve the automation of further arriving business data analytics. The second approach is mainly based on the use of ontologies [6] as a data store explicating the whole knowledge needed to implement the automatic assistance tool.

The rest of the paper is organized as follows: Section 2 discusses the closely related works in respect of ML-based data analytics in industry 4.0. Section 3 derives the goals that proposed framework is expected to achieve in the industry 4.0 domain. Section 4 introduces the main components of the proposed framework and discusses how these components collaborate to achieve the pursued goals. We later on show, how to implement the framework for representative application scenarios, in the Sect. 4.2, and we discuss the prototypical implementation. Finally, the Sect. 5 concludes the paper and outlines future perspectives.

2 Related Works

This section contains an overlapped overview of two research areas: (1) the application areas of data analytics in manufacturing industry, and (2) the challenges of applying ML algorithms as well as knowledge engineering in Industry 4.0 area. Likewise, we go through the limitations of previous works to motivate the work described in this paper.

2.1 Application Areas of ML-Based Data Analytics in Manufacturing

In manufacturing industry, ML techniques have been applied across three different levels that can be briefly described by the terms: *plan*, *make*, and *maintain*. These terms map the fundamental stages composing the manufacturing projects, which are summarized as below:

Plan is a set of tasks that deal with demand forecasting and jobs scheduling activities, which are, in fact, the main elements of production planning processes. The *demand forecasting* mostly uses historical time-series to predict future demands. Especially in large supply chains, demand forecasting can lead to a mitigated bullwhip effect [7]. The *jobs scheduling* is concerned with the process of arranging, controlling, and optimizing work along with the workloads in a production process or manufacturing process. In manufacturing, the purpose of scheduling is to minimize the production time and costs, such that a production facility knows when to make a product, with which resources, and through which equipment [2].

Make   stage encompasses activities like *product inspections and process diagnostics* and *predictions*. These tasks aim to optimize the manufacturing processes with respect to quality, time, and cost criteria. The *product inspection* can use image data in order to analyze the product state (e.g. the surface quality of steel [8]). The *process diagnostics* and *predictions* take care of controlling and improving the quality of products. Adjusting the technological processes with respect to these parameters can improve the productivity and the quality of the final product as well as it may reduce the manufacturing costs [1].

Maintain   stage concerns mainly the activities like *diagnostic* and *predictive maintenance*. These two application areas are strongly correlated [2]. Their aim is to reduce the downtime of manufacturing machinery, tools, and equipment. In recent years, a new maintenance paradigm known as condition-based maintenance has gained considerable attention [2]. It utilizes online condition variables to detect, identify, and forecast potentially detrimental fault conditions. The current condition-based plans designated as the diagnostic maintenance assesses the current state of the system in real-time. It performs maintenance when a triggering event occurs, such as the growth of a fault frequency beyond a particular threshold. However, a major drawback of the diagnostic maintenance is its inability to anticipate a breakdown well ahead of time. To overcome this issue, the predictive maintenance is used. It monitors the current and past states of the system and expects its future operational load in order to predict the remaining life period [2]. Such temporal information helps maintenance engineers to proactively order important spare parts that might not be immediately available in case of a failure.

2.2   Challenges in Building ML Predictive Models with Big Manufacturing Data

The predictive modeling is crucial to transform large manufacturing datasets, or *big industrial data* into actionable knowledge for various industrial applications. Predictive models can guide manufacturing decision making as well as personalized manufacturing. For example, we can prevent shutdowns in the production lines by predicting the time of occurrences of machinery failures and the criticality of the failures [2].

As a major approach to predictive modeling, ML is aided with computer algorithms, such as Decision Trees, Support Vector Machine, and Neural networks [9]. These algorithms use past experiences to improve both predictions and decision making. It is important to note that, historically, most of ML algorithms have been criticized for providing no explanations for their prediction results. Nevertheless, new methods have been recently developed to automatically explain prediction results of machine learning models without losing accuracy [10]. Nowadays, providing

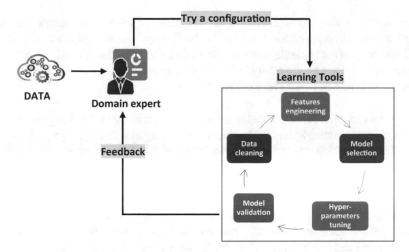

Fig. 1 The ML configuration tuning process

explanations is considered as a critical requirement of applying AI, specially for the real-time systems. This leads to the development of a new discipline called Explainable AI[1]. However, existing ML tools (such as AutoWeka [11] and TPOT [12]) are observed to present difficulties to use them efficiently in industrial area. In this regard, we discuss two major challenges in the following sections.

Challenge 1: efficiently and automatically selecting algorithms and hyperparameters values

All machine learning algorithms have two kinds of model parameters:

1. the ordinary parameters that are automatically learned and optimized during the training of the model;
2. the hyperparameters (categorical and continuous) which are manually set by the user of a machine learning tool before beginning the training of the model.

Given a modeling problem like, predicting whether an equipment failure will occur, an analyst manually builds an incremental model, as shown in Fig. 1. Once a learning problem is defined, the practitioner needs to find adequate learning tools to solve it. These tools can target different parts of the ML pipeline, i.e., features engineering, model selection, and hyperparameters tuning or optimization. To obtain a good learning performance, (s)he will try to set a configuration using personal experience or intuition about the underneath data and tools. Then, based on the feedback about how the learning tools performed, the practitioner will adjust the configuration hoping that the performance shall be improved. Such a trial-and-error process terminates once a desired performance is achieved or the computational budget runs out [9].

It is empirically proved in the available literature that the algorithms and the used hyperparameters values, affect the model accuracy [11]. Furthermore, the effective combination of an algorithm and the hyperparameter values varies with respect to the modeled problem. In the literature, some authors explore the automatic search of algorithms and hyperparameters values [3]. Evidently, it shows that automatic search methods can obtain equivalent or even better results than those resulting from the manual tuning done by the machine learning experts [9].

Challenge 2: Black-box nature of AutoML

As we mentioned before, there are several challenges that tackle the application of ML in the manufacturing space. Besides the difficulty of constructing a high-quality dataset, a much bigger issue arises, consisting of a lack of transparency regarding the decisions made by AutoML systems making them as black-boxes [10]. The lack of transparency leads ML experts and novices alike to question the results that were automatically obtained. If users cannot interpret the obtained results, they will not trust the AutoML system they are attempting to use and hence, they will hesitate to implement the model in critical applications, especially in manufacturing fields where interpretation and transparency of algorithms are a must for a system to be adopted into a workflow [1]. Another reason that justifies the low adoption rate of AutoML solutions in the industrial space is that the current methods for the ML pipeline optimization are inefficient on the large datasets originating from the manufacturing environment with various formats (i.e. images, sensors data) along with different quality levels [4].

3 Automated Machine Learning

Automated machine learning is the process of automating the tasks of applying ML to real-world problems. The main aim is to leverage the human expertise by allowing non-ML experts to define and parametrize data analytics solutions under predefined conditions on a target dataset. The core problem considered by AutoML can be formulated as follows: given a dataset, a machine learning task and a performance criterion; solve the task with respect to the dataset while optimizing the performance [3]. Finding an optimal solution is especially challenging due to the growing amount of available ML algorithms and their hyperparameters configurations [9].

Multiple approaches have been proposed to tackle the above problem. These approaches range from automatic features engineering [13] to automatic model selection [3, 9] and automatic hyperparameters tuning [12, 14]. Some approaches attempt to automatically and simultaneously choose a learning algorithm and optimize its hyperparameters. These approaches are also known as Combined Algorithm Selection and Hyperparameters optimization problem (CASH) [11, 12]. A solver for the CASH problem aims to pick an algorithm from a list of options and then tune it to give the highest validation performance amongst all the possible combinations of algorithms and hyperparameters.

The field of AutoML [3] is among the rapidly emerging subfields of ML that attempts to address the theoretical and algorithmic challenges as well as the development and deployment of systems that fully automate the ML process [10]. The goals of AutoML can be summarized as given below:

- Increasing the productivity by playing the role of a shield against methodological errors and over-estimations of performance.
- Democratizing the application of ML to non-experts of data analysis by providing them with "off the shelf" solutions.
- Enabling the knowledge practitioners to save time and effort.

Automation of analytics workflows or their parts have been studied and attempted actively over the past decade. As a result, there are various approaches to support data scientists and neophyte ML domain experts. Hereafter, we show two main approaches respectively based on the meta-learning and the use of ontologies.

3.1 Meta-Learning-Based Approach

The selection of the best algorithm for a given task can be investigated in a sub-area of ML known as meta-learning [5]. Meta-learning or learning to learn, aims to learn which algorithm shall work well for a dataset with certain characteristics, or which hyperparameters shall give a good performance. The idea of meta-learning for configuration recommendation is based on the following assumption: "*Algorithms show similar performance for the same configuration for similar problems*". It consists of generating a meta-model that maps characteristics of problems (i.e., meta-features) to the performance of algorithms that can be used to solve those problems.

In order to train a meta-model, a meta-dataset, for the learning purpose, is required. A meta-dataset contains information about machine learning experiments. For instance, it captures the knowledge about what combination of machine learning algorithm is used with which hyperparameters configurations, and how well the resulting model performed. In addition, for each experiment, it also describes the dataset on which the experiment was performed. The description of the dataset is done by meta-features that capture information about the data, such as the number of attributes, classes, kurtosis, skewness, etc.

3.2 Ontology-Based Approach

The ontology-based approaches for semantic data mining attempt to make use of formal ontologies in the data mining process. In contrast to the conventional data-driven meta-learning, semantic data mining is extensively driven by the knowledge of the data mining process as well as its components expressed in a data mining ontology or knowledge base. A well designed ontology can assist data analysts and neophyte ML domain experts to select appropriate modeling techniques. It may not only help to build specific models but it is also helpful to describe, in a number

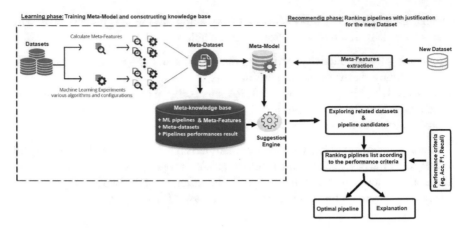

Fig. 2 The workflow of the proposed framework

of ways, the rationality behind the selection of the techniques and models [6]. By expressing the domain expertise in a formal structure, one can use logical reasoning to reduce the search space and hence, find the most predictive model for a given dataset.

Nural [6] gives a comparison of the meta-learning based approach with the Scala-Tion ontology-based suggestion approach [6] on a set of 114 datasets. Using Scala-Tion, each dataset is provided as an input to the suggestion engine and the suggested modeling technique is recorded. When predicting the top-1 performing technique, the ontology-based approach achieves an accuracy of 51% compared to 75% with meta-learning.

4 Framework and Methodology

The global architecture of the proposed framework is depicted in Fig. 2. The ML pipelines suggestion is provided by the *Suggestion Engine* through the use of a knowledge base (KB). This knowledge base is a collection of inductive meta-features that describes the datasets, the pipelines and their interdependencies. When a new dataset is presented to the system, the suggestion engine provides a recommendation of the most appropriate classifiers. This is achieved by combining the pipelines of the knowledge base with the morphological characteristics originating from the meta-model. It maps the characteristics of the given dataset to a label that describes similar datasets which are already processed in the meta-learning space.

Table 1 An example (sample) from the meta-learning space

Dataset	Metafeatures					
	nb. of Instances	nb. of Classes	nb. of Attr.	Class Entropy	Attr. Skewness	LD1
[8]	1941	2	34	0.93	3.65	0.72
[15]	45312	2	9	0.98	12.62	0.88
[16]	98050	2	29	1	1.36	0.89

4.1 Conceptual Description

The proposed framework is built upon the meta-learning concept. Therefore, it consists of two main phases: the learning phase and the recommendation phase.

In the *learning* phase, we generate and train multiple pipelines architectures on a large set of diverse industrial datasets. For this purpose, we run $\mathcal{N}$ classifiers, each classifier with $\mathcal{K}$ different hyperparameters configurations, over $\mathcal{D}$ datasets. The results of all runs are stored in a knowledge base where each record represents one run of a classifier C with hyperparameters configurations $\mathcal{K}$ over a dataset $\mathcal{D}$. In particular, each record stores the meta-features that model the dataset, the pipeline, and their interdependencies, as shown in Table 1. In addition to the results of the runs including the classifier performance metrics (e.g., Accuracy, F1-Score); we then use these meta-features to train a meta-model that is able to predict the optimal pipeline of a given dataset.

The main functionality of the meta-model can be formally defined as follows: given a set of learning algorithms space $C = \{C_1^{(1)}, \ldots, C_{\mathcal{N}}^{(\mathcal{K})}\}$, a dataset $\mathcal{D}$ divided into disjoint training D_{train}, and validation $D_{validation}$ sets, and an evaluation measure $\mathcal{M}$, the goal is to identify the classifier(s) $C^{(i)*}$ where $C^{(i)*} \in C$ and $C^{(i)*}$ is a tuned version of $C^{(i)}$ that maximizes or minimizes the $\mathcal{M}$ on $\mathcal{D}$.

$$C^{(i)*} \in \underset{C^{(i)} \in C}{Argmin} \; \mathcal{M}(C^{(i)}, D_{train}, D_{validation}) \tag{1}$$

In the *recommending* phase, the system receives a new unseen dataset and an evaluation metric. We first extract the meta-features that describe the dataset and then, we use the meta-model for suggesting the top modeling pipeline $C^{(i)*}$. The content of the knowledge base is then analyzed in order to produce a sorted list containing the candidate pipelines with an explanation of each suggestion based on the provided performance criteria.

We discuss in the below, some of the important concepts of the proposed framework:

Datasets A dataset represents a set of train/test data that we use during the offline phase to train the meta-learners. The input variables used for building machine learning models can come from various industrial sources, such as structured data in manufacturing administrative systems (Cyber-Physical Systems), manufacturing sensors (machine's condition parameters (temperature, vibration, etc.), operating parameters (speed, pressure, etc.), and environmental parameters (humidity, temperature, etc.)), etc.

Meta-features that characterize a dataset. Two main classes of meta-features have been proposed: (i) *General measures* that include general information related to the dataset at hand. To a certain extent, they are conceived to measure the complexity of the underlying problem. Some of them are: the number of attributes, classes, missing values, etc.(ii) *Statistical and information-theoretic measures* that describe statistics of the attributes and class distributions of a dataset sample. They include different summary statistics per attribute like mean, standard deviation, skewness, class entropy, etc.

Meta-knowledge base provides the means to store and search the knowledge issued from offline phase. The main functionality of the Meta-knowledge base can be divided into the two following categories: (i) Storing the knowledge (i.e. pipelines, performance results, meta-features) of the offline phase in a structured way (i.e. ontology or knowledge graph,). (ii) Searching and acquiring information.

An extensive usage of ontologies or knowledge graphs in our system is planned not only for better understanding the suggestion engine's queries and obtain information but also for driving the explanation acquisition process.

4.2 Prototypical Implementation and Use Cases

The learning phase is performed offline and consists of two main steps. In the first step, the meta-dataset is established. For each classification algorithm that is considered by the system for application, i.e., for the time being the proposed system supports 08 classification algorithms.

The meta-dataset is constructed by extracting 42 datasets characteristics (meta-features) using the PyMFE tool [17] for the general, statistical and info-theoretical categories, and by generating different measures on the performance of the classification algorithms on the datasets (e.g., predictive accuracy, precision, recall). Datasets characteristics and performance measures altogether are referred to as meta-data. The meta-knowledge base is provided as a data source for the suggestion engine. It is applied for metadata management which is built on top of the meta-knowledge base to implement the system and develop the graphical user interface. In the second step, meta-learning is performed on top of the meta-Knowledge base.

We implement a prototype of the proposed framework using the Python programming language. The used ML algorithms implementations are provided by the Sklearn data-mining library. These classifiers are *AdaBoost, SVM, Extra Trees, Gradient Boosting, Decision Tree, Logistic Regression, Random Forest, and Stochastic Gradient Descent (SGD)* classifiers. We have been performing an empirical study to evaluate the performance that can be achieved by using the proposed framework on various datasets. These data are gathered from state-of-the-art papers, mainly dealing with industry 4.0 related problems using ML solutions. They cover binary and mul-

Table 2 Comparative results of the effectiveness of the proposed system over domain experts (industrial researchers) configurations

Dataset	Task	Recommended config. result	Original paper result
[8]	Pipeline networks Failure risk analysis	93.74	85
[15]	RUL prediction	99.41	98.95
[16]	APS system failure prediction	99.10	92.56
[18]	Chatter prediction	97.06	95

ticlass classification problems from different industrial levels to ensure a meaningful evaluation of the prototype's capabilities.

As shown in Table 2, the obtained results are relatively more accurate than the results from the original papers. Thence, it can be observed that some ML oriented manufacturing works could be further improved, simply through the use of a better ML algorithm configurations.

Many use cases in manufacturing are compatible with the proposed AutoML based Framework as they often share similar logic. The usual goal is to help domain experts in identifying a particular category, e.g., of a root cause, error type, and the health of machinery. Such scenarios exist in any phase of a product lifecycle, e.g., during the control of the quality of a product, predictive maintenance of a machine or after sales customer support.

5 Conclusion and Future Work

The implication of advanced data analytics in industrial processes can significantly improve the manufacturing processes. There exists a wide range of IDA methods devoted to solve the problems in the manufacturing field. However, the practical implementation is still in its early stages. Among others, the major reason that impede the use of advanced data analytics in industrial contexts, is the lack of tools for decision makers in order to choose the suitable technique. In traditional approaches, the selection of ML algorithms is a tedious task because of trial-and-error strategy.

The industrial experts usually have limited computing expertise to develop ML predictive models. The presented work falls within a framework that assists the industry 4.0 actors and researchers to integrate the ML solutions with the help of past experiences and without having to perform extensive experiments. The proposed framework supports the whole process of iterative machine learning on big industrial data. It extracts manufacturing parameters, and then assists the users to build and evaluate the appropriate predictive models. The proposed approach provide means to enhance the usability of big industrial data to foster the smart factory discovery along with safety improvements.

In the future, we plan to further extend the scope of our framework by evaluating its effectiveness against state-of-the-art systems, and on different kinds of use cases with a particular focus on the predictive maintenance activity in the manufacturing industry. This may also require to add the support for further data formats and ML algorithms, as needed by the respective use cases.

Acknowledgements This work has been supported, in part, by Hestim, CNRST Morocco, and University of the Littoral Cote d'Opale, Calais France.

References

1. Garouani M, Ahmad A, Bouneffa M et al (2022) Towards big industrial data mining through explainable automated machine learning. Int J Adv Manuf Technol. https://doi.org/10.1007/s00170-022-08761-9
2. Sharp M, Ak R, Hedberg T (2018) A survey of the advancing use and development of machine learning in smart manufacturing. JMS 48:170–179. https://doi.org/10.1016/j.jmsy.2018.02.004
3. Reif M et al (2014) Automatic classfier selection for non-experts. Pattern Anal Appl 17(1):83–96. https://doi.org/10.1007/s10044-012-0280-z
4. Villanueva Zacarias AG, Reimann P, Mitschang B (2018) A frame-work to guide the selection and configuration of machine-learning-based data analytics solutions in manufacturing. In: Procedia CIRP, vol 72, pp 153–158. https://doi.org/10.1016/j.procir.2018.03.215
5. Garouani M et al (2021) Towards the automation of industrial data science: a meta-learning based approach. In: Proceedings of the ICEIS'23, pp. 709–716. https://doi.org/10.5220/0010457107090716
6. Nural MV (2017) Ontology-based semantics vs meta-learning for predictive big data analytics. University of Georgia, Athens, GA, USA
7. Wolf H et al (2019) Bringing advanced analytics to manufacturing: a systematic mapping. In: IFIP advances in information and communication technology, pp 333–340. https://doi.org/10.1007/978-3-030-30000-5_42
8. Mazumder RK, Salman AM, Li Y (2021) Failure risk analysis of pipelines using data-driven machine learning algorithms. In: Structural safety, vol 89, p 102047. https://doi.org/10.1016/j.strusafe.2020.102047
9. Luo G (2016) PredicT-ML: a tool for automating machine learning model building with big clinical data. In: HISS 4.1, pp 1–16. https://doi.org/10.1186/s13755-016-0018-1
10. Garouani M et al (2021) AMLBID: An auto-explained automated machine learning tool for big industrial data. In: SoftwareX, vol 16. https://doi.org/10.1016/j.softx.2021.100919
11. Thornton C et al (2013) Auto-WEKA: combined selection and hyperparameter optimization of classification algorithms. In: Proceedings of the ACM SIGKDD, pp 847–855. https://doi.org/10.1145/2487575.2487629
12. Olson RS, Moore JH (2019) TPOT: a tree-based pipeline optimization tool for automating machine learning, pp 151–160. https://doi.org/10.1007/978-3-030-05318-5_8
13. Katz G, Shin ECR, Song D (2017) ExploreKit: automatic feature generation and selection. In: Proceedings—IEEE international conference on data mining, ICDM, pp 979–984. https://doi.org/10.1109/ICDM.2016.176
14. Feurer M, Springenberg JT, Hutter F (2015) Initializing Bayesian hyperparameter optimization via meta-learning. In: Proceedings of the national conference on artificial intelligence, vol 2, pp 1128–1135

15. Benkedjouh T et al (2015) Health assessment and life prediction of cutting tools based on support vector regression. JIM 26(2):213–223. https://doi.org/10.1007/s10845-013-0774-6
16. Costa CF, Nascimento MA, IDA (2016) Industrial challenge: using machine learning for predicting failures. In: Advances in intelligent data analysis, vol XV, pp 381–386. https://doi.org/10.1007/978-3-319-46349-0_33
17. Alcobaca E et al (2020) MFE: towards reproducible meta-feature extraction. J Mach Learn Res 21(111):1–5
18. Saravanamurugan S et al (2017) Chatter prediction in boring process using machine learning technique. Int J Manuf Res. https://doi.org/10.1504/IJMR.2017.10007082

Multidimensional Data Visualization Based on the Shortest Unclosed Path Search

Oleg Seredin⬤, Egor Surkov⬤, Andrei Kopylov⬤, and Sergey Dvoenko⬤

Abstract The paper considers methods of multidimensional data visualization based on the search for the shortest unclosed path between the objects of the study sample and its mapping to a two-dimensional plane as an unclosed graph (chain), a columnar chart of the objects distribution along the found path or projection onto the path. The shortest unclosed path computing is executed with distance matrix between objects, which allows to use the method for both multidimensional data and data represented only by paired comparisons of objects functions. In this work an algorithm for greedy search of a quasi-shortest unclosed path is implements and also its modifications are proposed. The algorithms are tested on model and real-world data and also data obtained during the study of the problem of detecting falls using Microsoft Kinect V2.

Keywords Shortest unclosed path · Multidimensional data visualization · Pairwise similarity function

1 Introduction

Data visualization is a clearly information representation. One of the targets of data visualization is a simplifying their comprehension for the following analysis, i.e., a graphic demonstration of the certain object behavior. Visualization is also needed for large datasets analysis [1]. Visualization methods are applied to the processing of data arrays that are represented as real or complex numbers on a continual or discrete scale and derived in the process of conducting real or computing experiments. Also, during visualization, it is necessary to take into account the importance of interdisciplinary relation, when methods and tools of one subject field are successfully applied to solve

Supported by Ministry of Science and Higher Education of the Russian Federation within the framework of the state task FEWG-2021-0012.

O. Seredin · E. Surkov (✉) · A. Kopylov · S. Dvoenko
Tula State University, Tula, Russia
e-mail: eg-su@mail.ru

© The Author(s), under exclusive license to Springer Nature Switzerland AG 2022　　　279
N. H. T. Dang et al. (eds.), *Artificial Intelligence in Data and Big Data Processing*,
Lecture Notes on Data Engineering and Communications Technologies 124,
https://doi.org/10.1007/978-3-030-97610-1_23

a problem from another area of study. This is also due to the necessity of information extraction to obtain new characteristics of the study process or phenomenon and the issues of preparing data for better perception [2].

The complexity of data visualization appears when dimension of the studied objects feature space becomes more than three. In the simplest case, data visualization involves image of a certain function dependence of one or more parameters. This problem is trivial for two- and three-dimensional cases when it is sufficient to construct a chart of the corresponding dimension. However, in real tasks an object of the surrounding world is described by more than one pair of features to describe them, it is necessary to use tens and hundreds of factors influencing them [3]. Dependency visualizing with a large number of discriptive Resive features is quite problematic. Thus, when selecting the number of characteristics of an object, the following are taken into account, both the "simpleness" and clearness of visualization as well as the reliability and accuracy of the study [4].

Formally, the task of data visualization appears like this: it is necessary to find such a representation of objects from the original dimension space M into space of smaller dimension $M \leq 3$, that would minimize the loss in the object description quality. Such methods are called purposeful projection into a small-dimensional space [5].

In this work methods for visualizing multidimensional data based on the shortest unclosed path between the objects of the studied sample are proposed and demonstrated. The paper considers several criteria for the shortest unclosed path search and also proposes algorithms for its computation. The experimental results on model and real-world data, such as Iris Data Set [6] Abalone Data Set [7] are presented. The experiment was also conducted on data from the study of the fall detection task [8]. An experimental comparison of the shortest path search algorithms is performed and also, as part of this work, comparative tables of time spent on their calculation are presented.

2 Multidimensional Data Visualization Based on the Shortest Unclosed Path Search

The shortest unclosed path (SUP) is a graph of $N - 1$ edges that connects N objects of space in such a way that the total length of the edges is minimal [9]. The visualization of multidimensional data core idea in the work is visualization based on the search for the shortest unclosed path (SUP) between the sample objects and its mapping to a two-dimensional plane in the following form:

1. an unclosed graph (chain). This visualization method is displaying points on two-dimensional plane and connected them by edges in the order calculating by SUP algorithm. This approach is applicable only for two-dimensional data;

2. a column chart of objects distribution along the computed path. This is the basic approach of data visualization that is used in the work to represent data. This

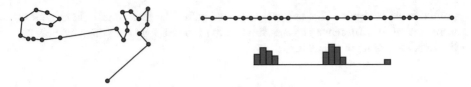

Fig. 1 Visualization by a column chart

visualization method is that path obtained by SUP calculating could be "stretch out" into a straight line, where objects will be located in the order found by the SUP. This a straight line could be visualized as a column chart. The path length is plotted along the abscissa axis, which can be divided into segments (their maximum number is equal to the objects number). Along the ordinate axis, the number of objects that fell into the bin is plotted (Fig. 1);

3. projection onto the path. The third visualization method is a projection onto the path, the distance between objects along the path is plotted along the abscissa axis, the Euclidean distance from the starting object to the current one is plotted along the ordinate axis. This visualization method allows to show the dependency of the position of relative to each other objects from any dimension space. The advantage of the second and third visualization approaches is that the algorithms is independent from the feature space dimension which describe objects. These algorithms could be implemented only on the matrix of distances between objects (further in the text we will use the term point, understanding the object as a point in multidimensional space).

2.1 Description of the SUP Search Algorithms

The paper considers two criteria for the SUP search:

1. The first criteria is to find the shortest unclosed path and is mathematically expressed as the minimization of the following functional:

$$J_1 = min \sum_{i=2}^{N} d_{i,i-1}, \tag{1}$$

where $d_{i,j}$—distance between i-th and j-th points, $i = 1, \ldots, N$—the order of elements in the path bypassing, N—number of elements. Note, if the objects are represented by a distance matrix A then the task of SUP searching on such objects is reduced to solving the problem of search such a distance matrix by permutation of corresponding rows and columns that the sum of elements above the main diagonal is minimal: $J_1 = min \sum_{i=1}^{N-1} A_{i,i+1}$, where $A_{i,j}$—distance between i-th and j-th points.

2. The second criteria is heuristic and is used when it is necessary to find the path with the shortest length and with the greatest distance between the terminal

points, thus "straighten" the shortest unclosed path. The second criteria provides a minimum of the difference between the length of the SUP and the distance between the terminal points of the path:

$$J_2 = min\left(\sum_{i=2}^{N} d_{i,i-1} - \alpha \cdot d_{1,N} \right),\tag{2}$$

where α—weight coefficient (in this work $\alpha = 1$).

The paper implements an algorithm for the shortest unclosed path search which is a brute-force of connecting points and choosing the shortest path, that is, one that's sum of the distances between the points when traversing is minimal (further algorithm A0). The complexity of this algorithm is that $N!$ operations of constructing different chains are performed. It takes $N - 1$ calls to the distance matrix to calculate the sum of the one chain elements. In case of there are symmetric chains, the complexity estimate will be halved and will be equal $\frac{(N-1)\cdot N!}{2}$. The shortest unclosed path can only be found by algorithm A0, however, this algorithm requires a lot of time and is implemented only for a small number of points (during experimental studies in this paper, it was found out that this is approximately 15 points). Therefore, its modifications were proposed—algorithms A1, A2, A3, A4, which are greedy and, in general, are not able to find the SUP, but they allow for an acceptable operating time and a quasi-shortest unclosed path.

Algorithm A1—its idea is that first you need to find a pair of points (terminal points), the distance between which will be minimal. Then save them to an array with isolated points (these are the points that are included in the path). To each of these terminal points, find the nearest of the remaining free points and compare these distances. Add the point which the distance from the terminal point is minimal to the isolated points. Further, for two terminal points (one of which has been updated) repeat the search for the nearest of the free points and save the one that turned out to be closer, and so repeat the iterations until all the free points become isolated. The complexity of this algorithm is equal to $N^2 - \frac{3N}{2}$. The disadvantage of this algorithm is that it finds only one path.

Algorithm A2—modification A1, is that for each beginning point there is the nearest one and these two points presented as starting segment. Then the actions are repeated according to the algorithm A1. Thus, this approach performs N operations of chain construction from which the one with the minimum length is selected The complexity of this algorithm is equal to $(N^3 - N^2 - N)/2$.

Algorithm A3—modification A1, is that the SUP graph is constructed from each pair of points. Thus, N^2 graph construction is performed. The complexity of this algorithm is equal to $\frac{N^4}{2} - N^3$.

Obviously, A2 and A3 will include the solution found when using A1, and the solution found with A2 is included in A3.

Algorithm A4—modification of the algorithm A3, is that initially one point is selected as starting point. Then all pairs are attached to the left and right of starting point, which become terminal for this triple. Further, the search is performed as in the algorithm A1. The complexity of this algorithm is estimated as $\frac{N^5}{4} - \frac{7N^4}{4} + \frac{9N^3}{2} - 5N^2 + 2N$.

Recursive variants of these algorithms A1R, A2R, A3R, A4R are also proposed, which consist in the fact that if several equidistant points are found when searching for the nearest point (from the free list) to the terminal point, then each option of such a path is considered. At the end, a path that provides the minimum of the criterion J_1 or J_2 is selected. The number of graph constructions depends on how many pairs of points with the same distance will be contained in the data. It takes more time to calculate a recursive greedy algorithm, however, it is assumed that this time is still several orders of magnitude less than for calculations with the A0 algorithm.

3 Experiments on Model and Real Data

3.1 Data Description

Dataset 1 (13 points). Figure 2a shows an example of the location of points on a two-dimensional plane forming a polyline. The specificity of this dataset is that there are several solutions to the problem of the shortest unclosed path search and the optimal solution for such a simple configuration is computed only by one of the proposed greedy algorithms (A4).

Dataset 2 (15 points). Figure 2b shows the dataset 2, which consists of 15 points which organized into three clusters. Dataset 3 (15 points). Figure 2c shows an example presented the location of points forming two circles. Dataset 4 (18 points). The points form two circles is shown on Fig. 2d.

Dataset 5. The real dataset of Fisher's irises is used as a dataset 5 [7]. The dataset contains 150 objects, each of which is described by four features. Fisher's irises is interesting due to a large number of equidistant objects and it is difficult to search a SUP on such a set of points, since there will be a corresponding number of identical potential SUP.

Dataset 6. The real dataset Abalone is used as a dataset 6 [10]. The dataset is used to predict the age of mollusks from physical measurements. The set contains 4177 objects, each of which is described by eight features, one of which is a prediction value (ring), therefore it will not be used to describe objects.

3.2 Results of Visualization of Model Data Sets by the Developed Algorithms

1. The first experiment was conducted on dataset 1. Table 1 presents a comparative result of the algorithms by the values that were obtained when searching for the shortest unclosed path on the data in the form of a polyline.

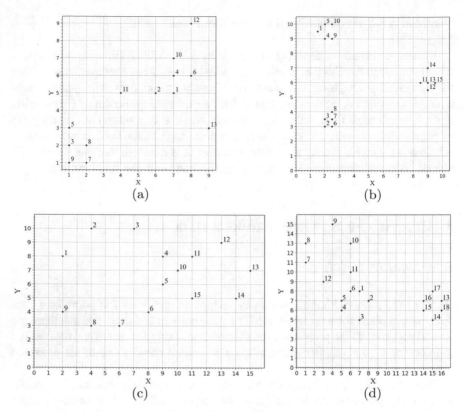

Fig. 2 Model datasets **a** the points form a polyline, **b** the points have a cluster structure, **c** the points have a cluster structure, **d** the points form three circles

Table 1 Results of algorithms for dataset 1

Algorithm	A0	A1	A1R	A2	A2R	A3	A3R	A4	A4R
Search by criterion J_1									
Path length	**20.832**	22.339	22.339	22.339	22.339	22.339	21.004	**20.832**	**20.832**
$d_{1,N}$	9.219	8.0	8.246	8.0	8.246	8.0	8.246	9.219	9.219
Path length—$d_{1,N}$	11.613	14.339	14.092	14.339	14.092	14.339	12.758	11.613	11.613
Search by criterion J_2									
Path length	20.832	22.339	22.339	22.339	22.339	23.217	21.69	20.832	20.832
$d_{1,N}$	10.630	8.0	8.246	8.0	8.246	9.219	9.219	9.219	10.63
Path length—$d_{1,N}$	**10.202**	14.339	14.092	14.339	14.092	13.998	12.470	11.612	**10.202**

Fig. 3 A variant of the SUP for dataset 1, found by algorithms A0, A4, A4R by minimizing criteria J_1 and J_2

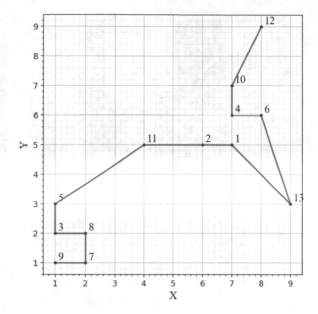

Algorithm A0 on this dataset finds three SUP with a length of 20.832. The solutions found when minimizing criteria J_1 and J_2 coincide. The solution with such a long path was found by the A4 and A4R algorithms, but all three variants of the shortest unclosed paths on this dataset could only be computed by the A4R algorithm. Figure 3 shows one of the three SUP options for dataset 1.

Figure 4a shows the visualization of the dataset 1 presented by a column chart for the path found with the algorithms A0, A4, A4R when optimizing the criterion J_1. Figure 4b shows the visualization of the dataset 1 as projection on the path found by algorithms A0, A4, A4R using the criterion J_1.

2. The second experiment was conducted on dataset 2. Table 2 presents a comparative result of the algorithms by the values that were obtained when searching for the shortest unclosed path on the data in the form of two circles.

The result of A1, A2, A3 are the same (Fig. 5a). The result of the algorithm A1, A2 is shown in Fig. 5b, A4 is shown in Fig. 5c. The result of A0, A4R, A3 is shown in Fig. 5d.

Figure 6a shows a visualization of dataset 2 in the form of a column chart for the path found by algorithms A0, A3R, A4R when optimizing J_1 criterion. Figure 6b shows the visualization of the dataset 2 as projection on the path found by the algorithm A0, A3R, A4R when optimizing the criterion J_1

3. The third experiment was conducted on dataset 3. Table 3 presents a comparative result of the algorithms by the values that were obtained when searching for the shortest unclosed path on the data in the form of two circles.

The results of the search for a solution by all the proposed algorithms while minimizing the criterion J_1 coincide (Fig. 7). The result of optimizing the J_2 criterion is shown in Fig. 7a, b. The algorithms find different paths at the same terminal vertices.

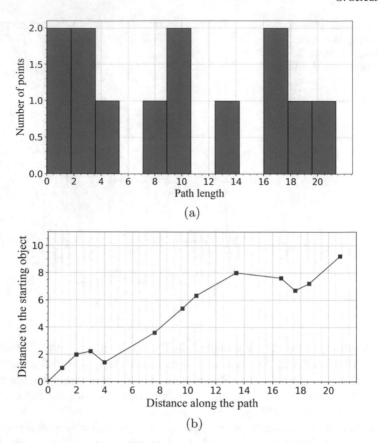

Fig. 4 Dataset 1 visualization **a** by a column chart **b** as a projection on the path

Table 2 Results of algorithms for dataset 2

Algorithm	A0	A1	A1R	A2	A2R	A3	A3R	A4	A4R
Search by criterion J_1									
Path length	**18.947**	19.330	18.948	19.330	18.948	19.330	**18.947**	19.328	**18.947**
$d_{1,N}$	7.159	7.159	7.159	7.159	7.159	7.159	7.159	7.159	7.159
Path length—$d_{1,N}$	11.788	12.17	11.789	12.17	11.789	12.17	11.788	12.170	11.788
Search by criterion J_2									
Path length	19.857	19.357	19.357	19.357	19.357	19.357	19.265	19.857	19.857
$d_{1,N}$	8.732	7.906	7.905	7.905	7.905	7.905	7.906	8.5	8.732
Path length—$d_{1,N}$	**11.125**	11.451	11.451	11.451	11.451	11.451	11.360	11.357	**11.125**

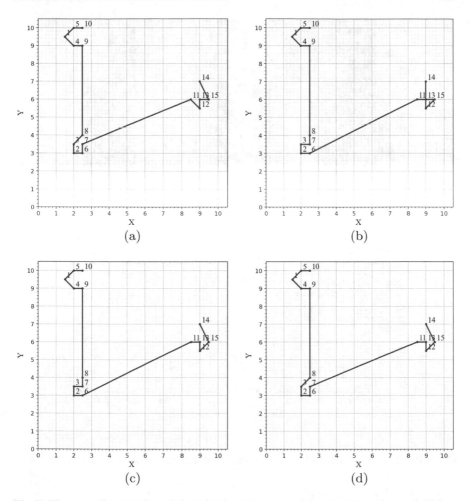

Fig. 5 The result of the SUP algorithms on the dataset 2 **a** algorithm A0, A3R, A4R, length 18.947, **b** algorithms A1, A2, A3 length 19.330, **c** algorithm A1R, A2R, length 18.948, **d** algorithm A4, length 19.328

None of the greedy algorithms could achieve the minimum of the J_2 criterion (see Fig. 8a). The recursive version of the A4 algorithm has found a solution very close to the optimal one.

For dataset 3, all algorithms calculated the same shortest unclosed path. Figure 8a shows a visualization of the dataset 3 in the form of a column chart for the path when optimizing criterion J_1. Figure 8b shows the visualization of the dataset 3 as projection on the path when optimizing the criterion J_1.

4. The fourth experiment was conducted on a dataset 4. Table 4 presents a comparative result of the algorithms by the values that were obtained when searching for the shortest open path on the data in the form of three circles (18 points).

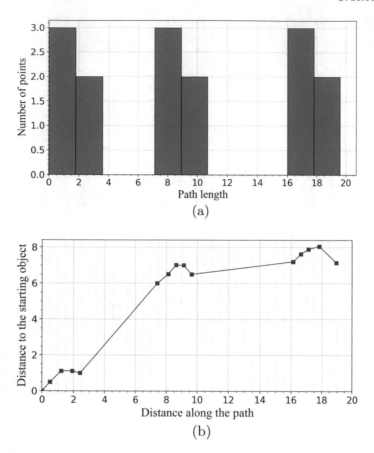

Fig. 6 Dataset 2 visualization **a** by a column chart, **b** as a projection on the path

Table 3 Results of algorithms for dataset 3

Algorithm	A0	A1	A1R	A2	A2R	A3	A3R	A4	A4R
Search by criterion J_1									
Path length	**32.730**	**32.730**	**32.730**	**32.730**	**32.730**	**32.730**	**32.730**	**32.730**	**32.730**
$d_{1,N}$	4.0	4.0	4.0	4.0	4.0	4.0	4.0	4.0	4.0
Path length—$d_{1,N}$	28.730	28.730	28.730	28.730	28.730	28.730	28.730	28.730	28.730
Search by criterion J_2									
Path length	36.024	33.828	33.828	33.828	33.828	33.828	33.724	33.724	36.128
$d_{1,N}$	11.180	8.602	8.602	8.602	8.602	8.602	8.602	8.602	11.180
Path length—$d_{1,N}$	**24.844**	25.226	25.226	25.226	25.226	25.226	25.121	25.121	24.948

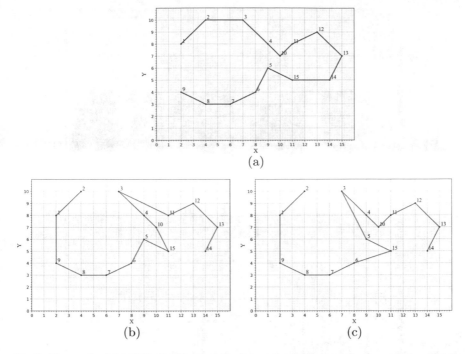

Fig. 7 The result of the SUP algorithms on the dataset 3 **a** during optimization criterion J_1 (all algorithms), length 32.73, **b** during optimizing the criterion J_2 by the algorithm A0, J_2=24.844, **c** during optimizing the criterion J_2 by the algorithm A4R, J_2=24.948

The omissions in the tables are explained by the fact that calculations using the A0 algorithm cannot be performed in an acceptable time. The results of calculations using the A3, A3, A4, A4R algorithms coincide and find the quasi-shortest unclosed path (Fig. 9a). The result of the calculation using algorithms A1, A2, A1, A2 is a longer path (Fig. 9b), however, the path calculated by the recursive versions of the algorithms has a greater distance between the terminal points (Fig. 9c).

Figure 10 shows the visualization of the dataset 4 in the form of a column chart for the path found by the greedy algorithms A3, A4, A3R, A4R when optimizing the criterion J_1. From the analysis of Fig. 10a, b, it can be seen that only the third cluster (objects 14–18) is well separated. Figure 10a shows the visualization of the dataset 4 as projection on the path found by the greedy algorithms A3, A4, A3R, A4R when optimizing the criterion J_1.

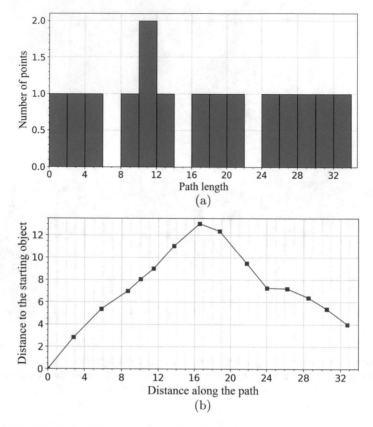

Fig. 8 Dataset 3 visualization **a** by a column chart, **b** as a projection on the path

Table 4 Results of algorithms for dataset 4

Algorithm	A0	A1	A1R	A2	A2R	A3	A3R	A4	A4R
Search by criterion J_1									
Path length	–	38.246	38.246	38.246	38.246	**36.627**	**36.627**	**36.627**	**36.627**
$d_{1,N}$	–	10.296	12.649	10.296	12.649	12.042	12.042	12.042	12.042
Path length—$d_{1,N}$	–	27.951	25.597	27.951	25.597	24.586	24.586	24.586	24.586
Search by criterion J_2									
Path length	–	38.246	38.246	38.246	38.246	36.627	36.627	38.58	38.58
$d_{1,N}$	–	10.296	12.649	12.649	12.649	12.042	12.042	15.231	15.231
Path length—$d_{1,N}$	–	27.951	25.597	25.597	25.597	24.586	24.586	**23.348**	**23.348**

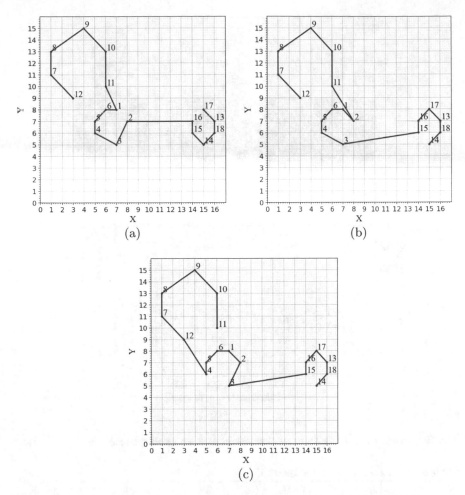

Fig. 9 The result of the SUP algorithms on the dataset 4 **a** algorithms A3, A4, A3R, A4R, length 36.627, **b** algorithms A1R-A2R, length 38.246, **c** algorithms A1, A2, length 38.246

3.3 Results of Real Dataset Visualization by the Developed Algorithms

5. The fifth experiment was conducted on Fischer irises (dataset 5). Table 5 presents a comparative result of the algorithms by the values that were obtained when searching for the shortest unclosed path. The omissions in this and following tables are explained by the fact that algorithm computing cannot be performed in an acceptable time.

Visualization of the constructed quasi-shortest path by the greedy A3R algorithm is shown in Fig. 11. To represent a four-dimensional object on a two-dimensional graph, the last two descriptive features were excluded.

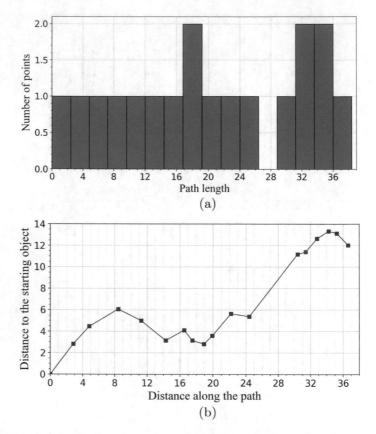

Fig. 10 Dataset 4 visualization **a** by a column chart, **b** as a projection on the path

Table 5 Results of algorithms for dataset 5

Algorithm	A0	A1	A1R	A2	A2R	A3	A3R	A4	A4R
Search by criterion J_1									
Path length	–	53.719	53.164	53.370	51.685	53.122	**51.627**	52.972	–
$d_{1,N}$	–	5.893	6.535	6.583	6.535	6.535	6.535	5.944	–
Path length—$d_{1,N}$	–	47.825	46.629	46.788	45.150	46.587	45.091	47.028	–
Search by criterion J_2									
Path length	–	53.719	53.164	53.370	51.685	53.1222	51.627	53.102	–
$d_{1,N}$	–	5.893	6.535	6.583	6.535	6.535	6.535	6.535	–
Path length—$d_{1,N}$	–	47.825	46.629	46.788	45.150	46.587	**45.091**	46.567	–

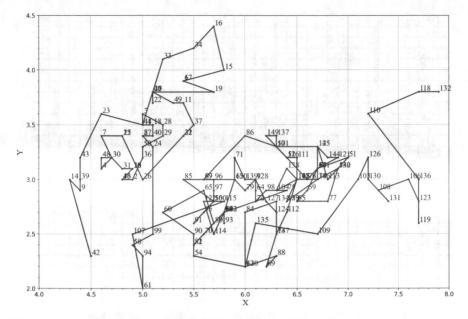

Fig. 11 The result of the SUP search with A3R algorithm on the dataset 5

Table 6 Results of algorithms for dataset 6

Algorithm	A0	A1	A1R	A2	A2R	A3	A3R	A4	A4R
Search by criterion J_1									
Path length	–	175.185	175.185	**173.392**	**173.392**	–	–	–	–
$d_{1,N}$	–	2.715	2.715	2.715	2.715	–	–	–	–
Path length—$d_{1,N}$	–	172.469	172.469	170.677	170.677	–	–	–	–
Search by criterion J_2									
Path length	–	175.185	175.185	173.392	173.392	–	–	–	–
$d_{1,N}$	–	2.715	2.715	2.715	2.715	–	–	–	–
Path length—$d_{1,N}$	–	172.469	172.469	**170.677**	**170.677**	–	–	–	–

Figure 12a shows a visualization of dataset 5 as a column chart for the path found by the A3R algorithm when optimizing the criterion J_1. Figure 12b shows the visualization of the dataset 5 as a projection on the path found by the A3R algorithm when optimizing the criterion J_1. Analysis of Figs. 11 and 12 shows good separability of the first class of colors, which is a well-known result of Iris data.

6. The sixth experiment was conducted on the Abalone dataset (dataset 6). Table 6 presents a comparative result of the computing with algorithms by the values that were obtained when searching for the shortest unclosed path.

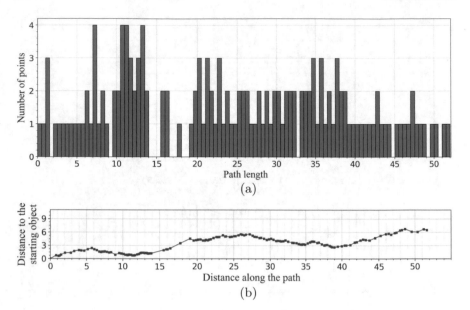

Fig. 12 Dataset 5 visualization **a** by a column chart, **b** as a projection on the path

Figure 13a shows a visualization of dataset 6 in the form of a column chart for the path found by the A2, A2R algorithms when optimizing the criterion J_1. Figure 13b shows the visualization of the dataset 6 as a projection on the path found by the algorithm A2, A2R when optimizing the criterion J_1.

According to analysis of the data demonstrated by charts in Figs. 3, 4, 5, 6, 7, 8, 9, 10, 11, 12 and 13 and in Tables 1, 2, 3, 4, 5 and 6 the following conclusions could be drawn. Algorithm A0 shows the best result, unfortunately it cannot be used on large samples with more than 15 objects. However, the same result was obtained using various modifications of the greedy algorithm (Tables 1, 2 and 3). Also, it's notable that recursive versions of algorithms allow to find shorter paths. However, their disadvantage is that their implementation requires a large amount of RAM and computing time. But it's still less than brute-force algorithm. It follows from the above tables that it is worth giving preference to the A4R algorithm.

4 An Experiment on Data Obtained During the Study of Fall Detection Task

4.1 Data Description

In the fall detection task [10, 11], the goal was set—to obtain a basic assembly of skeletal models for further comparison with the test sample [8]. The comparison is calculated using a pre-selected measure of dissimilarity between skeletal models.

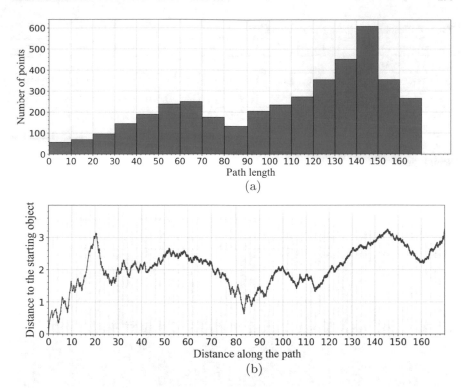

Fig. 13 Dataset 6 visualization **a** by a column chart, **b** as a projection on the path

The result of comparing the basic assembly and a test sample of objects is presented by a visualized distance matrix. It is assumed that the quality of the visualized matrix directly depends on the structure of the basic assembly of objects, exactly the order of skeletal models in the basic assembly. Firstly, the order of the rows in the distance matrix between the basic skeletal representations is not strictly defined. However, the structure of the basic assembly should reflect the similarity between objects of the basic assembly themselves. In the fall detection task, the searching for the shortest unclosed path between objects just yields an opportunity to arrange the objects of the basic assembly. Due to the fact that the elements of the basic assembly are represented by a distance matrix, will conduct an experiment with such a data set. Dataset 7 contains 136 elements, indeed.

4.2 Dataset Visualization Results

Table 7 presents a comparative result of the algorithms by the values that were obtained when searching for the shortest unclosed path on real data obtained during the study of fall detection task. The results of A4 and A4R are the same and provide the shortest path.

Table 7 Results of algorithms for dataset 7

Algorithm	A0	A1	A1R	A2	A2R	A3	A3R	A4	A4R
Search by criterion J_1									
Path length	–	6.000	6.000	5.984	5.984	5.971	5.971	**5.950**	**5.950**
$d_{1,N}$	–	0.455	0.455	0.415	0.415	0.420	0.420	0.420	0.420
Path length—$d_{1,N}$	–	5.545	5.545	5.569	5.569	5.551	5.551	5.529	5.529
Search by criterion J_2									
Path length	–	6.000	6.000	5.999	5.999	5.989	5.989	5.957	5.957
$d_{1,N}$	–	0.455	0.455	0.455	0.455	0.455	0.455	0.455	0.455
Path length—$d_{1,N}$	–	5.545	5.545	5.544	5.544	5.544	5.533	**5.502**	**5.502**

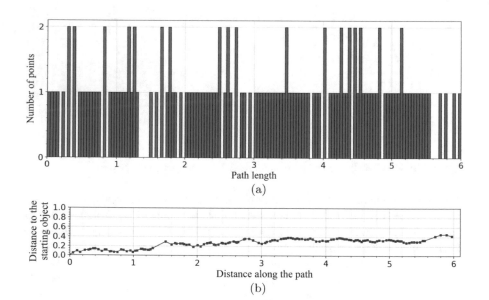

Fig. 14 Dataset 7 visualization **a** by a column chart, **b** as a projection on the path

Figure 14a shows a visualization of the dataset 7 in the form of a column chart for the path found by the A4, A4R algorithms using the criterion J_1. Figure 14b shows the visualization of the dataset 7 as a projection onto the path found by the A4, A4R algorithms for the criterion J_1.

According to analysis of Fig. 14, it could be seen that three clusters are visually separable in the data. Considering the basic assembly of objects in the fall detection task [8], we can make sure that the same data structure is observed on the graphs, the objects are clearly divided into skeletal models of three groups (standing, sitting, lying down).

Table 8 Estimate of algorithms computational time (hh:mm:ss.ms)

Points	A0	A1	A2	A3	A4	A1R	A2R	A3R	A4R
10	00:00:00.15	00:00:00.01	00:00:00.01	00:00:00.01	00:00:00.156	00:00:00.135	00:00:00.140	00:00:00.183	00:00:00.200
11	00:00:00.159	00:00:00.01	00:00:00.01	00:00:00.01	00:00:00.197	00:00:00.133	00:00:00.129	00:00:00.192	00:00:00.201
12	00:00:01.885	00:00:00.01	00:00:00.01	00:00:00.01	00:00:00.190	00:00:00.124	00:00:00.177	00:00:00.164	00:00:00.205
13	00:00:23.590	00:00:00.01	00:00:00.01	00:00:00.02	00:00:00.213	00:00:00.133	00:00:00.139	00:00:00.172	00:00:00.225
14	00:05:28.203	00:00:00.01	00:00:00.01	00:00:00.02	00:00:00.202	00:00:00.132	00:00:00.128	00:00:00.163	00:00:00.277
15	01:23:25.670	00:00:00.01	00:00:00.01	00:00:00.03	00:00:00.205	00:00:00.135	00:00:00.135	00:00:00 167	00:00:00.263
50	–	00:00:00.02	00:00:00.07	00:00:00.68	00:00:03.327	00:00:00.184	00:00:00.200	00:00:00.345	00:00:04.796
100	–	00:00:00.03	00:00:00.30	00:00:01.220	00:01:44.368	00:00:00.138	00:00:00.250	00:00:01.867	00:02:35.542
136	–	00:00:00.06	00:00:00.68	00:00:02.858	00:10:33.835	00:00:00.173	00:00:00.350	00:00:6.680	00:14:31.741
150	–	00:00:00.16	00:00:00.145	00:00:05.867	00:16:59.998	00:01:25.904	02:15:39.300	92:39:20:100	–
4177	–	00:00:22.760	37:25:00:01	–	–	00:00:22.760	37:25:00:01	–	–

5   Estimate of the Algorithms Computational Time

Table 8 shows the comparative results of the algorithm computational time on data of different volumes. The algorithms were implemented by means of the Java high-level programming language. The experiments were conducted on a computing device with an Intel core i7 8800HQ processor and 16 GB of RAM.

Analyzing the obtained time characteristics to draw the following conclusions: the algorithm A0 (brute-force), although it provides a search for the shortest path, but from the point of view of time costs, it is not realized for data that contains more than 15 objects. Algorithm A1 is the fastest, but its modifications A2, A3, A4 are able to find shorter paths. Among all the proposed versions of the greedy algorithm, it is the recursive ones that make it possible to find the shortest paths, meanwhile the time costs depend on the number of equidistant objects in the data. However, the time spent on computing recursive algorithms is still much less than on computing using the A0 algorithm. All versions of greedy algorithms are able to calculate the quasi-shortest path for 50 points in less than 5 s.

6   Conclusion

The main idea of visualization of multidimensional data, proposed and implemented in this paper, is the search for the shortest unclosed path between the studied objects and its mapping to a two-dimensional plane in the form of an unclosed graph (chain), a column chart of the objects distribution along the found path or projection onto the path. The paper proposes a method for searching SUP, such as a brute-force algorithm, a greedy algorithm and its various modifications. Tables of the SUP calculating results comparison by the algorithms proposed in the work, as well as estimates of the algorithms computational time are presented. The mapping method proposed in the paper allows to estimate the presence of local concentrations (clusters) in the data. The advantage of this visualization approach is that the algorithm undependable from the dimension of the feature space describing objects and could be implemented only on the distance matrix between objects.

References

1. Hofmann T, Buhmann J (1995) Multidimensional scaling and data clustering. In: Advances in neural information processing systems, pp 459–466
2. Naud A (2001) Neural and statistical methods for the visualization of multidimensional data. In: Dissertation katedra metod komputerowych umk
3. Pastizzo MJ, Erbacher RF, Feldman LB (2002) Multidimensional data visualization. Behav Res Methods Instruments Comput 34:158–162
4. Yakovlev S, Seredin O (2018) Multidimensional data visualization based on decision trees. Tidings of the Tula State University. Technical Science, Publishing house of TSU, Tula, pp 137–145

5. Jimenez LO, Landgrebe DA (1999) Hyperspectral data analysis and supervised feature reduction via projection pursuit. IEEE Trans Geosci Remote Sens 37(6):2653–2667
6. Fisher R (1936) The use of multiple measurements in taxonomic problems. In: Annual Eugenics, 7, Part II
7. Sam W (1995) Extending and benchmarking Cascade-Correlation. PhD thesis, Computer Science Department, University of Tasmania
8. Surkov E (2021) The study about basic assembly of skeletal representations in the fall detection task, Lomonosov-2021: collection of abstracts XXVIII International scientific conference of students, postgraduates and young scientists. Publishing department of the faculty of computational mathematics and cybernetics, Moscow, pp 82–83
9. Yu J, You X, Liu S (2021) Dynamic reproductive ant colony algorithm based on piecewise clustering. Appl Intell 51:8680–8700
10. Seredin OS et al (2019) A skeleton features-based fall detection using microsoft kinect v2 with one-class classifier outlier removal. In: International archives of the photogrammetry, remote sensing & spatial information sciences C, pp 189–195
11. Seredin OS, Kopylov AV, Surkov EE (2020) The study of skeleton description reduction in the human fall-detection task. Comput Optics 44(6):951–958. https://doi.org/10.18287/2412-6179-CO-753

Rider Churn Prediction Model for Ride-Hailing Service: A Machine Learning Approach

T. A. H. Tran and K. P. Thai

Abstract This paper presents a non-contractual churn prediction model for 4-wheel ride-hailing services. The purpose is to predict the churn tendency of the riders within a 3-month threshold. Using the transactional data of Uber and the KDD (Knowledge Discovery in Databases) methodology to conduct the empirical study with the support of python programming language. Three classifiers have been applied including SVM (Support Vector Machine), LR (Logistic Regression), and G-NB (Gaussian Naïve Bayes). The empirical results show that all classifiers achieve higher-than-90% accuracy. Depending upon the characteristics of ride-hailing service, LR is determined as the best model. This work also applies RFM (Recency, Frequency, Monetary Value) theory to segment churn values and its managerial implication is valuable for on-demand mobility companies.

Keywords Data mining · Non-contractual churn · Ride-hailing service

1 Introduction

Ride-hailing is an on-demand transportation service that provides an efficient travel mode by matching drivers and travelers via smartphone apps [1]. This industry has experienced significant growth in adoption since the introduction of Uber, in 2009 [2]. In the past, this used to be a taxi service, but now there are more types available such as car-hailing, bike/motorcycle hailing. Many enterprises (e.g., Grab, GoJek) located in Vietnam suffer from a loss of valuable customers to their competitors, this is known as "customer churn" [3, 4] but this paper calls it "rider churn" to better differentiate customers coming from ride-hailing services.

T. A. H. Tran (✉) · K. P. Thai
School of Business Information Technology, University of Economics, Ho Chi Minh City, Vietnam
e-mail: hongtran695.k44@st.ueh.edu.vn

K. P. Thai
e-mail: phungthk@ueh.edu.vn

© The Author(s), under exclusive license to Springer Nature Switzerland AG 2022   301
N. H. T. Dang et al. (eds.), *Artificial Intelligence in Data and Big Data Processing*,
Lecture Notes on Data Engineering and Communications Technologies 124,
https://doi.org/10.1007/978-3-030-97610-1_24

Before 2014, the riders in Viet Nam were provided with numerous traditional taxi services. The entry of multinational, on-demand mobility companies like Uber and Grab in 2014 shifted riders' commuting preference from regular taxi services due to the competitive pricing strategies of such companies. After Uber left in 2018, many ride-hailing apps were launched in Viet Nam to fill in the Uber space and compete with Grab [5]. For the 4-wheel ride-hailing service, several platforms (Grabcar, Becar, VATO, etc.) have competitive strategies to attract new customers and retain current customers. These platforms are having a huge impact on the traditional taxi companies and force them to change the way they are doing business. To avoid the loss of customers to competitors, many conventional taxi businesses, such as Mai Linh, Vinasun, Vinataxi, started app-based taxi services in the ride-hailing industry.

The aforementioned overview of the Vietnamese 4-wheel ride-hailing market proves that this industry is very dynamic and gives riders a lot of choices. Customers become more knowledgeable and less patient nowadays, which easily leads them to switch to competitors. Thus, churn becomes a very serious problem. To meet the need of surviving in this competitive industry, the retention of existing riders becomes a huge challenge. Because retaining an existing customer is a much lower cost than acquiring a new customer [6], properly detecting riders who are likely to be churned could help improve customer retention.

Over the last decade, there has been increasing interest in studies related to contractual churn prediction (e.g., telecom, insurance). In contrast, there is extremely little academic research to investigate non-contractual churn predictions. To propose a rider churn prediction model for ride-hailing services regarding non-contractual churn, the objectives of this study are: (i) Propose a classification model based on appropriate classifiers, research process and features of interest, (ii) Train, compare and evaluate machine learning models to identify the best one, (iii) Make predictions on the given dataset, (iv) Segment churners based on RFM theory, and (v) Give managerial implications using the results of prediction and analysis.

2 Literature Review

2.1 Customer Churn

"Churn" means the discontinuation of a contract [3], this definition is a recognized model for subscription-based companies (e.g., telecom, banking, etc.). For non-subscription-based companies (e.g., retail, e-commerce), churn can be determined broadly as the propensity of customers to terminate using the product/service of a company in a given period. In e-commerce, "customer churn" can be defined as when an enterprise loses existing customers to its competitor. Churn belongs in the last (fourth) stage of the customer life cycle (Fig. 1), thus, it is of great importance to maintain customers in the third stage (maintenance stage). Furthermore, customer churn has become a common issue for many enterprises globally. To overcome this

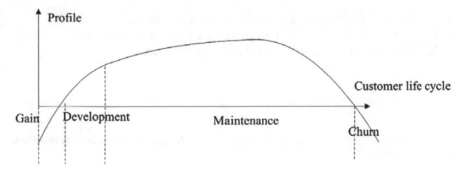

Fig. 1 Customer life cycle. *Source* Yu et al. [4]

challenge, several data mining techniques for churn analysis have been applied in different fields, e.g., finance, airline, telecommunications [4]. However, little attention has been paid to churn analysis regarding the e-commerce or m-commerce sector.

There are two concepts involved in customer churn:

- **Contractual and Non-contractual Relationships**: Contractual relationship [8] relates to subscription-based companies where the customer enters a "contract" with a business whereas a non-contractual relationship [8] is not bound by any agreement. Churners in a non-contractual relationship are more abstractly defined compared with contractual churners [7, 8]. The non-contractual relationship is typically in sectors (e.g., retail, e-commerce) that customers can freely start and stop services.
- **Intermittent Loss and Permanent Loss**: These concepts are usually concerned with non-contractual relationships. Intermittent Lost [8] means the customers did not buy products/services of a business during a specific time threshold, but they may make transactions beyond the time threshold, such customers are not completely lost. In contrast, Permanent Lost [8] means the customers stopped using the product/service of an enterprise completely for many reasons (e.g., the change of customer's spending habits, the change of the growth stage that the customer is no longer needs product, etc.).

2.2 RFM Theory

Hughes [9] considered the RFM model (Recency-Frequency-Monetary) as a behavior-based model used to analyze a customer's behavior and then make predictions based on the behavior in the database. That being said, RFM theory is commonly applied in studies related to predicting customer behavior in various fields (e.g., direct marketing) [6]. RFM variables were commonly defined as below [10] and can be used as segmenting variables [11] for several purposes, such as customer segmentation, finding the best customers, etc.

- R (Recency): the length of a period since the last purchase
- F (Frequency): the number of purchases made within a certain period
- M (Monetary): the money spent during a certain period

2.3 Classification Algorithm

Various classification models (classifiers) have been applied for churn prediction, this work applies three widely used classifiers as follows:

- **SVM (Support Vector Machine)**: The foundations of SVM have been developed by Vapnik and are gaining popularity due to many attractive features, and promising empirical performance [12]. SVM is a supervised learning model based on structural risk minimization and the performance of SVM can be improved by kernel function [13]. Four common kernels are linear, polynomial, RBF, and sigmoid [12].
- **LR (Logistic Regression)**: This is one of the most commonly utilized linear statistical models for discriminant analysis. Logistic Regression is a parameter-based model belonging to supervised learning and models the chance of an outcome that yields a continuous probability between 0 and 1, this transformation is called to Logistic (or Sigmoid) function [14].
- **NB (Naïve Bayes)**: NB is a simple learning algorithm based on Bayes' rule together with a strong assumption that the attributes are conditionally independent given the class [15]. There are three types of NB: Multinomial NB, Bernoulli NB [16], and Gaussian NB (G-NB) [17]. This study applies G-NB for the classification problems and can be defined that G-NB assumes a gaussian (normal) distribution of data if predictors take up a continuous value [17].

3   Research Methodology

3.1   The KDD Process

This work applies the methodology of Knowledge Discovery in Databases (or the KDD for short). The definition of the KDD process is found out from the work of Fayyad et al. [18] due to its popularity and comprehensiveness. The KDD process involves data selection, data preprocessing, data transformation, data mining, pattern evaluation, and knowledge representation; and each stage of the KDD process can involve significant iteration [18].

3.2 Empirical Research Design

The empirical research process is designed depending on KDD methodology as shown in Fig. 2:

- **(1) Data Selection**: This study makes use of transactional data of Uber Peru 2010 which was obtained from the Kaggle repository (URL: https://www.kaggle.com/marcusrb/uber-peru-dataset). This dataset records transactions of the riders at the time of ride request automatically by the app, including fare, distance, estimated travel time, current car availability (demand/offer balance), etc. Before the riders get picked up, the ride can be canceled by either the driver or the rider.
- **(2) Data Preprocessing**: Perform data reduction [19] by leaving out redundant attributes to obtain a reduced representation of the dataset. Then, two data cleaning techniques [19] are applied including filling missing values and handling outliers. Besides, the rider who didn't have any drop-off trip will be removed from the dataset.
- **(3) Data Transformation**: Aggregate the dataset on the customer level with features of interest. The obtained data will be applied the same data cleaning techniques described above, including filling missing values. The distribution of this dataset is skewed so normalize data by Z-Score [19] to ease the mathematical calculations as well as reduce the effect of larger features [20].
- **(4) Data Labeling**: In non-contractual churn prediction, many existing studies split periods to make churn labels within a time threshold for observed customers (Table 1).

Depending upon the characteristics of the ride-hailing industry, the further research suggested in Aleksandrova [21] and churn labeling method of Xia and He [8], this study splits the transactional data into the observation period (from 1/1/2010 to 9/30/2010) and the forecast period (from 10/1/2010 to 12/31/2010). If the riders do not have any trip (known as drop-off trip) during the forecast period, such riders will

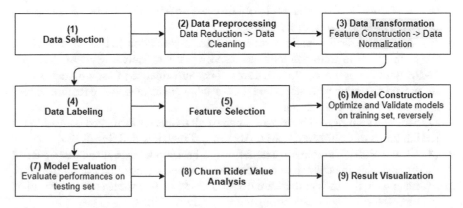

Fig. 2 Empirical research design

Table 1 A literature review on labeling method of non-contractual churn

Article	Market sector	Churn labeling method
Xia and He [8]	Online shopping	If the customer does not make any purchase in the forecast period, such customers are considered to be churn
Aleksandrova [21]	Concrete production	Customers are considered to be churn if they make no transaction in the next 6-month period
Miguéis et al. [22]	Retailing	Depending on periods of three months and less-than-30% spending to identify churners
Buckinx and Van del Poel [6]	FMCG retail	Depending on a period of five months to make churn label

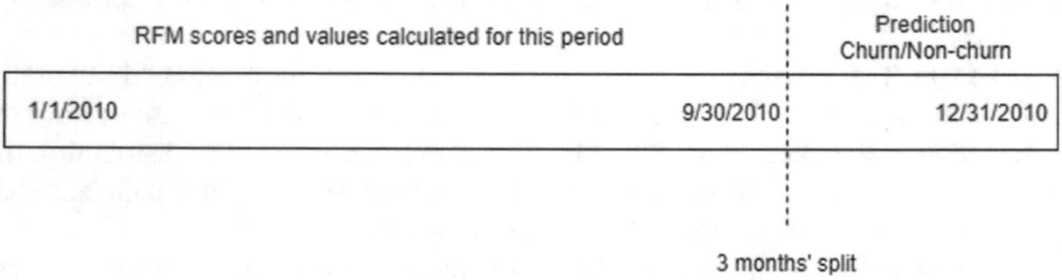

RFM scores and values calculated for this period	Prediction Churn/Non-churn
1/1/2010 9/30/2010	12/31/2010

3 months' split

Fig. 3 Date span and churn labeling by 3-month split

be labeled churn within the 3-month threshold, marked "1"; if the riders have at least one drop off trip during the forecast period, such riders will be labeled non-churn, marked "0" (Fig. 3).

- **(5) Feature Selection**: The dataset obtained from data preprocessing and data transformation stages is 80:20 split into train and test sets. ANOVA f-test [23] is then performed on a training set to obtain a well-performing subset of features as proposed input features for modeling.
- **(6) Model Construction**: Perform k-fold cross-validation [24] with k = 10 on the training set to train and validate three classifiers including SVM, LR, and G-NB with the proposed input features. The performance of each model is then optimized by tuning parameters and determining the most valuable features based on feature weight calculation.
- **(7) Model Evaluation**: The performance of all models is evaluated on testing set [21] in terms of Confusion Matrix, Accuracy, Precision, and Recall [25].
- **(8) Churn Rider Value Analysis**: After the best model is selected, apply RFM theory to divide the churners into different segments. Depending upon the work of Xia and He [8], the churners are divided based on two value states of F and M in comparison with corresponding average values of them.
- **(9) Result Visualization:** Visualize the results obtained in each stage.

4 Results

The date span of this dataset is from 1/1/2010 to 12/31/2010 and there are 23,111 instances with 28 attributes, but this work only makes use of 9 attributes that are the most relevant as shown in Table 2. After pre-processing and aggregating the data on customer level, the output is features of interest including 18 predictors and one target variable ("status" variable) for 989 riders. See Table 3 for descriptive statistics for the data after pre-processing and data transformation.

The obtained dataset was then 80:20 split into train and test sets. The distribution of churn class and a non-churn class of the dataset, training set, and testing set is depicted in Fig. 4 respectively. We can see that there are no notable differences between the two classes, thus the imbalanced class problem [26] does not appear.

ANOVA f-test [23] was performed to select the best features which were the most useful at predicting the target variable. Based on feature importance ranking, 12 out of 18 predictors were selected as proposed input features for modeling. With the given proposed input features, performing tenfold cross-validation on the training set to train and evaluate the three classifiers before generalized on the testing set. After model training, it's desirable to optimize models by tuning parameters and selecting the best features for each model, thus reducing the computational cost of modeling and improving the performance. By trial and error, parameter selection and predictors having high relevance value for each classifier are shown in Table 4. Three of the classifiers were then trained again with the corresponding input features listed in Table 4. The performance of all models was evaluated and shown in Table 5.

In Table 5, although with less than half the number of input features, all classifiers can achieve approximately good performance, i.e., greater-than-90% accuracy on

Table 2 Descriptions of attribute used

No	Attributes used	Descriptions
1	User_id	Each rider has a unique ID
2	Start_at	Time of booking, e.g., 16/11/2010 16:44
3	Journey_id	Each ride has a unique ID
4	End_state	The end state of the ride (*drop off*, *driver cancel*, *rider cancel*, *no show*, *not found*, *failure*)
5	Price	Price is determined based on distance and duration
6	Rider_score	A score of the driver rated by the corresponding riders for completed trips
7	Distance	Distance of a trip measured from pickup point to drop-off point
8	Duration	Duration of a trip measured from pickup point to drop-off point
9	Start_type	Types of booking: • *Reserved*—ride was scheduled in advance on the Uber app. • *asap*—the ride was booked and started as soon as possible at the time the rider requested on the Uber app. • *delayed*—ride scheduled in advance on Uber app showed up late.

Table 3 Descriptive statistics

	count	mean	standard deviation	iniu	25%	50%	75%	max
Recency	989	97.19818	98.39638	0	14	55	158	362
Frequency	989	17.04044	42.86638	1	1	5	14	447
Monetary	989	60559.21335	172065.02445	1300	5893	17260	52325	3674821
avg rider score	989	4.73161	0.63569	0	4.72727	4.93182	5	5
total distance	989	245904.11729	2104566.69924	0	15634	45280	129583	60740138
total duration	989	14182.35996	87399.06010	0	638	2377	8066	2554244
R quautile	989	3.02326	1.43216	1	2	3	4	5
F quantile	989	3	1.41564	1	2	3	4	5
M quautile	989	3	1.41564	I	2	3	4	5
RFM score	989	3.00775	1.32132	1	1.66667	3	4.33333	5
avg rides per week	989	0.32770	0.82435	0.01923	0.01923	0.09615	0.26923	8.59615
end state RiderCancel	989	3.71284	14.74714	0	0	1	3	365
start type asap	989	14.34783	44.55441	0	1	3	11	811
start type reserved	989	7.99090	24.76911	0	0	1	6	400
start type delayed	989	0.01921	0.16419	0	0	0	0	3
count fail trip	989	5.31749	17.33512	0	0	2	4	400
count charged fail trip	989	0.21436	0.93002	0	0	0	0	17
total cancel fee	989	475.92012	2435.93969	0	0	0	0	41803
status	989	0.42265	0.49423	0	0	0	1	1

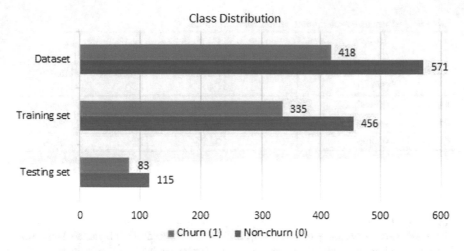

Fig. 4 Class distribution

Table 4 Model optimization

Classifier	Parameter	Input predictors	Predictor explanation
SVM	Kernel: Linear Σ = 0.10 C = 1.00	R_quantile, F_quantile, avg_rides_per_week, Frequency	*Frequency*: Total number of drop-off trips *R_quantile*: 1–5 score of Recency value. Larger the value, the smaller the score *F_quantile*: 1–5 score of Frequency value. Larger the value, the larger the score *RFM_score*: The average of individual R_quantile, F_quantile, M_quantile with equal weights *avg_rides_per_week*: The average number of drop-off trips per week *start_type_asap*: The number of drop-off trips whose start type was "asap" *end_state_RiderCancel*: The number of failed trips canceled by the riders *count_fail_trip*: The number of failed trips
LR	Solver: Liblinear	R_quantile, F_quantile, avg_rides_per_week, Frequency, start_type_asap	
G-NB	No configuration	R_quantile, RFM_score, count_fail_trip, Recency, end_state_RiderCancel	

Table 5 Performance evaluation measures for each model

	10-Fold cross-validation on the training set		Prediction result on the testing set		
	Est. Accuracy (%)	Std	Accuracy (%)	Precision (%)	Recall (%)
SVM	98.104	0.019	96.46	100	91.57
LR	98.862	0.019	96.97	98.73	93.98
G-NB	93.805	0.011	92.93	88.76	95.18

the dataset. Besides, there is neither overfitting [27] nor underfitting [27] due to no notable difference between accuracy on the training set and testing set. Thus applying any model can obtain reasonably good results in practice. The experiments also reflect the superiority of RFM variables in churn prediction because a customer is considered to be churn relying on various variables such as demographic variables. This study only makes use of transactional data to predict rider churn based on mainly RFM features, as a result, all the models are of good performance.

Uber is a mobility startup whose disruptive technology has resulted in the new business platform model. As such startups, consumer education strategy is a go-to choice at the early establishment, say, in the context of Peru market in 2010. Thus, the churn prediction model would matter how correctly several churners and non-churners are classified by the model. This study hence needs more robust metrics (Precision and Recall) besides Accuracy.

Precision helps when the costs of false positives are high and Recall helps when the cost of false negatives is high. The ability to have high values on Precision and Recall is always desired but, it is difficult to get that. Relying on the characteristics of ride-hailing service and the measurements (Accuracy, Precision, Recall), Logistic Regression was determined as the best model to predict binary churn tendency of the riders within the 3-month threshold. A summary of prediction results of Logistic Regression is shown by a confusion matrix illustrated in Table 6.

Having determined the best classification model, the values of churners were divided into different segments beyond simply classifying either churn or non-churn. The result of this task is shown in Table 7.

We can see that F_2M_2 presents the largest proportion of 39.23 percent among churners. Such proportion is a little bit strange because the higher the value of

Table 6 Confusion matrix of logistic regression

	Actual Positive Class Churner (1)	*Actual Negative Class* Non—churner (0)
Predicted Positive Class Churner (1)	True Positive = 78	False Positive = 1
Predicted Negative Class Non—churner (0)	False Negative = 5	True Negative = 114

Table 7 The classification of churn values using RFM theory

Churn Segment	Quantity	Rate (%)	Description
F_2M_2	164	39.2344	F_1: lower than the average frequency
F_1M_1	113	27.0335	F_2: higher than the average frequency
F_1M_2	72	17.2249	M_1: lower than the average spending M_2: higher than the average spending
F_2M_1	69	16.5072	

Frequency the more loyal are the customers of the company, similar for Monetary [10]. It could be due to the presence of extreme values and beyond the scope of this study. In general, the likelihood of churn tendency is significantly reduced as the frequency (F) and the spending (M) increase, since the rates of churn segments descend from F_1M_1 (27%), F_1M_2 (17,2%) to F_2M_1 (16,5%).

5 Conclusions

5.1 Managerial Implications

The managerial implications of this work provide an important guide for on-demand mobility companies to improve customer adhesiveness. Although the scope of this study was narrowed in the 4-wheel ride-hailing industry, the managerial implications can be widely applied for many types of ride-hailing services (e.g., two-wheel, four-wheel) since the characteristics are almost the same.

In the context of ride-hailing services, the relationship is non-contractual. Thus, it's difficult to judge the riders whether or not they will give up the enterprise completely in such a scenario, but we can judge intermittently loss of the riders within a time threshold (3 months in this study), such riders may make transactions beyond the time threshold. Therefore, it's desirable to analyze and understand churn segments for better retention solutions in the ride-hailing industry.

Depending upon RFM theory, the research findings revealed that the likelihood of churn tendency is significantly reduced as the frequency and the spending increase. For retention solutions, managers should start from Recency value first and then Monetary, try to push rider's activity more frequently and the spending as much as possible, this will effectively reduce churn rate in a dynamic and competitive business environment. Although low-value customers (F_1M_1 churn segment) are less valuable to the business and can be ignored from the traditional management concept, such customers should not be ignored and should be considered as a crucial challenge that mobility companies must be solved.

In the era of big data, several prevailing statistical techniques for customer churn management are replaced by more machine learning and predictive analysis techniques. Deep analysis from big data is the go-to method to understand the customer in this industry. Once churners have been identified with loss reasons, rapid action must be taken by the marketing department to prevent churn properly.

Furthermore, the state-of-the-art marketing concept focuses on relationship marketing instead of transactional marketing. That is marketing efforts should concentrate on building long-term relationships with customers (relationship marketing) rather than one-off sale transaction purpose (transactional marketing). Thus, the major double objective as regards the marketing aspect in the ride-hailing industry is not just to attract new customers, but also to retain them for the long term.

Finally, it will be a shortcoming if the role of drivers isn't involved in rider retention solutions. Because the drivers make a crucial role in delivering a smooth experience for the riders directly. Putting efforts in meeting the drivers' satisfaction and improving driver retention is as important as the riders. Thus, rider retention can be significantly improved from two sides of the drivers and the companies.

5.2   Limitations and Further Research

Limitations. Due to time and ability limited, there were still some limitations:

- This work treated outliers as interesting outliers that may contain valuable knowledge but doing nothing to study such outliers yet. This may result in missing deeper insights of the riders for retention solutions.
- Target variable was constructed without having a data sample of actual churners and loyal riders thus is hard to analyze hidden rules.
- Customers are judged to be churn depending on many types of features (RFM, demographic, purchase behavior). But this study mainly relied on RFM features as predictors due to the difficulty of collecting a desirable dataset.
- The data set is small, and a little old.
- Small data sets with high levels of accuracy suggest that models may be overfitting.

Further Research. Some areas need to be further explored as follows:

- Using ensemble method or hybrid method to combine models instead of a single classifier. These methods can take advantage of the various prediction models to increase the reliability and stability of the prediction results.
- Label churn by the method of RFM theory
- Study interesting outliers by reporting findings with and without outliers since this method can explain any difference in the results [26].

Acknowledgements This work was supported by School of Business Information Technology—University of Economics Ho Chi Minh City under bachelor's thesis.

References

1. Mao H, Deng X, Jiang H, Shi L, Li H, Tuo L, Shi D, Guo F (2021) Driving safety assessment for ride-hailing drivers. Accid Anal Prev 149:105574
2. Clewlow RR, Mishra GS (2017) Disruptive transportation: the adoption, utilization, and impacts of ride-hailing in the United States
3. Lazarov V, Capota M (2007) Churn prediction. Bus Anal Course TUM Comput Sci 33:34
4. Yu X, Guo S, Guo J, Huang X (2011) An extended support vector machine forecasting framework for customer churn in e-commerce. Expert Syst Appl 38(3):1425–1430
5. Mordor Intelligence. Vietnam Taxi Market (2020—2025). 2020 [cited 31 Oct 2021]. Available from: https://www.mordorintelligence.com/industry-reports/vietnam-taxi-market
6. Buckinx W, Van den Poel D (2005) Customer base analysis: partial defection of behaviourally loyal clients in a non-contractual FMCG retail setting. Eur J Oper Res 164(1):252–268
7. Bakkevig JK, Methi M. Non-contractual churn prediction using Hierarchical Temporal Memory (Master's thesis, Universitetet i Agder; University of Agder)
8. Xia G, He Q (2018) The research of online shopping customer churn prediction based on integrated learning. In: Proceedings of the 2018 international conference on mechanical, electronic, control and automation engineering (MECAE 2018). Qingdao, China, Mar 2018, pp 30–31
9. Hughes AM (1996) Boosting response with RFM. Mark Tools 4–8
10. Wang CH (2010) Apply robust segmentation to the service industry using kernel induced fuzzy clustering techniques. Expert Syst Appl 37(12):8395–8400
11. Allenby GM, Arora N, Ginter JL (1998) On the heterogeneity of demand. J Mark Res 35(3):384–389
12. Hsu CW, Chang CC, Lin CJ (2003) A practical guide to support vector classification
13. Kirui C, Hong L, Cheruiyot W, Kirui H (2013) Predicting customer churn in mobile telephony industry using probabilistic classifiers in data mining. Int J Comput Sci Issues (IJCSI) 10(2 Part 1):165
14. Kirasich K, Smith T, Sadler B (2018) Random forest vs logistic regression: binary classification for heterogeneous datasets. SMU Data Sci Rev 1(3):9
15. Webb GI, Keogh E, Miikkulainen R (2010) Naïve Bayes. Encycl Mach Learn 15:713–714
16. McCallum A, Nigam K (1998) A comparison of event models for naive bayes text classification. In: AAAI-98 workshop on learning for text categorization vol 752, no 1. 26 Jul 1998, pp 41–48
17. Moraes RM, Machado LS (2009) Gaussian naive bayes for online training assessment in virtual reality-based simulators. Mathware Soft Comput 16(2):123–132
18. Fayyad U, Piatetsky-Shapiro G, Smyth P (1996) From data mining to knowledge discovery in databases. AI Mag 17(3):37
19. Han J, Pei J, Kamber M (2011) Data mining: concepts and techniques. Elsevier, 9 June 2011
20. Hossein Javaheri S (2008) Response modeling in direct marketing: a data mining based approach for target selection
21. Aleksandrova Y (2018) Application of machine learning for churn prediction based on transactional data (RFM analysis). In: 18 International multidisciplinary scientific geoconference SGEM 2018: conference proceedings 2018, vol 18, no 2.1, pp 125–132
22. Miguéis VL, Camanho A, e Cunha JF (2013) Customer attrition in retailing: an application of multivariate adaptive regression splines. Expert Syst Appl 40(16):6225–32
23. Kumar M, Rath NK, Swain A, Rath SK (2015) Feature selection and classification of microarray data using MapReduce based ANOVA and K-nearest neighbor. Procedia Comput Sci 1(54):301–310
24. Anguita D, Ghio A, Ridella S, Sterpi D (2009) K-fold cross validation for error rate estimate in support vector machines. In: DMIN July 2009, pp 291–297

25. Hossin M, Sulaiman MN (2015) A review on evaluation metrics for data classification evaluations. Int J Data Min Knowl Manage Process 5(2):1
26. Japkowicz N, Stephen S (2002) The class imbalance problem: a systematic study. Intell Data Anal 6(5):429–449
27. Jabbar H, Khan RZ (2015) Methods to avoid over-fitting and under-fitting in supervised machine learning (comparative study). Comput Sci Commun Instrum Devices 163–72`

Online Hate Speech Detection on Vietnamese Social Media Texts In Streaming Data

Hanh Hong-Phuc Vo, Huy Hoang Nguyen, and Trong-Hop Do

Abstract This paper proposes a online hate speech detection system for streaming social media text. A combination of novel preprocessing techniques and state-of-the-art transfer learning model is proposed for social media hate speech detection. The experiment was conducted using the ViHSD dataset. The effectiveness of the proposed preprocessing method is verified as the proposed model achieves a significant improvement over the existing works on ViHDS dataset. An online hate speech detection system that detects and processes online hate speech in real-time was built to consolidate the effectiveness of the proposed system.

Keywords Hate speech detection · Social network text · Natural language processing · Transfer learning · Deep learning · Streaming data

1 Introduction

Social media platforms such as Facebook, TikTok and YouTube have been growing in popularity in recent years. There has been an increase in social media usage, particularly during the Covid-19 pandemic [8]. This resulted in a large amount of information, including hate speech. Many well-known platforms are demonstrating their efforts to address this issue through policies and technologies.

The difficulty is that it is hard to judge if the content is a violation or not without all the background information, such as other connected content or articles. That

H. Hong-Phuc Vo · H. H. Nguyen · T.-H. Do (✉)
University of Information Technology, Ho Chi Minh City, Vietnam
e-mail: hopdt@uit.edu.vn

H. Hong-Phuc Vo
e-mail: 18520275@gm.uit.edu.vn

H. H. Nguyen
e-mail: 18520842@gm.uit.edu.vn

Vietnam National University, Ho Chi Minh City, Vietnam

burning issue is even more difficult in the case of Vietnamese due to our language's varied vocabulary and complicated syntax, as well as the fact that most social media users utilize abbreviations and idiosyncratic, non-formal terms. Furthermore, the noise and scarcity of the imbalance dataset make this effort more difficult.

Detecting hate speech in streaming data is a difficult task. The truth is that most social media comments contain abbreviations and idiosyncratic, non-formal phrases, as well as weird characters. Furthermore, the volume of information, especially hate speech, grows rapidly by the second. This necessitates real-time processing and analysis of systems.

The contribution of the paper is three-fold. First, novel preprocessing strategies are proposed to increase the hate speech recognition model's accuracy. In these, using a Vietnamese abbreviation dictionary is a highlight approach. Second, PhoBER is the proposed model with highly effective hate speech recognition. Finally, a hate speech detection system for online social media is constructed by using spark and kafka.

The ViHSD dataset has been chosen for experimentation and evaluation. The proposed hate speed detection system achieves an F1-score of 0.6888, which is higher than that of existing works on the ViHSD dataset. PhoBERT is used in a streaming system to collect and analyze social media data in real time.

2   Related Works

In recent years, hate speech detection in Vietnamese has brought a lot of attention from many researchers. Huu et al. [2] proposed effective steps of text prepossessing combined with Logistic Regression to address the problem on the hate speech dataset provided by VLSP. Luu et al. [3] had a study on building the ViHSD dataset and hate speech detection in Vietnamese social media texts on this dataset. They experiment and evaluate their dataset on SOTA models. The highest performance is 0.6269 macro F1-score with m-bert cased.

Additionally, there have also been some research efforts for text classification tasks for Vietnamese social media texts. Nguyen [6] introduced different preprocessing techniques and key clause extraction with emotional context to improve the machine performance. Nguyen et al. [7] proposed PhoBERT model as well as using traditional machine learning models and neural network models. PhoBERT model achieves more outstanding performance than the other models. Hence, PhoBERT model is proposed in our system based on the related work.

3 Proposed Online Social Media Hate Speech Detection System

3.1 Data Preprocessing

Based on the dataset attributes of comments on social media, the technical solution for text preprocessing stage work are proposed by following:

- Remove duplicate and NULL. Social me dataset includes 2553 duplicated comments and 2 NULL comments. Additionally, the dataset is divided into train, dev and test data. Thus, train, dev, and test data are combined and removed duplicate and NULL comments. Then using train_test_split to split train, dev and test data following rate 7:1:2 in the paper of ViHSD [3]. Table 3b shows number of labels in each class after removing duplicate and NULL comments.
- Lowercase all comments.
- Convert all characters such as ":)", ":3", ":(" to emojis.
- Remove all URL, mail, hashtag, mention tag.
- Remove all mixed words and numbers, such as "10k", "5tr", "3km".
- Remove all punctuation, special characters, and numbers.
- Remove all emojis.
- Normalize all repeated words such as haaaa, okkk, vlll.
- Normalize abbreviations. Creating a Vietnamese abbreviation dictionary from the dataset, which is the most essential aspect of the preprocessing step. Most abbreviations in each comment are manually replaced by origin words in the abbreviation dictionary. This dictionary contains 1725 terms chosen from train, dev, and test data. Moreover, the abbreviation dictionary also corrects a few of spelling mistakes such as (correct: liêm sỉ),, sa sỉ, sịn (correct: xịn). Mistaken words are not included in Vietnamese dictionaries or hardly appear in comments on social media. Table 1a shows some examples of Vietnamese abbreviation dictionary.
- Predict abbreviations with different meanings but same form. Ridge is part of the Linear Regression family to be used as a model to predict abbreviations. Input is a comment, an abbreviation, start position and end position of the abbreviation in the comment. Output is the meaning of the abbreviation in the comment. Table 1b shows some examples with the same abbreviations but different meanings.
- Tokenize all comments. The space in Vietnamese is different to the space in English or common languages, is only the sign for separating syllables, not words. Tokenizing of texts is important, and it directly affects the results of models. Thus, ViTokenizer from the Pyvi library[1] is selected to tokenize all comments.

---

[1] https://pypi.org/project/pyvi/.

Table 1 Vietnamese abbreviation dictionary

(b) Some examples with the same form but different meanings

(a) Some examples of Vietnamese abbreviation dictionary

Abbreviations	Normalization
thíc, thick, thít, thik, thjx	thích
j, ji, rì, zì	gì
cutoe, dth, dthw, kewt	dễ thương

Abbreviations	Meaning
ah	à
	anh
bn	bao nhiêu
	bạn
bt	biết
	bình thường

3.2 Transfer Learning Model

After preprocessing, PhoBERT is implemented for hate speech detection due to its outperformance on Viet-namese NLP tasks. Furthermore, it has not yet been experimented with ViHSDdataset.

PhoBERT are large-scale monolingual language models pre-trained for Vietnamese. It was published by Nguyen et al., [5]. PhoBERT model achieves outstanding results on four Vietnamese NLP tasks, namely: Part-of-speech (POS) tagging, Dependency parsing, Named entity recognition (NER) word-level, and a syllable-level natural language inference. PhoBERT model pre-training approach is based on RoBERTa the model is optimized from BERT pre-training for better performance.

3.3 Traditional Machine Learning Models

Besides, traditional machine learning models that are commonly used in text classification tasks are also interested. Before training, TF-IDF is applied for feature extraction.

Feature Extraction TF-IDF is used to quantify words in a set of documents, which generally compute a score for each word to signify its importance in the document and corpus.

Traditional machine learning models

- **Support Vector Machine (SVM)** is a linear model for classification and regression problems. SVM tries to create a line or a hyperplane which separates the data into classes.
- **Logistic Regression (LR)** is a popular statistical technique to predict binomial outcomes. This is one of the basic and well-known methods of classification algorithms.
- **Ridge classifier (RC)** first converts the target values into {-1, 1} and then treats the issue as a regression task (multi-output regression in the multiclass case), optimizing the same objective as Ridge regression.

3.4 Deep Learning Models

In addition, deep learning models are also implemented for this task. Word embedding which helps learn representation for text is adopted before training models. **Word embedding** is a learned representation for text in which words with the same meaning are represented similarly. This approach to representing words and documents may be considered one of the key breakthroughs of deep learning on challenging natural language processing problems. FastText and PhoW2V are responsible for this stage.

- **fastText** is a multilingual pre-trained word vector, including Vietnamese, released by Grave et al. [1].
- **PhoW2V** is a pre-trained Word2Vec word embedding for Vietnamese, which is published by Nguyen et al. [4].

Deep learning models

- **Gated Recurrent Units (GRU)** is an advancement of recurrent neural networks. Compared to Long Short-Term Memory (LSTM), GRU is simpler but faster to train, and its performance is approximately equal to LSTM in several NLP tasks.
- **Bi-directional Recurrent Neural Networks (BiRNN)** is a combination of two RNNs - one RNN moves forward, beginning from the start of the data sequence, and the other, moves backward, beginning from the end of the data sequence. Due to this mechanism, BiRNN can get information from the past (forward) and future (backward) states at the same time.
- **Bi-directionalGRU (BiGRU)** have the same mechanism as BiRNN but replacing RNN cells by GRU cells with BiGRU.
- **Convolutional Neural Networks (Text-CNN)** is a class of deep learning methods which has been dominating in computer vision tasks. CNN is composed of multiple building blocks but mostly are convolution layers, pooling layers, and fully connected layers.

3.5 Real-Time Hate Speech Detection System

Streaming data is the continuous flow of data generated by various sources, for example, from applications, networking devices, and server log files, to website activity, social media data, banking transactions, etc. By using stream technology, data streams can be processed, stored, analyzed, and acted upon as they are generated in real-time.

With instances of social media hate speech continuing to rise, the need for real-time monitoring of posts to quickly take action on the offensive is more important than ever. Under that pressure, social media platforms need to create tools to be able to process the huge volume of data created quickly and efficiently, especially to prevent hate speech.

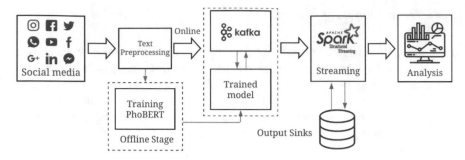

Fig. 1 The architecture of the online social media hate speech detection system.

In this research, Apache Kafka and Apache Spark are implemented which provide scalable, trusted, and low latency streaming platforms.

The architecture of the online social media hate speech detection system in this paper is illustrated in Fig. 1. Data is collected from social networks such as comments and classified using a proposed system for hate speech detection. Data after being classified is put into topics using Kafka produce. In this system, Kafka is used as a messaging system to connect components. Kafka Producer is responsible for writing data to topics, which is stored in distributed storage called Brokers. Spark structured streaming executes streaming data from Kafka topics to output sinks, which has a few formats like parquet, Kafka, console or memory, etc. Eventually, those classified comments are analyzed to gain insights, helping prevent hate speech.

4   Experiment

4.1   Dataset

In this paper, a few experiments are conducted on ViHSD dataset [3]. This dataset contains 33,400 comments about entertainment, celebrities, social issues, and politics gathered from Vietnamese pages and YouTube videos. Each comment is labeled with one of three labels: CLEAN, OFFENSIVE, or HATE, which are labeled 0, 1, 2 correspondingly. The examples of dataset are describes in Table 2. The number of labels in each class is shown in Table 3a. According to [3], the meaning of 3 labels are explained below:

CLEAN   : There is no harassment in the comments.
OFFENSIVE   : The comment contains abuse and even profanity, yet it does not target any specific object.
HATE   : The comments are harassing and abusive in nature, and are directed against an individual or group of people based on their traits, religion, and ethnicity.

Table 2 The examples in ViHSD dataset

No.	Comment	Label
1	Sao t gửi đc bây *(Why am I able to send it?)*	0
2	Ý thức còn ít hơn cả số tiền trong túi t *(Your level of civility is lower than the amount in my wallet)*	1
3	Quá ngu lồn đi =))) *(F*cking idiot =))))*	1
4	Im mẹ đi thằng mặt lon *(Shut the fuck up, you're a fucking c*nt)*	2
5	Bóng dơ *(Dirty queer)*	2

Table 3 Number of labels in each class

(a) Before remove duplicate and NULL

Dataset	0	1	2
Training	19886	1606	2556
Development	2190	212	270
Test	5548	444	688

(b) After remove duplicate and NULL

Dataset	0	1	2
Training	17685	1506	2401
Development	2551	223	311
Test	5040	435	695

ViHSD dataset has certain properties because it was crawled directly from user's comments on social networks:

- Having duplicate and NULL comments.
- There is a big skew between 3 classes.
- There are plenty of abbreviations or emojis, numbers and characters mixed in almost all comments.
- In comments, special characters such as &, % and # can be found.
- Some foreign language words can be found in the comments.
- Break marks such as a dot or a semicolon are not used consistently and clearly.
- Some comments contain unaccented letters.

4.2 Experimental Results

Table 4a shows the final experimental results for each model. The parameters of each model are shown in Table 4b. For traditional machine learning models, Grid-SearchCV technique is used to find suitable parameters for models. To assess the performance of all models, the macro F1-score is used as an evaluation metric. With an F1-score of 0.6888, PhoBERT outperforms the other models. The results with word embedding are not significantly different for deep neural network models. When using PhoW2V, CNN achieved its best performance with a 0.6720 F1-score,

Table 4 The experimental results and parameters of each model

(b) The experimental parameters of each model

Model	Parameters
LogisticRegression	C=10
SVM	C=10, gamma=0.1, kernel='rbf'
GRU+fastText	learning_rate=1e-3, drop_out=0.4, layer=1, units=150
GRU+PhoW2V	learning_rate=1e-3, drop_out=0.4, layer=1, units=150
BiGRU+fastText	learning_rate=5e-4, drop_out=0.4, layers=2, units=[150, 50]
BiGRU+PhoW2V	learning_rate=5e-4, drop_out=0.4, layers=3, units=[300, 150, 50]
CNN+fastText	learning_rate = 12e-4 filter_sizes = [2,3,5,6] num_filters = 64, drop_out = 0.4
CNN+PhoW2V	learning_rate = 12e-4 filter_sizes = [2,3,5,6] num_filters = 64, drop_out = 0.4
PhoBERT	epochs=2, batch_size=16, learning_rate=5e-5

(a) The experimental results of each model

Model	F1-score
Logistic Regression	0.6286
SVM	0.6261
GRU + fastText	0.6569
GRU + PhoW2V	0.6631
BiGRU + fastText	0.6540
BiGRU + PhoW2V	0.6716
CNN+fastText	0.6561
CNN+PhoW2V	0.6720
PhoBERT	**0.6888**

and Bi-GRU came in second with a 0.6716 F1-score. Furthermore, rather than fast-Text, PhoW2V embedding outperforms all deep neural network models. Moreover, Logistic Regression achieves the highest result with 0.6286 F1-score among traditional machine learning models.

4.3 Compare with Previous Study

Except for SVM, the majority of model results outperform the best results of existing works on the ViHSD dataset. Although removing duplicates in the dataset makes the new test set more difficult than the original test set, positive results are obtained. Thanks to generating Vietnamese abbreviation dictionary significantly improves the performance of the models. Our best result is greater than the 0.0619 F1-score of m-bert cased by the existing works on ViHDS dataset.

4.4 Error Analysis

Table 5 presented some wrong prediction samples in ViHSD dataset. Firstly, the comments number 1 had "déo", "ghê" having both negative and positive meanings in Vietnamese and our model can not understand the context of the comments. Additionally, the comment number 3 had "thằng" is a pronoun that causes confusion between labels 1 and 2. Following is the comment number 4 due to problematic data quality. The predicted result is correct, but the label is annotated incorrectly.

Table 5 Some error cases

#	Comments	Predict	True
1	\<person name\> bị thế đéo nào đu'Ợ'c *(How the hell did this happen)*	1	0
2	ghê vãi *(how disgusting)*	0	1
3	nhu' thằng điên *(mad guy)*	2	1
4	xạo ồn *(c*nt liar)*	1	2
5	già mất nết quá *(D*mn that spoiled old man)*	1	2
6	con gia nay chuyen gioi ho nhin ghê quá *(how disgusting, biddy is transgender)*	0	2

Because the inter-annotator agreement for the dataset is just K=0.52 approximately threshold 0.5. Next the comment number 5, "già" here refers to old people in a compact and irregular form, not an adjective old this is a complexity of the Vietnamese social media texts. The comment number 6 almost does not have accented letters. Furthermore, as mentioned earlier, ViHSD is an imbalanced dataset, having as a result influenced significantly on the predicted results. Figure 2 shows the confusion matrix of PhoBERT. The correct prediction rate of label 0 is much higher than that of labels 1 and 2.

Fig. 2 Confusion matrix

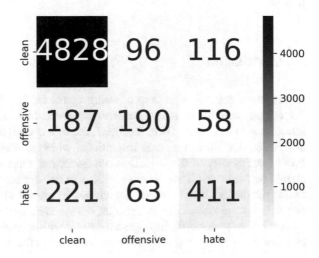

	user	timestamp	raw_comment	sentiment
0	Tiềm Doan	2021-10-13 23:04:45	Họ đá bóng bằng óc	OFFENSIVE
1	Tiềm Doan	2021-10-13 23:03:48	Thi thoang ho lai như lam Việt Nam hi vong va ...	HATE
2	Tiềm Doan	2021-10-13 23:02:21	Nhìn Trung Quốc đá. Cho vn bóng chạy từ bên nà...	HATE
3	Đức hải	2021-10-12 08:15:10	êu gọi Bộ Chính trị Đảng Cộng sản Việt Nam dừn...	HATE
4	cò cô ma	2021-10-11 04:35:28	thương ghê ai ghét cái lưỡi bò giơ tay	OFFENSIVE
5	Nguyễn Nhựt Minh	2021-10-10 06:00:10	Chán đéo muốn nói	OFFENSIVE

(a) Statistics of hate speech comments

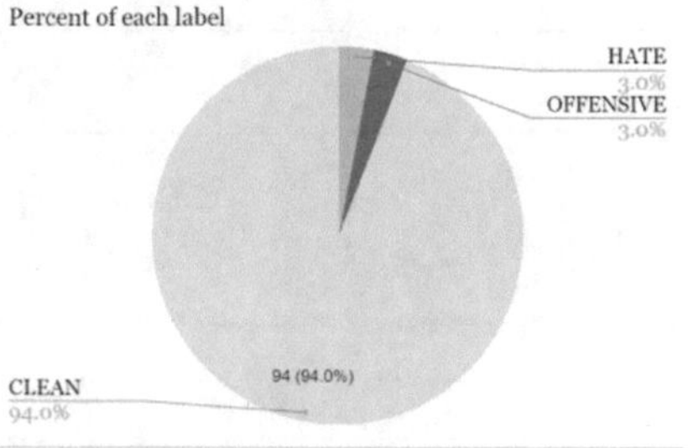

(b) Percentage of each label

	user	most_hate_speech
0	Tiềm Doan	3
1	Nguyễn Nhựt Minh	1
2	Đức hải	1
3	cò cô ma	1

(c) Users with the most hate speech

Fig. 3 Some analyses using real-time hate speech detection system.

4.5 Real-Time Analysis from Hate Speech Detection Results

Figure 3 shows statistics of 100 comments collected from a video on social media. The system detects hate speech immediately and provides useful information.

5 Conclusion

In summary, the effective preprocessing methods, PhoBERT and the hate speech detection system for streaming data using spark and kafka on online social media are introduced in this study. With PhoBERT, the best overall macro F1-score is achieved at 0.6888. Our work has some limitations, which will be discussed further below. Firstly, the quality of the ViHSD is not good; the dataset's accuracy of annotation should be improved if we want to build a better system for real-world tasks. Secondly, the dataset is unbalanced, which has an impact on predicted results. Finally, due to language knowledge limitations, data is unable to adequately preprocess.

A set of stop words for hate speech dataset will be generated in the future, as well as solve the imbalance problem between classes in the dataset.

References

1. Grave E, Bojanowski P, Gupta P, Joulin A, Mikolov T (2018) Learning word vectors for 157 languages. arXiv preprint arXiv:1802.06893
2. Huu QP, Trung SN, Pham HA (2019) Automated hate speech detection on vietnamese social networks. Tech. rep, EasyChair
3. Luu ST, Nguyen KV, Nguyen NLT (2021) A large-scale dataset for hate speech detection on vietnamese social media texts. In: Fujita H, Selamat A, Lin JCW, Ali M (eds) Advances and trends in artificial intelligence. Artificial Intelligence Practices. Springer International Publishing, Cham, pp 415–426
4. Nguyen AT, Dao MH, Nguyen DQ (2020) A pilot study of text-to-sql semantic parsing for vietnamese. arXiv preprint arXiv:2010.01891
5. Nguyen DQ, Nguyen AT (2020) Phobert: pre-trained language models for vietnamese. arXiv preprint arXiv:2003.00744
6. Nguyen KPQ, Van Nguyen K (2020) Exploiting vietnamese social media characteristics for textual emotion recognition in vietnamese. In: 2020 International Conference on Asian Language Processing (IALP). IEEE, pp 276–281
7. Nguyen LT, Van Nguyen K, Nguyen NLT (2021) Constructive and toxic speech detection for open-domain social media comments in vietnamese. Lect Notes Comput Sci 572–583. https://doi.org/10.1007/978-3-030-79457-6_49
8. Wiederhold BK (2020) Social media use during social distancing

Comprehensive Study of Deep Regression Models for Weather Forecasting

Hoang Dung Nguyen⑩, Van Thinh Phan⑩, Hoang Phu Dinh⑩,
Van Thuan Nguyen⑩, and Thanh-Hai Tran⑩

Abstract Weather forecasting aims to predict climate parameters at certain time frames and places. Accurate short-term prediction plays an important role in many fields such as agriculture and industry. Recently, several digital weather forecasting models (NWP) are being used because of their ability to quantitatively forecast meteorological factors in detail in space and time. But the results still contain many errors because the atmospheric system is dominated by disturbances of small spatial and temporal scales. One of the new research in weather forecasting that improves prediction results is the use of artificial intelligence with large databases. In this paper, we will deploy a series of outstanding weather regression models such as RNN, LSTM, CNN-LSTM, and TCN. We then determine the correlation between the weather parameters and analyze the weather forecasting performance of the models in three use cases: Single parameter prediction from a single input (SISO—Single Input Single Output), Single parameter prediction from many inputs (MISO—Multiple Inputs Single Output), single parameter prediction from correlated inputs (CISO—Correlated Inputs Relevant Single Output—CISO). We ran our tests on four weather data sets and got promising results. This gives important and effective recommendations for designing suitable models for future realistic short-term weather forecasting applications.

H. D. Nguyen · V. T. Phan · H. P. Dinh · V. T. Nguyen · T.-H. Tran (✉)
School of Electrical and Electronic Engineering, Hanoi University of Science and Technology, Hanoi, Vietnam
e-mail: hai.tranthithanh1@hust.edu.vn

H. D. Nguyen
e-mail: dung.nguyenhoang@hust.edu.vn

V. T. Phan
e-mail: thinh.pv163930@sis.hust.edu.vn

H. P. Dinh
e-mail: phu.dh163166@sis.hust.edu.vn

V. T. Nguyen
e-mail: thuan.nv163958@sis.hust.edu.vn

© The Author(s), under exclusive license to Springer Nature Switzerland AG 2022   327
N. H. T. Dang et al. (eds.), *Artificial Intelligence in Data and Big Data Processing*,
Lecture Notes on Data Engineering and Communications Technologies 124,
https://doi.org/10.1007/978-3-030-97610-1_26

Keywords Regression model · Long-short term memory · Correlation · Convolutional neural network · Weather forecasting

1 Introduction

Weather forecasting is a process that aims to predict the state of the atmosphere shortly. Humans have attempted to predict the weather informally for many millennia, and official weather forecasting systems began in the nineteenth century. Weather forecasting is accomplished by collecting data on the current state of the atmosphere and applying a scientific understanding of atmospheric processes to predict the evolution of the atmosphere. Numerical Weather Prediction (NWP) is an advanced forecasting technology that increasingly supports today's weather forecasting. The observed data are used as input for the numerical weather forecasting models. The models are commonly a combination of complex mathematical expressions that simulate physical processes to predict the behavior of the atmosphere. The final outputs are forecasts of wind, temperature, humidity, precipitation, and other meteorological factors.

Weather forecasting using deep learning (WFDL) belongs to the data-driven weather forecasting approach that has taken advantage of outstanding achievement in deep learning recently. The original datasets are fed into deep learning models, which aim to find underlying patterns or relationships of data and capture the characteristics of weather changes through large amounts of annotated data. Recently, several methods have been proposed for weather forecasting [2, 4, 12]. The methods can be divided into two main categories. The first category learns a model that takes a single input (a parameter of weather such as temperature) and predicts a single output (the temperature). This category considers parameters of weather as independent variables and processes them independently. The second category considers the impact of one weather parameter (e.g. humidity) on another parameter (e.g. temperature). As a consequence, the methods of this category take multiple parameters (temperature, humidity, rainfall, etc.) into account to predict a single parameter (temperature for instance). Certainly, there exists a correlation among parameters. Exploring this correlation to improve the prediction model is an important task. On the one side, it helps to increase the reliability of forecasting compared to the use of a single parameter. On the other side, it can reduce the computational time because only correlated parameters are considered instead of all parameters. So it emerges a new question. How helpful is the correlation among weather parameters for improving prediction accuracy?

In this paper, we will deploy a range of the latest deep regression models for weather forecasting (Recurrent Neural Network—RNN, Long-Short Term Memory—LSTM, Convolutional Neural Network CNN-LSTM, and Temporal Convolutional Neural Network—TCN). We then investigate the correlation of weather parameters. Finally, we analyze empirically the performance of these models in three use cases: Single Input Single Output (SISO-Input and output are the same weather parameter),

Multiple Inputs Single Output (MISO-The output weather parameter is one of the input weather parameters and the input is all the measured weather parameters). In addition, to analyze the performance of correlated parameters, we investigate MISO models that take only correlated parameters as inputs. We name it CISO (Correlated Inputs, Single Output). Besides, we will study the impact of window size (the number of samples to be fed into the models to predict the new one) on the performance of prediction. We conduct our experiments on four datasets among which three datasets are publicly available and one dataset is built ourselves. Experimental results show that the combination of the TCN model with SISO achieves the best performance and the window size doesn't have much impact on the model's performance. In summary, our contribution is threefold: (i) investigation of correlation of weather inputs that would be better for weather prediction; (ii) deployment of a range of CNN models for weather prediction to find out the best model to be used in the future; (iii) implementation and evaluation of different architectures (SISO, MISO, CISO) on three datasets.

2 Related Works

Weather forecasting using deep learning (WFDL) can overcome the challenges of conventional NWP forecasting models including low power consumption to compute, ease of understanding, ease of installation, and high portability. According to the characteristics of meteorological data, we reduce the weather forecasting problem to Time Series Analysis and analyze the most suitable deep learning models for the problem such as RNN, LSTM, TCN.

A recurrent neural network (RNN) was introduced into the weather forecasting model in 2015 [10]. RNN was compared with two other models, the Conditional Restricted Boltzmann Machine (CRBM) and the Convolutional Network (CN). After comparing the RMSE, the researchers concluded that the RNN could be applied in precipitation prediction with an appropriate level of accuracy. Unfortunately, with the increasing number of samples, RNN can no longer remember and learn more.

Long Short Term Memory networks (LSTM) are a special type of RNN. LSTM has the ability to learn long-term dependencies. LSTM was introduced by Hochreiter and Schmidhuber [6], and then it has been improved and popularized by so many researchers. LSTM model is widely used in time series problems [1, 8]. In addition to using Basic LSTM (Vanilla LSTM) for weather forecasting, there are other variations of it such as Stacked LSTM [7], Bidirectional LSTM [3].

Besides, it is also possible to combine the LSTM with CNN as the model produced by Roesch and Günther [9]. The model of Fu et al. [3] has been proposed for weather forecasting. In addition to RNN and LSTM, the temporal convolutional neural network (TCN) is also the model introduced for weather forecasting by Hewage et al. [5]. This model is evaluated by predicting and tracking weather parameters quite accurately over a period of several hours.

In the above works, the proposed methods have been evaluated on some weather datasets. Unfortunately, these datasets are not publicly available which makes them difficult to compare. We, therefore, utilized several weather databases that are publicly downloadable on weather websites. Then, we deploy significant deep regression models RNN, LSTM, CNN-LSTM, and TCN for weather forecasting problems. We will further study the correlation of weather parameters and recommend the most suitable option to implement such a system in practical application.

3 Proposed Framework for Weather Forecasting

3.1 General Framework

Our proposed framework for weather forecasting is illustrated in Fig. 1. In this study, we are interested in different types of inputs to machine learning models: SISO, MISO and, CISO. Different deep regression models will be deployed including RNN, LSTM, CNN-LSTM, and TCN. In the following, we will describe the computation and analysis of correlation among parameters and re-visit the studied regression models

3.2 Correlation of Weather Parameters

Correlation analysis(r) is a test statistic that measures the relationship or association among dependent variables. In this time series problem, the correlation coefficient measures the degree of the linear relationship between two weather parameters. The correlation coefficient has no units and is in the range $[-1, 1]$ in which $r > 0$ is a positive correlation while $r < 0$ is a negative correlation and $r = 0$ is uncorrelated.

In this paper, we conduct a Pearson correlation analysis to evaluate the linear correlation between a pair of weather parameters. The Pearson correlation formula is calculated as follows:

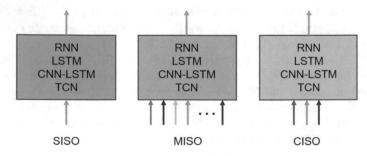

Fig. 1 Proposed architectures for weather forecasting: SISO (left), MISO (middle), CISO (right)

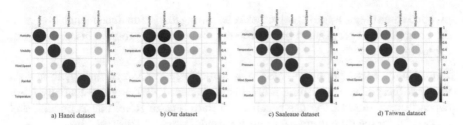

Fig. 2 Correlation analysis on four experimented datasets: **a** Hanoi dataset, **b** Our dataset, **d** Taiwan dataset; **c** Saaleaue dataset

$$r_{xy} = \frac{\sum_{i=1}^{n}(x_i - x')(y_i - y')}{\sqrt{\sum_{i=1}^{n}(x_i - x')^2 \sum_{i=1}^{n}(y_i - y')^2}} \tag{1}$$

where r_{xy} is the correlation coefficient between two parameters to be calculated, x_i is the value of the first weather parameter, y_i is the value of the second weather parameter captured at the same time, (x', y') are the average values of those two weather parameters in the data set.

Our aim when analyzing the correlation of parameters in the weather dataset is to evaluate whether a machine learning model consisting of only correlated parameters gives better results than the machine learning models using all parameters or a single parameter as input. Figure 2 illustrates the correlation of weather parameters on four datasets utilized in this work. The color of the circles in each cell represents the correlation between a pair of parameters (x, y) while its size shows the significance of correlation (the value r_{xy}. Bigger the size the more correlated the pair of parameters. We observe that the correlation among a pair of parameters may vary from dataset to dataset. In the following, two parameters x, y will be considerably correlated if their value r_{xy} is bigger than a threshold.

3.3 Revisit of Deep Regression Models

In this section, we will revisit regression models we will utilize for evaluating weather prediction on various datasets. The models range from a very basic one such as RNN to deal with time series data to more complex ones with the use of CNN, LSTM and till the latest model TCN for prediction problem.

Recurrent Neural Network—RNNs are neural networks that contain loops instead of just transmitting input information directly over the network. The RNN layer also receives its previous output as input. RNNs have memory and are therefore suitable for time-dimensional problems. However, standard RNNs do not perform well in practice when they are used to solve tasks containing long-term dependencies because the error gradient propagated back through the network is prone to disappearing or exploding.

Long-Short Term Memory—LSTM is a modified version of the cyclic neuron to avoid the gradient vanishing problems of standard RNNs and potentially learn dependency limits. LSTM has become a certain type of RNN that is popular and successful for the time it runs. It trains the model using backpropagation.

CNN-LSTM is an LSTM architecture specifically designed for sequence prediction problems with spatial inputs, such as images or video. The CNN-LSTM architecture involves the use of Convolutional Neural Networks (CNN) layers for feature extraction on input data combined with LSTMs to support sequence prediction.

Temporal Convolutional Neural Network—TCN [11] is a very effective machine learning model in many different problems. It has advantages such as (1) processing the input sequence as a whole rather than sequentially like RNN, (2) the ability to better control the memory size of the model, ease to adapt to different applications and different domains, (3) TCN avoids RNN problems like exploding/vanishing gradients. The main feature of TCN is that it is complex, causal, and TCN can map input of any length to an output of the corresponding length. Assume that the input is a 1-D sequence input $x \in R^n$ and a filter f: $\{ 0,1 \ldots ,k - 1 \} \to R$, the dilated convolution operation F on element s of the sequence is defined as:

$$F(s) = (x *_d f)(s) = \sum_{i=0}^{k-1} f(i).p_{s-d.i} \tag{2}$$

where d is the dilation factor, k is the filter size, and s-d.i stands for the direction of the past. Figure 3 depicts the architecture of TCN with configurable expansion factors d = 1, 2, 4. The dilation introduces a fixed step between each adjacent filter press. The greater expansion and larger filter size k allow for efficient expansion of the receptive profile. An exponential increase in k usually increases the depth of the network.

Weather data is time-series data. The weather prediction problem is also the sequential prediction problem. Machine learning models such as RNN, LSTM, CNN-LSTM, or TCN have been used a lot in this problem [2–4, 7, 10]. In particular, CNN-LSTM has been proven to be the most effective model in regression models with time series data many years ago [8]. However, in a few recent studies, TCN has proven to be very effective and outstanding in many different data types including weather data [5]. Therefore, we decided to deploy these machine learning models in our study to evaluate the effectiveness of these machine learning models. We implement a variation of each machine learning model as presented in the next sub-section.

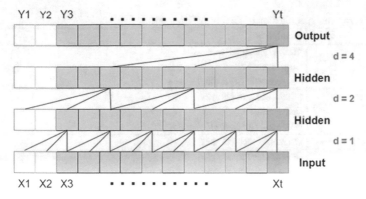

Fig. 3 A typical TCN layer (adapted from [5])

3.4 Implementation of Deep Learning-Based Regression Models

In this section, we will show how to implement our deep learning models. For each model type, we design various configurations. The parameters of each layer in the machine learning models are presented in Table 1. The implementation code was written in Python 3.4 on Google Colab—Platform Intel Xeon Processor with two core @ 2.3 GHz, 16 GB RAM. The deep learning models were implemented using Keras, Numpy, Pandas libraries. Callbacks in Keras are used to save the best weights for each training session. The correlation models were trained for 20 epochs, and the other models were trained for 50 epochs. All experiments were run with Earlystopping to prevent unnecessary overfitting. They use adaptive moment estimation Adam with a batch size equal to 128. Using a mean-squared loss function, the Adam algorithm ensures that the learning steps during the training process are scale-invariant relative to the parameter gradients. In the LSTM layers, we use activation Rectified Linear Unit (Relu) to avoid linearity.

4 Experiments

4.1 Datasets

The weather data used in this study consists of four datasets captured in four different locations: Taiwan dataset[1] is extracted from an air quality file in northern Taiwan from 0:00 on January 1, 2015, to 12:00 on December 31, 2015, by the Environmental

---

[1] https://www.kaggle.com/nelsonchu/air-quality-in-northern-taiwan.

Table 1 Parameter specification of all forecasting models

Models	Description
RNN	RNN layer with 128 units
	Dropout rate 0.2
	Fully connected layer with 1 neuron
LSTM Vanilla	LSTM layer with 128 units
	Dropout rate 0.2
	Fully connected layer with 1 neuron
LSTM Stack	LSTM layer with 128 units
	Dropout rate 0.2
	LSTM layer with 64 units
	Dropout rate 0.2
	Fully connected layer with 1 neuron
LSTM Bidirection	Bidirection of LSTM layer with 64 units
	Dropout rate 0.2
	Fully connected layer with 1 neuron
CNN-LSTM 1	Convolutional layer with 64 filters of size
	Max pooling layer
	Flatten layer
	Fully connected layer with 64 neurons
	Dropout rate 0.2
	LSTM layer with 64 units
	Fully connected layer with 1 neuron
CNN-LSTM 2	Convolutional layer with 64 filters of size
	Convolutional layer with 128 filters of size
	Max pooling layer
	Flatten layer
	Fully connected layer with 128 neurons
	Dropout rate 0.2
	LSTM layer with 128 units
	Dropout rate 0.2
	LSTM layer with 64 units
	Fully connected layer with 32 neurons
	Fully connected layer with 1 neuron
CNN-LSTM 3	Convolutional layer with 64 filters of size
	Convolutional layer with 128 filters of size
	Max pooling layer
	Flatten layer
	Dropout rate 0.2
	LSTM layer with 128 units
	Fully connected layer with 32 neuron
	Fully connected layer with 1 neuron
TCN	TCN layer with 128 filters, dropout rate 0.2
	Kernel size = 3
	Learning rate = 0.02, Dilations = 32
	Fully connected layer with 1 neuron

Table 2 The parameters used in our study and their corresponding units

Parameters and units	Hanoi dataset	Taiwan dataset	Saaleaue dataset	Our dataset
Temperature(°C)	[10.5:39.7]	[10:37]	[−4.77:36.55]	[24:54.1]
Rainfall(mm)	[0:98]	[0:65]	[0:8.12]	na
Humidity(%)	[25.52:100]	[13:99]	[23.2:100]	[7:97]
Wind Speed (m/s)	[0:9.81]	[0:11]	[0:37.22]	[0:14.88]
Pressure(hPa)	na	na	[965:1022]	[999:1010]
Visibility(km)	[0.4:17.9]	None	na	na
UV Index	na	[0:12]	na	[0:6.69]
Sampling rate	1 h	1 h	10 min	15 s
Total samples	8784	8760	26,333	90,906

Table 3 Correlation between weather parameters

Dataset	Parameters	Correlation coefficient(r)	Level
Ha Noi	Visibility, humidity	−0.50	Medium
Taiwan	Humidity, UV index	−0.54	Medium
Saaleaue	Temperature, humidity	−0.70	Strong
	Temperature, Pressure	0.77	Strong
Our dataset	Temperature, humidity	−0.94	Strong
	Temperature, UV index	0.72	Strong
	Humidity, UV index	−0.64	Medium

Protection Administration, Executive Yuan, Taiwan; Saaleaue dataset[2] is taken from 0:00 on July 1, 2020, to 0:00 on January 1, 2021; Hanoi dataset was taken from 0:00 on January 1, 2020, to 12:00 on December 31, 2020, taken from the website and our dataset collected by our self-designed system at the West Lake from June 5, 2021, to June 24, 2021, at a location near West Lake in Hanoi.

In two datasets Hanoi and Taiwan, the weather parameters are measured every 1 h, while the Saaleaue dataset is measured every 10 min and our dataset is measured most densely every 15 s. All four datasets have five weather parameters but are not the same type. In addition, the correlation between parameters is different (see Fig. 2). Hanoi dataset provides temperature, humidity, visibility, rainfall, and wind speed. Taiwan dataset includes parameters such as temperature, humidity, UV index, rainfall, and wind speed. Saaleaue includes parameters such as temperature, humidity, pressure, wind speed, and rainfall. Meanwhile, the West Lake dataset includes parameters such as temperature, humidity, UV index, pressure, and wind speed.

---

[2] https://www.bgc-jena.mpg.de/wetter.

In each dataset, we conduct linear interpolation to fill in the missing values then normalize the parameters to the range [0:1]. The datasets are split into three training, validation, and test sets with a ratio of 8:1:1. Table 2 shows the range of values of weather parameters and their units in four datasets while Table 3 and Fig. 2 present the correlation of the weather parameters in each dataset. However, we will only be interested in weather parameters with the correlation coefficient higher than average ($|r| \in [0.5 : 1]$). For evaluation, we compute RMSE (Root Mean Square Error) between the predicted value and the ground truth one. The smaller RMSE, the better the model is.

4.2 Experimental Results

Performance analysis on datasets:

In the following, we show the results (RMSE) obtained by machine learning models on each experimented dataset. On the Hanoi dataset, MISO gets the highest RMSE for almost all models, CISO sometimes achieves lower RMSE than SISO but on average, SISO is the best with the smallest RMSE (2.14) compared to MISO (2.49) and CISO (2.44). On the Taiwan dataset, MISO gets higher RMSE too for almost all machine learning models except for LSTM_2 and TCN. On the Saaleaue dataset, the result is quite inconsistent according to models and use cases. SISO-TCN is the best model on this dataset. On our dataset, the best use case is SISO for all of the models with the average RMSE = 0.49 following are MISO (average RMSE = 0.79) and CISO (average RMSE = 1.09) (Fig. 4).

Comparison of models:

Table 4 recapitulates the comparative results obtained by each machine learning model according to three use cases: SISO, MISO, and CISO. Among the machine learning models we tested, TCN is the best machine learning model. This is understandable because TCN has most of the advantages of RNN, LSTM, or CNN machine learning models, even TCN is more optimal than those machine learning models. Therefore, we recommend using this model to predict weather parameters.

Analysis performance according to sliding window size:

We evaluate the effect of the sliding window size on the model's prediction results. A temporary sliding window is used to prepare the data. The input sample is time steps from 1 to 24, the prediction time is immediately following the time step. The distance between two sliding windows is a 1-time step. We use a function to split the data into input and output then divide them into 3 datasets: training, evaluation,

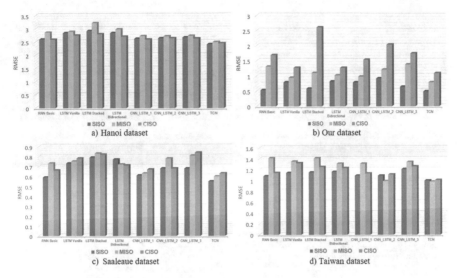

Fig. 4 Average RMSE value of studied models on datasets: **a** Hanoi dataset, **b** Our dataset, **c** Saaleaue dataset, **d** Taiwan dataset

Table 4 Average RMSE value of studied models on the experimented datasets

Model	Hanoi dataset			Taiwan dataset			Saaleaue dataset			Our dataset		
	SISO	MISO	CISO	SISO	MISO	CISO	SISO	MISO	CISO	SISO	MISO	CISO
RNN	2.60	2.86	2.59	1.08	1.41	1.14	0.59	0.73	0.66	0.53	1.31	1.69
LSTM Vanilla	2.84	2.89	2.75	1.14	1.35	1.32	0.73	0.75	0.78	0.74	0.94	1.27
LSTM Stacked	2.92	3.22	2.80	1.15	1.41	1.25	0.79	0.83	0.82	0.58	1.10	2.61
LSTM Bidi-rectional	2.84	2.98	2.70	1.16	1.31	1.23	0.77	0.72	0.71	0.82	1.02	1.27
CNN-LSTM 1	2.62	2.72	2.59	1.09	1.31	1.13	0.61	0.63	0.67	0.79	0.98	1.54
CNN-LSTM 2	2.64	2.71	2.64	1.09	0.99	1.11	0.68	0.78	0.68	0.92	1.21	2.04
CNN-LSTM 3	2.67	2.73	2.63	1.21	1.34	1.26	0.68	0.81	0.84	0.64	1.38	1.75
TCN	**2.41**	**2.49**	**2.44**	**1.00**	**0.98**	**1.01**	**0.55**	**0.60**	**0.63**	**0.49**	**0.79**	**1.09**

and testing. It is noticed that the selected model in this experiment is TCN SISO. Because the TCN SISO model gives the best RMSE results.

The RMSE value corresponding to each sliding window from 1 to 24 time steps of each weather parameter in each data set is shown in Fig. 5. However, unlike the finding was given by Hewage [5] on nonweather data, RMSE and sliding window size have a linear relationship. In this study, we observe a fairly consistent performance along with various window sizes. It can be explained that weather parameters too far in the past do not affect the prediction results for the present time.

So the sliding window might not be a good factor too much affecting the results of the machine learning model with weather data. Even with humidity data increasing window size sometimes makes the prediction results worse.

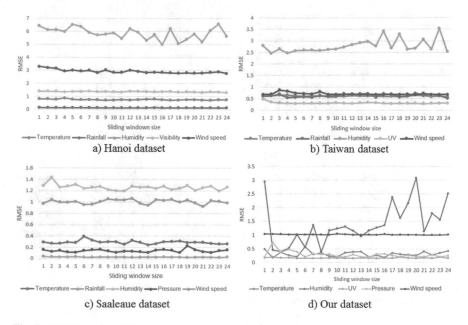

Fig. 5 Variation of RMSE in terms of window size using TCN model

Discussions:

Our experiments show that among different regression models, TCN achieved the best performance over three kinds of inputs. Regarding the correlation analysis, we found that weather parameters correlate together but in different manners. The use of correlated inputs sometimes gives a better performance than using all parameters or a single parameter while requiring less computational cost (on the Hanoi dataset). Otherwise, SISO and MISO may achieve better results, depending on the datasets. Overall, TCN SISO gives the highest accuracy. In terms of window size, the accuracy is quite stable for the case of temperature, rainfall but not humidity. Besides, thicker sampling for short-term prediction also seems to improve prediction error. Specifically, with the wind data shown in Fig. 5, the thicker sampling makes the prediction results better than other data sets with less sampling.

5 Conclusions

In this paper, we have investigated four neural networks (RNN, LSTM, CNN-LSTM, TCN) and their variations for weather forecasting problems. We also analyzed the correlation among weather parameters and experimented with three inputs to predict a weather output: single input, multiple inputs, correlated inputs. We validate our framework on four datasets which contains multiple parameters captured at various sampling rates, locations, and times and found that TCN applied in Single Input Single Output achieved the best results and is recommended for being used in practical applications. In the future, we will improve the TCN model with a transformer architecture or apply variational inference to capture inter-correlation among samples in a window and uncertainty of the data.

References

1. Chung J, Gulcehre C, Cho K, Bengio Y (2014) Empirical evaluation of gated recurrent neural networks on sequence modeling. arXiv preprint arXiv:1412.3555
2. Fente DN, Singh DK (2018) Weather forecasting using artificial neural network. In: 2018 second international conference on inventive communication and computational technologies (ICICCT). IEEE, pp 1757–1761
3. Fu Q, Niu D, Zang Z, Huang J, Diao L (2019) Multi-stations–weather prediction based on hybrid model using 1d cnn and bi-lstm. In: 2019 Chinese Control Conference (CCC). IEEE, pp 3771–3775
4. Hewage P, Behera A, Trovati M, Pereira E (2019) Long-short term memory for an effective short-term weather forecasting model using surface weather data. In: IFIP international conference on artificial intelligence applications and innovations. Springer, Heidelberg, pp 382–390
5. Hewage P, Behera A, Trovati M, Pereira E, Ghahremani M, Palmieri F, Liu Y (2020) Temporal convolutional neural (tcn) network for an effective weather forecasting using time-series data from the local weather station. Soft Comput 24(21):16453–16482
6. Hochreiter S, Schmidhuber J (1997) Long short-term memory. Neural Comput 9(8):1735–1780
7. Karevan Z, Suykens JA (2018) Spatio-temporal stacked lstm for temperature prediction in weather forecasting. arXiv preprint arXiv:1811.06341
8. Livieris IE, Pintelas E, Pintelas P (2020) A cnn-lstm model for gold price time-series forecasting. Neural Comput Appl 32(23):17351–17360
9. Roesch I, Günther T (2019) Visualization of neural network predictions for weather forecasting. In: Computer graphics forum, vol 38. Wiley Online Library, pp 209–220
10. Salman AG, Kanigoro B, Heryadi Y (2015) Weather forecasting using deep learning techniques. In: 2015 international conference on advanced computer science and information systems (ICACSIS). IEEE, pp 281–285
11. Shaojie Bai J, Zico Kolter VK (2018) An empirical evaluation of generic convolutional and recurrent networks for sequence modeling. arXiv preprint arXiv:1803.01271v2
12. Zhang Z, Dong Y (2020) Temperature forecasting via convolutional recurrent neural networks based on time-series data. Complexity

Sentiment Analysis of COVID-19 Tweets: Leveraging Stacked Word Embedding Representation for Identifying Distinct Classes Within a Sentiment

Aakash Bhandari⑩, Vivek Kumar⑩, Pham Thi Thien Huong, and Dang N. H. Thanh⑩

Abstract The pandemic that arose due to the novel Corona Virus Disease of 2019 (COVID) has become the biggest challenge of all time. The entire world's population has stormed social media to express their opinions, emotions, and sentiments. This manuscript implements classical machine and deep learning approaches with static and stacked word embeddings to identify the sentiments of the COVID-19 tweets extracted from *Twitter*. The problem we have tackled in this manuscript is the multi-class classification problem for three and five classes, respectively. Our proposed deep learning model with stacked word embeddings has outperformed the individual static pre-trained embeddings representation, classical machine, and deep learning approaches altogether. The proposed model has proven useful in complex classification tasks such as identifying classes belonging to the same group of sentiments namely Extremely Negative and Negative, Extremely Positive and Positive. The experimental results also show the superior performance of stacked word embeddings for the peculiar contextual semantic comprehension from small tweets and dealing with the unbalancedness of the experimental dataset. We achieved the accuracy with stacked embeddings with accuracy being 73.01% and 84.25% for three and five classes, respectively.

Vivek Kumar provided the manuscript's conceptualization, problem formulations, and experimental design. This research work has received funding from University of Economics Ho-Chi-Minh City.

A. Bhandari
Deerwalk Institute of Technology, Tribhuvan University (TU), Kirtipur, Nepal

V. Kumar (✉)
Marie Sklodowska-Curie Researcher, University of Cagliari, Cagliari, Italy
e-mail: vivekkumar0416@gmail.com

P. Thien Huong · Dang N. H. Thanh
University of Economics Ho-Chi-Minh-City (UEH), Ho-Chi-Minh-City, Vietnam
e-mail: huongpham.192118006@st.ueh.edu.vn

Dang N. H. Thanh
e-mail: thanhdnh@ueh.edu.vn

© The Author(s), under exclusive license to Springer Nature Switzerland AG 2022
N. H. T. Dang et al. (eds.), *Artificial Intelligence in Data and Big Data Processing*,
Lecture Notes on Data Engineering and Communications Technologies 124,
https://doi.org/10.1007/978-3-030-97610-1_27

Keywords COVID-19 · Emotion detection · Sentiment analysis · Machine
learning · Deep learning · Stacked word embeddings

1   Introduction

COVID-19 has become a major health concern for people around the world. The
virus has brought not only physical or psychological effects but also caused economic crises around the globe. The sudden outbreak of COVID-19 made government, institutions, and people express their opinions, feelings, and thoughts through
social media platforms such as *Reddit, Twitter*, etc. The lifestyle within the four walls
and the false and misleading information floating around lead to stress, anxiety, and
depression. The latest updates from the World Health Organization indicates the
current situation of COVID-19 in terms of which is shown in Fig. 1.

With the advent of new intelligent technologies, the use of Artificial Intelligence
in executing all sorts of tasks related to text, speech, image, audio, video, etc. in [5–8,
10, 15] has exponentially increased, and sentiment analysis [4, 16] is one the most
sought after research domains in Natural Language Processing (NLP). COVID-19
has thus drawn the attention of researchers to generate valuable insights from the
information available in social media to tap the different sentiments expressed by
people.

Sentiment analysis has become instrumental in understanding the people's sentiment and developing tools and measures to minimize the collateral damage. However, machine intelligence has proven its worth; they are yet to achieve the human
brain's innate ability to understand the emotions expressed in the complex context.
To contribute towards bridging this gap of contextual understanding, we have pro-

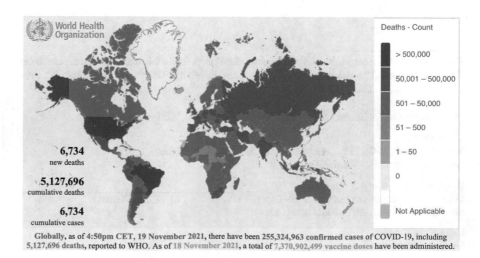

Globally, as of 4:50pm CET, 19 November 2021, there have been 255,324,963 confirmed cases of COVID-19, including
5,127,696 deaths, reported to WHO. As of 18 November 2021, a total of 7,370,902,499 vaccine doses have been administered.

Fig. 1 World Health Organization report on COVID-19

posed a Deep Learning(DL) model utilizing Bidirectional Long-Short Term Memory(BiLSTM) network with stacked word embeddings in this paper. Our work focuses on precisely identifying the sentiments associated with the tweets about COVID-19. In this paper, we have used Classical Machine Learning(CML), DL with bag-of-words, and static and stacked pre-trained word embedding representations.

The rest of the manuscript is organized as follows. Section 2 provides the overview of the related work. Section 3 presents the problem statement we are tackling, the dataset description, and the pre-processing applied to the experimental dataset. Section 4 describes the architecture of proposed models and experiments conducted. Section 5 presents the results and discussion. Finally, the conclusion and future work are summed up in Sect. 6.

2 Related Work

In the duration of the pandemic, several research works have surfaced which exploit sentiment analysis to generate insights from COVID-19 tweets on *Twitter*. In the manuscript [3] the authors did analysis of two types of tweets (pre-pandemic and during the pandemic) gathered in the duration of pandemic. The outcomes suggested that no useful words are found in WordCloud or computation using word frequency in tweets. Hence, the study concluded that related tweets failed to guide on COVID-19 Pandemic. Moreover, the authors have explored the Gaussian membership function based on the fuzzy rule to identify sentiments from tweet data correctly. Another work [13] provides two essential machine learning classification methods (textual analytics) and compares their effectiveness in classifying Corona Virus tweets. They observed 91% accuracy for short tweets by Naive Bayes Method whereas reasonable accuracy of 74% with logistic regression classification method. However, both methods demonstrated weaker performance for longer tweets. One work [12] used *Twitter* dataset using IDs provided by IEEE data port to perform COVID-19 sentiment analysis. The pre-processing techniques and features extraction techniques used by them to train the machine learning model takes the 80% data (6022 Tweets) and evaluate its performance using the remaining 20% data (1506 tweets). Tweets are classified into three categories, i.e., positive, neutral or negative. The Extra Trees Classifier outperformed all the other models with an accuracy score of 0.93 using their proposed concatenated features set. The study in [2] proposed a new fusion model that uses a large-scaled *Twitter* dataset representing eight highly affected countries from COVID-19 for sentiment analysis. The model proposed is a hybrid fusion model that uses five DL classifiers and employs a meta-learning method to improve the final output. The outcome of the study suggested that the sentiment in their tweets were related to the events and news in their countries. It also included the number of total deaths, number of recoveries, and newly infected cases.

A novel recent approach of using multiple pre-trained word embeddings together also called stacked embeddings is introduced in the manuscript [1]. The advantage of using stacked embeddings over individual pre-trained word embeddings is that it combines the overall representation of each word embedding that helps in better embeddings representation semantically. But so far, the use of stacked word embeddings is under-explored, and only a few research works are available in the literature [14]. The manuscript [1] presented a multipurpose unified framework called *FLAIR*[1] for to tackle several NLP tasks. One of the novelties of our work is that we have also employed stacked word embedding representations for the sentiment analysis task at hand.

3 Problem Statement, Dataset and Pre-processing

This section explains the problem statements we tackled and provides insights about the dataset used for conducting the experiments.

3.1 Problem Statement

In this manuscript, we aim to identify sentiments from the tweets related to COVID-19. The problem we are addressing first is a multi-class classification problem with five target classes. Since the data is non-uniformly distributed across the classes, we have transformed the first problem statement to a multi-class classification problem with three target classes. The average length of tweets is just twenty-five words, which makes identifying subtly close sentiments difficult, such as Extremely Negative-Negative and Extremely Positive-Positive in our case. Therefore to study the effect on the classification model in identifying semantically similar classes, we have merged the original five classes into three classes, namely `Negative`, `Neutral` and `Positive`.

3.2 Dataset Description

The dataset we used for our experiments is essentially comprised of tweets related to COVID-19 extracted from *Twitter*. The extracted data has following attributes: `ScreenName`, `TweetAt`, `Location`, `UserName`, `OriginalTweet` and `Sentiment`. To avoid any privacy concerns `UserName` and `ScreenName` have been replaced by codes. `Sentiment` attribute represents the labels associated with the tweets. The tweets are manually annotated in five classes, namely *Negative, Extremely Negative, Neutral, Extremely Positive* and *Positive*. The distribution of data for each class is presented in Table 1. The dataset used in this work is available

[1] https://github.com/flairNLP/flair.

Table 1 Distribution of class instances in train and test datasets

	Original classes				Merged classes				
Dataset	Extremely negative	Negative	Neutral	Positive	Extremely positive	Negative	Neutral	Positive	Total
Train	5481	9917	7713	11422	6624	15398	7713	18046	41157
Test	592	1041	619	947	599	1633	619	1546	3798

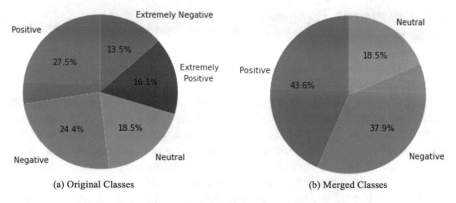

(a) Original Classes   (b) Merged Classes

Fig. 2 Cumulative distribution of train and test data for original and merged classes

at *Kaggle*.[2] Post merging the classes, the data is more uniformly distributed for the classes `Negative` and `Positive` in the resulting dataset thus generated. For better representation, we have presented the total distribution of the dataset (test and train) for original and merged classes in Fig. 2. To prepare the input data for the CML and DL models, we have performed standard steps to preprocess the data [9, 11]. The pre-processing steps implemented are mentioned below:

1. We converted text from each tweet into lowercase so that a single token could represent the words with the same spelling (e.g., *what* and *What* will have a common token, *what*).
2. We expanded the contractions used in the text into their normal form (e.g. *we'll* into *we will*).
3. Punctuations are heavily used in the dataset, which does not provide important information about sentiment. Thus, we have removed all the punctuation for the tweet data.
4. Whitespaces and newlines do not add much value to sentiment analysis; we simply removed them from the texts.

---

[2] https://www.kaggle.com/datatattle/covid-19-nlp-text-classification.

4 Classification Models and Experiments

This section presents the classification models used for the sentiment analysis and explains the approaches employed to perform the experiments.

4.1 Experiments Performed

For our work, we implemented CML and DL, classification models. The CML models employed bag-of-words representation while DL models were used with pre-trained static and stacked word embeddings representations. For the static word embeddings we have used pre-trained *GLoVe*[3] and *fastText*[4] word embeddings with DL models. For the stacked word embeddings, we have used the *FLAIR* framework to use the combination of GLoVe and fastText pre-trained word embeddings. *FLAIR* is a simple framework for state-of-the-art NLP downstream tasks developed by *Zalando Research*[5] . It is built on top of *PyTorch*[6] and is based on the concept of contextual string embeddings. It provides a number of pre-trained embeddings including *GLoVe* and *fastText*. The rationale behind using the stacked embedding for our experiments is to harness the advantages of each word embeddings representation and use it for superior contextual text interpretation. All the experiments are performed with datasets of five and three classes. As mentioned in the problem formulation section, the classes belonging to the same set of sentiment require explicit details to be distinguished. For our dataset, initially with five classes, determining Extremely Negative from Negative and Extremely Positive from Positive is more tedious than identifying *Positive*, *Neutral* and *Negative*. Therefore, to study the effect of unbalanced data in identifying classes with similar sentiment polarity, we have merged the original five classes into three classes, namely Negative, Neutral and Positive. Reducing the classes has improved the class distribution, and we performed all the experiments with the dataset with three classes. The computational resource used to run our experiments is depicted in Table 2.

---

[3] https://nlp.stanford.edu/projects/glove/.

[4] https://fasttext.cc/docs/en/english-vectors.html/.

[5] https://research.zalando.com/.

[6] https://pytorch.org/.

Table 2 Computational resource used for experiments

Item	Specification
Processor	Intel Core i3-7100 (-HT-MCP-) CPU @ 3.90 GHz
GPU	NVIDIA GeForce GTX 1080, 8 GB memory
Graphic Card	NVIDIA graphic driver version 418.87.00
CUDA	Version 10.1
Python	Version 3.6.6
Operating System	Ubuntu 17.10

4.2 Classification Models

For CML, we have used Support Vector Machine(SVM),[7] Naive Bayes(NB)[8] and Random Forest(RF)[9] classifiers available at Sci-Kit[10] Learn library. For DL approaches, we have used BiLSTM based sequential Deep Neural Network classification model, the architecture of which is presented in Fig. 3.

The proposed DL model consists of an embeddings layer at the top initialized by embedding matrix, followed by two BiLSTM layers, and one dense layer. The inputs to the embedding layer are the static and stacked embeddings. For our experiments static embedding matrix are generated by *GloVe* and *fastText* while *FLAIR* is used to generate the stacked embeddings which is essentially the combination *GloVe* and *fastText*. Finally the output layer yields the classes for the multi-class classification task. We have set the dropout rate to 0.20, and the dimension used for the pre-trained word embeddings matrix is 300.

5 Results and Discussion

This section presents the results achieved by the performed experiments and provides a detailed discussion over them. The metrics which we have used to evaluate the performances of employed approaches are Recall, Precision, and and F-1 scores given by by the formulas below:

$$Precision = \frac{TruePositive}{TruePositive + FalsePositive} \tag{1}$$

---

[7] https://scikit-learn.org/stable/modules/svm.html.

[8] https://scikit-learn.org/stable/modules/naive_bayes.html.

[9] https://scikit-learn.org/stable/modules/generated/sklearn.ensemble.RandomForestClassifier.html.

[10] https://scikit-learn.org/stable/.

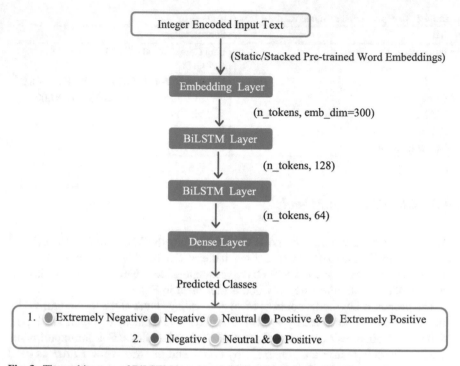

Fig. 3 The architecture of BiLSTM based model for multi-class classification

$$Recall = \frac{True\,Positive}{True\,Positive + False\,Negative} \tag{2}$$

$$F1 = 2 \times \frac{Precision \times Recall}{Precision + Recall} \tag{3}$$

To sum up, Tables 3 and 4 present the results of CML and DL models for classification task for 5 and 3 classes respectively. The confusion matrix results for static and stacked word embeddings are shown in Fig. 4. The CML approaches signify SVM, NB, and RF classifiers and Static and Stacked embeddings are pre-trained word embedding used with DL approaches. For our experiments, the baseline is CML classifiers. Our experimental results show that CML has performed poorly for the tasks at hand, and overall the DL approach with stacked word embeddings has outperformed the DL approaches with static word embeddings and the baseline CML and approaches. The best accuracy score is achieved by the DL model used with stacked word embeddings which are **73.01** for five classes and **84.25** for three classes classification problems, respectively. Our results also strengthen the assumption of reducing the classes from 5 to 3 to tackle the subtle difference between the classes of the same polarity. The class reduction to balance the dataset distribution has significantly improved the performance of all the CML and DL approaches for each class.

For instance, the overall accuracy of DL model using stacked word embeddings has improved by **11.24%**, i.e., from **73.01** to **84.25**. On the other hand, the difference in accuracy between DL approaches with static and stacked word embeddings for five classes is **1.52%** and for three classes **0.21%**. These outcomes underline the efficacy of using stacked word embeddings in classifying the critical classes which essentially belong to the same group are sentiments that are often difficult to be identified semantically for small input text length.

6 Conclusion and Future Work

Social media is a powerful platform for communication and accessing information about the ongoing and *Twitter* is one of them floating with abundant text data. People express all sorts of views and opinions driven by different sentiment aspects on *Twitter* by means of tweets. Unfortunately, the sentiment expressed in tweets is not always positive, and it impacts the well-being of associated people as well. Tweets full of *Violent, Racist, Abusive, Derogatory, Threatening* content can stir fear and depres-

Table 3 Results of CML and DL classification models used with 5 classes

| Classes | Metrics | CML Approaches | | | Static Embeddings | | Stacked Embeddings |
		SVM	Naive Bayes	Random Forest	fastText	GloVe	GloVe+fastText
Extremely Negative	Precision	0.3076	0.1645	0.2105	0.7409	**0.7698**	0.7432
	Recall	0.0135	0.2584	0.0473	0.7246	0.6891	**0.7483**
	F1	0.0258	0.2010	0.0772	0.7327	0.7272	**0.7457**
Negative	Precision	0.2692	0.2852	0.2729	0.6682	0.6634	**0.6870**
	Recall	0.1978	0.1517	0.3198	0.6695	0.6493	**0.6916**
	F1	0.2281	0.1981	0.2945	0.6689	0.6563	**0.6893**
Neutral	Precision	0.4181	0.2418	0.3700	0.8055	0.7931	**0.8431**
	Recall	0.1938	0.3004	0.1954	**0.7964**	0.7802	0.7899
	F1	0.2649	0.2680	0.2558	0.8009	0.7866	**0.8156**
Positive	Precision	0.2452	0.2598	0.2462	**0.6716**	0.6437	0.6706
	Recall	0.7043	0.1256	0.4857	0.6652	**0.7307**	0.7117
	F1	0.3637	0.1694	0.3268	0.6684	0.6844	**0.6905**
Extremely Positive	Precision	0.0000	0.1582	0.2000	0.7444	**0.7946**	0.7855
	Recall	0.0000	0.2871	0.0834	**0.7779**	0.7495	0.7462
	F1	0.0000	0.2040	0.1177	0.7608	**0.7714**	0.7654
Overall	Accuracy	0.2635	0.2074	0.2611	0.7148	0.7130	**0.7301**
	Precision	0.2510	0.2329	0.2608	0.7148	0.7169	**0.7326**
	Recall	0.2635	0.2074	0.2611	0.7148	0.7130	**0.7301**
	F1	0.2004	0.2037	0.2345	0.7147	0.7137	**0.7310**

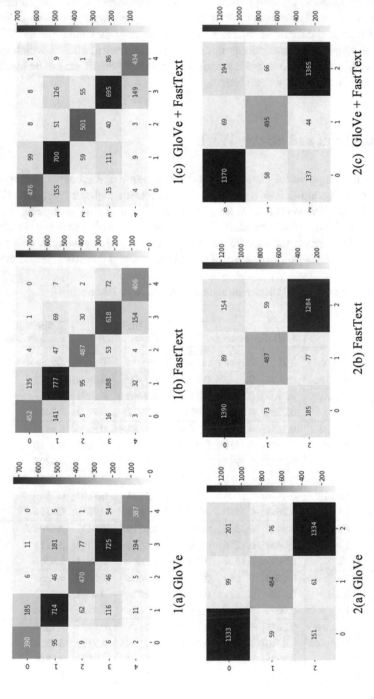

Fig. 4 Confusion matrix results for 5 and 3 classes classification respectively

Table 4 Results of CML and DL classification models used with 3 classes

| Classes | Metrics | CML Approaches | | | Static Embeddings | | Stacked Embeddings |
		SVM	Naive Bayes	Random Forest	fastText	GloVe	GloVe+fastText
Negative	Precision	0.4231	0.4513	0.4566	0.8410	0.8510	**0.8810**
	Recall	0.2663	0.4090	0.3674	**0.8550**	0.8470	0.8140
	F1	0.3269	0.4291	0.4071	0.8480	**0.8490**	0.8460
Neutral	Precision	0.0000	0.2381	0.4529	0.8200	0.8010	**0.8440**
	Recall	0.0000	0.2827	0.0856	0.7350	**0.7740**	0.7590
	F1	0.0000	0.2584	0.1440	0.7750	0.7870	**0.7990**
Positive	Precision	0.3873	0.3998	0.4152	0.8360	**0.8450**	0.8080
	Recall	0.6940	0.4094	0.6358	0.8550	0.8600	**0.9060**
	F1	0.4972	0.4046	0.5024	0.8450	0.8530	**0.8550**
Overall	Accuracy	0.3970	0.3886	0.4307	0.8350	0.8404	**0.8425**
	Precision	0.3396	0.3956	0.4392	0.8353	0.8402	**0.8455**
	Recall	0.3970	0.3886	0.4307	0.8357	0.8404	**0.8425**
	F1	0.3429	0.3913	0.4030	0.8351	0.8402	**0.8419**

sion in common people, especially in the dire situation caused due to COVID-19. Therefore, the need for measures to identify tweets' nature and sentiment to prevent unfortunate situations is very much the need of the hour. In this paper, we have aimed to identify the sentiments associated with the tweets related to COVID-19. The sentiments associated with the tweets belong to five classes, namely *Extremely Negative, Negative, Neutral, Positive* and *Extremely Positive*. We have leveraged CML, and DL approaches, bag-of-words, pre-trained static and stacked word embedding representations to classify the COVID-19 tweets. Our findings show that the proposed DL model used with stacked embeddings has achieved promising results and can be used to analyze the sentiments of tweets accurately and more effectively. The two primary outcomes and novelty of this study are: (i) Stacked word embeddings can play a significant role in the precise contextual interpretation of the text to identify the subtle difference between the sentiments belonging to the same group of emotions. (ii) Stacked word embeddings can also help mitigate the effect of the unbalancedness of the experimental dataset in classification tasks. As future work, we aim to use dynamic (contextual) word embeddings such as ELMo, BERT, etc., in stacked embeddings for better semantic interpretation of text for the given context.

Acknowledgements This research is funded by University of Economics Ho Chi Minh City (UEH), Ho Chi Minh City, Vietnam.

References

1. Akbik A, Bergmann T, Blythe D, Rasul K, Schweter S, Vollgraf R (2019) Flair: an easy-to-use framework for state-of-the-art nlp. In: Proceedings of the 2019 conference of the North American chapter of the association for computational linguistics (demonstrations), pp 54–59
2. Basiri ME, Nemati S, Abdar M, Asadi S, Acharrya UR (2021) A novel fusion-based deep learning model for sentiment analysis of covid-19 tweets. Knowl-Based Syst 228:107242
3. Chakraborty K, Bhatia S, Bhattacharyya S, Platos J, Bag R, Hassanien AE (2020) Sentiment analysis of covid-19 tweets by deep learning classifiers-a study to show how popularity is affecting accuracy in social media. Appl Soft Comput 97:106754
4. Khanna P, Sasikumar M 920110 Recognizing emotions from human speech. In: Thinkquest 2010. Springer, Heidelberg, pp 219–223
5. Kumar V, Kalitin D, Tiwari P (2017) Unsupervised learning dimensionality reduction algorithm pca for face recognition. In: 2017 international conference on computing, communication and automation (ICCCA). IEEE, pp 32–37
6. Kumar V, Mazzara M, Messina A, Lee J (2018) A conjoint application of data mining techniques for analysis of global terrorist attacks. In: International conference in software engineering for defence applications. Springer, Heidelberg, pp 146–158
7. Kumar V, Mishra BK, Mazzara M, Thanh DN, Verma A (2020) Prediction of malignant and benign breast cancer: a data mining approach in healthcare applications. In: Advances in data science and management. Springer, Heidelberg, pp 435–442
8. Kumar V, Recupero DR, Riboni D, Helaoui R (2020) Ensembling classical machine learning and deep learning approaches for morbidity identification from clinical notes. IEEE Access 9:7107–7126
9. Kumar V, Verma A, Mittal N, Gromov SV (2019) Anatomy of preprocessing of big data for monolingual corpora paraphrase extraction: source language sentence. Emerging Technol Data Mining Inf Security 3:495
10. Kumar V, Zinovyev R, Verma A, Tiwari P (2018) Performance evaluation of lazy and decision tree classifier: a data mining approach for global celebrity's death analysis. In: 2018 international conference on Research in Intelligent and Computing in Engineering (RICE). IEEE, pp 1–6
11. Riboni D (2020) Tf-idf vs word embeddings for morbidity identification in clinical notes: an initial study
12. Rustam F, Khalid M, Aslam W, Rupapara V, Mehmood A, Choi GS (2021) A performance comparison of supervised machine learning models for covid-19 tweets sentiment analysis. Plos One 16(2):e0245909
13. Samuel J, Ali G, Rahman M, Esawi E, Samuel Y et al (2020) Covid-19 public sentiment insights and machine learning for tweets classification. Information 11(6):314
14. Schweter S, März L (2020) Triple e-effective ensembling of embeddings and language models for ner of historical german. In: CLEF (Working notes)
15. Thanh DN, Erkan U, Prasath VS, Kumar V, Hien NN (2019) A skin lesion segmentation method for dermoscopic images based on adaptive thresholding with normalization of color models. In: 2019 6th international conference on Electrical and Electronics Engineering (ICEEE). IEEE, pp 116–120
16. Tiwari P, Mishra BK, Kumar S, Kumar V (2020) Implementation of n-gram methodology for rotten tomatoes review dataset sentiment analysis. In: Cognitive analytics: concepts, methodologies, tools, and applications. IGI Global, pp 689–701

Gather Android Application Information on Google Play for Machine Learning Based Security Analysis

Nguyen Tan Cam, A Nguyen Thi Yen Nhi, Nguyen Vuong Thinh, Nghi Hoang Khoa, and Van-Hau Pham

Abstract With the development of smartphones, Android is the most popular mobile operating system. Android ecosystem is very various. There are plenty of Android app stores such as Google Play, APKPure, etc. The information of applications which are available in these app stores is used for many different research fields (e.g., Android malware detection, repackaged application detection, detect inconsistencies between application metadata and application behavior). In this paper, we propose the system which allows collecting the information of Android applications on Google Play to provide the dataset for machine learning based security analysis. The experimental results reveal a lot of potential applications.

Keywords Android application statistic · Android security · Android dataset · Google Play · Machine learning

1 Introduction

In recent years, Android is the most well-known operating system. According to IDC [1], Android operating system always accounts for more than 80% as compared

N. T. Cam (✉)
Hoa Sen University, Ho Chi Minh City, Vietnam
e-mail: camnt@uit.edu.vn

N. T. Cam · A. N. T. Y. Nhi · N. V. Thinh · N. H. Khoa · V.-H. Pham
University of Information Technology, Ho Chi Minh City, Vietnam
e-mail: 18521198@gm.uit.edu.vn

N. V. Thinh
e-mail: 18520367@gm.uit.edu.vn

N. H. Khoa
e-mail: khoanh@uit.edu.vn

V.-H. Pham
e-mail: haupv@uit.edu.vn

Vietnam National University, Ho Chi Minh City, Vietnam

with the other mobile operating systems (Fig. 1). Android smartphones when sold to users, they are pre-installed some applications. Besides that, the user can install lots of applications from app stores such as Google Play, APKPure, APKMirror, etc.

Google Play was released in 2008. It is the official market of Google. It also provides a large number of applications which belongs to many different categoties including Movies, Apps, Books, Entertainments. According to Statista [2], 96.9% of all Android applications on Google Play were available for free (Fig. 2).

Each application when uploaded on Google Play contains much information such as APK file, application title, application description, required Android version, developer, appId, version, etc. These information are given to not only the end user but also experts who analyse the security of Android ecosystem. For example, APK files can be used for malware classification systems. While the information about description can be applied to extract behaviors of applications by using natural language processing (NLP) [3].

The studies related to Android malware classification by using machine learning usually consider apps in Google Play as benign app set in their dataset. Thus, building a system that allows gathering information of Android applications on Google Play can be helpful to create datasets.

Meanwhile, lots of study which focuses on detecting inconsistencies in the description of application's behaviors in privacy policy and the actual behaviors of applications also require a large amount of information about Android apps on Google Play. Once more, gathering these information will help the related works save time for creating datasets.

In this research, we propose a system that allows collecting information about Android applications on Google Play for machine learning based security analysis. Particularly, the information that we extract includes the application's metadata such as application title, description, required Android OS version, developer, appId,

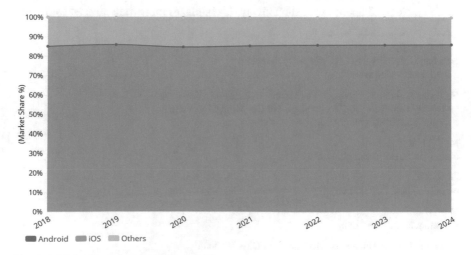

Fig. 1 Mobile OS market share forecast [1]

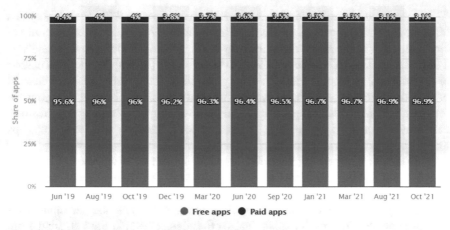

Fig. 2 The rate of free apps on Google Play of October 2021

application version, etc. In addition, the system integrates pre-processing modules of machine learning algorithms that are used to classifiy Android malware. This integration allows the proposed system to extract features for related works. This approach helps in reducing the cost of creating datasets for related works.

The main contributions of this paper include:

- Building a system that can be used to download APK files of applications on Google Play as well as downloading applications' information (metadata).
- Testing the system with 2,357,000 apps on Google Play.
- Building a system that is intergrated with feature extraction modules of some Android malware detection systems based on machine learning.
- Listing related information such as the top permissions that are the most common, the most requested Android operating system versions, etc.

The paper has five main sections. Some related works are disscussed in Sect. 2. Section 3 presents our proposed system. Section 4 explains the evaluation. Conclusions are discussed in Sect. 5.

2 Related Works

There are many studies conducted analyzing a large number of applications on Google Play. There are many studies carrying out analyzing numerous applications on Google Play. Haoyu et al. [4] proposed the system that is used for gathering the information from app markets such as Google App and Chinese App markets. They use these information to detect fake apps, malicious apps, expired apps. However, their systems cannot extract features for malware detection systems using machine learning.

Another study is AndroZoo [5]. AndroZoo is proposed by Allix et al. It checks a large number of applications on Google Play by using VirusTotal. Like the study of Haoyu et al., it is not capable of extracting features for machine learning models.

Similar to AndroZoo, the solution of Wang et al. [6] gathers the information of applications on Google Play. They only use VirusTotal to scan malware instead of providing the ability to extract information and characteristics of applications for related studies about detecting Android malware by using machine learning. Besides, they only downloaded apps in 2014, 2015, and 2017. Thus, we need to update apps that are colleted from Google Play.

With the development of image classification algorithms, some research on using image classification algorithms to classify Android malware is more and more popular. IMCEC [7] was proposed by Danish et al. It is used to classify Android malware by using image classification algorithms. With APK files as input, their system generates images corresponding to the features which are extracted from these files. The classification of Android malware is also considered as the classification of the images which are created from APK files. In addition, the cost for feature extraction and corresponding image generation is high. Therefore, we need to a system that allows undertake these tasks as a pre-processing step of related works.

Like IMCEC, Halil et al. [8] proposed the system that can be used to classify Android malware by using image classification algorithms. Particularly, they converted the information in the Android application's source code into grayscale images, and then used some algorithms such as Decision Tree, k-nearest neighbors, Random forest, AdaBoost, Gradient Boost to classify Android malware.

AndroPyTool [9] integrates tools like AndroGuard [10], FlowDroid [11], DroidBox [12], AVClass [13], Strace [14], and VirusTotal [15]. However, it also needs to be integrated with modules that can extract application's features for Android malware classification by using machine learning.

3   Proposed System

In this study, we propose the system that is used to download Android applications and related information on Google Play. It is also used to extract features of applications that are compatible with Android malware detection solutions that use machine learning. The architecture of proposed system is presented in Fig. 3.

3.1   Metadata Collector Module

Metadata collector module gathers metadata of Android applications on Google Play. There are 43 kinds of information including application title, application description,

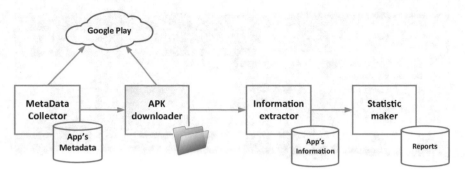

Fig. 3 The architecture of the proposed system

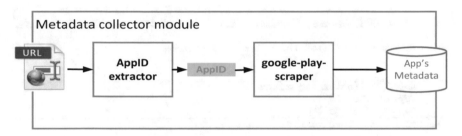

Fig. 4 The main tasks of Metadata collector module

required Android OS version, developer, appId, application version, etc. The information about AppID is used to download APK files of Android application on Google Play. The Fig. 4. shows the main tasks of Metadata collector module.

The Fig. 5 presents an example of the HTML code of Google Play website. In this study, we reply on HTML's information to extract AppIDs of apps on Google Play.

The Fig. 6 shows the algorithm of AppID Collector, which is used to collect the AppID of an application from the application's HTML on Google Play website.

In the example of the HTML code of Google Play website in Fig. 5, the string < a href = "/store/apps/details?id = com.facebook.katana" aria-hidden = "true" tabindex = "-1" class = "poRVub" > < /a > shows the AppID of the application is *com.facebook.katana*. After having the AppID, we use google-play-scraper [16] to gather the information of application. In this study, we gather 43 types of information which are related to applications.

3.2 APK Downloader Module

This module is designed to download APK files of Android apps on Google Play. It considers AppID as input data in order to download APK files. The module uses

Fig. 5 An example of the HTML code of Google Play website

AppID Collector Algorithm
Input: URL
Ouput: Package ID

```
1.   function linkExtracter(startUrl):
2.     | Rule(allow="app/details", callback=parseLink)
3.     | Rule(allow="app", callback=linkExtracter)
4.   function parseLink(url):
5.     | html = select(url)
6.     | idPackage = htm.select("href format")
7.     | return idPackage
8.   function main()
9.     | allowDomain = URL
10.    | linkExtracter(allowDomain)
```

Fig. 6 The algorithm of AppID Collector

Google Play API and Google accounts to interact with Google Play and download APK files. The use of many Google accounts and multi-threading techniques are a solution to increase the speed of APK files download. The following figure shows the algorithm of the APK downloader module (Fig. 7).

```
APK downloader Algorithm
Input:  Application  ID:  appID,  Google  account:  gsfId,
authSubToken
Output: APK file
1.  from gpapi.googleplay import GooglePlayAPI
2.  server = requests.get('url api token'')
3.  authSubToken gsfId = server.response
4.  api = GooglePlayAPI(locale, timezone)
5.  api.login(gsfId, authSubToken)
6.  api.download(appID)
```

Fig. 7 The algorithm of APK downloader

3.3 Information Extractor Module

In addition to the information that is extracted from metadata collector module, this module extracts the properties and characteristics of the application. The properties of the application is extracted by AndroPyTool [9]. Meanwhile, the application's features are extracted using the feature extractor modules of related machine learning-based Android malware detection solutions [7, 8] and our previous works [17]. In this study, we extract the types of features which are presented in Table 1.

Figure 8 depicts an example of the result from application's feature extraction. This result can be used for methods which are used to detect Android malware by using machine learning.

There are many related studies using image classification algorithms to classifiy Android malware [7]. Therefore, in this study, we integrated their features extractor modules and their images convertor modules. This approach is useful for creating pre-processed data of the datasets which use Android apps on Google Play such as benign application set. Figure 9 shows some examples of converting from application

Table 1 The feature categories are extracted in this study

Feature name	Description
Package name	The package name
Permissions	The list of Permissions
API Calls	The list of System calls
Opcodes	The list of Dalvik opcodes
System Command	The list of System command
Activities	The list of Activity components
Services	The list of Service components
Receivers	The list of Receivers
Flowdroid	The list of Data flows by using FlowDroid

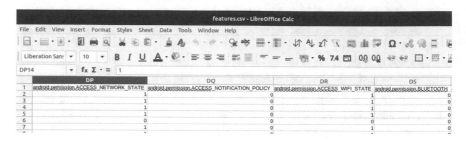

Fig. 8 An example of the result from application's feature extraction

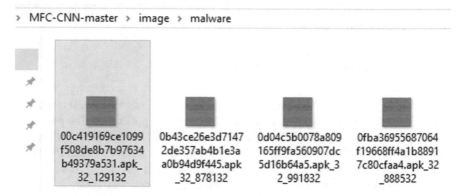

Fig. 9 Some examples of converting from application features into images

features into images.

3.4 Statistic Maker Module

Because the system can extract the information from many Android applications on Google Play, the information needs to be managed effectively. In the proposed system, we develop Statistic Maker module to aim at statistics of the information of the applications. Valuable statistics like most commonly used permissions list, common size of an application, most commonly requested Android OS version, etc.. Besides, we design this module in a modular way. This approach allows the module to be easily extended to deal with other data types in the future.

4 Evaluation

The proposed system is experimented at the Information Security laboratory. It is one of the main laboratories at the University of Information Technology, Vietnam National University Ho Chi Minh City. Table 2 presents the hardware configuration of the computer system that we used in this experiment.

During the evaluation process, we collect the information of 2,357,000 Android apps on Google Play. In addition, the proposed system extracts the specific information of 12,549 applications that are compatible with machine learning-based Android malware detection methods [7, 8]. Table 3 shows some statistical results from testing the proposed system.

Figure 10 shows the results of metadata extraction from Tiktok Lite application on Google Play. This metada includes 43 types of information. Some of the long text in this image is omitted.

One of the important information is the list of permissions that are required by the application. In our study, we discovery the top 10 most popular permissions. Figure 11. shows the top 10 most requested permissions. The results in this figure shows that Internet access is the most commonly requested permission. Its rate is 98%. Among the apps we analyzed, the rates of permission which is related to SD card are 65% and 45% respectively.

When developing applications, programmers require the lowest version of operating system (requested Android OS version) so that their application can be compatible. Statistics of the requested Android OS version which is required by Android apps on Google Play are helpful in limiting the scope of related works. The analysis results show that the lowest version of the most requested operating system is 4.1 (Fig. 12). This result answers related research questions such as "which Android

Table 2 The configuration setting of the test system

Component	Value
Operating System	POP! OS
Memory	64 GB
CPU	Intel(R) Xeon(R) v4 E5-2660, 2.00 GHz 32 core
Hard disk	500 GB

Table 3 Some statistical results from testing the proposed system

Criteria name	Value
Download time per 10,000 applications	3141 s (52.35 min)
Maximum application size	389.59 Megabytes
Smallest application size	0.00150967 Megabytes
Average application size	35.2321 Megabytes

```
1.{
2.     "title":"TikTok Lite",
3.     "description":"TikTok is <<omitted>> @tiktok",
4.     "descriptionHTML":"TikTok is <<omitted>> @tiktok",
5.     "summary":"TikTok Lite <<omitted>>.",
6.     "installs":"100,000,000+",
7.     "minInstalls":100000000,
8.     "maxInstalls":252862099,
9.     "score":4.3590918,
10.    "scoreText":"4.4",
11.    "ratings":1410748,
12.    "reviews":642745,
13.    "histogram":{
14.        "1":146391,
15.        "2":35932,
16.        "3":57701,
17.        "4":95393,
18.        "5":1075329
19.    },
20.    "price":0,
21.    "free":true,
22.    "currency":"USD",
23.    "priceText":"Free",
24.    "offersIAP":false,
25.    "IAPRange":"undefined",
26.    "size":"53M",
27.    "androidVersion":"4.1",
28.    "androidVersionText":"4.1 and up",
29.    "developer":"TikTok PTE.ltd.",
30.    "developerId":"5299725221209516149",
31.    "developerEmail":"feedback@tiktok.com",
32.    "developerWebsite":"undefined",
33.    "developerAddress":"10 ANSON ROAD #10-08<<omitted>>",
34.    "privacyPolicy":"https://www.tiktok.com/legal/privacy-policy",
35.    "developerInternalID":"5299725221209516149",
36.    "genre":"Video Players & Editors",
37.    "genreId":"VIDEO_PLAYERS",
38.    "familyGenre":"undefined",
39.    "familyGenreId":"undefined",
40.    "icon":"https://play-lh. <<omitted>>",
41.    "headerImage":"https://play-lh.google<<omitted>>",
42.    "screenshots":[
43.        "https://play-lh.googleuser<<omitted>> ", <<omitted>>
44.    ],
45.    "video":"undefined",
46.    "videoImage":"undefined",
47.    "contentRating":"Teen",
48.    "contentRatingDescription":"undefined",
49.    "adSupported":false,
50.    "released":"undefined",
51.    "updated":1611584087000,
52.    "version":"17.8.3",
53.    "recentChanges":"undefined",
54.    "comments":[
55.        "Best experience <<omitted>>
56.    ],
57.    "editorsChoice":false,
58.    "appId":"com.zhiliaoapp.musically.go",
59.    "url":"<<omitted>>/apps/details?id=com.zhiliaoapp.musically.go <<omitted>>"
60.}
```

Fig. 10 The results of metadata extraction from Tiktok Lite application on Google Play

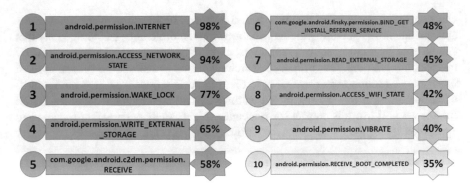

Fig. 11 The top 10 most requested permissions

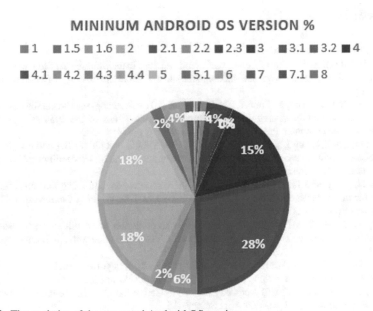

Fig. 12 The statistics of the requested Android OS version

operating system versions should we limit our research scope to?" "Are the analysis of the operating systems which have the version that is lower than 8.0 meaningful?".

5 Conclusion

There are a large number of Android apps on Google Play. The types of information of every application are diverse. Creating the system which can extract the information of many applications on Google Play is useful because these types of information can become datasets of related works in the future. In this study, we proposed the system that allows gathering the information from Android apps on Goolge Play that can be used by security analysis solutions, especially solutions that analyze security by machine learning. Our proposed system is tested to extract the information of more than 2 million apps from Google Play. In addition, we extracted the features of 12,549 applications. The extracted features can be used by related studies. In the future, we will use this dataset for relevant projects such as detecting repackaged apps, identifying the relations between applications on Google Play, etc.

Acknowledgements This research is funded by Hoa Sen University under grant number CS202111.

References

1. IDC (2021) Smartphone Market Share
2. Statista (2021) Distribution of free and paid Android apps in the Google Play Store as of January 2021. Available: https://www.statista.com/statistics/266211/distribution-of-free-and-paid-and roid-apps/
3. Qu Z, Rastogi V, Zhang X, Chen Y, Zhu T, Chen Z (2014) Autocog: measuring the description-to-permission fidelity in android applications. In: Proceedings of the 2014 ACM SIGSAC conference on computer and communications security, pp 1354–1365
4. Wang H, Liu Z, Liang J, Vallina-Rodriguez N, Guo Y, Li L et al (2018) Beyond google play: a large-scale comparative study of chinese android app markets. In: Proceedings of the internet measurement conference 2018, pp 293-307
5. Allix K, Bissyandé TF, Klein J, Le Traon Y (2016) Androzoo: collecting millions of android apps for the research community. In: 2016 IEEE/ACM 13th Working Conference on mining software repositories (MSR), pp 468–471
6. Wang H, Li H, Guo Y (2019) Understanding the evolution of mobile app ecosystems: A longitudinal measurement study of google play. In: The world wide web conference, pp 1988–1999
7. Vasan D, Alazab M, Wassan S, Safaei B, Zheng Q (2020) Image-based malware classification using ensemble of CNN architectures (IMCEC). Comput Secur 92:101748
8. Ünver HM, Bakour K (2020) Android malware detection based on image-based features and machine learning techniques. SN Appl Sci 2:1–15
9. Martín A, Lara-Cabrera R, Camacho D (2018) A new tool for static and dynamic Android malware analysis. In: Data science and knowledge engineering for sensing decision support, pp 509–516
10. Desnos A (2020) AndroGuard, 20 May 2020. Available: https://github.com/androguard/and roguard
11. Arzt S, Rasthofer S, Fritz C, Bodden E, Bartel A, Klein J et al (2014) FlowDroid: precise context, flow, field, object-sensitive and lifecycle-aware taint analysis for Android apps. In: Presented at the proceedings of the 35th ACM SIGPLAN conference on programming language design and implementation, Edinburgh, United Kingdom
12. Lantz (2015) DroidBox. Available: https://code.google.com/p/droidbox/
13. Sebastián M, Rivera R, Kotzias P, Caballero J (2016) Avclass: a tool for massive malware labeling. In: International symposium on research in attacks, intrusions, and defenses, pp 230–253
14. Chaykovsky V (2020) strace—linux syscall tracer. Available: https://strace.io/
15. Virustotal.com (2021) Virustotal. Available: https://www.virustotal.com

16. PyPI (2021) Google Play Scraper, 11 Nov 2021. Available: https://pypi.org/project/google-play-scraper/
17. Hoang Khoa N, Tan Cam N, Pham V-H, Nguyen AG-T (2021) Detect Android malware by using deep learning: experiment and evaluation. In: 2021 The 5th international conference on machine learning and soft computing, pp 129–134

Image Processing and Computer Vision

Self-knowledge Distillation: An Efficient Approach for Falling Detection

Quang Vu Duc[ID]**, Trang Phung**[ID]**, Mai Nguyen**[ID]**, Bao Yen Nguyen**[ID]**, and Thu Hien Nguyen**[ID]

Abstract Self-knowledge distillation, the related idea to knowledge distillation, is a novel approach to avoid training a large teacher network. In this paper, we propose an efficient self-knowledge distillation approach for falling detection. In our approach, the network shares and learns the knowledge distilled via embedded vectors from two different views of a data point. Moreover, we also present the lightweights yet robust network to address this task based on (2+1)D convolution. Our proposed network uses only 0.1M parameters that reduce hundreds of times compared to other deep networks. To demonstrate the effectiveness of our proposed approach, two standard datasets such as the FDD and URFD datasets, have been experimented. The results illustrated state-of-the-art performance and outperformed that compared to independent training. Moreover, with 0.1M parameters, our network demonstrates easy deployment on edge devices, e.g., phones and cameras, in real-time without GPU. The code is available at https://github.com/vdquang1991/self_KD_falling_detection.

Keywords Falling detection · Knowledge distillation · Deep learning · Convolutional neural network

1 Introduction

Falling detection is one of the popular problems in the field of human activity recognition, attracting a lot of attention from scientists. This is an important problem and has great significance for the protection of human health. This problem aims to propose accurate and real-time predictions when a person falls to minimize lying on the floor and waiting for help.

Q. Duc
National Central University, Taoyuan City, Taiwan

Q. Duc · M. Nguyen · B. Nguyen · T. Nguyen
Thai Nguyen University of Education, Thai Nguyen, Vietnam

T. Phung (✉)
Thai Nguyen University, Thai Nguyen, Vietnam
e-mail: phungthutrang.sfl@tnu.edu.vn

© The Author(s), under exclusive license to Springer Nature Switzerland AG 2022
N. H. T. Dang et al. (eds.), *Artificial Intelligence in Data and Big Data Processing*,
Lecture Notes on Data Engineering and Communications Technologies 124,
https://doi.org/10.1007/978-3-030-97610-1_29

Recently, numerous deep models have been proposed with many different architectures to accomplish the mentioned goal. For example, Nunez et al. proposed an approach to utilize the pre-trained model VGG-Net from ImageNet [20]. The authors adopted the optical flow images to predict the falling instead of conventional RGB images. Besides, their model is first to pre-trained on large-scale annotation datasets e.g., ImageNet and UCF101, and then fine-tuned to falling detection. Far apart from [20] that focus on the image domain as an input, Cameiro et al.perform the issue with input as videos [2]. The authors used three types of input videos, including RGB frames, optical flow frames and pose feature frames. Finally, the authors demonstrated that using 3D Convolutional Neural Networks (CNN) hybrid with RGB frames or optical flow frames can achieve state-of-the-art (SOTA) performance. However, it also presented several drawbacks, for example (i) 3D CNNs require a large model size and computational cost than 2D CNNs. (ii) using optical flow frames and/or pose feature frames require computationally expensive to extract from RGB frames. Therefore, 3D CNNs with input as optical flow frames or pose feature frames have proved robust models. Although these models can boost performance for many computer vision tasks. But it is very hard to deploy them on embedded and mobile devices due to their model size. Thus, these models may not be opportune for falling detection that requires real-time detection.

Knowledge distillation [8] is one of the most popular approaches to boost the performance of lightweight models (i.e., student models) by learning the knowledge distilled from deep models (i.e., teacher models). Various distillation-based approaches have been proposed recently, such as response-based knowledge, relation-based knowledge, feature-based knowledge, etc. [5]. However, the performance of the student network may be degraded since the gap between the teacher and student networks are large [19]. Self-knowledge distillation have been proposed to address this limitation (see Fig. 1). Various Self-knowledge distillation approaches have been introduced to replace conventional knowledge distillation and achieved SOTA per-

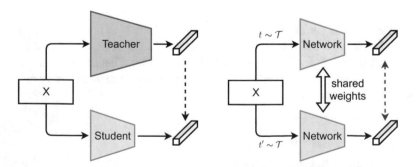

Fig. 1 The comparison between knowledge distillation (left) and self-knowledge distillation (right). Knowledge distillation-based methods always require one or more large-scale teacher networks to instruct the training for the student network. Meanwhile, there is no teacher in the self-knowledge distillation approaches. There is only one network in the self-knowledge distillation, and it can distill the knowledge by itself

formance such as image classification [12, 14], action recognition [24], etc. In this paper, we presented the self-knowledge distillation-based approaches for fall detection. There is only one network in our approach and it shares and learns knowledge from each other during the training process via embedded vectors. Moreover, we also proposed an efficient lightweight model based on (2+1)D convolution for falling detection. With the lightweight backbone, our model can easily deploy on embedded or mobile devices without GPU support. Our contributions are summarized as follows:

(1) We have proposed an efficient approach based on self-knowledge distillation for falling detection. To our best knowledge, this is the first approach focus on self-knowledge distillation to address this task.
(2) Inspired by (2+1)D convolution in [21], we have introduced a new robust lightweight model. With only 0.1M parameters, our network can easily deploy on edge devices.
(3) Experiment results have shown that our approach outperforms that compared to independently training. Besides, it also achieved SOTA performance in terms of accuracy, sensitivity, specificity with only 0.1M parameters.

2 Related Work

Currently, deep learning is one widespread technique in computer vision. It has achieved high performance in various image recognition problems such as handwriting recognition, iris recognition, traffic sign recognition,...In the most recent statistics on Deep learning systems for fall action recognition. Milon Islam et al. reviewed 37 progressive systems. Most of the studies used CNN architecture (21 systems - 56.8%), only 12 systems (32.4%) used LSTM architecture, and four systems (10.8%) used auto-encoder architecture [11].

The CNN architecture is good for finding patterns and shapes in specific images. So, fall detection systems often leverage the power of CNNs in classification or image detection [1, 17, 25]. If the one dimensional (1D) CNN method has been introduced for both Sensor-device [4] and Vision-device [23], 3D CNN [10, 13, 16] and Feedback Optical Flow Convolutional Neural Network (FOF CNN) [9] mainly developed for Vision device. The experimental comparison results show that 3D CNN or CNN with 10-fold cross-validation [17] is best among the reviewed systems [11].

Far apart from the above approaches that focus only on performance, our proposed approach targets building a robust network for edge devices. Therefore, our method goals are both performance and model size. To reduce the model size, we utilize (2+1)D convolution to build a suitable lightweight network. To achieve SOTA performance, we proposed an efficient self-knowledge distillation via embedded vectors. The details of our approach are presented in Sect.3.

3 Methodology

In this section, we introduce the Self-Knowledge Distillation (SKD) to efficiently falling detection. We first briefly discuss the problem setup and then describe SKD framework and its components in detail.

3.1 Problem Setup

Given a set of N training samples in the form of $\mathcal{D} = \{(x^{(1)}, y^{(1)}), (x^{(2)}, y^{(2)}), \dots,$ $(x^{(N)}, y^{(N)})\}$ such that $x^{(i)}$ is a video clip and $y^{(i)}$ is its label (i.e., falling or not). Let $\mathcal{B} = \{(x^{(i)}, y^{(i)})\}_{i=1}^{n}$ denote a mini-batch input from $\mathcal{D}$. Note that $(n \ll N)$. We denote the binary cross-entropy (BCE) between two probability distributions p and q as $BCE(p, q)$. And the mean square error (MSE) is utilized to compared the difference between two vector a and b as $MSE(a, b)$.

3.2 Self-knowledge Distillation

Our proposed SKD approach is illustrated in Fig. 2. In particular, for a mini-batch input B, we perform data augmentation operators (e.g. flip, contrast adjustment, etc.) for each $x^{(i)}$ in B. Let $x_1^{(i)}, x_2^{(i)} = \mathcal{T}(x^{(i)})$ denote the two randomly augmented views from the original video $x^{(i)}$ where $\mathcal{T}(\cdot)$ is the set of data augmentation operators. Note that $x_1^{(i)}, x_2^{(i)}$ have the same the label $y^{(i)}$. Both views $x_1^{(i)}$ and $x_2^{(i)}$ are then passed into the model $\mathcal{F}_\theta$, where θ denotes the set of parameters of $\mathcal{F}$. We assume the

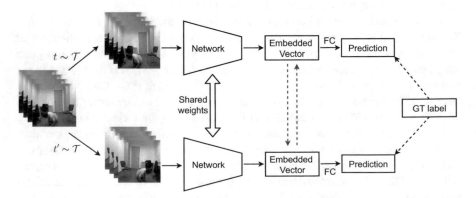

Fig. 2 Overview of the SKD approach. The black line is the forward path; the black dashed line indicates the binary cross-entropy (BCE) loss; the blue dashed line denotes self-knowledge distillation loss. FC means the fully connected layer. GT label is the ground-truth label of the input video

outputs of the model correspond to embedded vector $z_1^{(i)}$ and $z_2^{(i)}$, i.e., $z_1^{(i)} = \mathcal{F}_\theta(x_1^{(i)})$ and $z_2^{(i)} = \mathcal{F}_\theta(x_2^{(i)})$. The model $\mathcal{F}_\theta$ shares weights between the two views. A fully connected (FC) layer with a sigmoid activation function is adopted to generate the logit prediction from $\mathbf{z}_1^{(i)}$ and $\mathbf{z}_2^{(i)}$ as follows:

$$
\begin{aligned}
p_1^{(i)} &= sigmoid(FC(z_1^{(i)})) \\
p_2^{(i)} &= sigmoid(FC(z_2^{(i)}))
\end{aligned}
\tag{1}
$$

where $p_1^{(i)}$ and $p_2^{(i)}$ represents the logit value for the falling class. The standard BCE loss between the logit value p and ground truth label y is calculated by:

$$
\mathcal{L}_{sup}\left(y^{(i)}, p_1^{(i)}, p_2^{(i)}\right) = BCE\left(y^{(i)}, p_1^{(i)}\right) + BCE\left(y^{(i)}, p_2^{(i)}\right)
\tag{2}
$$

where $\mathcal{L}_{sup}$ is the canonical supervised learning loss between the ground truth label $y^{(i)}$ and the logit value $p^{(i)}$ from the model. The proposed SKD method performs the self-distillation via embedded vectors as follows:

$$
\mathcal{L}_{skd}\left(z_1^{(i)}, z_2^{(i)}\right) = MSE\left(stop\_grad(z_1^{(i)}), z_2^{(i)}\right) + MSE\left(stop\_grad(z_2^{(i)}), z_1^{(i)}\right)
\tag{3}
$$

where $\mathcal{L}_{skd}$ denotes the self-knowledge distillation loss between two embedded vectors of the same input data point (i.e., the same input video). The $stop\_grad$ is the stop-gradient operator [3]. This means that $z_1^{(i)}$ (or $z_2^{(i)}$) is treated as a constant in this term. By integrating Eqs. 2 and 3, we can construct the final optimization objective for the entire network as follows:

$$
\mathcal{L}_{net}\left(y^{(i)}, p_1^{(i)}, p_2^{(i)}, z_1^{(i)}, z_2^{(i)}\right) = \frac{1}{n}\sum_{i=1}^{n}\mathcal{L}_{sup}\left(y^{(i)}, p_1^{(i)}, p_2^{(i)}\right) + \lambda\mathcal{L}_{skd}\left(z_1^{(i)}, z_2^{(i)}\right)
\tag{4}
$$

where λ is the loss weight for the self-distillation loss $\mathcal{L}_{skd}$. The pseudo-code of our proposed SKD approach is illustrated in Algorithm 1.

3.3　Network Architecture

In 2015, He et al. proposed an efficient network, namely ResNet [7]. One of the most contributions is the skip connection technique. ResNet can avoid the vanishing gradient issue without reducing the network's performance. That helps the deep layers at least no worse than the shallow ones. Moreover, the upper layers get more information directly from the lower layers so it adjusts the weight more efficiently with ResNet architecture. After the success of ResNet, various model architectures have been introduced based on the ResNet backbone. Experiments have shown that these architectures can be trained with CNN models with depths of up to thousands

---

Algorithm 1: The pseudo-code SKD framework for Falling Detection

---

    Input: Training set $\mathcal{D}$
           $\mathcal{F}_\theta$: the 3D deep network
           T: the set of data augmentation operators
           λ: loss weight factor.
 1 Initialize parameters θ
 2 **repeat**
 3 | Sample a batch (x, y) from the training set $\mathcal{D}$
 4 | $x_1, x_2 = T(x)$
 5 | $z_1, z_2 = \mathcal{F}_\theta(x_1), \mathcal{F}_\theta(x_2)$
 6 | $p_1, p_2 = sigmoid(FC(z_1)), sigmoid(FC(z_2))$
 7 | Calculate the loss $\mathcal{L}_{net}\left(y, p_1, p_2, z_1, z_2\right)$ by Eq. 4
 8 | Update parameters θ by computing the gradient of $\mathcal{L}_{net}$
 9 **until** *convergence*
10 **return** $\mathcal{F}_\theta$

---

of layers. ResNet has quickly become the most popular architecture in deep learning and computer vision.

Hara et al. have proposed a 3D ResNet model to perform the action recognition [6]. However, the models are both very deep and complex, and they are trained on large datasets. Therefore, these 3D ResNet models are not suitable for fall detection in this work. To reduce the model size and complexity of the 3D CNNs, Tran et al.proposed (2+1)D CNN to replace 3D CNN [21]. Specifically, the temporal and spatial dimensions of a filter are separated in (2+1)D CNNs. By such factorization, 3D conv operator is replaced with a 2D conv operator (in space, i.e., spatial domain) followed by a 1D conv operator (in time, i.e., temporal domain). Figure 3a illustrates the difference between two architectures including 3D CNN and (2+1)D CNN. By experiment, the authors demonstrated that (2+1)D CNNs outperform conventional 3D CNN.

Figure 3b depicts the difference between two 3D conv and (2+1)D conv layers. In which, with the 3D conv layer, the kernel size is often used with $t \times d \times d$ of dimensions. Meanwhile, in (2+1)D conv, the 3D conv operator is split into two smaller convolutions with the first conv operator having a kernel size of $1 \times d \times d$ and the second conv operator having a kernel size of $t \times 1 \times 1$. Moreover, the number of parameters and computational cost in (2+1)D convolution layers are significantly reduced compared to conventional 3D convolution layers.

In this work, we introduce a lightweight network based on (2+1)D convolution that is denoted in Table 1. In which, Conv_1, Conv_2x, Conv_3x and Conv_4x are the convolution layer. The output of each convolution layer is passed into a batch normalize layer and a ReLU layer. In column "Specification", $7 \times 7 \times 7, 16$ (for example) denotes this convolution layer has kernel size is set to $7 \times 7 \times 7$ and there are 16 filter in this layer. The symbol s denotes the stride in conv operator and max-pooling (MaxPool) layers. The FC is the fully connected layer, in this layer, we use the sigmoid activation function to provide the classification prediction for input video.

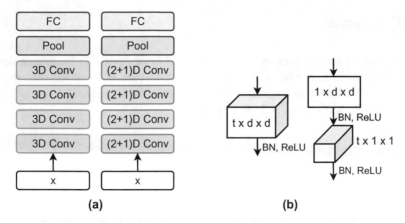

Fig. 3 Comparison between 3D CNN with (2+1)D CNN. a illustrates the 3D convolution layers are replaced by the (2+1)D convolution layers in the same backbone network. **b** denotes the difference between 3D convolution and (2+1)D convolution.

Table 1 Overall architecture of our proposed network

Layer	Specification	Output size
Input		$T \times 224 \times 224 \times 3$
Conv_1	$7 \times 7 \times 7, 16$ $s = (2, 2, 2)$	$\frac{T}{2} \times 112 \times 112 \times 16$
MaxPool	$3 \times 3 \times 3$ $s = (1, 2, 2)$	$\frac{T}{2} \times 56 \times 56 \times 16$
Conv_2a	$1 \times 3 \times 3, 32$ $s = (1, 2, 2)$	$\frac{T}{2} \times 28 \times 28 \times 32$
Conv_2b	$3 \times 1 \times 1, 32$ $s = (1, 1, 1)$	$\frac{T}{2} \times 28 \times 28 \times 32$
Conv_3a	$1 \times 3 \times 3, 64$ $s = (1, 2, 2)$	$\frac{T}{2} \times 14 \times 14 \times 64$
Conv_3b	$3 \times 1 \times 1, 32$ $s = (2, 1, 1)$	$\frac{T}{4} \times 14 \times 14 \times 64$
Conv_4a	$1 \times 3 \times 3, 128$ $s = (1, 2, 2)$	$\frac{T}{4} \times 7 \times 7 \times 128$
Conv_4b	$3 \times 1 \times 1, 128$ $s = (2, 1, 1)$	$\frac{T}{8} \times 7 \times 7 \times 128$
Global_Ave_Pool		$\frac{T}{8} \times 128$
Drop Out	Drop rate = 0.5	$\frac{T}{8} \times 128$
FC	Unit = 1, sigmoid	1

In which, FC denotes the fully connected layer. s illustrates the stride

4   Experiment

In this section, we first present the datasets used in this work. The implementation details and evaluation metrics also are introduced in this section. We then experimented and evaluated our approach against other SOTA methods in terms of sensitivity, specificity, accuracy, and model size.

4.1   Datasets and Evaluation Metrics

Datasets In this paper, we have used two standard datasets including Fall Detection Dataset (FDD) and UR Fall Detection (URFD), to evaluate the SKD's performance compared to SOTA methods.

The FDD dataset was published in 2013. This dataset contains videos recorded in two locations, the coffee room, and the home room. All videos in the dataset were recorded by a single camera with 25 fps and set to have a frame resolution of 320 240 pixels. The actors in each video perform normal activities at home and fall at different times, all actions are performed at random.

The URFD dataset was built by Bogdan Kwolek et al.in 2014 to identify people falling through different types of devices such as cameras, accelerometers, Microsoft Kinect (in this paper, we only use the videos captured from the camera in the dataset and without using information from other devices) [15]. The dataset consists of 70 videos with 30 videos containing different fall actions and the remaining 40 videos containing normal daily activities, such as: sitting, walking, bending, etc.

Metrics. The fall detection task can be seen as a supervised learning problem with binary classification. In particular, the model needs to decide whether an input video clip is labeled as a fall or not. To evaluate the performance of a classifier, sensitivity, specificity, and accuracy are common metrics. The three metrics are defined as follows:

$$sensitivity = \frac{TP}{TP + FN} \tag{5}$$

$$specificity = \frac{TN}{TN + FP} \tag{6}$$

$$accuracy = \frac{TP + TN}{TP + TN + FP + FN} \tag{7}$$

where TP is true positive i.e., the number of video clips labeled as falls is predicted correctly by the model. TN - true negative i.e., the number of video clips labeled as non-fall is predicted correctly by the model. FP—false positive i.e., the number of video clips labeled as non-fall is predicted incorrectly by the model. FN—false negative i.e., the number of video clips labeled as fall is predicted incorrectly by the model.

4.2 Implementation Details

Our network is trained from scratch with the stochastic gradient descent (SGD) optimizer. The training videos are divided into clips with 16 frames of length and each frame is $224 \times 224 \times 3$ of dimensions. The batch size is set to 32 clips. The learning rate is initialized to 0.001 and is reduced 10 times if the model does not improve the accuracy in 10 consecutive epochs. All models are trained with 100 epochs and the metrics are calculated on the test set. In this work, we utilize the data augmentation strategy set proposed by Vu et al. [24]. Each augmentation operator has a chosen probability of 0.5. To evaluate the network performance, we compare the results of the model with independently training (Baseline) and SOTA methods proposed in [2, 20, 22] in terms of sensitivity, specificity, accuracy, model size, and computational cost (FLOPs [18]).

4.3 Performance Comparison

Comparison with independently training. To evaluate the effectiveness of our proposed approach, we have compared the SKD method to the independently training (Baseline) method with the same network architecture on the FDD and URFD datasets. As can be seen in Tables 2 and 3, the SKD framework outperforms the baseline method. Specifically, the SKD achieves 96.7%, 99.8%, and 100% in terms of accuracy, sensitivity, and specificity that increase 1.7%, 0.8%, and 1.0% compared to the Baseline method, respectively.

In Table 3, our proposed SKD method outperforms the Baseline method in terms of accuracy and specificity by a margin of 0.3% and 1.0%, respectively. These results show that the SKD approach can improve generalization and performance compared to independently training. The improvement comes from self-knowledge distilled by the network via an efficient data augmentation strategy.

Table 2 The comparison of the SKD framework with the Baseline (independently training) method with the same network architecture on the FDD dataset

Method	Accuracy	Sensitivity	Specificity
Baseline	95.0	99.0	99.0
SKD (Ours)	**96.7**	**99.8**	**100.0**

All mechanisms are trained with the same network

Table 3 The performance of the SKD framework compares to the Baseline (independently training) method with the same network architecture on the URFD dataset

Method	Accuracy	Sensitivity	Specificity
Baseline	98.0	100.0	99.0
SKD (Ours)	**98.3**	100.0	**100.0**

All mechanisms are trained with the same network

Table 4 Comparison in terms of accuracy, sensitivity, specificity, model size, and FLOPs between our proposed SKD approach and SOTA methods on the FDD dataset

Method	Accuracy (%)	Sensitivity (%)	Specificity (%)	Model size (M)
VGG [20]	97.0	99.0	97.0	14.7
RGB [2]	80.52	**100.0**	79.02	14.7
OF [2]	96.43	99.9	96.17	14.7
PE [2]	63.01	100.0	60.15	14.7
OF&PE&RGB [2]	98.43	99.9	98.32	44.1
MobileNet3D [22]	**99.5**	99.0	99.8	0.2
SKD (Ours)	96.7	99.6	**100.0**	**0.1**

$M = 10^6$

Table 5 Comparison in terms of accuracy, sensitivity, specificity, model size, and FLOPs between our proposed SKD approach and SOTA methods on the URFD dataset

Method	Accuracy (%)	Sensitivity (%)	Specificity (%)	Model size (M)
VGG [20]	95.0	100.0	92.0	14.7
RGB [2]	96.99	100.0	96.61	14.7
OF [2]	96.75	100.0	96.34	14.7
PE [2]	93.24	94.41	93.09	14.7
OF&PE&RGB [2]	98.77	100.0	98.62	44.1
MobileNet3D [22]	**99.9**	100.0	99.9	0.2
SKD (Ours)	98.3	**100.0**	**100.0**	**0.1**

$M = 10^6$

Comparison with SOTA methods. We present our experiment of the SKD method against the SOTA methods on the FDD dataset in Table 4. As can be shown in Table 4, our proposed SKD approach obtains SOTA performance compared to existing methods meanwhile model size and computational cost are reduced significantly. In particular, our network only uses 0.1M parameters that reduce 147 times that compared to VGG-Net in [2, 20] and 2 times compared to MobileNet3D [22].

We compare the SKD framework with existing methods on the URFD dataset in Table 5. Our approach achieves the best performance in terms of sensitivity and specificity only with 0.1M parameters. It implies that our network and approach are remarkably effective for falling detection with any datasets. Especially, our method achieves 100% in terms of both sensitivity and specificity. This means our approach is suitable for the URFD dataset.

5 Conclusion

In this work, we have proposed a simple yet robust self-knowledge distillation approach for falling detection. Except training with ground truth labels, our proposed method is learned via knowledge distilled from embedded vectors of two different views of the same input video. The exchange and sharing of knowledge between two embedded vectors of the same data point help to enhance the generalization performance of the model. The experimental results have shown that our proposed method outperforms independently training. Only with 0.1M params, our network achieves competitive performance compared to state-of-the-art heavy networks. In future work, we will investigate and apply the proposed method to other computer vision tasks involving.

References

1. Adhikari K, Bouchachia H, Nait-Charif H (2017) Activity recognition for indoor fall detection using convolutional neural network. In: MVA. IEEE, pp 81–84
2. Cameiro SA, da Silva GP, Leite GV, Moreno R, Guimarães SJF, Pedrini H (2019) Multi-stream deep convolutional network using high-level features applied to fall detection in video sequences. In: IWSSIP. IEEE, pp 293–298
3. Chen X, He K (2020) Exploring simple siamese representation learning. arXiv preprint arXiv:2011.10566
4. Cho H, Yoon S (2019) Applying singular value decomposition on accelerometer data for 1d convolutional neural network based fall detection. Electronics Lett 55(6):320–322
5. Gou J, Yu B, Maybank SJ, Tao D (2021) Knowledge distillation: a survey. IJCV 129(6):1789–1819
6. Hara K, Kataoka H, Satoh Y (2018) Can spatiotemporal 3d cnns retrace the history of 2d cnns and imagenet? In: CVPR, pp 6546–6555
7. He K, Zhang X, Ren S, Sun J (2016) Deep residual learning for image recognition. In: CVPR, pp 770–778
8. Hinton G, Vinyals O, Dean J (2015) Distilling the knowledge in a neural network. arXiv preprint arXiv:1503.02531
9. Hsieh YZ, Jeng YL (2017) Development of home intelligent fall detection iot system based on feedback optical flow convolutional neural network. Ieee Access 6:6048–6057
10. Hwang S et al (2017) Maximizing accuracy of fall detection and alert systems based on 3d convolutional neural network. In: IoTDI. IEEE, pp 343–344
11. Islam MM, Tayan O, Islam MR, Islam MS, Nooruddin S, Kabir MN, Islam MR (2020) Deep learning based systems developed for fall detection: a review. IEEE Access 8:166117–166137
12. Ji M et al (2021) Refine myself by teaching myself: Feature refinement via self-knowledge distillation. arXiv preprint arXiv:2103.08273
13. Kasturi S, Filonenko A, Jo KH (2019) Human fall recognition using the spatiotemporal 3d cnn. In: Proceedings IW-FCV, pp 1–3
14. Kim K, Ji B, Yoon D, Hwang S (2020) Self-knowledge distillation: a simple way for better generalization. arXiv preprint arXiv:2006.12000
15. Kwolek B, Kepski M (2014) Human fall detection on embedded platform using depth maps and wireless accelerometer. Comput Methods Programs Biomed 117(3):489–501
16. Li S, Xiong H, Diao X (2019) Pre-impact fall detection using 3d convolutional neural network. In: ICORR. IEEE, pp 1173–1178
17. Li X, Pang T, Liu W, Wang T (2017) Fall detection for elderly person care using convolutional neural networks. In: CISP-BMEI. IEEE, pp 1–6
18. Ma N, Zhang X, Zheng HT, Sun J (2018) Shufflenet v2: practical guidelines for efficient cnn architecture design. In: ECCV, pp 116–131
19. Mirzadeh SI et al (2020) Improved knowledge distillation via teacher assistant. In: AAAI, vol 34, pp 5191–5198

20. Núñez-Marcos A, Azkune G, Arganda-Carreras I (2017) Vision-based fall detection with con-
    volutional neural networks. WCMC
21. Tran D et al (2018) A closer look at spatiotemporal convolutions for action recognition. In:
    CVPR, pp 6450–6459
22. Trieu XH, Madrid VRM, Albacea EA (2021) A lightweight model for falling detection. In:
    RIVF, IEEE
23. Tsai TH, Hsu CW (2019) Implementation of fall detection system based on 3d skeleton for
    deep learning technique. IEEE Access 7:153049–153059
24. Vu DQ, Le N, Wang JC (2021) Teaching yourself: a self-knowledge distillation approach to
    action recognition. IEEE Access 9:105711–105723
25. Yhdego H, Li J, Morrison S, Audette M, Paolini C, Sarkar M, Okhravi H (2019) Towards
    musculoskeletal simulation-aware fall injury mitigation: transfer learning with deep cnn for
    fall detection. In: SpringSim. IEEE, pp 1–12

Improving Gastroesophageal Reflux Diseases Classification Diagnosis from Endoscopic Images Using StyleGAN2-ADA

Phuong-Thao Nguyen⓪, Thanh-Hai Tran⓪, Viet-Hang Dao⓪, and Hai Vu⓪

Abstract Gastroesophageal Reflux Disease (GERD) is a common gastrointestinal disease with an increasing trend in recent years in the world as well as in Vietnam. This disease becomes a great burden on the healthcare system due to its huge impact on the quality of life for patients. The diagnosis of GERD is based on typical symptoms while lesions detection and severity classification on upper endoscopy are necessary to confirm. However, during the procedure, the esophagitis erosions could be not evident to recognize and discriminate the severity levels by young endoscopists. This paper presents a method for automatic classification of Gastroesophageal Reflux Diseases from endoscopic images. Our proposed method utilizes the state-of-the-art deep neural network (i.e. ResNet-50) for feature extraction and classification. To improve the classification results, we investigate two data augmentation techniques: Affine Transformation and Generative Adversarial Network (i.e. StyleGAN2-ADA) on different color models (RGB, HSV). Our main contribution is to propose a framework that integrates StyleGAN2-ADA based data augmentation to enrich the training dataset for the CNN model. Experiments are conducted on our self-collected dataset of GERD with two classes (GERD A and GERD B). Experimental results show a significant improvement from 83.2% to 91.7% of accuracy when using StyleGAN-ADA on RGB channels compared to the original data.

This research is funded by Vietnam Ministry of Science and Technology under grant number KC-4.0-17/19-25 "Research and Develop Intelligent Diagnostic Assistance System for Upper GastroIntestinal Endoscopy Images".

P.-T. Nguyen · T.-H. Tran (✉) · H. Vu
School of Electrical and Electronic Engineering, Hanoi University of Science and Technology, Hanoi, Vietnam
e-mail: hai.tranthithanh1@hust.edu.vn

P.-T. Nguyen
e-mail: thao.np182977@sis.hust.edu.vn

H. Vu
e-mail: hai.vu@hust.edu.vn

V.-H. Dao
Institute of Gastroenterology and Hepatology, Hanoi Medical University Hospital, Hanoi, Vietnam
e-mail: daoviethang@hmu.edu.vn

Keywords Gastroesophageal reflux disease · Endoscopic images · Convolutional neural network · Generative adversarial network

1 Introduction

According to Montreal definition [16], Gastroesophageal Reflux Disease (GERD) occurs when stomach contents flow back into the esophagus causing troublesome symptoms and/or complications. GERD is an increasingly common disease in the world and according to Lyon consensus, pH 24 h measurement and upper gastrointestinal endoscopy (UGIE) are the two exploration tests to confirm the diagnosis. The aim of UGIE is to identify typical lesions such as erosions, complications of GERD such as esophageal stricture, ulcer, Barrett's esophagus and to exclude other diseases with mimic symptoms. In cases with typical erosions due to reflux, it is defined as erosive gastroesophageal reflux disease (ERD). To evaluate GERD, the Los Angeles classification (LA grade) system was established more than 20 years ago, validated in many studies and used commonly worldwide. However, the severity assessment is sometimes challenging as it depends on the experience of the doctors/endoscopists and some other technical factors such as cleansing level, inflation level, patients' breathing in case of UGIE without anesthesia. Nowadays, with the increasing of advanced deep learning, many CNN models have been applied in medical image analysis to assist the doctors for better diagnosis and treatment of the diseases (e.g. radiology image interpretation [4], obstructive pulmonary disease recognition in computed tomography [1], diabetic retinopathy screening [19], esophageal cancer endoscopic diagnosis [9], dysplasia in Barrett's esophagus, and detection of early gastric cancers [11]). Convolutional Neural Networks (CNNs) based methods try to learn models from large annotated datasets to extract hidden features from images for classification. This paper aims to support the doctors identifying the grades of reflux diseases severity on UGIE images using recent advanced deep neural networks. Given an input endoscopic image, we try to classify it into grades defined by the Los Angeles classification (LA grade). These grades are illustrated in Fig. 1, in which Grade A and B are the most common types. The most difference between these two grades is the length of mucosal breaks, which is no more than 5 mm for Grade A and more than 5 mm for Grade B.

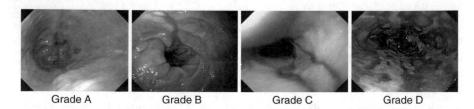

| Grade A   Grade B   Grade C   Grade D |

Fig. 1 Examples of four grades of reflux esophagitis diseases defined by Los Angeles criteria

In this paper, we will investigate the ResNet-50 model for grade classification from endoscopic images. To address the scarcity of dataset, we employ two data increment techniques such as Affine Transformation and Generative Adversarial Network (i.e. StyleGAN2-ADA) in different color models (RGB, HSV). We validate our method on a self-collected dataset of GERD Grade A and Grade B. In summary, the contribution of this paper is three-fold. First, we deploy the ResNet-50 model for the classification of GERD A and GERD B from endoscopic images. Second, we develop and study the effects of two data augmentation techniques. Finally, the performance of the model will be evaluated on our dataset in terms of accuracy, precision, recall, and F1-score.

In the remainder of this paper, we will present related works in Sect. 2. In Sect. 3, we describe our proposed framework with the presentation of color models and re-visiting of ResNet-50 and StyleGAN2-ADA. Experiments and conclusions are presented in Sects. 4 and 5 respectively.

2 Related Works

As aforementioned in the introduction section, the application of machine learning algorithms in medical image analysis has achieved impressive results. However, research on the analysis of GERD endoscopic images is still limited. Most of the research till now focused on Barrett's esophagus due to the risk of progression into esophageal cancer in case of a long segment with dysplasia area. However, in Asian countries including Vietnam, the prevalence of Barrett's esophagus is uncommon, and mainly are the short-segment type with low risk of neoplasm progression [13].

In a work by Yousefi et al. [20], they created a segmentation method which is a hybrid of a clustering method and a Level Set algorithm. The spatial fuzzy C-mean algorithm is designed to cluster pixels based on a Euclidean distance function, and the Level Set method served as a threshold to cut out the original curve created by the fuzzy C-mean algorithm. Compared to the manual segmentation by experts, their results show high accuracy with about 95% in all four cases.

Another work in Barrett's esophagus is a computer-aided detection system for early neoplastic lesions in Barrett's esophagus by Sommen et al. [15]. They developed the application with a combination of 3 methods and placed them as follows. First, the endoscopic images go through a region of interest detection and a Gabor bank filter simultaneously to create useful features; then they use Support Vector Machine - a method of machine learning, to learn those features and draw annotations on images. The result annotations showed that their system had the potential to approach the performance of experts when their system achieved 0.21–0.62 F-score in delineation similarity to gold standard annotations. This is an impressive result given that they trained their system on 100 images.

An automatic approach to Barrett's esophagus which utilized CNN was proposed by Mendel et al. [10] used ResNet50 [3]—one of the most popular CNN architecture, with transfer learning in analyzing precancerous condition. They sampled 100 images from the endoscopic vision challenge MICCAI 2015 to create 7823 image patches that were segmented and labeled as C0 (non-cancerous) or C1 (cancerous) by experts for training, and resize the same image, applying 50 pixel offset to create another 768 images for evaluation. They got the best F1 score of 0.91. In [17], the authors investigated color and texture based features for boundary delineation of reflux esophagitis lesions from endoscopic images.

Related to the problem of GERD grade classification, to the best of our knowledge, there exists a very limited number of works. In [12], the authors combined the QUestionario Italiano Diagnostico (QUID) questionnaire with an artificial neural network (ANN)-assisted algorithm to differentiate GERD patients from healthy individuals. In [5], the researchers proposed a hierarchical heterogeneous descriptor followed by a support vector machine (HHDF-SVM) method for GERD diagnosis from conventional endoscopic images. Recently, the VGG deep neural network was deployed for the classification of GERD images into three classes (Grade A+B, Grade C+D, and normal) [18].

3   Proposed Methods

3.1   General Framework

In this paper, we propose a framework for the classification of GERD grades from endoscopic images using deep learning. The proposed framework is illustrated in Fig. 2. It consists of two main phases:

– Training phase: To deal with the scarcity of dataset, we first enhance the annotated data with two techniques: Affine Transformation and StyleGAN2-ADA [7]. Next, the enriched data undergoes two color models that are RGB and HSV, then a combination of RGB and HSV. Finally, the ResNet-50 model will be trained with this data.

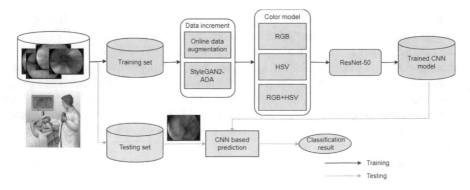

Fig. 2 Our proposed framework for GERD classification from endoscopic images

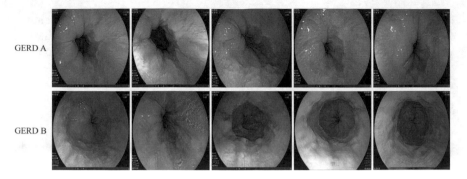

Fig. 3 Examples of GERD images generated by StyleGAN2-ADA

- Testing phase: The endoscopic images after being fed to the trained CNN model are classified into GERD grades of severity. This phase attempts to assess the performance of the model in terms of accuracy, precision, recall, and F1-score. In our work, we investigate only Grade A or Grade B because of the availability of the data. However, our proposed method can be applied to classify the whole four grades according to the Los Angeles grading whenever the data are collected.

3.2 StyleGAN2-ADA for Data Augmentation

Generative Adversarial Networks (GANs) [2] were first introduced in 2014. Its aim was to understand image distributions and generate new images from these distributions. GAN consists of two neural networks that operate in parallel. The first neural network is a generator, which generates new images by learning the distribution of training images. The second neural network is a discriminator and its goal is to classify between real images and generated images by generator. StyleGAN2 [7] was developed based on this theory, which is a GAN with a style-based generator. It inherited the progressive training from ProGAN [6], starting by training the generator and discriminator with a very low-resolution layer (e.g. 4×4) to capture coarser features of the images, then higher resolution layers are added gradually so that the network can learn more detailed features.

Unlike traditional GAN, StyleGAN2 introduces a mapping network that comprises eight fully connected layers to encode an input latent vector into an intermediate latent vector. A problem with training GAN is that it requires large datasets to avoid overfitting, whereas, in reality, those datasets are difficult to obtain. Therefore, Adaptive Discriminator Augmentation (ADA), was proposed to address the lack of data problem. ADA controls the augmentation strength p by incrementing/decrementing p according to the level of overfitting. Some examples of generated GERD images are shown in Fig. 3. We notice that the StyleGAN2-ADA has successfully generated images of endoscopic images of two grades. Validation of these generated images by the doctors will be a future work. In the following, we will investigate if the augmented data helps to improve the classification results.

3.3 ResNet-50 for Classification

One problem of training deep neural networks with small datasets is that the model is likely to be overfitting because it "remembers" every training example instead of learning the generalization of the data. Additionally, our dataset is still small but very challenging. As a result, we choose ResNet-50, an adequately deep network for learning discriminant features between two classes while avoiding overfitting.

ResNet stands for Residual Network, which is a CNN architecture proposed in [3]. By adding residual blocks to the network, ResNet tackles the degradation problem, which has been exposed when the network goes deeper and begins to converge, accuracy becomes saturated and declines quickly. A residual block is a stack of layers in such a way that the output of a layer is taken and added to another layer deeper in the block. In image classification tasks, ResNet has been shown to outperform several other CNN models. ResNet-50 is ResNet with 50 layers. The architecture of ResNet-50 consists of five sequential layers, a fully connected layer, and a final softmax layer. The first layer is a 7×7 convolutional layer, followed by a 3×3 max pooling. Each layer in the next four layers contains three, four, six, three residual blocks, respectively, which means that ResNet-50 has totally sixteen residual blocks. A residual block comprises a 3×3 convolutional layer between two 1×1 layers, each layer followed by a batch normalization layer. The architecture of ResNet-50 used in this paper is shown in Fig. 4.

3.4 Training the CNN Models

To generate images, we used the transfer learning techniques with a pre-trained StyleGAN2-ADA model that was trained with the FFHQ dataset, which is a high-quality face dataset. As higher the resolution of the images, the more time it takes

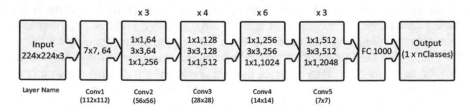

Fig. 4 Architecture of ResNet-50

to train the network, we resize the input images to 510×510. With one point in the latent space, the network generates one image.

To classify the images, we took advantage of transfer learning techniques. We trained the ResNet-50 model using the Tensorflow 2 deep learning framework. Adam, a stochastic optimization algorithm is used since it works well on small datasets [8]. We fine-tune all layers of the network, with an initial learning rate of 0.001 and decay by 0.5 when validation loss has stopped improving. Images were resized to 224×224 to make sure that they were compatible with ResNet-50. The batch size was 32 and we trained the network for 50 epochs.

4 Experiments

4.1 Datasets

The dataset was collected from patients in Vietnam's hospitals. It comprises 779 images, with 470 images of GERD grade A and 309 images of GERD grade B. The default image resolution is 1280×1024. Those images were labeled by doctors into two grades according to the difference in the length of mucosal breaks. Figure 5 shows some example images in our dataset and their corresponding abnormalities highlight marks. The mucosal breaks on grade A and grade B are located around the Z-line landmark. At an early stage like grade A and grade B, these abnormalities are hard to be distinguished from the normal parts. Additionally, the two grades are different only in the length of the mucosal breaks, but the length varies in different images according to the endoscopic camera angles, making it difficult to classify correctly.

To evaluate the proposed method, we apply K-fold validation with K = 10. That means we split the data into ten subsets, each time we take nine subsets for training and one remaining subset for testing. The reported result is the average performance of ten folds on testing sets. For training sets, as the obtained dataset is small and slightly unbalanced, therefore we first flipped a part of GERD grade B images in training set horizontally so that each class had the same instances. Then we applied some online data augmentation (DA) techniques (Affine Transformation). Besides, StyleGAN2-ADA is utilized to generate more 256 training images for each class. We conducted separate experiments to investigate whether Affine Transformation or StyleGAN2-ADA produces better results.

4.2 Experimental Results

To perform the experiments, we first trained the ResNet-50 network with original dataset (Origin), then with data enriched by online augmentation techniques (DA), and finally with StyleGAN2-ADA (GAN) in RGB, and HSV. With the combination of

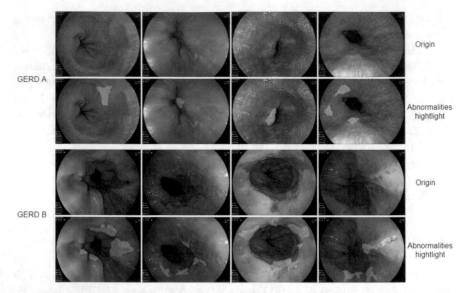

Fig. 5 Examples of GERD images in our dataset and their corresponding abnormalities highlight marks

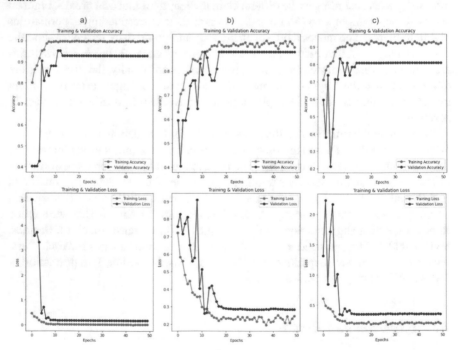

Fig. 6 Process of training CNN models with RGB color model. **a** With StyleGAN2-ADA, **b** with online data augmentation, **c** original dataset

RGB and HSV color models, we only trained the network with the original dataset and the augmented data generated by StyleGAN2-ADA. We skip the online data augmentation because it exceeded the hardware power. We compare the performance of ResNet-50 with different color models and data generation techniques in terms of accuracy, precision, recall, and F1-score. 10-fold cross-validation was performed so that the network can be more generalized (Fig. 6).

Table 1 shows the comparison of average accuracy, precision, recall and F1-score between original dataset and different data generation techniques under different color models. Overall, ResNet-50 gives better classification results when it was trained with images generated by different techniques in the RGB color model. We saw that the result when StyleGAN2-ADA was applied increased remarkably compared to original dataset and online data augmentation usage, under RGB and HSV color model. As shown in Fig. 3, the generated images look quite authentic as StyleGAN2-ADA network was able to capture detailed characteristics of each class, therefore it contributed a large number of images to the training set. However, the result of StyleGAN2-ADA was lower than original dataset in the combination of RGB and HSV. With online data augmentation techniques, the result was higher than original dataset utilization in RGB color model, but it decreased in HSV color model. As shown in Table 1, the result with StyleGAN2-ADA application in RGB color model outperforms other results, with accuracy, precision, recall, and F1-score are 91.7%, 92.2%, 91.7%, and 91.7%, respectively. The process of network training with RGB color model is shown in Fig. 6. The detailed classification result of each class is illustrated in Fig. 7a. We noticed that ResNet-50 performed better in predicting GERD grade A than grade B in terms of precision, recall, and F1-score.

As there were no available datasets of GERD images for the classification task, we can not produce a fair comparison between our proposed method with existing ones. However, we summarize some relevant works [5, 12, 18] in Table 2 that may help the readers to compare with our proposed method, dataset, and performance. It is noticed that the method in [12] used questionnaire data instead of image data. Comparing to [5, 18], our work is the first research about GERD grades A and B classification. The works in [5] only classified between normal and GERD cases while [18] did not separate grade A and B but grouped the data as grade A+B, C+D, and normal cases, then classified them.

Table 1 Comparative results of ResNet-50 performance with different color models and data generation techniques

	RGB			HSV			RGB + HSV	
	Original data	DA	StyleGAN-ADA	Original data	DA	StyleGAN-ADA	Original data	StyleGAN-ADA
Accuracy (%)	83.2	86.3	**91.7**	83.4	80.9	88.2	85.5	84
Precision (%)	86	87.2	**92.2**	82	81.4	88.7	86.9	87.7
Recall (%)	83.2	86.3	**91.7**	83.4	80.9	88.2	85.5	84
F1-score (%)	82.1	86.2	**91.7**	81.6	80.9	88.1	85.2	82.4

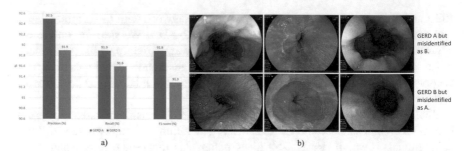

Fig. 7 **a** Classification result of each class when StyleGAN2-ADA was used in RGB color model, **b** some misidentified images

Table 2 Comparison of different AI systems for prediction of GERD

Task	Algorithm	Dataset	Evaluation method	Overall accuracy
Binary classification	Machine learning (ANN) [12]	QUID[a] questionnaire (577 GERD patients, 94 normal cases)	hold-out	99.2%
Binary classification	Machine learning (HHDF-SVM) [5]	147 RGB images (39 GERD patients, 108 normal cases)	10-fold cross-validation	93.2%
Three-class classification	Deep learning (GERD-VGGNet) [18]	671 NBI[b] images (GERD A–B: GERD C–D: normal = 244:229:198)	10-fold cross-validation	98.9% ± 1%
Binary classification	Deep learning + StyleGAN2-ADA (proposed method)	779 RGB images (470 GERD A, 309 GERD B)	10-fold cross-validation	91.7%

[a] QUestionario Italiano Diagnostico
[b] Narrow-band image

For better understanding of ResNet-50 performance, we apply the Grad-CAM [14] to different testing images. Grad-CAM is a visualization method, which employs gradients of any target class (i.e. GERD A and GERD B) that flow into the final convolutional layer to provide a coarse localization highlighting the essential regions in the images for prediction. By looking at the regions that the network considered as important, we can know whether the model learned the correct features of the images, i.e. learned the abnormal patterns. The Grad-CAM visualization of ResNet-50 trained with original data vs data generated by different data increment methods in the RGB color model is illustrated in Fig. 8. We can clearly see that the Grad-CAM result of when the model trained with data augmented by StyleGAN2-ADA is better

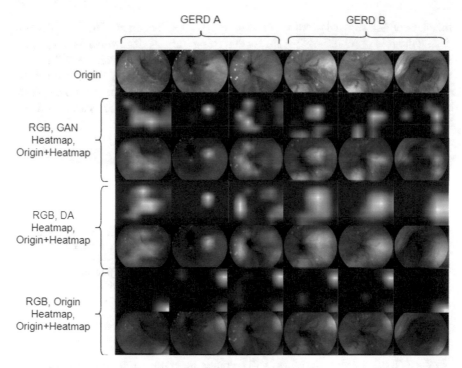

Fig. 8 Grad-CAM visualization of ResNet-50 network trained with original data versus data generated by different data increment methods in RGB color model

than other methods as it converged around the Z-line landmark and highlighted the abnormal regions in the images.

The ResNet-50 model still made some mistakes in diagnosing GERD grade A and grade B. Figure 7b shows examples of misidentified images by the network. According to the Los Angeles classification criteria, the difference between grade A and B is only the length of mucosal breaks with the cut off of 5mm, which is difficult for the network to exactly point out. Moreover, the captured images could contain some unwanted artifacts, such as blurring and scattered light on abnormal parts of the esophagus, affecting the performance of ResNet-50.

5 Conclusions

In this paper, we presented a method that consisted of two stages. The first stage utilized StyleGAN2-ADA to generate more images for training the model. The second stage deployed a state-of-the-art CNN (ResNet-50) for classification. To validate our proposed framework, we have built a small dataset of GERD endoscopic images with two grades according to Los Angeles criteria. We have experimented our method on this dataset and showed promising results of 91.7% of accuracy. These results confirmed the role of StyleGAN2-ADA for enriching training data, which is a necessary step to deal with the lack of large annotated datasets. In the future, we will test our method on a bigger dataset with more levels of disease severity. We will verify the generated images by StyleGAN2-ADA with doctors and retrain the model on the clean augmented dataset. We will also study a regression model to assist the doctors to make the final decision based on the prediction of our system.

References

1. Das N, Topalovic M, Janssens W (2018) Artificial intelligence in diagnosis of obstructive lung disease: current status and future potential. Curr Opinion Pulmonary Med 24(2):117–123
2. Goodfellow I, Pouget-Abadie J, Mirza M, Xu B, Warde-Farley D, Ozair S, Courville A, Bengio Y (2020) Generative adversarial networks. Commun ACM 63(11):139–144
3. He K, Zhang X, Ren S, Sun J (2016) Deep residual learning for image recognition. In: Proceedings of the IEEE conference on computer vision and pattern recognition, pp 770–778
4. Hosny A, Parmar C, Quackenbush J, Schwartz LH, Aerts HJ (2018) Artificial intelligence in radiology. Nat Rev Cancer 18(8):500–510
5. Huang CR, Chen YT, Chen WY, Cheng HC, Sheu BS (2015) Gastroesophageal reflux disease diagnosis using hierarchical heterogeneous descriptor fusion support vector machine. IEEE Trans Biomed Eng 63(3):588–599
6. Karras T, Aila T, Laine S, Lehtinen J (2017) Progressive growing of gans for improved quality, stability, and variation. arXiv preprint arXiv:1710.10196
7. Karras T, Aittala M, Hellsten J, Laine S, Lehtinen J, Aila T (2020) Training generative adversarial networks with limited data. arXiv preprint arXiv:2006.06676
8. Kingma DP, Ba J (2014) Adam: A method for stochastic optimization. arXiv preprint arXiv:1412.6980
9. Kumagai Y, Takubo K, Kawada K, Aoyama K, Endo Y, Ozawa T, Hirasawa T, Yoshio T, Ishihara S, Fujishiro M et al (2019) Diagnosis using deep-learning artificial intelligence based on the endocytoscopic observation of the esophagus. Esophagus 16(2):180–187
10. Mendel R, Ebigbo A, Probst A, Messmann H, Palm C (2017) Barrett's esophagus analysis using convolutional neural networks. In: Bildverarbeitung für die Medizin 2017. Springer, Heidelberg, pp 80–85
11. Mori Y, Kudo SE, Mohmed HE, Misawa M, Ogata N, Itoh H, Oda M, Mori K (2019) Artificial intelligence and upper gastrointestinal endoscopy: Current status and future perspective. Digestive Endoscopy 31(4):378–388
12. Pace F, Riegler G, de Leone A, Pace M, Cestari R, Dominici P, Grossi E, Group ES et al (2010) Is it possible to clinically differentiate erosive from nonerosive reflux disease patients? A study using an artificial neural networks-assisted algorithm. Eur J Gastroenterol Hepatol 22(10):1163–1168
13. Quach DT, Pham QT, Tran TL, Vu NT, Le QD, Nguyen DT, Dang NL, Le HM, Le NQ, Sharma P et al (2020) Prevalence, clinical characteristics, and risk factors of barrett esophagus in vietnamese patients with upper gastrointestinal symptoms. Medicine 99(34)
14. Selvaraju RR, Cogswell M, Das A, Vedantam R, Parikh D, Batra D (2017) Grad-cam: visual explanations from deep networks via gradient-based localization. In: Proceedings of the IEEE international conference on computer vision, pp 618–626
15. van der Sommen F, Zinger S, Curvers WL, Bisschops R, Pech O, Weusten BL, Bergman JJ, Schoon EJ et al (2016) Computer-aided detection of early neoplastic lesions in Barrett's esophagus. Endoscopy 48(7):617–624

16. Vakil N, Van Zanten SV, Kahrilas P, Dent J, Jones R et al (2006) The montreal definition and classification of gastroesophageal reflux disease: a global evidence-based consensus. Official J Am College Gastroenterol—ACG 101(8):1900–1920
17. Vu DH, Nguyen LT, Nguyen VT, Tran TH, Dao VH, Vu H (2021) Boundary delineation of reflux esophagitis lesions from endoscopic images using color and texture. In: 2021 international conference on Multimedia Analysis and Pattern Recognition (MAPR). IEEE, pp 1–6
18. Wang CC, Chiu YC, Chen WL, Yang TW, Tsai MC, Tseng MH (2021) A deep learning model for classification of endoscopic gastroesophageal reflux disease. Int J Environ Res Public Health 18(5):2428
19. Wong TY, Bressler NM (2016) Artificial intelligence with deep learning technology looks into diabetic retinopathy screening. Jama 316(22):2366–2367
20. Yousefi-Banaem H, Rabbani H, Adibi P (2016) Barrett's mucosa segmentation in endoscopic images using a hybrid method: spatial fuzzy c-mean and level set. J Med Signals Sens 6(4):231

A (2+1)D Attention Convolutional Neural Network for Video Prediction

Trang Phung⬡, Van Truong Nguyen⬡, Thi Hong Thu Ma⬡, and Quang Vu Duc⬡

Abstract Video prediction is a challenging generative modeling task. The objective of algorithms is to generate future scenes from a given image sequence. However, instead of giving a ground-truth prediction, these methods often give blurry predictions. This is a major challenge to predictive models. The causes of the blurry predictions are these models didn't capture the features of the context and the variation of motion from the input. In this paper, we introduce (2+1)D Convolutional Neural Network (CNN) combined with two attention modules named spatial and temporal attention. The (2+1)D CNN helps the model learn the spatial-temporal features better than 2D CNN. And two attention modules are to boost the representation power of CNNs. The experiment results have shown that our approach achieved state-of-the-art performance in video prediction on standard datasets in terms of both PSNR and SSIM compared to existing methods.

Keywords Video prediction · Convolutional neural network · Attention

1 Introduction

Video prediction is one of the fundamental problems in vision tasks. The goal of this task is to predict future frames given past video frames. The predicted future frames may be in the form of RGB images [6, 12] and/or optical flow [15, 28]. These future frames can be used for various tasks such as action prediction [27], video coding [17],

T. Phung
Thai Nguyen University, Thai Nguyen, Vietnam

V. T. Nguyen · Q. V. Duc
Thai Nguyen University of Education, Thai Nguyen, Vietnam

T. H. T. Ma (✉)
Tan Trao University, Tuyen Quang, Vietnam
e-mail: thutq7@gmail.com

Q. V. Duc
National Central University, Taoyuan, Taiwan

© The Author(s), under exclusive license to Springer Nature Switzerland AG 2022
N. H. T. Dang et al. (eds.), *Artificial Intelligence in Data and Big Data Processing*,
Lecture Notes on Data Engineering and Communications Technologies 124,
https://doi.org/10.1007/978-3-030-97610-1_31

video surveillance [5], autonomous driving [29], etc. This issue has been researched in a long time and recently, various deep learning-based methods have significantly boosted the video prediction performance. In which, most of these methods used CNN or Long short-term memory (LSTM) or its variation Convolutional LSTMs (ConvLSTMs) [20].

The most difficult problem in the video prediction task is uncertainty because there are many future reasonable outcomes for a given input frame sequence. Therefore, the output of the models rapidly declines over time as uncertainty increases [22]. There have been a lot of studies proving the effectiveness of CNNs to predict future frames [6, 10, 12, 15, 28]. Usually, these methods have proposed different auto-encoder model architectures to encode the given frames and generate future frames. For example, the method in [12] can predict two-way prediction (both future and past frames) using the retrospective cycle constraints. In [10], the authors proposed the multi-level wavelet analysis method that combined in a CNN model to uniformly address with spatial and temporal information. The model in [3] is trained a ConvL-STM network to get features from an image sequence and a convolutional network is initialized by the output of the ConvLSTM network for generating the next frames. Besides architectural design, various training strategies such as the VAEs [3] and/or GANs [10, 12] have been introduced and used very common.

However, there always exists a distinct gap between the ground truth and their predictions. That is a major challenge to predictive models because it requires mastering not only the visual context abstraction but also the various motions of each object over time. Instead of giving a ground-truth prediction, these methods often give blurry predictions. This represents the uncertainty of models. The cause of the blurry predictions is that the models do not fully capture relevant information from the past. Inspired by the success of the attention module on image and video data such as SENet [9], CBAM [26], 3D STC-ResNet [4]. In this paper, we propose an efficient CNN architecture to produce future frame(s) that has achieve state-of-the-art (SOTA) performance for video prediction. Our contributions can summarize as follows:

- First, we propose to utilize the (2+1)D convolution ((2+1)D conv) to extract spatial-temporal features from the given frame sequence. To our best knowledge, this is the first work focus on (2+1)D conv for video prediction. 2D conv are limited by the intra-frame and lack of explicit inter-frame modeling capabilities. Meanwhile, 3D conv have been proposed as a promising method to replace recurrent modeling [19]. Various video prediction approaches in [1, 25] utilized 3D conv to learn temporal consistency. In [24], Du et al. proposed a method called (2+1)D conv to split a 3D conv operation into two separate parts, first is a 2D conv to get spatial features and a 1D conv to learn temporal information. The authors demonstrated that (2+1)D convolution has more advantages than 3D convolution on action recognition.
- Second, we propose two attention modules called spatial attention and temporal attentions to boost the representation power of CNNs. Attention has been widely applied to image and text objects. In video objects, they are also used in some tasks such as action recognition [4]. In our approach, spatial and temporal

attention modules are suggested to increase the learnability of spatial-temporal features from the input image sequence. In the experiment, we found that spatial and temporal attention modules significantly improve the predictive power of the model.

2 Related Work

In this section, we present related studies using deep neural networks to video prediction. VAEs and GANs are two common training strategies for this task.

VAE-based approaches. These methods usually build a Variational AutoEncoder (VAE) model to address this task. In which, Kulback-Leibler divergence loss use to the latent layer as a standard Gaussian. In [18], the authors proposed a new loss function called image gradient loss and a multi-scale model, that significantly reduces blurring artifacts. Byeon et al. [2] presented a new architecture named Parallel Multi-Dimensional LSTM. Moreover, the authors introduced the blending units instead of ConvLSTM to learn both past and current contexts, their model achieved SOTA performance on a few of the datasets. Lotter et al. [17] presented a deep neural network to perform two tasks including predicting the object movement and learning internal representation.

GAN-based approaches. Generative Adversarial Networks (GAN) was the first introduced by Goodfellow et al. [7] and has achieved a lot of great success in deep learning and AI. There are two networks in GAN including generator and discriminator networks. While the discriminator is used to distinguish real image and fake image (generated by generator), the generator is trained to mislead the discriminator. Various GAN methods have been introduced for video prediction. For examples, Liang et al. [15] presented a new approach that contains two generators and two discriminators to increase the performance of frame prediction and real/fake classification. Especially, their method utilizes both RGB image and optical flow image domains. The predictions of the future RGB frame and future optical flow form a closed loop. They generate and share features with each other to boost the performance of video prediction. Kwon and Park [12] proposed a CNN to predict both future and past frames and two discriminator models classify the output image and image sequence as real or fake. In [23], the authors proposed a model called PreCNet based on predictive coding to predict the next frame video. Liu et al. [16] proposed a new method for video prediction and compression via latent space. In [14], the authors proposed an efficient approach for long-term video prediction via semantic structures and modeling structures.

Hybrid approaches. All approaches in this group have used both Kulback-Leibler divergence loss (from VAE models) and discriminator networks (from GAN models) to address this task. For example, Lee et al. [13] proposed the video prediction model based on both Kulback-Leibler divergence and discriminator losses. An overview of video prediction can be found in [20].

3 Proposed Method

In this section, we first introduce the problem setup of the video prediction task. We then describe our proposed network including (2+1)D CNN and two attention modules. Finally, we describe the objective function for training our method.

3.1 Problem Setup

Given X context frames $X = (x_1, x_2, \ldots, x_m)$ being the input of length m where $x_i \in \mathbb{R}^{H \times W \times C}$ illustrates the ith frame. H, W, and C correspond with the height, width, and channel. Our task is to predict n future frame(s) $X' = (x_{m+1}, x_{m+2}, \ldots, x_{m+n})$ (next-frame prediction if $n = 1$). Thus, our objective is to learn a generative model based on given information from the input sequence and then predict future frame(s).

3.2 Network Architecture

We develop an CNN model for the generator network based on the architecture from [12]. The generator is illustrated in Fig. 1.

The generator network G is trained to generate the next frame from the input image sequence and the 2D discriminator and 3D discriminator networks are trained to classify between real or fake frame (clip) from output of G. In the encoder, (2+1)D convolution combine with two attention modules are applied to extract spatial-temporal features from the given frame sequence. Then, 2D transposed convolutions are utilized to upscale the feature maps from smaller H and W to larger H and W and predict

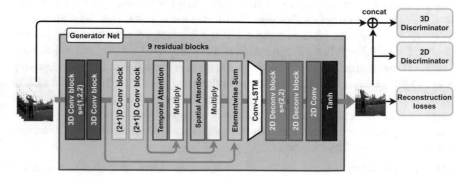

Fig. 1 Overview of (2+1)D attention CNN. In which, our method contains one generator network and two discriminator networks. Each Conv block includes an instance of a batch normalization layer and an activation layer with ReLU function followed by one Conv layer

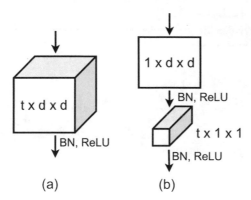

Fig. 2 The comparison between 3D conv and (2+1)D conv. **a** is a normal 3D conv that is performed by using a standard filter $t \times d \times d$ of dimension where t, d denote the temporal and width (height) extent, respectively. **b** is a (2+1)D conv block that splits the 3D conv operator into two parts including a spatial 2D conv operator and a temporal 1D conv operator.

the next frame. For multi-frame prediction, we can repeat the process by reusing the predicted frame as the input of the encoder network.

Figure 2 shows the difference between 3D conv and (2+1)D conv. Compared to 3D conv, (2+1)D conv uses a ReLU layer between the 2D and 1D conv. Therefore, it doubles the number of nonlinearities in the network. Moreover, Tran et al. [24] have demonstrated that the optimization process of (2+1)D conv is easier than conventional 3D conv due to separate spatial and temporal elements.

The spatial and temporal attention modules are shown in Fig. 3. We assume that the input of the temporal attention $\mathbf{A_t}$ is a tensor $\mathbf{F}$ with the dimension of (T, H, W, N) . After applying average spatial pooling, we have a new tensor with the dimension of $(T, 1, 1, N)$. Through two Conv layers, we get the output with $(T, 1, 1, N)$ of dimension. In spatial attention $\mathbf{A_s}$, we use average temporal pooling instead of average spatial pooling, and then through two Conv layers, we get the output with $(1, H, W, N)$ of dimension. The overall of both attention modules can be illustrated as follows:

$$
\begin{aligned}
\mathbf{F_1} &= \mathbf{A_t}(\mathbf{F}) \otimes \mathbf{F} \\
\mathbf{F_2} &= \mathbf{A_s}(\mathbf{F_1}) \otimes \mathbf{F_1}
\end{aligned}
\tag{1}
$$

where $\otimes$ denotes element-wise multiplication.

Temporal attention module. We create a temporal attention map by exploiting the feature inter-motion relationship. The temporal attention focuses on "what" is moving from an input frame sequence. To calculate temporal attention efficiently, we adopted spatial average pooling to squeeze the spatial dimension. Then, we applied the convolution layers to generate a temporal attention map.

Spatial attention module. Far apart from temporal attention, spatial attention focuses on "where" is an informative part. This module cares about the feature

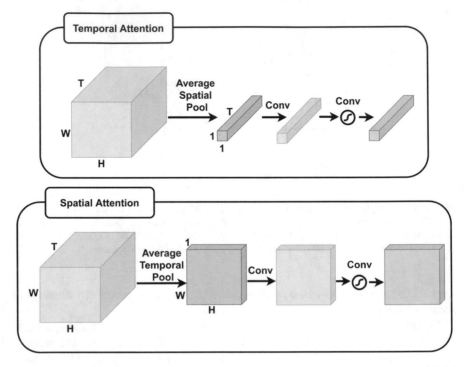

Fig. 3 The details of both attention modules. As illustrated, temporal module (spatial module) utilizes global spatial (temporal) pooling outputs and forward them to convolution layers. The final layer of each module is the sigmoid function

inter-spatial relationship. To calculate spatial attention, firstly, an average pooling is used along the time axis. Then, we apply the convolution layers to make spatial attention maps.

Given an input image sequence, two attention modules are placed in a sequential manner to compute complimentary attention. We will discuss experimental results between with and without attention modules in Sect. 4.

3.3 Objective Function

In the training process, we minimize the following objective function

$$
\begin{aligned}
\mathcal{L} = \lambda_{img}\mathcal{L}_{img} + \lambda_{gdl}\mathcal{L}_{gdl} + \lambda_{ps}\mathcal{L}_{ps} \\
+ \lambda_{adv}^{2D}\mathcal{L}_{adv}^{2D} + \lambda_{adv}^{3D}\mathcal{L}_{adv}^{3D}
\end{aligned}
\tag{2}
$$

where λ_{img}, λ_{gdl}, λ_{ps}, λ_{adv}^{2D}, and λ_{adv}^{3D} are hyper-parameters to trade-off among these different losses. In which, λ_{img} denotes the loss weight of $\mathcal{L}_{img}$ (Eq. 3), λ_{gdl} is the loss weight of the loss function $\mathcal{L}_{gdl}$ (Eq. 4), λ_{ps} is the loss weight of the loss function $\mathcal{L}_{ps}$ (Eq. 5), λ_{adv}^{2D} and λ_{adv}^{3D} denote the loss weight of the loss function $\mathcal{L}_{adv}^{2D}$ and $\mathcal{L}_{adv}^{3D}$ (Eq. 7), respectively.

Reconstruction losses. We assume that $X = (x_1, x_2, \ldots, x_m)$ is a given frame sequence. Let y and $\hat{y}$ be the ground truth frame and the predicted frame from **G**, respectively, i.e., $\hat{y} = \mathbf{G}(X)$. We used three reconstruction loss functions for training the generator **G** including $\mathcal{L}_{img}$, $\mathcal{L}_{gdl}$, and $\mathcal{L}_{ps}$. The first loss function ($\mathcal{L}_{img}$) is formulated by:

$$\mathcal{L}_{img} = \mathcal{L}_1(y, \hat{y}) \tag{3}$$

where $\mathcal{L}_1(\cdot, \cdot)$ is the standard L1 error between two frames (two images). The second loss function is Gradient Difference Loss (GDL) in [18] as follows:

$$\mathcal{L}_{gdl}(y, \hat{y}) = \sum_{i,j} \left| |y_{i,j} - y_{i-1,j}| - |\hat{y}_{i,j} - \hat{y}_{i-1,j}| \right|^\alpha + \left| |y_{i,j} - y_{i,j-1}| - |\hat{y}_{i,j} - \hat{y}_{i,j-1}| \right|^\alpha \tag{4}$$

where α is an integer greater or equal to 1 and $|\cdot|$ is the absolute value function operation. The third loss function is formulated is the perceptual losses [11]. Here, we utilize the output features of the mid-level CNN as information to estimate the perceptual similarity of two given inputs (y vs. $\hat{y}$). In particular, we use the output of the ReLU 4-3 layer from the VGG-16 for this operation as follows:

$$\mathcal{L}_{ps} = \mathcal{L}_2(VGG(y), VGG(\hat{y})) \tag{5}$$

Adversarial losses. Let the 2D and 3D discriminator model denote $\mathbf{D_{2D}}$ and $\mathbf{D_{3D}}$, where $\mathbf{D_{2D}}$ ($\mathbf{D_{3D}}$) learns to distinguish whether the output frame (clip) is from the real clip (video) or generated by **G**. We formulate the adversarial loss on the discriminator $\mathbf{D_{2D}}$ and $\mathbf{D_{3D}}$ as:

$$\begin{aligned} \mathcal{L}_{D_{2D}} &= -\log D_{2D}(y) - \log(1 - D_{2D}(\hat{y})) \\ \mathcal{L}_{D_{3D}} &= -\log D_{3D}(X \oplus y) - \log(1 - D_{3D}(X \oplus \hat{y})) \end{aligned} \tag{6}$$

The adversarial loss for the generator **G** as:

$$\begin{aligned} \mathcal{L}_{adv}^{2D} &= -\log D_{2D}(\hat{y}) \\ \mathcal{L}_{adv}^{3D} &= -\log D_{3D}(X \oplus \hat{y}) \end{aligned} \tag{7}$$

where $X \oplus y$ is the concatenate operation the frame y to the given frame sequence X.

4 Experiment

In this section, we first introduce two standard datasets used in this work. The implementation details also is presented in this section. We then experimented and evaluated our approach against other SOTA methods in terms of PSNR and SSIM.

4.1 Datasets and Implementations

In this paper, we have used three standard datasets including CalTech Pedestrian, KITTI and UCF101, to evaluate our approach's performance that compared to SOTA methods.

Car-mounted camera video. We adopt the KITTI dataset to train the model and the CalTech Pedestrian dataset is used for evaluation. Approximate 41K frames from "City", "Residential", and "Road" clips are sampled for training. Our experimental setup follows previous studies in [2, 12].

Human action videos. The UCF101 dataset [21] includes of 13K videos from 101 human action classes. For the fair comparison, we used 90% of sampled frames as the training set and the others for testing as in previous studies [2, 12].

Implementations. The input sequence length is set to 4 and all frames are normalized to be $[-1, 1]$. For data augmentation, we flipped frames and inversed input sequence with a probability of 0.5. We adopt the Adam optimizer with $\beta_1 = 0.5$ and $\beta_2 = 0.999$. The initial learning rate of 0.01 and linearly decay per every 100 epochs. Each epochs is defined as 12K iterations. For balancing different losses, we set $\lambda_{img} = 1, \lambda_{gdl} = 1$, $\lambda_{ps} = 1, \lambda_{adv}^{2D} = 0.005$, and $\lambda_{adv}^{3D} = 0.005$. We train our network on the PC with E5-2630 v3 CPU, 256 GB of RAM, dual 2080Ti GPUs. The result of model predict one next frame.

4.2 Quantitative Evaluation

The quantitative evaluation result of our approach and the SOTA methods is described in Table 1. Specifically, we used two quantitative evaluation metrics to perform including Peak Signal-to-Noise Ratio (PSNR) and Structural Similarity Index Measure (SSIM).

In the UCF101 dataset, most of the videos contain diversity in objects, actions, and backgrounds. We compare the performance of our approach to Copy-Last-Frame and existing SOTA methods. These results show that our methods achieved outperform the BeyondMSE [18] and CtrlGen [8] and ContextVP [2] in terms of both PSNR and SSIM. In particular, our method achieves 35.1 PSNR and 0.942 SSIM that

Table 1 Quantitative evaluation of video prediction algorithms on UCF101 and Caltech pedestrian datasets

Method	UCF101		Caltech pedestrian	
	PSNR	SSIM	PSNR	SSIM
Copy-last-frame	30.2	0.89	23.3	0.779
PredNet [17]	–	–	27.6	0.905
BeyondMSE [18]	32.0	0.92	–	0.881
ContextVP [2]	34.9	0.92	28.7	0.921
CtrlGen [8]	28.13	0.934	20.88	0.776
Retrospective cGAN [12]	35.0	0.94	**29.2**	0.919
DPG [6]	–	–	28.2	0.923
PreCNet [23]	–	–	28.4	**0.929**
Ours (w/o att)	33.7	0.913	27.8	0.911
Ours	**35.1**	**0.942**	**29.01**	**0.921**

outperforms 4.9 PSNR and 0.052 SSIM compared to Copy-Last-Frame. Besides, our method outperforms 4.9 PSNR and 0.052 SSIM compared to Retrospective cGAN (the second-best method).

In the CalTech Pedestrian dataset, the videos in this dataset are captured by a car-mounted camera, the videos record the moving vehicles with a lot of motions. We compare our approach with Copy-Last-Frame and six SOTA methods. Our network achieves 29.01 PSNR (increase 5.71 PSNR) and 0.921 (increase 0.142) compared to Copy-Last-Frame. As shown in Table 1, our approach near similar with the best SOTA method i.e., Retrospective cycle GAN [12] in terms of PSNR (29.1 vs. 29.2). For SSIM metric, our approach achieves approximate performance compared to PreCNet [23] (0.921 vs. 0.929).

To evaluate the efficiency of the two attention modules, we conduct a comparison of our network between using and without using the attention modules. Figure 4 shows the comparison between using and without using attention modules in video prediction on the UCF101 dataset. Meanwhile, we choose several actions with high speed (frames include large motions) for evaluation such as "Long Jump", "Archery", "Biking". The results have shown that the model with two attention modules produces much sharper results compared to the model without attention. For example. the action "Long Jump" (top left of the Fig. 4), the model with two attention modules generates the image of the leg of the athlete smoother and sharper compared to the network without the attention modules.

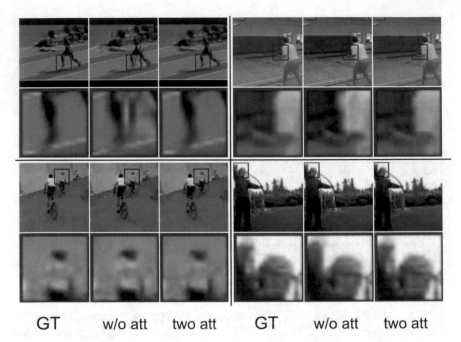

| GT | w/o att | two att | GT | w/o att | two att |

Fig. 4 A comparison of the predicted frame in the human action between using attention (two att) and without using attention (w/o att) modules. GT denotes the ground-truth frame

5 Conclusion

In this paper, we introduced a (2+1)D CNN model combined with spatial and temporal attention modules for video prediction. The temporal attention focus on the inter-motion relationship of features and the spatial attention focuses on the inter-spatial relationship of features. Through experiments, we demonstrate that two attention modules significantly increase the power of the predictive model. Comparing the result with related studies, our model achieved state-of-the-art performance in video prediction on several standard datasets. A limitation of the network is that the performance reduces significantly with multi-frame prediction compared to deep recurrent networks. So, in the future, we will research a new suitable model with multi-frame prediction and extend the experiment with other datasets.

Acknowledgements This research is funded by Thai Nguyen University, Tan Trao University and Thai Nguyen University of Education.

References

1. Aigner S, Körner M (2018) FutureGAN: anticipating the future frames of video sequences using spatio-temporal 3d convolutions in progressively growing GANs. arXiv preprint arXiv:1810.01325
2. Byeon W, Wang Q, Kumar Srivastava R, Koumoutsakos P (2018) ContextVP: fully context-aware video prediction. In: ECCV, pp 753–769
3. Castrejon L, Ballas N, Courville A (2019) Improved conditional VRNNs for video prediction. In: ICCV, pp 7608–7617
4. Diba A, Fayyaz M, Sharma V, Mahdi Arzani M, Yousefzadeh R, Gall J, Van Gool L (2018) Spatio-temporal channel correlation networks for action classification. In: ECCV, pp 284–299
5. Elafi I, Jedra M, Zahid N (2016) Unsupervised detection and tracking of moving objects for video surveillance applications. Pattern Recogn Lett 84:70–77
6. Gao H, Xu H, Cai QZ, Wang R, Yu F, Darrell T (2019) Disentangling propagation and generation for video prediction. In: ICCV, pp 9006–9015
7. Goodfellow I, Pouget-Abadie J, Mirza M, Xu B, Warde-Farley D, Ozair S, Courville A, Bengio Y (2014) Generative adversarial nets. In: Advances in neural information processing systems, vol 27
8. Hao Z, Huang X, Belongie S (2018) Controllable video generation with sparse trajectories. In: CVPR, pp 7854–7863
9. Hu J, Shen L, Sun G (2018) Squeeze-and-excitation networks. In: CVPR, June 2018
10. Jin B, Hu Y, Tang Q, Niu J, Shi Z, Han Y, Li X (2020) Exploring spatial-temporal multi-frequency analysis for high-fidelity and temporal-consistency video prediction. In: CVPR, pp 4554–4563
11. Johnson J, Alahi A, Fei-Fei L (2016) Perceptual losses for real-time style transfer and super-resolution. In: ECCV. Springer, Berlin, pp 694–711
12. Kwon YH, Park MG (2019) Predicting future frames using retrospective cycle GAN. In: CVPR, pp 1811–1820
13. Lee AX, Zhang R, Ebert F, Abbeel P, Finn C, Levine S (2018) Stochastic adversarial video prediction. arXiv preprint arXiv:1804.01523
14. Lee W, Jung W, Zhang H, Chen T, Koh JY, Huang T, Yoon H, Lee H, Hong S (2021) Revisiting hierarchical approach for persistent long-term video prediction. arXiv preprint arXiv:2104.06697
15. Liang X, Lee L, Dai W, Xing EP (2017) Dual motion GAN for future-flow embedded video prediction. In: ICCV, pp 1744–1752
16. Liu B, Chen Y, Liu S, Kim HS (2021) Deep learning in latent space for video prediction and compression. In: Proceedings of the IEEE/CVF conference on computer vision and pattern recognition, pp 701–710
17. Lotter W, Kreiman G, Cox D (2016) Deep predictive coding networks for video prediction and unsupervised learning. arXiv preprint arXiv:1605.08104
18. Mathieu M, Couprie C, LeCun Y (2015) Deep multi-scale video prediction beyond mean square error. arXiv preprint arXiv:1511.05440
19. Oprea S, Martinez-Gonzalez P, Garcia-Garcia A, Castro-Vargas JA, Orts-Escolano S, Garcia-Rodriguez J, Argyros A (2020) A review on deep learning techniques for video prediction. arXiv preprint arXiv:2004.05214
20. Rasouli A (2020) Deep learning for vision-based prediction: a survey. arXiv preprint arXiv:2007.00095
21. Soomro K, Zamir AR, Shah M (2021) A dataset of 101 human action classes from videos in the wild. In: Center for research in computer vision, vol 2
22. Srivastava N, Mansimov E, Salakhudinov R (2015) Unsupervised learning of video representations using LSTMs. In: ICML, pp 843–852
23. Straka Z, Svoboda T, Hoffmann M (2020) PreCNet: next frame video prediction based on predictive coding. arXiv preprint arXiv:2004.14878

24. Tran D, Wang H, Torresani L, Ray J, LeCun Y, Paluri M (2018) A closer look at spatiotemporal convolutions for action recognition. In: CVPR, pp 6450–6459
25. Wang Y, Jiang L, Yang MH, Li LJ, Long M, Fei-Fei L (2018) Eidetic 3d LSTM: a model for video prediction and beyond. In: ICLR
26. Woo S, Park J, Lee JY, So Kweon I (2018) CBAM: convolutional block attention module. In: ECCV, pp 3–19
27. Zeng KH, Shen WB, Huang DA, Sun M, Carlos Niebles J (2017) Visual forecasting by imitating dynamics in natural sequences. In: ICCV, pp 2999–3008
28. Zhang C, Chen T, Liu H, Shen Q, Ma Z (2019) Looking-ahead: neural future video frame prediction. In: ICIP. IEEE, pp 1975–1979
29. Zhang R, Isola P, Efros AA, Shechtman E, Wang O (2018) The unreasonable effectiveness of deep features as a perceptual metric. In: CVPR, pp 586–595

A Tri-valued Trimmed Mean Decision-Based Filter for Removal of Salt-and-Pepper Noise

Archit Sethi, Bharat Garg, and Rana Pratap Yadav

Abstract This paper presents a Tri-valued Trimmed Mean Decision-based De-noising Filter for Removal of Salt-and-Pepper noise. The proposed filter estimates noisy pixel value using the weighted median of noise-free pixels in the filter window. A fixed window size (3×3) is used at each iteration of the technique to maintain edge preservation and maximum correlation throughout the process. The proposed filter is evaluated on various natural and medical images with varying noise density (10–90%), where it provides high visual clarity and improved quality metrics. The simulation results show an average increase of 0.92 dB (2.6%) and 1.07 dB (2.8%) increased value of PSNR (Peak Signal-to-Noise Ratio) using grayscale and X-ray images, with varying noise density (10–90%).

Keywords Median filter · Salt-and-pepper noise · Image processing

1 Introduction

Digital images are corrupted by noise during data acquisition, by sensing elements and transmission, or by interference in the transmission channel. Salt-and-Pepper (SAP) noise is a type of Impulse noise, where corrupted pixels attain the maximum (Salt) or minimum (Pepper) value. The SAP noise is caused by random and sharp disturbances in the image signal, which makes it noisy and thus, degrade its quality. At high noise density (ND) $\geq$ 90%, the retention of information deteriorates, resulting in poor visual quality and lack of clarity. Non-linear filtering techniques are preferred for SAP noise removal since it restores the noise-free image closest to the original with clarity.

A. Sethi · B. Garg (✉) · R. P. Yadav
Thapar Institute of Engineering and Technology, Patiala, Punjab 147004, India
e-mail: bharat.garg@thapar.edu

A. Sethi
e-mail: asethi2_be16@thapar.edu

R. P. Yadav
e-mail: rpyadav@thapar.edu

© The Author(s), under exclusive license to Springer Nature Switzerland AG 2022
N. H. T. Dang et al. (eds.), *Artificial Intelligence in Data and Big Data Processing*,
Lecture Notes on Data Engineering and Communications Technologies 124,
https://doi.org/10.1007/978-3-030-97610-1_32

407

A common non-linear filtering technique, was standard median filter (MF) presented in [1]. It computed the median of pixel values in the filter window and it replaced the processed noisy pixel. When $ND \geq 60\%$, the details of image were not properly restored, thus, making it ineffective to be used at such noise densities. Decision-Based algorithm (DBA) for the median filter was presented in [16]. The processed a pixel either 0 or 255, i.e. minimum or maximum of the gray values and sorted the current filter window in a horizontal, vertical, and diagonal manner, taking out the central position pixel as the median value. If the pixel was still noisy, it restored the pixel by replacing it with the previously processed pixel. The limitation of this approach was the streaking effect, due to the continuous replacement of noisy pixels. Modified Decision-based Unsymmetric Trimmed Median (MDBUTM) filter [5] was proposed. It processed a noisy pixel and replaced it with the median of Noise Free Pixels (NFP) in the filter window. At high ND, the mean of all noisy pixels was computed. But, the image quality was decreased and resulted in faded filtered images.

A Fast Switching Based Mean-Median (FSBMM) filter in [18], was comprised of a combination of mean and median filters to estimate the value of processed Noisy Pixel (NP). The noisy pixel was restored by the nearest neighbor pixel when the result of the median was noisy. The NP on the boundary of the window was replaced by previously proposed pixels in their corresponding row or column vector when these boundary pixels could not be de-noised using the median. Hence, at high noise densities, the filtered images displayed repetition of pixels at the boundary, which resulted in poor visual filtered image quality. For higher noise densities, a Recursive Cubic Spline Interpolation (RSI) filter was presented in [17], which incorporated a numerical analytical method of cubic spline interpolation technique to estimate the value of NP. The RSI utilizes a global image information-based window, which improved the quality of the de-noised image. Also, it required at least two non-extreme pixels for interpolation, which are unlikely to locate at $ND \geq 90\%$.

For better visual clarity and coherence of filtered image from a highly corrupted image, a Probabilistic Decision-Based Median (PDBM) filter in [2] was proposed. PDBM estimates the noisy pixel by either trimmed median (TM) [5] or patched median (PM) [18]. But, at higher ND, the visual clarity drops drastically. Another algorithm, a Difference Applied Median Filter (DAMF) [4], used an iterative method of pre-processed pixels. Recently, a filtering technique was based on pixel density (BPD) [3]. It determined whether the currently processed pixel was noisy or noise-free, then decided the adaptive filter window. In a Three-Value Weighted Approach (TVWA) in [12], the NFPs of the current filter window were segmented into three groups based on their closeness to a minimum, maximum, and middle value. Weighted values were determined by the number of NFPs in a particular group corresponding to the total number of NFPs in the current filter window. At $ND \geq 80\%$, however, performance reduced greatly and resulted in blurred de-noised images. An Adaptive Switching Weighted Median (ASWM) [6] filter classified processed pixel as either noise-free or noisy, then de-noised using the median of the adaptive weighted window.

Recently, other techniques de-noised images at higher noise densities were presented in [8, 9]. An Adaptive Min-Max value-based Weighted Median (AMMWM) filter [7] computes highly correlated groups of NFPs using minimum and maximum values of the current filter window to restore the processed NP. It determined the weighted medians of these two groups. The study revealed that the approach was adaptive because it increased the filter window size by 2 when there were no NFPs. However, the filtering technique could restore NP at higher noise densities but, suffered from high computational complexity. A Convolution Neural Network inspired algorithm, Min-Max Average Pooling based (MMAP) filter was presented in [13]. This approach segmented the noisy image into two parts, which were passed through multiple layers of max and min pooling after pre-processing. The filter combined processed images from previous procedures and performed average pooling to reduce all residual noise. A new filtering approach, Significance Driven Inverse Distance Weighted (SDIDW) filter in [10] estimated the inverse distance weighted value using a minimum number of nearest original pixels. The weight of the original pixels was assigned based on the squared distance from the noisy pixel. However, at $ND \geq 90$, the output filtered image was reconstructed with poor visual clarity.

A novel Multilayer Decision-based Iterative (MDI) filter was presented in [15]. The approach divided the de-nosing process into pre-edge filtering, iterative filtering, and post-smoothing. Two copies of the noisy image were stored in α and β. For α, the Median-Mean-Mean sequence was followed and for β Mean-Median-Median sequence adhered. Finally, the post-smoothing step enhanced the quality of the reconstructed image. The Adaptive Weighted Min-Mid-Max Value-based (AWM3F) [14] computed the minimum, maximum and middle values along with their weights based on the number of NFPs. Hybrid Decision-based filtering (HBDF) in [11] de-noised image using a combination of linear, non-linear, and probabilistic techniques. In addition, to enhance the quality, a universal add-on was added, which used edge-detection and smoothening techniques to brush out fine details from filtered images.

Based on the observation, it was evident that PSNR values drop at $ND \geq 60\%$, thus, it was a challenge to maintain the visual coherency and clarity. Also, in adaptive filtering techniques, no matter what method was implemented, the blurring effects were quite obvious and had an overall impact on the clarity of filtered images. Hence, this paper presents a technique, that implements a decision-based Tri-Valued Length (TVL) and Trimmed Mean (TM) approach to counter blurring of images and preserve edges.

The paper is organized as follows: Sect. 2 presents the proposed TVTM flowchart with its algorithm, whereas Sect. 3 represents an illustration of the proposed TVTM. Section 4 presents simulation results and analysis on various benchmark images with varying noise density. Finally, Sect. 5 conclusions are given.

2 Proposed TVTM Filtering Algorithm

This algorithm considers two important aspects i.e. tri-valued length and trimmed mean. It improves the PSNR values as noise density increases along with edge preservation. First section presents preliminaries, followed by flow chart and algorithm of proposed TVTM filter and finally presents an example under certain noise density.

2.1 Preliminaries

Noise Model The minimum and maximum values of the gray level representing salt and pepper noise, respectively, are considered as noisy values. Noise vector (Γ) contains two extreme quantities (i.e. $\Gamma = [0, 255]$). The processed pixel encounters only noisy pixels and neglects the noise-free ($i P_{i,j}^n \in \Gamma$).

Minimum and Maximum group computation The pixels in current filter window $i P_{i,j}^n \geq 0$ and $i P_{i,j}^n \leq 255$ are stored in array (Δ). The minimum (Δ_{min}) and maximum (Δ_{max}) pixel values are computed from this array. The two groups A_{min} and A_{max} are created, which corresponds to B_{min} and B_{max} respectively. Finally, each pixel of Δ is assigned either B_{min} or B_{max} based on approximate closeness to A_{min} or A_{max}, respectively. These two groups are used to estimate the noisy pixel value in proposed algorithm.

Filter size The objective of this algorithm is to maintain the visual clarity and reduce blurring of image as much possible. Therefore, the filter window $\omega_{(2\kappa+1)(2\kappa+1)}^n$ is kept constant at 3×3, where $\kappa = 1$. As value of k increases, it results in blurred de-noised images with loss of edge preservation.

2.2 Flowchart of the Proposed Filter

The flowchart of the proposed filter is shown in Fig. 1. The algorithm is divided into two parts which are (a) tri-valued length and (b) trimmed mean. The window

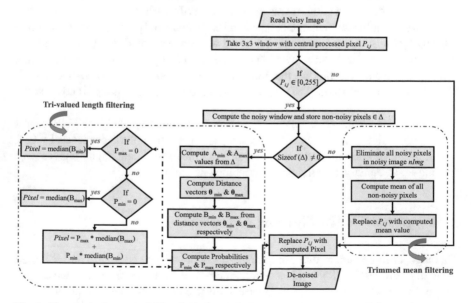

Fig. 1 Flowchart of proposed filter depicting various steps

Algorithm 1 TVTM $(nImg, OutImg)$

1: **Input** $nImg$ ▷ *Noisy Image of size M x N*
2: **Output** $OutImg$ ▷ *Output filtered Image*
3: **Initialize**: $Pixel \leftarrow 0$;
4: **for** each pixel $i P_{i,j}^n \in nImg$ **do**
5:      **Initialize**: $\Gamma \leftarrow [0, 255]$;   ▷ *Set containing noisy pixels*
6:      **Initialize**: $\kappa \leftarrow 1$;   ▷ *Set filter size*
7:      Compute $\omega_{(2\kappa+1)(2\kappa+1)}^n$ centered at $i P_{i,j}^n$;
8:      **for** each $\omega_{l,m}^n \notin \Gamma$ **do**
9:          Compute $\Delta \leftarrow [\Delta, \omega_{l,m}^n]$;   ▷ Δ *stores NEP of* $\omega_{l,m}^n$
10:      **end for**
11:      $\lambda_\Delta \leftarrow \text{length}(\Delta)$;   ▷ *Number of NEP*
12:      **if** $\lambda_\Delta \neq 0$ **then**
13:          $Pixel \leftarrow \text{TVL}(\Delta)$;   ▷ *Call TVL procedure*
14:      **else**
15:          $Pixel \leftarrow \text{TM}(nImg)$;   ▷ *Call TM procedure*
16:      **end if**
17:      $OutImg \leftarrow Pixel$;
18: **end for**
19: **Return** $OutImg$
20: **End**   ▷ *End of algorithm*

Algorithm 2 Function TVL (Δ)

1: $A_{min} \leftarrow \min(\Delta)$; $A_{max} \leftarrow \max(\Delta)$;   ▷ *Minimum and Maximum value of* Δ
2: $\theta^{min} \leftarrow |A_{min} - \Delta|$;   ▷ *Absolute difference between min value and each element of* Δ_i
3: $\theta^{max} \leftarrow |A_{max} - \Delta|$;
4: $B_{min} \leftarrow \phi$; $B_{max} \leftarrow \phi$;
5: **for** each Δ_i **do**   ▷ *Check closeness of* Δ_i *to* B_{min}/B_{max}
6:      **if** $\theta_i^{max} \leq \theta_i^{min}$ **then**
7:          $B_{max} \leftarrow [B_{max}, \Delta_i]$;
8:      **else**
9:          $B_{min} \leftarrow [B_{min}, \Delta_i]$;
10:      **end if**
11: **end for**
12: $\lambda_{min} \leftarrow \text{length}(B_{min})$; $\lambda_{max} \leftarrow \text{length}(B_{max})$;   ▷ *Number of NFP in* B_{min} *and* B_{max}
13: $P_{min} \leftarrow \lambda_{min}/\lambda_\Delta$; $P_{max} \leftarrow \lambda_{max}/\lambda_\Delta$;   ▷ *Probability of NFP in* B_{min} *and* B_{max}
14: **if** $P_{max} == 0$ **then**
15:      $Pixel \leftarrow \text{median}(B_{min})$;   ▷ *Compute median of* B_{min}
16: **else if** $P_{min} == 0$ **then**
17:      $Pixel \leftarrow \text{median}(B_{max})$;
18: **else**
19:      $Pixel \leftarrow P_{max}*\text{median}(B_{max}) + P_{min}*\text{median}(B_{min})$;
20: **end if**
21: **Return** $Pixel$;

---

size is kept constant $\kappa = 1$ to avoid blurring of the image and preserve the edges. To estimate the effective value of noisy pixels, $i P_{i,j}^n \geq 0$ and $i P_{i,j}^n \leq 255$ are considered, which are then stored in an array (Δ). The two sub-algorithms have been encircled with dashed lines as tri-valued length and trimmed mean, shown in Fig. 1.

Algorithm 3 Function TM $(nImg)$

1: **Initialize**: $N \leftarrow 0$; $G \leftarrow 0$; $M \leftarrow 0$;
2: **for** each pixel $nImg \notin \Gamma$ **do**
3:     $G \leftarrow G + nImg$;   ▷ *Eliminates all noisy pixels*
4:     $N \leftarrow N + 1$;
5: **end for**
6: $M \leftarrow G/N$;   ▷ *Computes mean of noise-free pixels*
7: $Pixel \leftarrow M$;
8: **Return** $Pixel$;

When $\Delta \neq 0$, then minimum and maximum values respectively from Δ are computed and distance vector (θ) corresponding to A_{min} and A_{max} are calculated ($\theta_{min} = |\Delta_i - A_{min}|, \theta_{max} = |\Delta_i - A_{max}|$). From these distance vectors, B_{min}, B_{max} are created. Next, step is checking the closeness of noise-free pixels in Δ_i, if $\theta^{max} \leq \theta^{min}$, then they are added in B_{max}, else, B_{min}. Probabilities $P_{max} = \lambda_{max}/\Delta$ and $P_{min} = \lambda_{min}/\Delta$ are computed. Finally, the noisy pixel is replaced by $Pixel$ obtained from Eq. 1.

$$Pixel = P_{max} * median(B_{max}) + P_{min} * median(B_{min}) \qquad (1)$$

The sub-algorithm TM, eliminates all the noisy pixels $\notin \Gamma$ and mean of remaining are computed, which would replace with current $Pixel$.

2.3 Algorithm of the Proposed Filter

The pseudo-code of the proposed algorithm given in Algorithm 1. The algorithm initializes variable $Pixel$ and window parameter (κ). The current filter window (3×3) on the noisy image and the number of noise-free pixels are stored in Δ. If the length of the array (Δ) $\notin 0$, then the TVL procedure is computed as shown in Algorithm 2. Minimum and maximum values are stored in B_{min} & B_{max} respectively, and distance parameters θ^{min} & θ^{max} and B_{min} & B_{max} are initialized, respectively. The two distance vectors are used to determine the closeness based on maximum and minimum value respectively, thus, added to B_{max} or B_{min} respectively. Then, the probabilities of each group are computed as $P_{max} = \lambda^{max}/\Delta$ and $P_{min} = \lambda^{min}/\Delta$ respectively. Finally, the weighted median of each group determines the estimated value of processed pixel $i P_{i,j}$. If the length of the array $\Delta \in 0$, then the TM procedure is followed as shown in Algorithm 3, where the noisy pixels are eliminated, and the mean of all noise-free are computed and replaced with $Pixel$.

3 Illustration of the Proposed Algorithm

The de-noising process of the proposed algorithm is demonstrated using an illustration of a filter window ($\kappa \leftarrow 1$) traversing the noise-matrix to produce a noise-free pixel value, as shown in Fig. 2. It illustrates the noise-free pixel estimation using TVL and TM. Firstly, B_{min} & B_{max} are computed from Δ. Corresponding distance vectors θ^{min} & θ^{max} are computed, by computing the absolute difference between

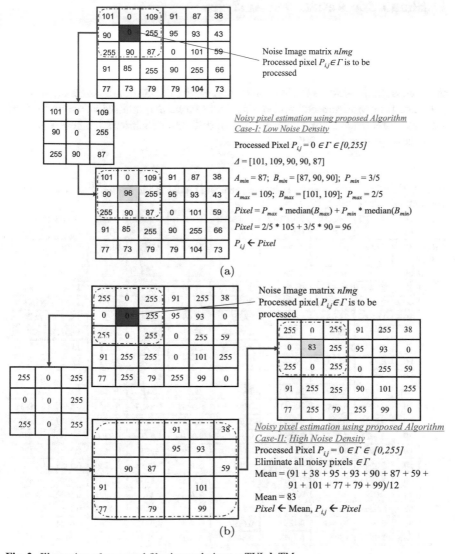

Noise Image matrix $nImg$
Processed pixel $P_{i,j} \in \Gamma$ is to be processed

Noisy pixel estimation using proposed Algorithm
Case-I: Low Noise Density

Processed Pixel $P_{i,j} = 0 \in \Gamma \in [0,255]$

$\Delta = [101, 109, 90, 90, 87]$

$A_{min} = 87;\ B_{min} = [87, 90, 90];\ P_{min} = 3/5$

$A_{max} = 109;\ B_{max} = [101, 109];\ P_{max} = 2/5$

$Pixel = P_{max} * median(B_{max}) + P_{min} * median(B_{min})$

$Pixel = 2/5 * 105 + 3/5 * 90 = 96$

$P_{i,j} \leftarrow Pixel$

(a)

Noise Image matrix $nImg$
Processed pixel $P_{i,j} \in \Gamma$ is to be processed

Noisy pixel estimation using proposed Algorithm
Case-II: High Noise Density
Processed Pixel $P_{i,j} = 0 \in \Gamma \in [0,255]$
Eliminate all noisy pixels $\in \Gamma$
Mean = (91 + 38 + 95 + 93 + 90 + 87 + 59 + 101 + 77 + 79 + 99)/12
Mean = 83
$Pixel \leftarrow Mean,\ P_{i,j} \leftarrow Pixel$

(b)

Fig. 2 Illustration of proposed filtering technique **a** TVL **b** TM

minimum/maximum values and Δ_i respectively. B_{min} & B_{max} are defined, initially with ϕ. Then closeness of Δ_i is checked, if $\theta^{max} \leq \theta^{min}$, then B_{max} is stored, otherwise, B_{min}. Finally, probabilities are computed i.e. P_{max} and P_{min}. If $P_{max} = 0$, then pixel is replaced with median (B_{min}) or vice-versa. When length of array (Δ) is equal to 0, then the noise matrix with all pixel values $\in \Gamma$, are eliminated and mean of NFPs are computed and replaced with $Pixel$.

4 Simulation Results and Analysis

The efficiency of the proposed algorithm is analyzed over the existing techniques using different benchmark images (Pirate, Living-room, Pepper, and medical image X-ray (Chest) each of size 512×512). The quality metrics for quantitative analysis are Peak Signal-to-Noise Ratio (PSNR) and Structural Similarity Index (SSIM) in [19]. The following subsections provide quantitative and qualitative analysis of the proposed filter over existing algorithms with different benchmark images by varying noise density from 10 to 90%.

Table 1 Quality metrics of filtered pirate image using proposed and existing SAP removal filters with varying noise density

	ND (%)	MF [1]	DBAMF [16]	MDBUTM [5]	FSBMMF [18]	RSIF [17]	PBDM [2]	DAMF [4]	BPDM [3]	Proposed TVTM
PSNR (dB)	10	37.85	38.17	31.86	38.77	38.53	25.51	31.89	23.03	38.72
	20	31.33	34.05	31.26	34.61	34.14	20.57	31.28	19.80	34.89
	30	25.46	31.91	30.66	32.28	31.90	17.97	30.67	17.96	32.80
	40	20.57	29.83	29.89	30.35	30.11	16.29	29.90	16.66	30.87
	50	16.71	28.03	28.79	28.85	28.69	15.11	28.78	15.57	29.71
	60	13.43	26.48	26.8	27.62	27.31	14.50	26.79	14.67	28.77
	70	10.8	24.61	23.57	26.45	25.96	13.58	23.55	13.63	27.84
	80	8.73	22.69	19.82	25.14	24.25	12.24	19.82	12.60	26.15
	90	6.91	19.67	15.62	22.88	21.89	9.77	15.61	10.84	23.16
	Avg	**19.09**	**28.39**	**26.48**	**29.67**	**29.20**	**16.18**	**26.48**	**16.09**	**30.33**
SSIM	10	0.9826	0.9830	0.8941	0.9845	0.9837	0.8425	0.8946	0.5870	0.9891
	20	0.9379	0.9587	0.8860	0.9624	0.9585	0.5949	0.8864	0.4219	0.9659
	30	0.8224	0.9313	0.8766	0.9364	0.9310	0.4086	0.8769	0.3356	0.9428
	40	0.6032	0.8950	0.8616	0.9049	0.8994	0.2834	0.8620	0.2791	0.9178
	50	0.3570	0.8485	0.8336	0.8680	0.8627	0.2058	0.8336	0.2355	0.8788
	60	0.1744	0.7931	0.7624	0.8273	0.8183	0.1706	0.7625	0.2001	0.8476
	70	0.0793	0.7172	0.6097	0.7802	0.7636	0.1302	0.6094	0.1530	0.8068
	80	0.0395	0.6229	0.3971	0.7230	0.6904	0.0962	0.3972	0.1055	0.7446
	90	0.0175	0.4542	0.1857	0.6191	0.5719	0.0543	0.1860	0.0524	0.6364
	Avg	**0.4660**	**0.8005**	**0.7008**	**0.8451**	**0.8311**	**0.3097**	**0.7010**	**0.2634**	**0.8589**

4.1 Comparative Analysis with Gray Scale Images

Benchmark images include three grayscale (Pirate, Living-room, Pepper), each of size 512 × 512. Table 1 summarizes the PSNR and SSIM values of reconstructed Pirate images with varying the ND from 10 to 90%. The quality metrics shown demonstrate that the proposed filter is superior to the existing filters. Similarly, the comparative analysis for the other images Living-room and Pepper images are shown in Figs. 3 and 4, respectively. Comparative analysis of graphical plots exhibits the superiority of the proposed algorithm over existing filters in both quality metrics at each sample of noise density. There is an increase in average PSNR by at least 2.6% over other filters. Figure 5 shows the de-noised images of benchmark image data set in comparison with existing filters at 80% noise density. The computational complexity of the proposed algorithm is also computed by computing filtering time.

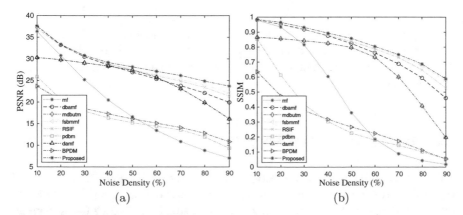

Fig. 3 Comparative analysis of **a** PSNR and **b** SSIM of living-room filtered image at varying noise density (10–90%)

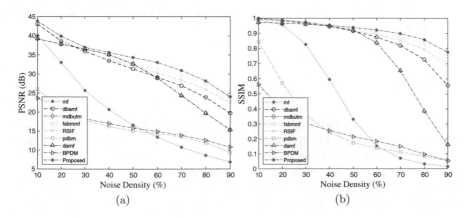

Fig. 4 Comparative analysis of **a** PSNR and **b** SSIM of pepper filtered image at varying noise density (10–90%)

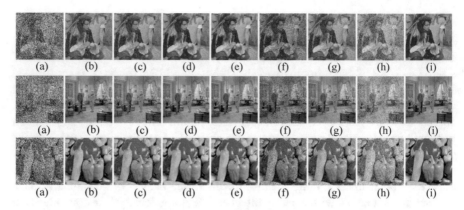

Fig. 5 Benchmark images at 80% ND filtered via **a** MF **b** DBAMF **c** MDBUTM **d** FSBMMF **e** RSIF **f** PDBM **g** DAMF **h** BPDM and **i** proposed filters

The proposed algorithm on an average has 1.78 s computation time. The algorithm is comparatively faster than other existing algorithms during comparative analysis. The proposed filter demonstrates better visual clarity and could preserve edges at high noise densities.

4.2 Analysis on Medical Images

This subsection includes the simulation results on medical X-ray (Chest) image of size 512×512. Table 2 shows that with varying noise density from 10 to 90%, the proposed filter exhibits superior and appealing results over existing filters, with an average increase of 2.89% in PSNR values. The graphical plot of quality metrics in Fig. 6 represents the high performance of the proposed filter. Figure 7 shows the visual coherency of the X-ray image at 90% noise density, with less blurring and edge preservation over other existing filters.

5 Conclusion

This paper presented a decision-based algorithm for SAP noise removal that effectively restores processed noise pixels using TVL and TM. Tri-valued length, comprised of a weighted average of median and probabilities values of current filter, for estimation of the noise-free pixel. The window size is kept constant to prevent blurred images at high noise densities, with significant preservation of edges. The trimmed mean function computes the mean of the noisy image by eliminating noisy pixels and replacing them with a processed pixel. The quality metrics on various

Table 2 Quality metrics of filtered X-ray images with varying noise density

	ND (%)	MF [1]	DBAMF [16]	MDBUTM [5]	FSBMMF [18]	RSIF [17]	PBDM [2]	DAMF [4]	BPDM [3]	Proposed TVTM
PSNR (dB)	10	43.33	44.21	34.83	44.63	40.32	25.53	34.72	23.94	45.40
	20	33.00	41.46	30.93	43.08	36.15	20.28	30.81	20.61	44.02
	30	25.97	39.54	28.33	40.08	33.16	17.79	28.23	18.71	41.37
	40	20.34	34.31	26.62	38.10	32.53	16.12	26.50	17.41	39.13
	50	16.36	35.68	24.81	36.82	30.09	15.07	24.71	16.31	37.95
	60	12.95	32.94	22.78	35.69	28.85	14.18	22.68	15.29	36.46
	70	10.36	31.02	20.21	34.28	26.92	13.28	20.14	14.25	35.42
	80	8.17	27.41	16.98	32.36	25.51	11.38	16.93	12.58	33.73
	90	6.35	23.60	13.45	27.99	23.18	8.56	13.41	10.33	29.19
	Avg	**19.65**	**34.47**	**24.33**	**37.01**	**30.75**	**15.80**	**24.24**	**16.61**	**38.08**
SSIM	10	0.9930	0.9807	0.9397	0.9796	0.9704	0.8314	0.9400	0.5738	0.9931
	20	0.9574	0.9725	0.9313	0.9756	0.9480	0.5412	0.315	0.4142	0.9853
	30	0.8333	0.9600	0.9249	0.9619	0.9354	0.3468	0.9252	0.3355	0.9742
	40	0.5738	0.9418	0.9145	0.9485	0.9197	0.2268	0.9147	0.2952	0.9655
	50	0.2985	0.9242	0.8858	0.9345	0.9027	0.1658	0.8858	0.2565	0.9556
	60	0.1195	0.8921	0.7884	0.9065	0.8826	0.1276	0.7884	0.2211	0.9447
	70	0.0445	0.8492	0.5746	0.8807	0.8562	0.1069	0.5743	0.1752	0.9275
	80	0.0176	0.7764	0.2923	0.8540	0.8142	0.0872	0.2920	0.1066	0.8890
	90	0.0084	0.6318	0.1042	0.7799	0.7425	0.0705	0.1037	0.0709	0.7948
	Avg	**0.4273**	**0.8810**	**0.7062**	**0.9135**	**0.8858**	**0.2783**	**0.7062**	**0.2722**	**0.9367**

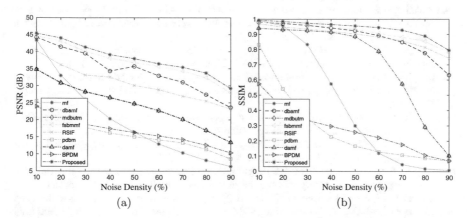

Fig. 6 Comparative analysis of **a** PSNR and **b** SSIM of X-Ray (chest) filtered image at varying noise density (10–90%)

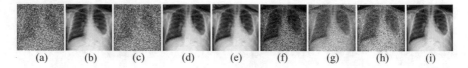

Fig. 7 X-ray (chest) image at 90% ND filtered via **a** MF **b** DBAMF **c** MDBUTM **d** FSBMMF **e** RSIF **f** PDBM **g** DAMF **h** BPDM and **i** proposed filters

benchmark image data set and medical images presented an average increase of at least 2.5% which is 1.07 dB higher than previous existing algorithms. Therefore, this algorithm could be effectively utilized for de-noising medical images.

References

1. Astola J, Kuosmanen P (2020) Fundamentals of nonlinear digital filtering. CRC Press
2. Balasubramanian G, Chilambuchelvan A, Vijayan S, Gowrison G (2016) Probabilistic decision based filter to remove impulse noise using patch else trimmed median. AEU-Int J Electron Commun 70(4):471–481
3. Erkan U, Gökrem L (2018) A new method based on pixel density in salt and pepper noise removal. Turk J Electr Eng Comput Sci 26(1):162–171
4. Erkan U, Gökrem L, Enginoğlu S (2018) Different applied median filter in salt and pepper noise. Comput Electr Eng 70:789–798
5. Esakkirajan S, Veerakumar T, Subramanyam AN, Prem Chand C (2011) Removal of high density salt and pepper noise through modified decision based unsymmetric trimmed median filter. IEEE Signal Process Lett 18(5):287–290
6. Faragallah OS, Ibrahem HM (2016) Adaptive switching weighted median filter framework for suppressing salt-and-pepper noise. AEU-Int J Electron Commun 70(8):1034–1040
7. Garg B (2020) An adaptive minimum-maximum value-based weighted median filter for removing high density salt and pepper noise in medical images. Int J Ad Hoc Ubiquitous Comput 35(2):84–95
8. Garg B (2020) Restoration of highly salt-and-pepper-noise-corrupted images using novel adaptive trimmed median filter. Signal Image Video Process 14:1555–1563
9. Garg B, Arya K (2020) Four stage median-average filter for healing high density salt and pepper noise corrupted images. Multimedia Tools Appl 79(43):32305–32329
10. Garg B, Rana PS, Rathor VS (2021) Significance driven inverse distance weighted filter to restore impulsive noise corrupted x-ray image. J Ambient Intell Humanized Comput 1–12
11. Ghumaan RS, Sohi PJS, Sharma N, Garg B (2021) A novel hybrid decision-based filter and universal edge-based logical smoothing add-on to remove impulsive noise. Turk J Electr Engi Comput Sci 29(4):1944–1963
12. Lu CT, Chen YY, Wang LL, Chang CF (2016) Removal of salt-and-pepper noise in corrupted image using three-values-weighted approach with variable-size window. Pattern Recogn Lett 80:188–199
13. Satti P, Sharma N, Garg B (2020) Min-max average pooling based filter for impulse noise removal. IEEE Signal Process Lett 27:1475–1479
14. Sharma N, Sohi PJS, Garg B (2021) An adaptive weighted min-mid-max value based filter for eliminating high density impulsive noise. Wireless Pers Commun 1–18
15. Sharma N, Sohi PJS, Garg B, Arya K (2021) A novel multilayer decision based iterative filter for removal of salt and pepper noise. Multimedia Tools Appl 1–15

16. Srinivasan K, Ebenezer D (2007) A new fast and efficient decision-based algorithm for removal of high-density impulse noises. IEEE Signal Process Lett 14(3):189–192
17. Veerakumar T, Esakkirajan S, Vennila I (2014) Recursive cubic spline interpolation filter approach for the removal of high density salt-and-pepper noise. Signal Image Video Process 8(1):159–168
18. Vijaykumar V, Mari GS, Ebenezer D (2014) Fast switching based median-mean filter for high density salt and pepper noise removal. AEU-Int J Electron Commun 68(12):1145–1155
19. Wang Z, Bovik AC, Sheikh HR, Simoncelli EP (2004) Image quality assessment: from error visibility to structural similarity. IEEE Trans Image Process 13(4):600–612

On Ranking of Image Quality Metrics

Sergey Dvoenko, Mikhail Kurbakov, and Manh Tran X.

Abstract In different problems of image processing, it needs to evaluate the quality of images. For a ground-truth case, the result is evaluated in comparison with the original image. Different metrics realize different ideas of what a more quality image is. On the other side, the human quality evaluation of images isn't so precise numerically. Therefore, may be better to use measurements in ordinal scales. Based on this approach, the known Kemeny's median is proposed here to coordinate image rankings from different metrics. In this paper, the modification of Kemeny's algorithm is developed. The coordination of the median ranking with rankings by a human is statistically tested for images from the public database. As a result, it can be recommended to use some of the preferred metrics from the given set to compare the results of different processing algorithms.

Keywords Image processing · Kemeny's median · Quality metrics

1 Introduction: Related Works and Problem Formulation

In different problems of image analysis, it needs to evaluate the quality of processing. The different approaches are used to do it. They are known as full-, and no-reference image quality assessments with intermediate reduced-reference ones [1–3]. These approaches are based on a different understanding of how to evaluate image quality. It seems the evaluation with the reference image can be implemented most simply. By the present, different approaches to full-reference image visual quality assessment have been proposed to build such estimations known as image quality metrics [4–9].

Different metrics are developed for usually different conceptions of what does it mean the better quality of an image. Therefore, the image quality is some sort of an ambiguous concept. For example, ground-truth metrics of difference squared based, in a case of improved quality can be interpreted by a human as not so good visually. On the contrary, other images noised by some specific distortions are interpreted

S. Dvoenko (✉) · M. Kurbakov · M. Tran X.
Tula State University, Tula, Russia

© The Author(s), under exclusive license to Springer Nature Switzerland AG 2022　　　421
N. H. T. Dang et al. (eds.), *Artificial Intelligence in Data and Big Data Processing*,
Lecture Notes on Data Engineering and Communications Technologies 124,
https://doi.org/10.1007/978-3-030-97610-1_33

as more contrast and accurate while their quality is reduced. As a result, different metrics were developed to pay attention to the human peculiarities of image sensing. Today, the problem of visual perception of image quality under the quality evaluation by metrics remains to be the key one, since in practice it needs to compare processing results by different algorithms.

From the other side of the problem of the quality metrics comparing, images and their processing tasks themselves are rather various ones. It is easy to see, the quality of images of various classes (types, etc.) is changed differently under different distortions, after processing by different algorithms, etc. Then, it needs to use different quality metrics. As a result, evaluation scores of the quality of images of different classes are usually different too.

That is why under modern conditions, comparing the quality of processing results appears to be not so a simple problem. It should be noted that it is a special case of the general problem of real-world image recognition [10]. As a rule, it needs to get by different metrics evaluation scores of images of different classes based on collections in public databases. Finally, it needs to demonstrate that some processing techniques can reach a sufficiently stable quality result for some sufficient set of specified images based on some sufficient set of metrics.

Let an average score of some set of metrics be used. Nevertheless, we fundamentally can face here the general multi-criteria problem. For this reason, it needs before to understand, can such an average evaluation score to be acceptable or not. It is known, that the set of criteria is contradictory usually, which leads to the Pareto problem. This means that the best evaluation scores usually correspond to a set of different decisions as to the Pareto set. In this case, the averaging is the attempt to avoid the Pareto problem based on the so-called "super-criterion" as a final combination of partial criteria with equal weights.

Nevertheless, it needs to find the optimal linear combination of weights of partial criteria in general. Unfortunately, the super-criterion isn't a continuous function also. As a result, small weight changes can lead to sharp changes in the optimal decision and the super-criterion appears to be a non-stable function.

2 The Proposed Comparing of Quality Metrics

In an attempt to avoid the Pareto problem, it needs to take into account that an image quality recognized by a human is not a numerical quality score. Therefore, it is suitable to use measurements in non-numerical scales like ordering ones, and use rankings. Specifically, it needs to order a set of images by scores of some quality metric. As it is known, positions of images as points on the ordering axis can be varied without the mutual positions having been changed.

On the other side, in a case of different rankings have been coordinated (e.g. as Kemeny's median [11–13]), such the averaged ranking appears to be the arithmetic mean of individual rankings by quality metrics. Finally, such the group ranking realizes the ranking of images by their quality like on the axis of a super-criterion discussed above. It needs to note, such the "super-criterion" provides less sensitivity to small score changes of partial criteria, i.e. quality metrics.

Based on image rankings, we can compare different processing algorithms. After processing, the quality evaluation by some metric given builds the ordering of images and specifies the best one, i.e. specifies the best processing algorithm. Nevertheless as pointed above, the quality evaluation is better to be based on different metrics, which produce their specific image rankings and can specify their own best images. Therefore, image rankings coordinated in the form of the Kemeny's median specify some algorithm with the best processing quality based on evaluation scores of the given set of quality metrics.

In the case the processing algorithms have not been used, the following approach can be proposed. First, let an image be noised by some set of noises (types of distortion). After the quality of its noisy variants has been evaluated by some quality metric, all noisy images become ordered by the metric scores. It is evident, the more sensitive this metric is to this type of distortion, the lower the quality evaluation score of the noisy image is. Therefore, the worst image by quality specifies the type of noise which most strongly degrades the image quality. Different metrics produce different rankings of images and specify their images with the worst quality. By Kemeny's median used as a coordination principle, it can be concluded that this type of noise most degrades this image based on evaluation scores of the given set of quality metrics.

In other words, the quality of this image is degraded the most by this type of distortion from the specified set according to the scores of the given set of metrics.

On the other side, let a set of images be noised by some type of distortion. The quality evaluation of noisy images by some metric orders them by evaluation scores of this metric. It is evident, the more sensitive this metric is to this distortion, the lower the quality evaluation score of noisy images is. Therefore, the worst image by quality specifies the type (class), which is most strongly degraded by this type of distortion. Different metrics produce different rankings of images and specify their images with the worst quality. By Kemeny's median used as a coordination principle, it can be concluded that this class of images is most degraded by this noise based on evaluation scores of the given set of quality metrics.

In other words, the quality of this image from the set is degraded the most by this distortion according to evaluation scores of the given set of quality metrics.

As a result, an opportunity appears for reasonable recommendations on using preferred metrics to evaluate the quality of images.

Nevertheless, it needs to preliminary evaluate the mutual coordination of quality metrics. It means that the similarity of rankings from different quality metrics needs to be evaluated. The coordination of rankings can be evaluated in different ways, e.g. statistically based on rank correlations with testing the corresponding null hypothesis about independent rankings. It needs to remark, in a case of the low coordination level it needs to find conflicting groups of quality metrics [14]. But this problem is not discussed in this paper.

The important peculiarity of this investigation consists in comparing numerical quality scores with the visual perception of image quality by a human as a part of the general perception problem [10]. In this paper, the result of the above proposed two cases of the quality comparing is discussed.

3 Experimental Data

The image set for testing is the public resource at the address [8]. This database contains 24 reference images from the Kodak database [9] with one synthetic image added (with patterns and textures). All 25 images have the size of 512×384 pixels. Reference images are obtained by cropping from original images as 8-bit RGB ones without any compression and other transformations (Fig. 1). Each image has been subjected to 17 types (Table 1) of distortion with four levels for each type. Types of distortion are named by developers of this database [7]. Distortion levels were specified based on human's perception of distortions only.

In our experiments, only level 2 of distortion (level 2 is not so intensive) has been used. The image quality was evaluated by 18 metrics with one result of expert survey added as MOS (Mean Opinion Score—expert survey), therefore by 19 metrics as a total: MSSIM, VIF, VIFP, VSNR, PSNR-HVS, PSNR-HVS-M, SSIM, NQM, UQI,

Fig. 1 Reference color images (1–24) from the TID2008 database with the 25th synthetic image represented here in a grayscale

Table 1 Types of distortion [7]

N	Type of distortion	Noise	Noise2	Safe	Hard	Simple	Exotic	Exot2
1	Additive Gaussian	•	•	•		•		
2	Additive in color		•					
3	Spatially correlated	•	•	•	•			
4	Masked		•		•			
5	High frequency	•	•	•				
6	Impulse	•	•	•				
7	Quantization	•	•		•			
8	Gaussian blur	•	•	•	•	•		
9	Image denoising	•			•			
10	JPEG compression			•		•		
11	JPEG2000 compress			•		•		
12	JPEG transmit errors				•			•
13	JPEG2000 trans. err				•			•
14	Non-eccentricity				•		•	•
15	Local block-wise						•	•
16	Mean (intensity) shift						•	•
17	Contrast change						•	•

XYZ (inverted), LINLAB (inverted), IFC, WSNR, DCTUNE (inverted), SNR, MSE (inverted), PSNR, PSNRY, MOS [7, 8].

It needs to remark, the evaluation scores of the MOS "metric" is the result of not yet another numerical metric calculation. It is the result of the special procedure of respondent questioning with the special processing of their answers to represent in a numerical form.

Initial data are used as two data cubes of the $17 \times 25 \times 19$ size in the form of standard ranks of 25 images ordered in each of 19 tables of the 17×25 size as 25 rankings-columns (as the type R17) and as 17 rankings-lines (as the type R25).

In the case of the ordered metric values having differed by no more than 0.001, then neighboring places have been assigned to them, and the corresponded standard rank has been calculated.

4 Kemeny's Median as a Coordination Principle

Let images be considered as an unordered set of alternatives $A = \{a_1, ..., a_N\}$, where according to the size of data cubes the $N = 17$ is for columns and $N = 25$ is for lines. The ordering of images by scores of some metric is considered as a ranking P of alternatives presented by a relation matrix $M(N, N)$ with elements

$$m_{ij} = \begin{cases} 1, & a_i \succ a_j \\ 0, & a_i \sim a_j \\ -1, & a_i \prec a_j \, . \end{cases}$$

It is known, distances between rankings as binary relations in ordinal scales are metric values [7]. Based on the same numeration in A, the distance between rankings P_u and P_v presented by matrices M_u and M_v is calculated as

$$d(P_u, P_v) = \frac{1}{2} \sum_{i=1}^{N} \sum_{j=1}^{N} |m_{ij}^u - m_{ij}^v|.$$

In this paper, two sets of n rankings by scores of quality metrics under investigation are considered, where $n = 25$ is for R17 rankings and $n = 17$ is for R25 rankings. It needs to find a group relation P coordinated with relations $P_1, ..., P_n$. It is well-known, that Kemeny's median P^* represents the group transitive relation as the ranking with the least distance to others

$$P^* = \arg\min_P \sum_{u=1}^{n} d(P, P_u).$$

The optimal decision is found based on the loss matrix $Q(N, N)$[12–16]. The total distance from a group ranking P to other rankings is calculated as

$$\sum_{u=1}^{n} d(P, P_u) = \frac{1}{2} \sum_{i=1}^{N} \sum_{j=1}^{N} \sum_{u=1}^{n} d_{ij}(P, P_u),$$

where partial distances d_{ij} are defined under conditions $m_{ij} = 1$ as

$$d_{ij}(P, P_u) = |m_{ij} - m_{ij}^u| = \begin{cases} 0, & m_{ij}^u = 1 \\ 1, & m_{ij}^u = 0 \\ 2, & m_{ij}^u = -1 \, . \end{cases}$$

Hence, the element $q_{ij} = \sum_{u=1}^{n} d_{ij}(P, P_u)$ of the loss matrix Q defines the total mismatch losses of the preference $a_i \succ a_j$ in the unknown ranking P relative to corresponding known preferences in rankings $P_1, ... P_n$.

The Kemeny's median is built as follows. At the current step, the row (and the corresponding column) with the minimal loss sum is removed from Q [12]. The corresponding alternative is added to the last place in the current ranking. Sometimes, it needs to calculate standard ranks for alternatives in the same place. The result is denoted as the ranking P_1 [12]. In a reversed scanning, it is checked the correspondence between binary relations $a_i \succ a_j$ in the ranking and penalties $q_{ij} \leq q_{ji}$ in the loss matrix. If it is not so, the current pair of alternatives is reversed. The result is denoted as the ranking P_{II}. This is Kemeny's median $P*$.

Is it evident, the final ranking is dependent on a discrete character of the loss matrix Q, because of not so "natural" partial distances d_{ij}. The discrete loss matrix more affects the result with a large number of alternatives, since indistinguishable alternatives appear more often. It makes building the ranking P_{II} difficult. To meet these conditions better, the ranking P_{II} is built as follows.

The reversed scanning gives the ranking R [15]. If no violations, then $R = P_I$, and Kemeny's median is $P* = R$. In the other case $R \neq P_I$, and backward scanning of the ranking R is repeated until it stops changing. In the case of "long" rankings, it is difficult to specify binary relations in the R ranking after paired permutations with a mixture of relationships like preference (better than) and indistinguishability (the same). Pair relations in R are restored as follows.

First, rows and columns in Q are reordered based on the permutation caused by the ranking R. The corresponding loss matrix is Q_R.

Second, the permutation R determines only the predefined order of the loss matrix Q_R reduction in contrast to the reduction of the loss matrix Q for the ranking P_I. Specifically, the row (and the corresponding column) with a non-minimal sum can be removed. If a few rows with the same sum are removed simultaneously, they are put in last places in the ranking with the same standard rank. The final ranking is Kemeny's median $P* = P_{II}$.

5 Experiments and Discussion

Experiment 1. Following the proposed approach to metrics comparing, the result consists of 25 Kemeny's medians (it is MK17 type as the ranking of 17 noisy variants of a single reference image) built for the coordination of 18 rankings for 18 metrics without the MOS. Medians are shown as diagrams (Fig. 2), where the horizontal axis shows distortion types by numbers from Table 1, the vertical axis shows values of standard ranks of alternatives (noisy images) in each MK17 after distortion of each reference image by all 17 types of noise. Also, 25 rankings by MOS (it is MOS17 type as the ranking by humans of the same 17 noisy variants) are shown as diagrams (Fig. 3) in the same way. It is rather obvious that many of MK17 and MOS17 rankings show both a tendency to be concordant mutually.

As stated above, we try to find the most strongly degraded images in MK17 and MOS17 rankings. Figure 4 shows the allocation of noise types according to the total numbers of appearance of noisy images in the last three (15–17) places represented by dark color and in the preceding three (12–14) places represented by light color.

It is obvious that in MK17 rankings at least three types of noise (15–17) most degrade this set of 25 reference images based on evaluation scores of the given set of quality metrics. And next, at least four types of noise (3, 11, 13, and 14) noticeably degrade this set of images too. In MOS17 rankings, types 3, 13, 15 of noise most

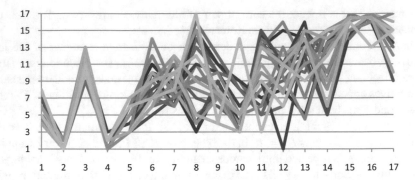

Fig. 2 The 25 Kemeny's medians MK17 as diagrams for 25 reference images. The X-axis is distortion numbers from Table 1, Y-axis is the values of standard ranks of noisy images as alternatives. Diagrams tend to be coordinated and look almost regular

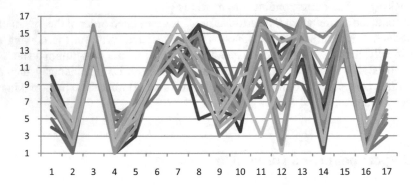

Fig. 3 The 25 MOS17 (human's) rankings as diagrams for 25 reference images. The X-axis is distortion numbers from Table 1, Y-axis is the values of standard ranks of noisy images as alternatives. Diagrams tend to be coordinated and look almost regular

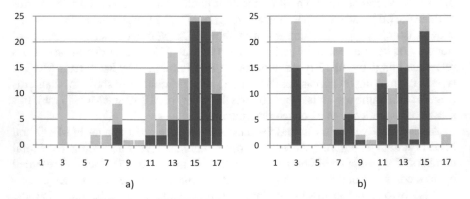

Fig. 4 Allocation of noisy images as alternatives: a) in MK17 rankings, b) in MOS17 rankings. The X-axis is distortion numbers from Table 1, Y-axis is the total number of placements of noisy images in 15–17 places (dark color) and 12–14 places (light color)

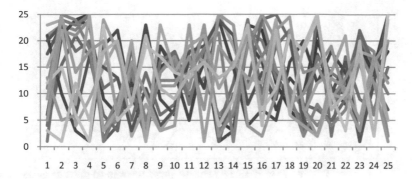

Fig. 5 The 17 Kemeny's medians MK25 as diagrams for 17 types of noise. The X-axis is image numbers from Fig. 1, Y-axis is the values of standard ranks of noisy images as alternatives. Diagrams are poorly coordinated and look chaotic

degrade 25 reference images according to visual human's evaluations. And types 6–8, 11 of noise noticeably degrade these 25 images according to visual human's evaluations also. Hence, at least four types (3, 11, 13, 15) of noise demonstrate concordance between quality metrics scores and human evaluations.

According to Table 1, it is interesting to compare some types of noise, specifically 10 with 11 and 12 with 13 in Fig. 4. It becomes obvious that the more advanced JPEG2000 compression makes more intensive distortions into images.

Experiment 2. The result consists of 17 Kemeny's medians (it is MK25 type as the ranking of 25 reference images degraded by a single type of noise) built for the coordination of 18 rankings for 18 metrics without the MOS. Medians are shown as diagrams (Fig. 5), where the horizontal axis shows image numbers from Fig. 1, the vertical axis shows values of standard ranks of alternatives (degraded images). Each of the 17 diagrams represents degraded images after distortion of 25 reference images by only one noise type. Also, 17 rankings by MOS (it is the MOS25 type as the ranking by humans of the same 25 degraded variants) are shown as diagrams (Fig. 6) in the same way. It is easy to see that concordance between MK25 medians is very weak. The same can be said about rankings by the MOS25 metric.

Figure 7 shows the same situation. In each ranking where reference images have been distorted by one type of noise, some of them are most degraded. But in contrast to the first experiment, it is not possible to specify the type of noise which most degrades a reference image in MK25 and MOS 25. Almost every image appears in three last and three before them places only in a third of all cases in general. Such heterogeneous reference images are ordered ambiguously by quality metric scores as is.

In contrast, the same situation for the MOS metric scores shows that it is difficult for a human to sort heterogeneous images of different types according to the degree of distortion. It is interesting, that some images have never been ranked after 20th place. These are images 9, 10, 21, 24 in the MK25 and the 10th image in the MOS25. Nevertheless, it can't be understood visually what is the basis for it.

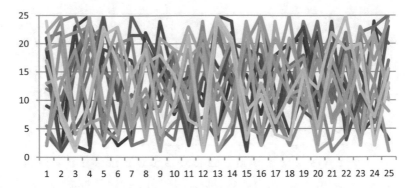

Fig. 6 17 MOS25 (human's) rankings as diagrams for 17 types of noise. The X-axis is image numbers from Fig. 1, Y-axis is the values of standard ranks of noisy images as alternatives. Diagrams are poorly coordinated and look chaotic

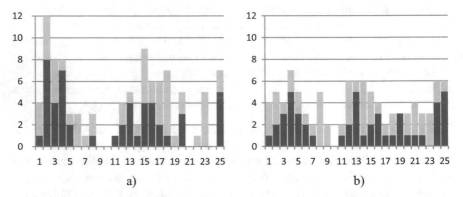

Fig. 7 Allocation of noisy images as alternatives: a) in MK25 rankings, b) in MOS25 rankings. The X-axis is reference image numbers from Fig. 1, Y-axis is the total number of placements of alternatives in 23–25 places (dark color) and 20–22 places (light color)

Statistical testing. The consistency of rankings is evaluated statistically based on testing the null hypothesis about the independence of corresponding rankings.

In the first experiment, the paired consistency of MK17 medians and MOS17 rankings for each of 25 reference images is tested by the Kendall rank correlation coefficient τ for the one-side significance level $\alpha = 0.05$ and the critical threshold $\tau_0 = 0.294$. All pairs of rankings are consistent with each other except for two of them for 20 and 23 reference images. In these cases, observed values $\tau = 0.273$ and $\tau = 0.288$ have not exceeded the critical threshold. For each of 25 reference images the group consistency is tested by the Kendall concordance coefficient W for a) 18 rankings R17 for the significance level $\alpha = 0.05$ and the critical threshold $W_0 = 0.0913$, b) 18 rankings R17 and corresponding medians MK17 for the significance level $\alpha = 0.05$ and the critical threshold $W_0 = 0.0865$, c) 18 rankings R17 and corresponding rankings MOS17 for the same $\alpha = 0.05$, $W_0 = 0.0865$. All groups of rankings are consistent.

In the second experiment, the results of the consistency testing of rankings are contradictory and not entirely unambiguous. The paired consistency of MK25 medians and MOS25 rankings for each of the 17 types of noise is tested by the Kendall rank correlation coefficient τ for the one-side significance level $\alpha = 0.05$ and the critical threshold $\tau_0 = 0.23479$. Only pairs for noises 7–9, 11–13, 16 out of 17 pairs, are consistent. That is, visual perception scores do not well agree with quality metric estimations in general.

For each of 17 types of noise the group consistency is tested by the Kendall concordance coefficient W for a) 25 rankings R25 for the significance level $\alpha = 0.05$ and the critical threshold $W_0 = 0.0843$, b) 25 rankings R25 and corresponding medians MK25 for the significance level $\alpha = 0.05$ and the critical threshold $W_0 = 0.0798$, c) 25 rankings R25 and corresponding rankings MOS25 for the same $\alpha = 0.05$ and $W_0 = 0.0798$. The test showed group consistency for all types of noise except type 16.

6 Conclusion

Experiments show that 18 metrics under investigation are not contradicted each other. Hence, both individual rankings of images by each metrics and group ones by medians become sufficiently coordinated in general. Specifically, if there are no contradictions, the set of metrics points as a whole to the same noise types, that cause most distortions in reference images in both experiments. It can be said the same about the noise types that cause the least distortions in images.

It should be noted, 18 quality metrics demonstrate sufficient coordination with human visual perception (with MOS metric) in the first experiment, but it is not so in the second one. The explanation consists of the following.

It becomes obvious that a human evaluates easier visual distortions based on comparing different modifications of the same reference image. But it also becomes obvious that a human makes more significant mistakes in a visual evaluation of distortions based on comparing noisy images from the different classes.

The result of the first experiment shows that it needs to believe in quality evaluation by metrics under investigation. The result of the second experiment shows that it needs to use quality evaluation by these metrics in a case of visual troubles.

Here, it is showed what types of noise most distort referencing images from the set under investigation. Therefore, in the case of different processing algorithms, the best of them should most effectively eliminate types of noise with the most distortions in images from this set, at least.

It is planned to use new types of Kemeny's median [14–16] and the modern version of the database [7, 8].

It is known, Kemeny's algorithm is the local procedure. Nevertheless, the decisions regarding the best and worst alternatives are usually the same in locally optimal rankings. Differences are often concentrated within rankings. And the more alternatives are included in rankings, the more differences in rankings to be. Such differences become important in a more detailed investigation of the set of alternatives. Kemeny's metric median [14–16] realizes the idea of immersing a set of rankings in a metric space. The single arithmetic mean is calculated for all rankings as points in it. Kemeny's metric median as a point in the space coincides with the arithmetic mean of all rankings. This property can reduce significantly the set of possible decisions. Such the ranking clarifies mutual positions of alternatives and reduces the number of indistinguishable pairs.

Acknowledgements This research is supported by the Russian Foundation for Basic research (RFBR) Grants 20-07-00055, 20-07-00441, 19-07-01178, by the Ministry of Science and Higher Education of RF within the framework of the state task FEWG-2021-0012.

References

1. Chandler DM (2013) Seven challenges in image quality assessment: past, present, and future research. ISRN Signal Processing. Article ID 905685. https://doi.org/10.1155/2013/905685
2. Mittal A, Moorthy AK, Bovik AC (2012) No-reference image quality assessment in the spatial domain. IEEE Trans Image Proc 21(12):4695–4708
3. Soundararajan R, Bovik AC (2012) RRED indices: reduced reference entropic differencing for image quality assessment. IEEE Trans Image Proc 21(2):517–526. https://doi.org/10.1109/TIP.2011.2166082
4. Wang Z, Bovik A, Sheikh H., E. Simoncelli, E (2004) Image quality assessment: from error visibility to structural similarity. IEEE Trans Image Proc 13(4):600–612
5. Sheikh H, Bovik A (2006) Image information and visual quality. IEEE Trans Image Proc 15(2):430–444
6. Chandler D, Hemami S (2007) VSNR: a wavelet-based visual signal-to-noise ratio for natural images. IEEE Trans Image Proc 16(9):2284–2298
7. Ponomarenko N, Lukin V, Zelensky A, Egiazarian K, Carli M, Battisti F (2009) TID2008—a database for evaluation of full-reference visual quality assessment metrics. Adv Mod Radioelectron 10:30–45
8. TID2008 Archive. http://www.ponomarenko.info/tid/tid2008.rar
9. Kodak Lossless True Color Image Suite. http://r0k.us/graphics/kodak/
10. Duin RPW (2021) The origin of patterns. Front Comput Sci 3:747195, 1–16. https://doi.org/10.3389/fcomp.2021.747195
11. Kemeny J, Snell J (1963) Mathematical models in the social sciences. MIT Press, Blaisdell, NY
12. Litvak D (1982) Expert information: methods of acquisition and analysis. Radio I Svyaz, Moscow (in Russian)
13. Bury H, Wagner D (2003) Application of Kemeny's median for group decision support. In: Yu X, Kacprzyk J (eds) Applied decision support with soft computing. Studies in Fuzziness and Soft Computing, vol 124. Springer, Berlin, Heidelberg, pp 235–262. https://doi.org/10.1007/978-3-540-37008-6_10
14. Dvoenko SD, Pshenichny DO (2021) Rank aggregation based on new types of the Kemeny's median. Pattern Recognit Image Anal 31(2):185–196. https://doi.org/10.1134/S105466182102061
15. Dvoenko SD (2021) A developing of the Kemeny median: new types and algorithms. Pattern Recognit Image Anal 31(3):376–380. https://doi.org/10.1134/S105466182103007X

16. Dvoenko S, Pshenichny D (2021) On new Kemeny's medians. In: Del Bimbo A et al (eds) ICPR 2021 workshops, LNCS, vol 12665. Springer, Cham, pp 97–102. https://doi.org/10.1007/978-3-030-68821-9_9

Exponentially Weighted Mean Filter for Salt-and-Pepper Noise Removal

Serdar Enginoğlu⬤, Uğur Erkan⬤, and Samet Memiş⬤

Abstract This paper defines an exponentially weighted mean using an exponentially decreasing sequence of simple fractions based on distance. It then proposes a cutting-edge salt-and-pepper noise (SPN) removal filter—i.e., Exponentially Weighted Mean Filter (EWmF). The proposed method incorporates a pre-processing step that detects noisy pixels and calculates threshold values based on the possible noise density. Moreover, to denoise the images operationalizing the calculated threshold values, EWmF employs the exponentially weighted mean (ewmean) in 1-approximate Von Neumann neighbourhoods for low noise densities and k-approximate Moore neighbourhoods for middle or high noise densities. Furthermore, it ultimately removes the residual SPN in the processed images by relying on their SPN densities. The numerical and visual results obtained with MATLAB R2021a manifest that EWmF outperforms nine state-of-the-art SPN filters.

Keywords Image denoising · Noise removal · Nonlinear filter · Salt-and-pepper noise · Exponentially weighted mean

S. Enginoğlu (✉)
Department of Mathematics, Faculty of Arts and Sciences, Çanakkale Onsekiz Mart University, Çanakkale, Turkey
e-mail: serdarenginoglu@gmail.com

U. Erkan
Department of Computer Engineering, Faculty of Engineering, Karamanoğlu Mehmetbey University, Karaman, Turkey
e-mail: ugurerkan@kmu.edu.tr

S. Memiş
Department of Computer Engineering, Faculty of Engineering and Natural Sciences, İstanbul Rumeli University, İstanbul, Turkey
e-mail: samettmemis@gmail.com

© The Author(s), under exclusive license to Springer Nature Switzerland AG 2022   435
N. H. T. Dang et al. (eds.), *Artificial Intelligence in Data and Big Data Processing*,
Lecture Notes on Data Engineering and Communications Technologies 124,
https://doi.org/10.1007/978-3-030-97610-1_34

1 Introduction

During the acquisition or transmission of a digital image, some distortion occurs in the image, which is called noise. These noises are divided into several types, including "Impulse Noise", "Gaussian Noise", and "Speckle Noise" [1]. This study considers the salt-and-pepper noise (SPN), a kind of impulse noise, in which some pixel values are distorted with probability p and change to 0 or 255. That is, p is the probability that a randomly selected pixel turns into 0 or 255 in a salt-and-pepper noisy image. These pixels, whose real value is degraded, are referred to as noisy pixels. Noise-free pixels are called regular pixels. Thus, regular pixels' probability of occurrence in an image with p-density SPN is $1 - p$.

The act of replacing the value of a potentially noisy pixel with a viable value is called noise removal. This value is commonly obtained by exploiting the values of the pixels near the pixel in question, namely neighbour pixels' values. To do so, a square sub-matrix (window) of order $(2k + 1) \times (2k + 1)$ is used, where k is a positive integer and contains the considered noisy pixel in its centre. At low noise density, regular pixels in a 3×3 window offer sufficient workable information about the central pixel's possible value. However, when the noise density increases, it is necessary to enlarge the window's size until it acquires an adequate number of regular pixels. Here, the inclusion of the pixels distantly located from the central pixel in the calculation increases the margin of error depending on the size of the window. Therefore, it is essential to determine a usable window size for a given noisy pixel by considering the noise density.

Studies on noise removal are crucial since it is a pre-treatment needed to promote image processing success. Among the current noise removal methods are the filters processing each pixel in a given image regardless of whether it is noisy or not (blind filters) [2, 3] or processing only possibly noisy pixels, the filters using fixed or adaptive windowing, and linear, non-linear, single-stage, and multi-stage filters. For example, k-Approximate Frequency Median Filter [3], a blind filter, uses the fixed window size $(2k + 1) \times (2k + 1)$ while Adaptive Frequency Median Filter [4] considers solely the possibly noisy pixels and employs an adaptive window size. This diversity of the parameters affects the performance and running time of noise removal filters. For instance, a filter working on possibly noisy pixels can run faster than a filter working on all the pixels in question [5–7].

Similarly, a single-stage filter can work faster than a multi-stage filter [8–10]. However, in this case, the noise removal performance of the used method is adversely affected. Moreover, iterative, recursive, or multi-stage filters tend to exhibit increased running time [11].

Adaptive Weighted Mean Filter (AWMF) [12] is an improved version of Adaptive Median Filter (AMF) [5], which utilizes adaptive windowing and the mean value of the regular pixels in the window. AWMF successfully denoises high-density SPN but poses some drawbacks, such as high-processing time and failure to denoise low-density SPN. Different Adaptive Median Filter (DAMF) [13] is a more effective tool intended for low- and middle-density SPN than AWMF but not as effective as AWMF for high-density SPN. To deal with these two drawbacks, Adaptive Riesz Mean Filter (ARmF) [14], which uses a weighted mean based on a sequence of proper fractions whose denominators are increasing polynomial values, has been proposed. ARmF outperforms AWMF and DAMF at any SPN density and runs faster than the two.

Besides, Iterative Mean Filter (IMF) [15] employs a 3×3 window with an iterative procedure and performs better than AWMF and DAMF. However, it suffers from a prolonged processing time. Moreover, Efficient Weighted-Average Filter (EWAF) [16] is designed for high-denoising performance and real-time denoising. It avails of 3×3 interpolation and distance transformation. The other two efficient weighted filters are Three-Values-Weighted Approach Filter (TVWAF) [17] and Unbiased Weighted Mean Filter (UWMF) [18]. In contrast to the aforesaid filters, using deep learning tools—i.e., min, max, and average pooling—Min–Max Average Pooling Filter (MMAPF) [19] is developed. MMAPF consists of three procedures and several complex steps.

This paper proposes Exponentially Weighted Mean Filter (EWmF). The proposed filter utilizes an exponentially decreasing sequence of simple fractions to weight the values of the neighbouring pixels of a possibly noisy pixel. By doing so, EWmF achieves the highest denoising performance compared to nine state-of-the-art SPN filters. The major contributions of the paper are highlighted as follows:

- A new type of weighted mean, namely exponentially weighted mean, is defined, which requires no complex procedures.
- Using the exponentially weighted mean, EWmF, which outperforms nine state-of-the-art SPN filters, is proposed.
- EWmF minimizes the errors caused by the operationalization of distantly neighbouring pixels.

The second section of the current study provides some basic concepts. It then presents the proposed salt-and-pepper noise removal filter, namely EWmF, explaining the authors' motivation and the filter's working principles. The third compares EWmF with the other considered methods according to their running time and noise removal performances at low, medium, and high densities. The concluding section briefly discusses the relevance of EWmF for SPN removal and the need for further research.

2 Preliminaries and the Proposed SPN Filter

This section firstly presents some of the basic notions. Throughout this paper, let $A := \left[a_{ij}\right]_{m \times n}$ be an image matrix (IM) such that a_{ij} is an unsigned integer number and $0 \leq a_{ij} \leq 255$. If $a_{ij} = 0$ or $a_{ij} = 255$, then a_{ij} is called a possibly noisy pixel of A; otherwise, a_{ij} is called a regular pixel of A. If there exists at least one possibly noisy entry in A, then A is called a possibly noisy image matrix or briefly a noisy image matrix (NIM).

Definition 2.1 Let A be an NIM. Then, the matrix $\ddot{A} := [\ddot{a}_{ij}]_{m \times n}$ is called binary matrix (noise mask) of A where

$$\ddot{a}_{ij} := \begin{cases} 0, & a_{ij} \text{ is a possibly noisy entry of } A \\ 1, & \text{otherwise} \end{cases}$$

Definition 2.2 Let $A = [a_{ij}]_{m \times n}$ be an image matrix (IM) and $t \in \{1, 2, \ldots, \min\{m, n\}\}$. Then, the matrix $\overline{\overline{A}}_{tsim} := [\overline{\overline{a}}_{rs}]_{(m+2t) \times (n+2t)}$ called t-symmetric pad matrix of A is defined as follows:

$$\begin{bmatrix}
a_{tt} & \cdots & a_{t1} & a_{t1} & a_{t2} & \cdots & a_{tn} & a_{tn} & \cdots & a_{t(n-t+1)} \\
\vdots & \ddots & \vdots & \vdots & \vdots & \ddots & \vdots & \vdots & \ddots & \vdots \\
a_{1t} & \cdots & a_{11} & a_{11} & a_{12} & \cdots & a_{1n} & a_{1n} & \cdots & a_{1(n-t+1)} \\
a_{1t} & \cdots & a_{11} & a_{11} & a_{12} & \cdots & a_{1n} & a_{1n} & \cdots & a_{1(n-t+1)} \\
a_{2t} & \cdots & a_{21} & a_{21} & a_{22} & \cdots & a_{2n} & a_{2n} & \cdots & a_{2(n-t+1)} \\
a_{3t} & \cdots & a_{31} & a_{31} & a_{32} & \cdots & a_{3n} & a_{3n} & \cdots & a_{3(n-t+1)} \\
\vdots & \ddots & \vdots & \vdots & \vdots & \ddots & \vdots & \vdots & \ddots & \vdots \\
a_{mt} & \cdots & a_{m1} & a_{m1} & a_{m2} & \cdots & a_{mn} & a_{mn} & \cdots & a_{m(n-t+1)} \\
a_{mt} & \cdots & a_{m1} & a_{m1} & a_{m2} & \cdots & a_{mn} & a_{mn} & \cdots & a_{m(n-t+1)} \\
\vdots & \ddots & \vdots & \vdots & \vdots & \ddots & \vdots & \vdots & \ddots & \vdots \\
a_{(m-t+1)t} & \cdots & a_{(m-t+1)1} & a_{(m-t+1)1} & a_{(m-t+1)2} & \cdots & a_{(m-t+1)n} & a_{(m-t+1)n} & \cdots & a_{(m-t+1)(n-t+1)}
\end{bmatrix}$$

Example 2.1 Let $A = \begin{bmatrix} 11 & 255 & 13 \\ 0 & 22 & 23 \\ 31 & 32 & 0 \end{bmatrix}$. Then,

$$\overline{\overline{A}}_{2sim} = \begin{bmatrix}
22 & 0 & 0 & 22 & 23 & 23 & 22 \\
255 & 11 & 11 & 255 & 13 & 13 & 255 \\
255 & 11 & \mathbf{11} & \mathbf{255} & \mathbf{13} & 13 & 255 \\
22 & 0 & \mathbf{0} & \mathbf{22} & \mathbf{23} & 23 & 22 \\
32 & 31 & \mathbf{31} & \mathbf{32} & \mathbf{0} & 0 & 32 \\
32 & 31 & 31 & 32 & 0 & 0 & 32 \\
22 & 0 & 0 & 22 & 23 & 23 & 22
\end{bmatrix}_{7 \times 7}$$

Definition 2.3 Let $A = [a_{ij}]_{m \times n}$ and $\overline{\overline{A}} = [\overline{\overline{a}}_{rs}]_{(m+2t) \times (n+2t)}$ be a t-symmetric pad matrix of A. Then, the matrix

$$\begin{bmatrix} \overline{\overline{a}}_{(i+t-k)(j+t-k)} & \cdots & \overline{\overline{a}}_{(i+t-k)(j+t+k)} \\ \vdots & \overline{\overline{a}}_{(i+t)(j+t)} & \vdots \\ \overline{\overline{a}}_{(i+t+k)(j+t-k)} & \cdots & \overline{\overline{a}}_{(i+t+k)(j+t+k)} \end{bmatrix}_{(2k+1)\times(2k+1)}$$

is called k-approximate matrix of a_{ij} and is denoted by A_{ij}^k. Here, $k \in \{1, 2, \ldots, t\}$.

Example 2.2 Let us consider Example 2.1. Then,

$$A_{21}^1 = \begin{bmatrix} 11 & 11 & 255 \\ 0 & 0 & 22 \\ 31 & 31 & 32 \end{bmatrix}$$

Definition 2.4 Let $A = \left[a_{ij}\right]_{m \times n}$ be an IM, the sum of A is defined as follows:

$$\sum A := \sum_{i=1}^{m} \sum_{j=1}^{n} a_{ij}$$

Definition 2.5 Let $A = \left[a_{ij}\right]_{m \times n}$ be an IM. Then, possibly noise density of A, denoted by $\mathrm{nd}(A)$, is defined as follows:

$$\mathrm{nd}(A) := \mathrm{round}\left(100 - \frac{100 \sum \ddot{A}}{mn}\right)$$

Here, $\ddot{A}$ is the binary matrix of A and round($\cdot$) is the rounding function to the nearest integer. (For $\sum \ddot{A}$, see Definition 2.4).

Secondly, this section discusses nine noise removal filters and offers a new SPN filter referred to as EWmF. Deriving the new value of a possibly noisy pixel from its neighbour pixels' values is the first procedure that comes to mind. Utilizing the median or mean of the neighbour pixels' values is applicable at low noise density but not at medium and high noise densities. This happens due to the fact that the neighbour pixels' values occur farther from the centre pixel's possibly correct value when the number of possibly noisy pixels increases. As in AWMF, weighting the values of possibly noisy neighbour pixels by 0 and the others by 1 increases a filter's noise removal success. Moreover, ARmF's relying on a weighting process using a sequence of proper fractions whose denominators are increasing polynomial values (for example, $n \geq 2$, $\frac{1}{n^2}$ or $\frac{1}{(\varepsilon+n)^4}$) significantly increases the success.

Optimizing the rate of convergence to zero of such sequences, used to weight pixels, ensures that the pixels' values farther from the centre less affect the centre pixel's new value. This study conducts a weighting using an exponentially decreasing sequence of simple fractions ($r, n \geq 2$, $\frac{1}{r^n}$) based on distance. In this way, we further minimize the errors caused by neighbour pixels with values far from the possible value of the central pixel in colour transitions. The proposed method takes into account the noise density of an image as in EWAF. This produces a threshold value that determines the rate of convergence to zero of the sequence based on noise density. By determining the image's noise density, specific steps can be used for different noise densities. For example, considering only the values of the neighbour pixels closest to the central pixel called four neighbourhoods enhances the noise removal performance at low noise density. Besides, EWmF incorporates a final stage, as in DAMF, that handles potentially unprocessed pixels at medium and high noise density, which increases success at high noise density and improves visual output.

Definition 2.6 Let $I_n = \{1, 2, \ldots, n\}$. Then, th : $I_{99} \rightarrow I_5$ is called threshold function such that

$$\text{th}(x) := \text{round}\left(\frac{20}{(50x - 1)^{1/3}}\right)$$

Here, round($\cdot$) is the rounding function to the nearest integer.

Definition 2.7 Let $A = [a_{ij}]$ be an IM. Then, the matrix $W^k := [w^k_{st}]_{(2k+1)\times(2k+1)}$ is called exponentially weight matrix (ewm) such that

$$w^k_{st} := \left(\frac{1}{\text{th}(\text{nd}(A)) + 1}\right)^{|k+1-s|+|k+1-t|}$$

Here, th is the threshold function (see Definition 2.6). Particularly, ewm for Von Neumann neighbourhood (4-neighbourhood) is as follows:

$$V^1 := V \odot W^1$$

where $\odot$ is the Hadamard product and $V = \begin{bmatrix} 0 & 1 & 0 \\ 1 & 1 & 1 \\ 0 & 1 & 0 \end{bmatrix}$.

Definition 2.8 Let $A = [a_{ij}]$ be an IM, A^k_{ij} be a k-approximate matrix of a_{ij}, $\ddot{A}^k_{ij}$ be the binary matrix of A^k_{ij}, and W^k (or V^1) be an ewm. Then, the exponentially weighted mean (ewmean) of A^k_{ij} is as follows:

$$\text{ewmean}\left(A^k_{ij}, \ddot{A}^k_{ij}, W^k\right) := \frac{\sum\left(A^k_{ij} \odot W^k \odot \ddot{A}^k_{ij}\right)}{\sum\left(W^k \odot \ddot{A}^k_{ij}\right)}$$

Here, $\odot$ is the Hadamard product.

The proposed method, unlike EWAF and IMF, employs adaptive windowing. It also uses a number of regular pixels equal to or greater than the threshold value when calculating the new value of the possibly noisy pixel. Thus, the proposed EWmF prevents the increase in error margin and eliminates likely not-a-number (NaN) cases. Finally, we present the algorithm of EWmF.

3 Simulation Results

In this section, we compare the proposed EWmF with AFMF [4], AWMF [12], DAMF [13], ARmF [14], IMF [15], EWAF [16], TVWAF [17], UWMF [18], and MMAPF [19] in terms of SSIM [20] values. We then carry out several simulations to compare EWmF with AFMF, AWMF, DAMF, IMF, ARmF, EWAF, TVWAF, UWMF, and MMAPF using 40 test images in TESTIMAGES Database [21] with 600×600 and 200 test images in UC Berkeley Database [22] with 321×481. We perform the simulations by utilizing MATLAB R2021a and a laptop with I(R) Core (TM) CPU i5-4200H@2.8 GHz and 8 GB RAM.

3.1 Image Quality Metric

Let $O := [o_{ij}]$ and $R := [r_{ij}]$ be the original image and restored image, respectively. SSIM [20] is defined by $\mathrm{SSIM}(O, R) := \frac{(2\mu_O\mu_R+C_1)+(2\sigma_{OR}+C_2)}{(\mu_O^2+\mu_R^2+C_1)+(\sigma_O^2+\sigma_R^2+C_2)}$ where $\mu_O, \mu_R, \sigma_O,$ σ_R, and σ_{OR} are the average intensities, standard deviations, and cross-covariance of images O and R, respectively. Additionally, $C_1 := (K_1 L)^2$ and $C_2 := (K_2 L)^2$ are two constants such that $K_1 = 0.01$, $K_2 = 0.03$, and $L = 255$, for 8-bit grayscale images.

Algorithm 1. EWmF(A)

Pre-Stage	**Input:** IM $A := [a_{ij}]_{m \times n}$ such that $\min\{m, n\} \geq 4$
	Output: Restored IM $R := [r_{ij}]_{m \times n}$ of A
	Convert A to double form
	Set $a_{ij} \leftarrow 0$ if $a_{ij} = 255$
	Set $t \leftarrow \text{th}(\text{nd}(A))$ (see Definition 2.5 and 2.6)

Main Denoising Stage

Obtain $W^1, W^2, W^3, W^4, V,$ and V^1 (see Definition 2.7)
Obtain $\bar{\bar{A}}_{4sim}$ (see Definition 2.2)
for all i and j,
    if $\bar{\bar{a}}_{(i+4)(j+4)} = 0$, **then**
        for k from 1 to 4 **do**
            Obtain A_{ij}^k for a_{ij} (see Definition 2.3)
            Compute $\ddot{A}_{ij}^k$ (see Definition 2.1)
            if $k = 1$ and $\sum(V \odot \ddot{A}_{ij}^1) \geq 3$, **then**
                $r_{ij} \leftarrow \text{ewmean}(A_{ij}^1, \ddot{A}_{ij}^1, V^1)$
                Break
            elseif $\sum \ddot{A}_{ij}^k \geq t$, **then**
                $r_{ij} \leftarrow \text{ewmean}(A_{ij}^k, \ddot{A}_{ij}^k, W^k)$
                Break
            end
        end
    else
        $r_{ij} \leftarrow \bar{\bar{a}}_{(i+4)(j+4)}$
    end
end

Final Denoising Stage for Residual SPN

Obtain $\bar{\bar{R}}_{2sim}$ (see Definition 2.2)
  for all i and j, **then**
    if $\text{nd}(A) < 80$ and $\bar{\bar{a}}_{(i+4)(j+4)} = 0$, **then**
        Obtain R_{ij}^1 for r_{ij} (see Definition 2.3)
        Compute $\ddot{R}_{ij}^1$ (see Definition 2.1)
        $r_{ij} \leftarrow \text{ewmean}(R_{ij}^1, \ddot{R}_{ij}^1, V^1)$
    elseif $\text{nd}(A) < 90$ and $\bar{\bar{a}}_{(i+4)(j+4)} = 0$, **then**
        Obtain R_{ij}^1 for r_{ij}
        Compute $\ddot{R}_{ij}^1$
        $r_{ij} \leftarrow \text{ewmean}(R_{ij}^1, \ddot{R}_{ij}^1, W^1)$
    elseif $\bar{\bar{a}}_{(i+4)(j+4)} = 0$, **then**
        Obtain R_{ij}^2 for r_{ij}
        Compute $\ddot{R}_{ij}^2$
        $r_{ij} \leftarrow \text{ewmean}(R_{ij}^2, \ddot{R}_{ij}^2, W^2)$
    end
  end
Convert R to uint8 form

3.2 Experimental Results on 40 Test Images in TESTIMAGES Database

Table 1 presents the mean SSIM values for the 40 test images from TESTIM-AGES Database with SPN densities ranging from 10 to 90%. The proposed filter

Table 1 Mean SSIM values of the denoising results of 40 images with various SPN densities

SPN filter	10%	20%	30%	40%	50%	60%	70%	80%	90%	Mean
AFMF	0.9627	0.9560	0.9462	0.9326	0.9126	0.8831	0.8367	0.7595	0.6147	0.8671
AWMF	0.9808	0.9748	0.9669	0.9567	0.9428	0.9233	0.8950	0.8509	0.7669	0.9176
DAMF	0.9910	0.9814	0.9696	0.9553	0.9377	0.9159	0.8869	0.8427	0.7566	0.9152
IMF	0.9914	0.9826	0.9722	0.9595	0.9436	0.9225	0.8931	0.8478	0.7673	0.9200
ARmF	0.9926	0.9852	0.9763	0.9652	0.9505	0.9300	0.9007	0.8555	0.7699	0.9251
EWAF	0.9924	0.9844	0.9751	0.9643	0.9512	0.9341	0.9099	0.8723	0.8000	0.9315
TVWA	0.9892	0.9806	0.9706	0.9585	0.9430	0.9221	0.8928	0.8460	0.6129	0.9017
UWMF	0.9800	0.9821	0.9723	0.9594	0.9448	0.9236	0.8964	0.8545	0.7798	0.9214
MMAPF	0.9914	0.9828	0.9724	0.9601	0.9468	0.9300	0.9074	0.8729	0.8036	0.9297
EWmF	**0.9928**	**0.9863**	**0.9783**	**0.9683**	**0.9558**	**0.9393**	**0.9158**	**0.8790**	**0.8075**	**0.9359**

The best performances are shown in bold

outperforms nine state-of-the-art SPN filters at any SPN density the aforesaid SSIM values. Moreover, the mean SSIM values for all the SPN densities reveal that EWmF is the best performing SPN filter.

3.3 Experimental Results on 200 Test Images from the UC Berkeley Database

Table 2 provides the mean SSIM values for the 200 test images from the UC Berkeley Database with SPN densities ranging from 10 to 90%. At all the SPN densities, EWmF performs better than the others by exhibiting an improvement in the SSIM value, corresponding to around 0.02. Besides, their mean SSIM results show that the proposed filter is more efficacious than the others.

3.4 Visual Outputs for an Image from UC Berkeley Database

Figure 1 shows the denoising results of the filters for the test image with ID: 198054 from the UC Berkeley Database. The test image is noised with 70% SPN in Fig. 1b. It can be observed that Fig. 1c has low quality. In Fig. 1d, e, the edge details are degraded. Figure 1f has white points in the zoomed-in hair details of the face. Figure 1g–i blur the details around the eyes. Figure 1j corrupts the hair details of the face and produces black speckles. Figure 1k blurs the face, especially the eye details. Figure 1l presents the denoising results of the proposed filter. The edges and

Table 2 Mean SSIM values of the denoising results of 200 images with various SPN densities

SPN filter	10%	20%	30%	40%	50%	60%	70%	80%	90%	Mean
AFMF	0.9215	0.9089	0.8958	0.8770	0.8506	0.8142	0.7625	0.6854	0.5576	0.8082
AWMF	0.9657	0.9525	0.9365	0.9171	0.8925	0.8605	0.8178	0.7580	0.6608	0.8624
DAMF	0.9788	0.9615	0.9410	0.9169	0.8884	0.8540	0.8110	0.7519	0.6537	0.8619
IMF	0.9790	0.9623	0.9431	0.9204	0.8929	0.8591	0.8155	0.7565	0.6680	0.8663
ARmF	0.9812	0.9670	0.9505	0.9305	0.9052	0.8721	0.8277	0.7653	0.6649	0.8738
EWAF	0.9795	0.9637	0.9458	0.9254	0.9009	0.8707	0.8312	0.7763	0.6876	0.8757
TVWA	0.9774	0.9609	0.9423	0.9206	0.8942	0.8609	0.8174	0.7560	0.5430	0.8525
UWMF	0.9759	0.9686	0.9520	0.9308	0.9047	0.8715	0.8295	0.7710	0.6800	0.8760
MMAPF	0.9792	0.9626	0.9435	0.9210	0.8964	0.8660	0.8276	0.7751	0.6887	0.8733
EWmF	**0.9829**	**0.9702**	**0.9546**	**0.9354**	**0.9107**	**0.8804**	**0.8404**	**0.7822**	**0.6906**	**0.8831**

The best performances are shown in bold

details of the test image are preserved with less blurring. These results can be easily observed in the hair details on the face. The SSIM result of the EWmF is 0.8095, while those of the others are below 0.8000.

4   Conclusion

This paper proposes EWmF for SPN removal. It consists of pre-processing, main, and final denoising phase. In the pre-processing step, all pixels in the given image are masked as noisy or regular, and a threshold value relying on the SPN density for the image is determined using the proposed threshold function. In the main step, the proposed weighted mean (ewmean), having based on an ewm and the threshold value, removes the SPN from the image. In the final step, the SPN density-based procedure of ewmean removes all the residual SPN from the image. The EWmF's greatest advantage is its operationalizing ewmean, threshold values, and different threshold-based weighting matrices so that it outperforms the nine state-of-the-arts SPN filters.

In future studies, ewmean can be combined with interpolation and distance transformation, focusing on real-time SPN denoising. On the other hand, each weighted mean has an advantage on different images and SPN densities. A fusion approach based on ewmean and the other weighted means can be proposed to combine them.

Fig. 1 Denoising results of the test image with ID: 198054 from UC Berkeley Database: (**a**) original image, (**b**) noisy image (0.0447), (**c**) AFMF (0.7281), (**d**) AWMF (0.7857), (**e**) DAMF (0.7811), (**f**) IMF (0.7704), (**g**) ARmF (0.7965), (**h**) EWAF (0.7991), (**i**) TVWA (0.7881), (**j**) UWMF (0.7949), (**k**) MMAPF (0.7899), and (**l**) EWmF (0.8095)

References

1. Russ JC, Russ JC (2008) Introduction to image processing and analysis. CRC Press Taylor & Francis Group, New York, USA
2. Tukey JW (1977) Exploratory data analysis, Reading. Addison-Wesley, MA
3. Erkan U, Gökrem L, Enginoğlu S (2019) k-Approximate frequency median filter in salt-and-pepper noise. In: 2nd International conference on science and technology; Engineering science and technology. Association of Kutbilge Academicians, Prizren, pp 395–405
4. Erkan U, Enginoğlu S, Thanh DNH, Hieu LM (2020) Adaptive frequency median filter for the salt and pepper denoising problem. IET Image Process 14:1291–1302

5. Hwang H, Haddad RA (1995) Adaptive median filters: new algorithms and results. IEEE Trans Image Process 4:499–502
6. Erkan U, Gökrem L, Enginoğlu S (2019) Adaptive right median filter for salt-and-pepper noise removal. Int J Eng Res Dev 11:542–550
7. Enginoğlu S, Erkan U, Memiş S (2020) Adaptive cesáro mean filter for salt-and-pepper noise removal. El-Cezerî J Sci Eng 7:304–314
8. Erkan U, Kılıçman A (2016) Two new methods for removing salt-and-pepper noise from digital images. ScienceAsia 42:28–32
9. Erkan U, Gökrem L (2018) A new method based on pixel density in salt and pepper noise removal. Turkish J Electr Eng Comput Sci 26:162–171
10. Erkan U, Thanh DNH, Enginoğlu S, Memiş S (2020) Improved adaptive weighted mean filter for salt-and-pepper noise removal. In: 2nd International conference on electrical, communication and computer engineering. IEEE Press, Istanbul, pp 1–5
11. Erkan U, Enginoğlu S, Thanh DNH (2019) A recursive mean filter for image denoising. In: International artificial intelligence and data processing symposium. IEEE Press, Malatya, pp 1–5
12. Zhang P, Li F (2014) A new adaptive weighted mean filter for removing salt-and-pepper noise. IEEE Signal Process Lett 21:1280–1283
13. Erkan U, Gökrem L, Enginoğlu S (2018) Different applied median filter in salt and pepper noise. Comput Electr Eng 70:789–798
14. Enginoğlu S, Erkan U, Memiş S (2019) Pixel similarity-based adaptive Riesz mean filter for salt-and-pepper noise removal. Multimedia Tools Appl
15. Erkan U, Thanh DNH, Hieu LM, Enginoğlu S (2019) An iterative mean filter for image denoising. IEEE Access 7:167847–167859
16. Hosseini H, Hessar F, Marvasti F (2015) Real-time impulse noise suppression from images using an efficient weighted-average filtering. IEEE Signal Process Lett 22:1050–1054
17. Lu CT, Chen YY, Wang LL, Chang CF (2016) Removal of salt-and-pepper noise in corrupted image using three-values-weighted approach with variable-size window. Pattern Recognit Lett 80:188–199
18. Kandemir C, Kalyoncu C, Toygar Ö (2015) A weighted mean filter with spatial-bias elimination for impulse noise removal. Digit Sig Process 46:164–174
19. Satti P, Sharma N, Garg B (2020) Min-max average pooling based filter for impulse noise removal. IEEE Signal Process Lett 27:1475–1479
20. Wang Z, Bovik AC, Sheikh HR, Simoncelli EP (2004) Image quality assessment: from error visibility to structural similarity. IEEE Trans Image Process 13:600–612
21. Asuni N, Giachetti A (2014) TESTIMAGES: a large-scale archive for testing visual devices and basic image processing algorithms. In: Smart tools and apps for graphics. The Eurographics Association, Cagliari, pp 63–70
22. Martin D, Fowlkes C, Tal D, Malik J (2001) A database of human segmented natural images and its application to evaluating segmentation algorithms and measuring ecological statistics. In: Proceedings of the IEEE international conference on computer vision, vol 2. IEEE Press, Vancouver, , pp 416–423

An Enhanced Algorithm of Bees Colony for Finding Best Threshold in Medical Images

Trong-The Nguyen, Trinh-Dong Nguyen, Thi-Kien Dao, and Vinh-Tiep Nguyen

Abstract This study investigates a solution to the status of the threshold medical images segmentation using an enhanced algorithm of bees colony (EABC). In medical image processing, effectively segmenting images into recognized objects is critical for classification and object recognition. On a given image, the fractional image threshold segmentation approach is modeled into N-level thresholds and then optimized using the EABC. For the fractional image order error threshold, the proposed scheme results are compared to those of other algorithms in the literature. In comparison to existing methods, the proposed scheme is more efficient in a variety of situations.

Keywords Image segmentation · Multi-level thresholds · Enhanced algorithm of bees colony · Fitness function · Variable coefficient

1 Introduction

The analysis of medical images based on simulation and measurement software combined with the image segmentation principle has become an integral part of the current pharmaceutical industry [1]. Healthy vision is the essential basis for the diagnosis of disease by physicians. Increasing the graphic quality of the medical picture helps increase the diagnosis rate for patients, improve the standard of current

T.-T. Nguyen · T.-K. Dao
Fujian Provincial Key Lab of Big Data Mining and Applications, Fujian University of Technology
Fujian, Fuzhou, China
e-mail: thent@uit.edu.vn

T.-T. Nguyen · T.-D. Nguyen (✉) · V.-T. Nguyen
University of Information Technology, VNU-HCM, Ho Chi Minh, Vietnam
e-mail: dongnt@uit.edu.vn

V.-T. Nguyen
e-mail: tiepnv@uit.edu.vn

Vietnam National University, Ho Chi Minh City, Vietnam

© The Author(s), under exclusive license to Springer Nature Switzerland AG 2022
N. H. T. Dang et al. (eds.), *Artificial Intelligence in Data and Big Data Processing*,
Lecture Notes on Data Engineering and Communications Technologies 124,
https://doi.org/10.1007/978-3-030-97610-1_35

medical care, and encourage more professional advancement [2]. Segmentation of the image is to define the picture as a series of related regions such that the image features are different in the different areas and display similarities in the same area. The widely used image segmentation methods currently include the threshold method [3], edge detection method [4, 5], zone method [6], morphological watershed method, etc. [7].

The multi-threshold image segmentation is a typical optimization problem that combines the maximum between-class variance method and some optimization algorithms to determine the appropriate threshold [2]. The metaheuristic algorithm is one of the efficient ways to determine the proper image thresholds [8]. It has the advantages included in the straightforward structure and low computational complexity and is suitable for solving the problem of multi-threshold image segmentation [9].

A new metaheuristic algorithm, the algorithm of bees colony (ABC) [10], is a generation of swarm intelligence optimization algorithm, which imitates bees' foraging behavior according to the division of labor; they carry out different activities share information in the process of operations. The ABC algorithm has some advantages in searching for the optimal solution in the solution space, e.g., fast coverage, robustness, and is easy to implement. The testing target of the traditional image segmentation method often is limited in two-dimensional, with no significant implementation spectrum [11]. The seven image features derive from the color space and use the process of seed area development in conjunction with heuristic optimization methods to produce the effects of image segmentation. In terms of time efficiency and accuracy error, the output of the proposed scheme is reasonably balanced. Nonetheless, the number of calculations increases due to the unnecessary extraction of the function, which makes the running time is longer [12].

This paper proposes a new enhanced algorithm of bees colony (EABC) algorithm for the multilevel segmentation technique used in medical images. The search formula of the improved algorithm makes it approach the optimal global value faster in the initialization population and enhances its ability to escape from the local optimum in the later stage of the algorithm. It is not only to avoid the ABC algorithm drawback of being trapped the local optimum but also is to improve the performance [10]. Then, the combined ABC algorithm with the fractional order image threshold segmentation method is applied to process medical images. The simulation results show that the suggested scheme offers better performance than other schemes for image threshold segmentation.

2   Fractional Order Image Thresholding Method

Multistage segmentation provides an effective method for image process and analysis. Furthermore, the automatic selecting parameters of an n-dimensional optimal threshold in image segmentation has always been a challenge. A solution to this problem is presented in this section [2, 11]. A given image with RGB elements of red, green, blue component is measure by a variable intensity levels as L, and the range of these levels is set at $\{0, 1, 2, ..., L-1\}$ that ϵ arange in $0 < i < L-1$, in this way, we can define it as following equation:

$$p_i^C = \frac{h_i^C}{N}, \quad \sum_{i=1}^{N} p_i^C = 1, \quad C = \{R, G, B\} \tag{1}$$

where C is the component of the image, that is, $C = \{R, G, B\}$; N is the image's total number of pixels; h_i^C is the number of pixels corresponding to intensity level i in component C; p_i^C is the image histogram h_i^C of each component C. The total mean of components of the image can be calculated as follows:

$$\mu_T^C = \sum_{i=1}^{L} i p_i^C, \quad C = \{R, G, B\} \tag{2}$$

If we want to calculate the N-level threshold, then the $N-1$ dimension threshold level t_j^C, $j = 1, \ldots, n-1$ is necessary. The calculation formula is as follows:

$$F^C(x, y) = \begin{cases} 0, & f^C(x, y) \le t_1^C \\ \frac{1}{2}(t_1^C + t_2^C), & t_1^C \le f^C(x, y) \le t_2^C \\ \quad \vdots & \\ \frac{1}{2}(t_{n-2}^C + t_{n-1}^C), & t_{n-2}^C \le f^C(x, y) \le t_{n-1}^C \\ L, & f^C(x, y) > t_{n-1}^C \end{cases} \tag{3}$$

where x and y in $f(x, y)$ are the pixel size H $\times$ W, (W and H mean the width and height) of the image, respectively. In this case, the pixels of a given image are classified into n classes $D_1^C \cdots D_n^C$, that are multiple objects or object-specific functions (for example, its topology or contained features) [11]. The simplest and most effective way to obtain the optimal threshold is to maximize the variance between classes, which is generally defined as

$$\mu_B^{C^2} = \sum_{j=1}^{L} w_j^C \left(\mu_j^C - \mu_T^C \right)^2, \quad C = \{R, G, B\} \tag{4}$$

where j is a specific number of the class with its the probability and mean value are w_j^C and μ_j^C in the class. The probability of the occurrence of w_j^C in $D_1^C \cdots D_n^C$ is expressed as follows.

$$
w_j^C = \begin{cases} \sum\limits_{i=1}^{t_j^C} p_i^C, & j = 1 \\ \sum\limits_{i=t_{j-1}^C+1}^{t_j^C} p_i^C, & 1 < j < n \\ \sum\limits_{i=t_{j-1}^C+1}^{L} p_i^C, & j = n \\ C = \{R, G, B\} \end{cases} \tag{5}
$$

In this case, the mean μ_j^C can be defined as

$$
w_j^C = \begin{cases} \sum\limits_{i=1}^{t_j^C} \frac{i p_i^C}{w_j^C}, & j = 1 \\ \sum\limits_{i=t_{j-1}^C+1}^{t_j^C} \frac{i p_i^C}{w_j^C}, & 1 < j < n \\ \sum\limits_{i=t_{j-1}^C+1}^{L} \frac{i p_i^C}{w_j^C}, & j = n \\ C = \{R, G, B\} \end{cases} \tag{6}
$$

If an objective function is constructed for each RGB component, then the problem of finding the n-dimensional threshold of an image can be simplified as the problem of finding the optimal threshold t_j^C to maximize the three objective functions $\sigma_B^{C^2}$ of each RGB component.

$$
\varphi^C = 1 < t_1^C < \cdots < t_{n-1}^C < L^{\sigma_B^{C^2} t_j^C}, \quad C = \{R, G, B\} \tag{7}
$$

As the threshold level increases, computing this optimization problem involves more computational effort. Many methods have been proposed in reference [10], which method should be used to solve the real-time application of this optimization problem. However, recently, the swarm intelligence optimization algorithm has become the most efficient alternative analysis method and has been used to solve such optimization problems.

3 Algorithm of Bees Colony (ABC)

Karabog, in Erciyes University, developed the artificial bees colony algorithm (algorithm of bees colony-ABC algorithm), Turkey [10], which is a swarm intelligence optimization algorithm for the global optimization issues. The algorithm of bees colony is a metaheuristic intelligent algorithm generated by simulating the foraging behavior of bees colonies. The ABC has attracted a large number of scholars for research because of its simple concept, few control parameters, easy implementation, and good optimization effect, and has gradually entered various application fields [13]. Significant progress has been made in the design and leaf constrained minimum spanning tree problem, vehicle routing problem.

3.1 Basic Principle of Artificial Bees Colony Algorithm

The ABC algorithm includes three types of bees: employment bee, reconnaissance bee, and follower bees. The employment bee mainly looks for food sources, collects relevant information, and then transmits the relevant food information to the following bee. The follower bee gets the relevant food from the employment bee. When the number of iterations of the bee colony algorithm is reached or certain conditions are met. Still, the amount of honey source can not be further improved, the honeybee becomes a scout bee and continues to search for a honey source near the nest. The original formula is expressed for updating bees' velocity and location as follows.

$$y_{i,j} = x_{i,j} + \emptyset_{i,j}(x_{i,j} - x_{k,j})$$
$$y_{i,j} = x_{i,j} + \emptyset_{i,j}(x_{i,j} - x_{k,j}) + \varphi_{i,j}(x_{best,j} - x_{i,j}) \qquad (8)$$

where $x_{i,j}$, $x_{k,j}$ and $x_{best,j}$ represent the position vectors for the employment bee, reconnaissance bee, and follower (best) bees, respectively; $y_{i,j}$ is the velocity vector, $\emptyset_{i,j}$ and $\varphi_{i,j}$ are the random numbers in [0, 1.5]. In the algorithm of bees colony, each honeybee corresponds to a feasible solution in the problem space. The number of honey sources is equal to the number of employed bees.

3.2 Enhanced Algorithm of Bees Colony (EABC)

The improved algorithm is implemented by adding equations for modifying update equations.

$$a = \left(1 + \frac{iter.}{\max(iiter.)}\right), b = \left(1 - \frac{iter.}{\max(iter.)}\right) \qquad (9)$$

where $iter.$ is the number of iterations of the artificial bees colony algorithm, the modification is that in the initial stage of the ABC algorithm, the bees search toward the current global optimal honey source. The increased iteration of search times, the influence of the current global optimal honey source on all bees colony search directions would be decreased in the later stage, and the random search's step size is increased so as to enhance the later stage. The coefficient set is used for the two coefficients to enhance the algorithm's ability to escape from the local optimum as following the formula.

$$\emptyset_{i,j} = 2 \times (rand - 0.5) \tag{10}$$

$$\varphi_{i,j} = 1.5 \times rand \tag{11}$$

In the formula, $\emptyset_{i,j}$ denotes the random number with the value range of $[-1, 1]$ and $\varphi_{i,j}$ denotes the random number with the value range of $[0, 1.5]$.

3.3 Fractional Order Image Thresholding Using EABC Algorithm

We use a fractional-order technique of image histogram combined with the location matrix coded solution of the EABC algorithm for threshold segmentation, as presented following in detail in this subsection. The problem of threshold selection in the fractional image segmentation method is transformed into the optimization problem of φ^C maximization by artificial bees colony algorithm as follows.

$$S_{\text{best}} = \max(\varphi^C) \tag{12}$$

The corresponding threshold is the best segmentation threshold through the steps of threshold optimization of fractional order image are carried out as follows.

Step 1. Prepare the images: The degree of high-quality image segmentation is calculated by using the method of high-quality image segmentation.

Step 2. Initialize the algorithm parameters: In the search range, half of the population of the algorithm is randomly placed, and the new honeybee is searched and the initial labeled honey source is determined;

Step 3. Update the local location: The following bees with the same number of employed bees search for new honey sources according to the improved search formula, and calculate the value of the high-quality degree of the honey sources searched. Compared with the previous honey sources, the high value of φ^C replaces the low value of φ^C as the initial honey source for the next generation of bee colony search.

Step 4. Update the global solution: Whether there is a honeybee searching for a honeybee remains unchanged after a certain number of iterations. If there is a honeybee, it is necessary to generate a new location to replace the corresponding honey source.

Step 5. Termination condition: check whether meet it or not, if it is not condition meet, it repeats steps (3) and (4). Otherwise, the threshold corresponding to the optimal solution S_{best} is outcome as the best threshold of image threshold segmentation.

Figure 1 shows a flowchart of the EABC for obtaining image threshold the optimal solution.

Fig. 1 A flowchart of the EABC for obtaining image threshold the optimal solution

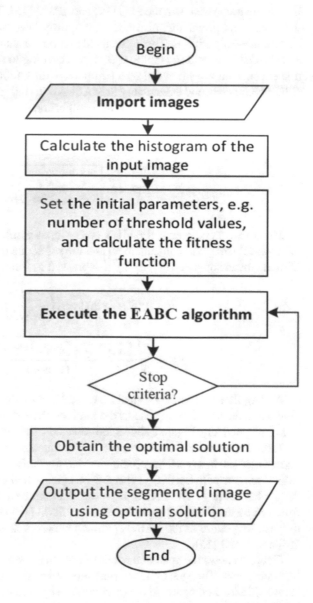

4 Experimental Results

The EABC algorithm is tested through modeling the fitness function for fractional-order image thresholding function with six images. The obtained results from the EABC are compared with the PSO [14] and GWO [15]. To avoid the randomness of the results and probability distribution, inappropriate statistical measures effectiveness of these algorithms. Therefore, all algorithms execute 25 times for each graph, and the test threshold th is set to 2, 3, 4, 5 according to reference. All algorithms in this experiment are simulated on a computer with Intel (R.) Core (TM.) i7-6700hq CPU of 2.60 GHz and 8G memory in MATLAB 7.10.0 (r2010a) programming environment. The stopping criterion of each experiment is 150 iterations, NP = 50. To verify stability, the standard deviation (STD.) is calculated at the end of each test according to the fitness function [16].

$$STD. = \sqrt{\sum_{i=1}^{\text{Max\_iter}} \frac{(\theta_i - \varepsilon)^2}{\text{Max\_iter}}} \tag{13}$$

Moreover, a noise ratio (PSNR)'s peak signal would compare the similarity of the images' (image's segmentations) and the reference images (the original images) with the mean square error (MSE.) is expressed as follows.

$$PSNR = 20 \log_{10}\left(\frac{255}{MSE}\right) \tag{14}$$

$$MSE. = \sqrt{\frac{\sum_{i=1}^{ro} \sum_{j=1}^{co} \left(I_0^a(i, j) - I_{th}^a(i, j)\right)}{ro \times co}} \tag{15}$$

Among them, I_0^a is the original images; I_{th}^a is the image's segments; a depends on the image; ro and co represents the total numbers of the rows and columns of the image respectively. Figure 2 shows the selected original images.

Table 1 displays the obtained experimental results of the EABC for the first four chosen medical images. Figure 3 shows a visually derived comparison of the proposed EABC scheme with the PSO [14] and GWO [15] schemes for images 01 and 06 with thresholds set at 4. The EABC algorithm has the best convergence and accuracy, followed by the PSO [14] algorithm and the GWO [15] algorithm. It can be seen that the resulting curve of the EABC approach achieves faster convergence of the PSO [14] and WGO [15].

Figure 4 shows the operation result of the EABC scheme with the images 01 and 06 histograms. The spine nuclear magnetic resonance primitive and segmentation results. Table 2 compares the experimental segmentation results of the proposed EABC scheme with GWO [15] and PSO [14]. The methods need to run 25 times for each image. The objective function obtained with the fraction order function is

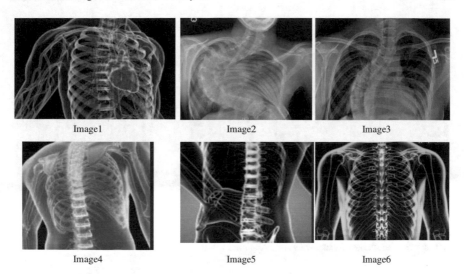

Fig. 2 The chosen medical scan images

Table 1 Experimental results of the EBAC for the chosen medical images

No. images	k	No. thresholds	PSNR	STD	MSE
Image01	2	8, 89	15.267	0.713	15.211
	3	24, 45, 125	20.809	1.125	20.123
	4	23, 52,104, 145	19.696	1.453	24.973
	5	39,76,121,145,189	22.404	1.753	29.348
Image02	2	24, 121	13.421	1.103	14.232
	3	64, 121, 165	16.865	0.953	20.254
	4	43,75,121,172	19.897	1.234	25.121
	5	45,89,123,156,187	21.234	1.523	29.342
Image03	2	67, 142	15.451	0.723	16.232
	3	35,101,151	18.862	0.453	22.254
	4	41, 76, 102,165	20.897	0.834	26.121
	5	23,56, 87, 121, 186	22.234	0.952	29.042
Image04	2	8, 89	14.267	0.712	14.211
	3	24, 45, 125	20.809	1.123	20.123
	4	23, 52,104, 145	19.696	1.453	24.973
	5	39,76,120,145,189	22.404	1.653	29.648

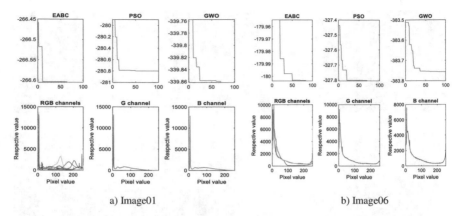

a) Image01   b) Image06

Fig. 3 Visually derived comparison of the proposed EABC scheme with the PSO and GWO schemes for the images 01 and 06 with thresholds set at 4

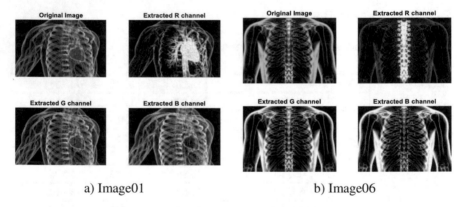

a) Image01   b) Image06

Fig. 4 The operation result of the EABC scheme with the images 01 and 06 histograms

calculated for each image for the average received values: MSE, PSNR, and STD—the standard deviation. To sum up, the effect of threshold segmentation with the fractional-order image method based on the EABC scheme is better than the other schemes, e.g., the PSO [14] and GWO [15] schemes.

5 Conclusion

This research proposed an enhanced algorithm of bees colony (EABC) for the solution to the state of the threshold medical image segmentation. As image segmentation's adaptability, the precision of the automatic selection can affect the accuracy and detection of target recognition, the image segmentation research has substantial

Table 2 Comparison of the segmentation results of the EBA with GA and PSO schemes

No. images	k	EABC			PSO [14]			GWO [15]		
		PSNR	STD	MSE	PSNR	STD	MSE	PSNR	STD	MSE
01	2	1.34E+01	6.67E−01	1.34E+01	1.11E+01	1.18E−01	1.43E+01	1.14E+01	9.27E−02	1.49E+01
	3	1.94E+01	1.05E+00	1.89E+01	1.38E+01	1.87E−01	1.92E+01	1.41E+01	1.02E−01	1.96E+01
	4	1.85E+01	1.36E+00	2.35E+01	1.60E+01	2.66E+01	2.26E+01	1.67E+01	1.85E−01	2.32E+01
	5	2.11E+01	1.55E+00	2.76E+01	1.84E+01	3.29E−01	2.64E+01	1.94E+01	2.53E−01	2.63E+01
02	2	1.26E+01	1.03E+00	1.34E+01	1.15E+01	4.57E−03	1.50E+01	1.14E+01	3.06E−03	1.57E+01
	3	1.59E+01	4.26E−01	2.00E+01	1.40E+01	1.02E−01	1.95E+01	1.40E+01	3.61E−02	1.90E+01
	4	1.87E+01	1.16E+00	2.36E+01	1.59E+01	2.42E−01	2.20E+01	1.65E+01	1.67E−01	2.26E+01
	5	2.00E+01	1.43E+00	2.76E+01	1.82E+01	2.84E−01	2.53E+01	1.89E+01	2.02E−01	2.55E+01
03	2	1.45E+01	6.80E−01	1.53E+01	1.13E+01	5.28E−02	1.53E+01	1.12E+01	7.14E−03	1.56E+01
	3	1.77E+01	4.26E−01	2.09E+01	1.37E+01	1.48E−01	1.96E+01	1.39E+01	7.51E−02	1.95E+01
	4	1.96E+01	7.84E−01	2.46E+01	1.58E+01	1.65E−01	2.35E+01	1.70E+01	7.90E−02	2.26E+01
	5	2.09E+01	8.95E−01	2.73E+01	1.83E+01	2.59E−01	2.76E+01	1.87E+01	1.76E−01	2.87E+01
04	2	1.34E+01	6.69E−01	1.34E+01	9.75E+00	8.12E−02	1.44E+01	1.31E+01	2.31E−03	1.31E+01
	3	1.96E+01	1.05E+00	1.89E+01	1.08E+01	1.88E−01	1.87E+01	1.14E+01	1.74E−01	1.79E+01
	4	1.85E+01	1.36E+00	2.35E+01	1.24E+01	2.42E−01	2.23E+01	1.32E+01	2.29E−01	2.33E+01
	5	2.11E+01	1.55E+00	2.79E+01	1.47E+01	3.70E−01	2.55E+01	1.51E+01	2.61E−01	2.50E+01
05	2	1.36E+01	1.03E+00	1.34E+01	1.26E+01	6.75E−02	1.73E+01	1.25E+01	1.11E−03	1.73E+01
	3	1.68E+01	4.26E−01	2.00E+01	1.41E+01	1.45E−01	2.16E+01	1.39E+01	4.91E−02	1.23E+01
	4	1.88E+01	1.16E+00	2.36E+01	1.48E+01	2.51E−01	2.50E+01	1.47E+01	1.32E−01	2.54E+01
	5	2.09E+01	1.43E+00	2.76E+01	1.58E+01	8.36E−01	2.87E+01	1.51E+01	4.62E−01	2.89E+01

(continued)

Table 2 (continued)

No. images	k	EABC			PSO [14]			GWO [15]		
		PSNR	STD	MSE	PSNR	STD	MSE	PSNR	STD	MSE
06	2	1.45E+01	6.80E−01	1.53E+01	1.33E+01	7.60E−02	1.75E+01	1.33E+01	1.85E−04	1.72E+01
	3	1.87E+01	4.26E−01	2.19E+01	7.70E+00	1.85E−01	2.17E+01	1.57E+01	1.60E−01	2.16E+01
	4	2.06E+01	7.84E−01	2.36E+01	1.40E+01	2.51E−01	2.47E+01	1.70E+01	2.29E−01	2.54E+01
	5	2.09E+01	8.95E−01	2.64E+01	1.77E+01	9.07E−01	2.87E+01	1.84E+01	5.71E−01	2.90E+01

practical utility. The fractional-order image threshold segmentation method of the EABC algorithm can effectively segment medical images. Based on inheriting the advantages of the ABC algorithm and then the fractional image threshold model, the proposed scheme shows that the segmentation effect, time efficiency. Compared with the other schemes, the EABC scheme is better than other fractional-order image threshold segmentation methods. In future work, the threshold optimization for image segmentation will be implemented with a variety the metaheuristic algorithms, e.g., improved flower pollination [2], parallel compact bat algorithm [17], and soccer league competition algorithm [7].

Acknowledgements This work is supported by the Vietnam National University, Ho Chi Minh City (VNU-HCM) under grant number D1-2022-03.

References

1. Dao T-K, Wang H-J, Yu J, Nguyen H-Q, Ngo T-G, Nguyen T-T (2021) An optimizing multilevel thresholding for image segmentation based on hybrid swarm computation optimization BT—advances in intelligent information hiding and multimedia signal processing. Presented at the (2021)
2. Nguyen T-T, Wang H-J, Dao T-K, Pan J-S, Ngo T-G, Yu J (2020) A scheme of color image multithreshold segmentation based on improved moth-flame algorithm. IEEE Access. 8:174142–174159. https://doi.org/10.1109/ACCESS.2020.3025833
3. Zhao M, Lin H-Y, Yang C-H, Hsu C-Y, Pan J-S, Lin M-J (2015) Automatic threshold level set model applied on MRI image segmentation of brain tissue. Appl Math Inf Sci 9:1971
4. Chu SC, Dao TK, Pan JS, Nguyen TT (2020) Identifying correctness data scheme for aggregating data in cluster heads of wireless sensor network based on naive Bayes classification. Eurasip J Wirel Commun Netw 2020(52):1–16. https://doi.org/10.1186/s13638-020-01671-y
5. Nguyen T-T, Jiang S-J, Dao T-K, Ngo T-G, Nguyen T-T-T, Do T-V (2021) An enhancing grasshopper optimization for efficient feature selection BT—advances in engineering research and application. In: Sattler K-U, Nguyen DC, Vu NP, Long BT, Puta H (eds) CERA 2020. Lecture notes in networks , Springer Chcm, vol 178. Springer International Publishing, Cham, pp 161–172
6. Gu F, Lu ZM, Pan JS (2005) Multipurpose image watermarking in DCT domain using subsampling. In: Proceedings—IEEE international symposium on circuits and systems, pp 4417–4420. https://doi.org/10.1109/ISCAS.2005.1465611
7. Qiao Y, Dao TK, Pan JS, Chu SC, Nguyen TT (2020) Diversity teams in for wireless sensor network deployment problem. Symmetry (Basel) 12:445. https://doi.org/10.3390/sym12030445
8. Horng M-H (2011) Multilevel thresholding selection based on the artificial bee colony algorithm for image segmentation. Expert Syst Appl 38:13785–13791. https://doi.org/10.1016/j.eswa.2011.04.180
9. Ngo T-G, Nguyen T-T, Ngo Q-T, Nguyen D-D, Chu S-C (2016) Similarity shape based on skeleton graph matching. J Inf Hiding Multimed Signal Process 7:1254–1264
10. Karaboga D, Basturk B (2007) A powerful and efficient algorithm for numerical function optimization: artificial bee colony (ABC) algorithm. J Glob Optim 39:459–471. https://doi.org/10.1007/s10898-007-9149-x
11. Jamil N, Sembok TMT, Bakar ZA (2011) Digital archiving of traditional songket motifs using image processing tool. In: 5th Recent researches in chemistry biology and environment, ICAC, pp 33–39

12. Du C-J, Sun D-W (2004) Recent developments in the applications of image processing techniques for food quality evaluation. Trends food Sci Technol 15:230–249
13. Karaboga D, Gorkemli B, Ozturk C, Karaboga N (2014) A comprehensive survey: artificial bee colony (ABC) algorithm and applications. Artif Intell Rev 42:21–57. https://doi.org/10.1007/s10462-012-9328-0
14. Chander A, Chatterjee A, Siarry P (2011) A new social and momentum component adaptive PSO algorithm for image segmentation. Expert Syst Appl 38:4998–5004
15. Kapoor S, Zeya I, Singhal C, Nanda SJ (2017) A grey wolf optimizer based automatic clustering algorithm for satellite image segmentation. Procedia Comput Sci 115:415–422
16. Yao S, Lin W, Ong E, Lu Z (2005) Contrast signal-to-noise ratio for image quality assessment. In: IEEE international conference on image processing. IEEE, pp I–397
17. Nguyen TT, Pan JS, Dao TK (2019) A novel improved bat algorithm based on hybrid parallel and compact for balancing an energy consumption problem. Information 10:194. https://doi.org/10.3390/info10060194

Fast Channel-Dependent Transmission Map Estimation for Haze Removal with Localized Light Sources

Andrei Filin, Inessa Gracheva, Andrei Kopylov, and Oleg Seredin

Abstract The presence of haze, dust, suspensions of various kinds on images significantly degrades visibility, which makes it difficult to analyze the scenes. Obtaining an accurate transmission map is crucial for eliminating the effect of such particles. In this work, the transmission map was refined considering the dependence of scattering on the wavelength and using the soft lightening mask. According to this assumption, each of the RGB light components corresponds to separate coefficients characterizing the model. The coefficients were obtained by maximizing the total structural similarity index (SSIM) when comparing the reconstructed images with the haze-free images from the I-HAZE dataset. Experiments were carried out on datasets with natural (O-HAZE and I-HAZE) and synthesized (SOTS outdoor and indoor) haze; the PSNR and SSIM metrics were calculated. The results obtained demonstrate an improvement in the quality of haze removal by the proposed method compared to the basic methods.

Keywords Haze removal · Channel-dependent transmission map · Dark channel prior · Point-light sources · Lightning weights map

1 Introduction

The presence of haze and fine dust can significantly degrade the visibility of scenes in computer vision, and leads to negative impacts on the results of image and video analysis. Therefore, haze removal techniques have attracted increased interest over the past two decades. On the one hand, a lot of effective methods have appeared that demonstrate impressive results for standard outdoor scenes, and on the other hand, the growing complexity of scenes, lighting conditions, increasing image resolution, transition from image to video processing impose additional requirements on robustness and computational complexity of haze removal algorithms.

A. Filin (✉) · I. Gracheva · A. Kopylov · O. Seredin
Institute of Applied Mathematics and Computer Science, Tula State University, Tula, Russia

© The Author(s), under exclusive license to Springer Nature Switzerland AG 2022   461
N. H. T. Dang et al. (eds.), *Artificial Intelligence in Data and Big Data Processing*,
Lecture Notes on Data Engineering and Communications Technologies 124,
https://doi.org/10.1007/978-3-030-97610-1_36

This paper focused on the single image dehazing methods [1], which are the most applicable for most real-world applications, as they do not need additional hardware like depth sensors or interaction with the user, which is especially important for video processing.

The dehazing methods are substantially based on the hazy image formation model, proposed by Cozman and Krotkov [2] and have two main problems in processing of complex scenes, obtained in low light conditions with the presence of localized light sources. The first problem is that the atmospheric light is no longer the sole and brightest lighting source, and thus an evaluation of the atmospheric light becomes challenging. The second problem is that the mentioned above optical model doesn't take into account attenuation and blending the reflected light, beamed by point-light sources, it does only for airlight.

Many efforts were taken to overcome these difficulties. To prevent significant color distortions and correct estimation of atmospheric light, usage of the most haze-opaque pixels from the 0.1% of the brightest pixels in the dark channel, was proposed in work [3]. Method [4] by Berman et al. is based on the observation, that colors on the haze-free image form dense clusters in RGB space, and each color cluster becomes a line, which is called the Haze-Line, in the hazy conditions. These lines intersect at the air-light position. Haze-lines, extracted by the Hough transform, are then used for atmospheric color estimation. Then Zhu et al. [5] proposed the color-plane model for air-light estimation, which is the combination of the color-line and haze-line models. In [6], each color channel of the original image is partitioned into 8×8 patches, each patch contains a pixel with a minimum brightness value. Then, among all the found minimum values of intensities, the maximum value is found in each color channel separately, forming the final atmospheric light estimation.

Another idea is to calculate the distributed estimation of atmospheric light for each pixel or patch instead of the single value. In the method of Li et al. [7] the glow model is embedded into the slightly modified standard haze model. The original image is decomposed to the sum of two images: a night haze image (a layer with large gradients preserved, estimated using the Laplace filter) and a glow image (a smooth layer). Glow images are formed by active light sources whose intensity is convoluted with the atmospheric point-spread function to produce a glowing effect on the image. Then a transmission map is calculated as a minimum of the ratio of the night haze image to atmospheric light in the 15×15 patch. The transmission map is estimated and refined using a guided filter. Then, all data is substituted into the standard dehazing model. Acuti et al. [8] proposed to estimate the airlight on image patches of multiple sizes, that are then merged together. Laplacian of the original image is used as supplementary input to emphasize fine details. These inputs are filtered according to the three quality weight maps—local contrast, saturation, and saliency. On the final step, a multi-scale fusion based on Laplacian pyramid decomposition of the inputs and a Gaussian pyramid of the normalized weights is applied.

Deep learning dehazing methods [9–11] are also affected by the disadvantages mentioned above. Difficulties in obtaining an image with and without haze, taken at the same time, lead to the fact that the vast majority of pairs of images in publicly available datasets, e.g. [12, 13], used for training, are synthesized based on the same Koschmieder's Optical Model. Therefore, twilight or nighttime images with localized light sources are often missed during the training process.

Another approach [14, 15] is based on the idea to form a hard or soft lightning mask, that is used to remove regions, affected by the localized light sources, from the estimation of the transmission map. This makes it possible to hold the classical hazy image formation model and preserve the low computational complexity of the haze removal method. A support vector data description method, modified on the basis of the ideas of Pattern Recognition in Interrelated Data, is applied to obtain the lightning mask [16].

This work is a further development of the method based on joint transmission map estimation and atmospheric-light extraction [15] with two main improvements. Instead of using the lightning mask, which excludes pixels, classified as affected by the localized light source, from the dark channel estimation, we propose here to form the lightning weights map, to weight corresponding values of the dark channel. This allows excluding localized light sources from the most hazy-opaque region, used for atmospheric light estimation by the He et al. method [3]. And avoid overexposure of these areas during the restoration process keeping the haze formation model valid.

The second improvement is founded on the fact that atmospheric lighting components propagate differently in a haze. Following this, separate coefficients were estimated for each of the RGB color components in the revised haze density estimation model [14]. The coefficients have been estimated using the supervised learning strategy on the I-HAZE real images dataset [17], so the model can more closely estimate the density of haze and therefore build an efficient reconstruction. Experimental results of the proposed model demonstrate parity compared the baseline methods on tested datasets in terms of quantitative and subjective evaluations and outperform them in terms of computation complexity.

In Sects. 2 and 3 the proposed approach is presented in detail. In Sect. 4, we present the experimental results of comparing our method with existing haze removal methods using real and synthetic haze images datasets. We also compare our method with existing haze removal methods and perform qualitative and quantitative analyzes. In Sect. 5 the conclusions are presented.

2 Enhanced Universal Atmospheric-Light Extractor and Dark Channel Rectification

In this research, we employ the widespread model [18], which takes into account two scattering effects—attenuation of incident light and mixing of atmospheric light and reflected light.

$$I_c(\mathbf{s}) = J_c(\mathbf{s})T(\mathbf{s}) + A_c(1 - T(\mathbf{s})), \tag{1}$$

where $I_c(\mathbf{s})$—the intensity of the hazy image at the position $\mathbf{s} \in S$ of color channel $c \in \{R, G, B\}$ of the RGB color space, $J_c(\mathbf{s})$—the intensity of the haze-free image at the position s and color channel c, $T(\mathbf{s})$—medium transmission map, $S = \{\mathbf{s} = (s_1, s_2) : s_1 = 1, \ldots, N_1, s_2 = 1, \ldots, N_2\}$—discrete pixels grid, A_c—atmospheric light of color channel c.

In this research we used method [3] to determine the transmission map, which suggests that the dark channel prior can be used for this purpose:

$$T_c(\mathbf{s}) = 1 - dark(I_c(\mathbf{s})), \tag{2}$$

where the dark channel of the haze image $dark(I_c(\mathbf{s}))$ is specified as:

$$dark(I(c, \mathbf{s})) = \min_{y \in \Omega(\mathbf{s})} \min_{c \in \{r,g,b\}} (I(c, \mathbf{s})), \tag{3}$$

where $\Omega(\mathbf{s})$ is the patch of the image located at a position $\mathbf{s}$, c—one of RGB color component.

Haze formation model (1) takes into account a single global illumination source. The Universal Atmospheric-Light Extractor (UALE) was proposed in [14] to eliminate the localized light sources and to keep Eq. (1) valid. The UALE is based on one class classifier as pixels, belonging to localized light sources, form the training set which represents just one class. But local decisions about pixel class need to be reconciled according to the concept of Pattern Recognition in Interrelated Data [16] to increase the robustness and accuracy. From the other side, Eq. (3) gives a rather rough estimation of haze density, and the next step for all dark channel prior based haze removal methods is to rectify it. In our previous work [15], we combined the coordination of local decisions of the classifier and rectification of the dark channel into one procedure using the highly computationally effective algorithm with structure transferring properties based on the gamma-normal probabilistic model [19] and dynamic programming.

Here we propose to use the result of local decisions coordination, normalized to the range from 0 to 1, as a soft lightening mask $W(\mathbf{s})$. The values, which are close to 0, mean that haze density at the corresponding pixel is actually determined by the local illumination. Otherwise, values closed to 1, mean that the dark channel estimation is caused by the haze itself. This mask represents a distributed illumination caused by different illumination sources and plays a role similar to the glow image in the method of Li et al. [7]. The final rectified dark channel estimates are weighted by the value of the soft lightning mask (Fig. 1).

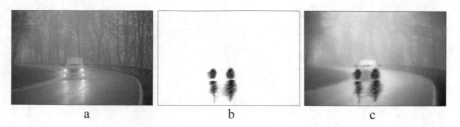

a   b   c

Fig. 1 **a** Initial hazy image, **b** lightening mask, **c** rectified dark channel

3 Channel-Dependent Transmission Map Estimation

According to the model [18], the transmission map is expressed as follow:

$$T(\mathbf{s}) = e^{-\beta d(\mathbf{s})} \tag{4}$$

It is assumed that the atmosphere is homogeneous (i.e., the scattering coefficient β is constant) and that the scattering of light does not depend on the wavelength. In this research, only the first assumption (about the homogeneity of the atmosphere) is observed. The assumption that scattering is independent of wavelength is rejected, so the coefficient becomes a vector, where each element is one of the RGB color components. According to these assumptions, the transmission map is expressed as follow:

$$T_c(\mathbf{s}) = e^{-\beta_c d(\mathbf{s})} \tag{5}$$

Since the transmission map in the atmospheric scattering model has an exponential dependence, the suggested transmission map also has an exponential dependence [15]. Thus, the estimated transmission map from Eq. (5) can be written as follows:

$$T(\mathbf{s}) = \beta_c (1 - \text{dark}(\mathbf{I}_c(\mathbf{s})) \times W(\mathbf{s}))^{\gamma_c} + \varepsilon_c \tag{6}$$

where ε_c is a prediction error of the model, β_c—scattering coefficient, γ_c—exponential factor, $\times$ means the Hadamard product.

In this research, we assume that representing the coefficients β_c, γ_c and ε_c as vectors, where each element denotes one of the color components, will improve the transmission map. To obtain the most accurate transmission map, the coefficients β_c, γ_c and ε_c are found using Nelder-Mead method [20] of objective function optimization. The structural similarity index (SSIM) [21] is adopted as the basis for the optimized function. After achieving the results of the last iteration, the coefficients β_c, γ_c and ε_c, showed maximum SSIM over all periods, were chosen.

We employ the I-HAZE [17] as the training set. This set contains 30 pairs of the real images of the same scenes, every pair-hazy and corresponding haze-free images.

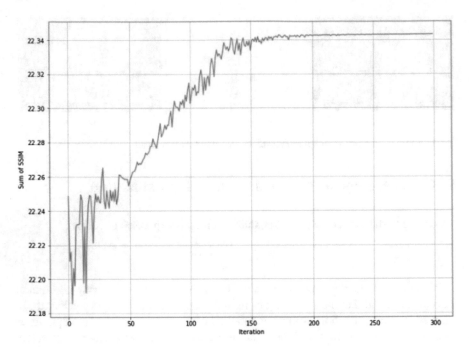

Fig. 2 Changing the SSIM in an iterative process

To perform experiments, the SciPy realization of the Nelder-Mead optimization method for python 3 is used. Figure 2 demonstrates changing of the sum of SSIM over images of the dataset during training iterations in our experiment. At the end of optimization, the following coefficients are obtained: $\beta_c = [0.92110, 0.93364, 0.90397]$, $\gamma_c = [0.81221, 0.84756, 0.73685]$, $\varepsilon_c = [-0.00019, 0, 0.00107]$. We use these values in our experiments.

4 Experimental Results

4.1 Synthetic Images

The data for the haze removal task must meet several conditions, which do not allow the collection of significant volumes of such datasets. The main problem is that both images in a pair haze/haze-free must get considering unchanged, except presence/absence of haze. Therefore, in the image dehazing task is often applied datasets, where hazy images have been synthesized via the atmospheric model and depth information. But the depth map is often inaccurate, which leads to distortion in the resulting images; moreover, such images do not fully represent the complexity of the internal processes. Nevertheless, image synthesizing makes it possible to produce much wider datasets for image dehazing tasks.

We used SOTS [13] dataset in our experiments. It includes 500 indoor and 500 outdoor synthetic images (1000 images in total) with different haze intensities.

4.2   Real Images

Datasets with real images are consist of pairs of images of the same scene, obtained in the same environment, the only difference is the presence/absence of haze.

As real datasets in this research we use:

- I-HAZE [17]—consists of 35 pairs of indoor images;
- O-HASE [22]—consists of 45 pairs of outdoor images.

In both datasets, the haze was obtained using a smoke generator.

4.3   Comparative Evaluation

The proposed algorithm compared with the method of Dhara [23], Qin [10], Berman [24], He [3], Zhu [25], also our method, proposed earlier [15]. The methods were compared in terms of average PSNR and SIMM [21] on both datasets with synthetic (SOTS indoor and SOTS outdoor) and real (I-HAZE and O-HAZE) images. All images are converted to a single 640 × 480 resolution before performing experiments.

Figure 3 demonstrates changing of the mean calculation time over 5 images for different resolutions (number of processed pixels). Figure 4 shows examples of the example result for the image from the test set. A quantitative comparison is shown in Table 1.

A quantitative assessment of the methods in Table 1 shows that the method, proposed in current research, demonstrates comparable results to the baseline methods; method is slightly superior to the method proposed in the previous article [15]. The neural network-based method (Qin et al. [10]) demonstrates higher results on the SOTS datasets because the indoor and outdoor models, which are used for our experiments, were trained using SOTS indoor and outdoor datasets, respectively.

The diagram in Fig. 3 demonstrates lower computation time growth rates compared to other considered methods.

Visually, it can be noted that in the shown example (Fig. 4), the proposed method darkens the image much less while perceptible removing haze from the image.

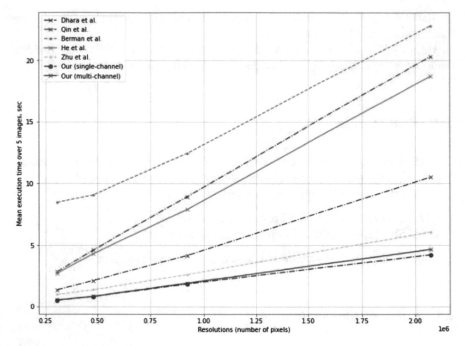

Fig. 3 Changing of the mean calculation time over 5 images for different resolutions. Note, that our single and multi-channel methods have similar computational complexity, therefore its' plots on the current figure's scale are overlaying

5   Conclusion

The experimental results demonstrate a parity of the proposed method and other the baseline method [15] while keeping the linear relation between computational complexity and the resolution of the image under processing. It's worth mentioning that the proposed method has the lowest computation time along with examined methods.

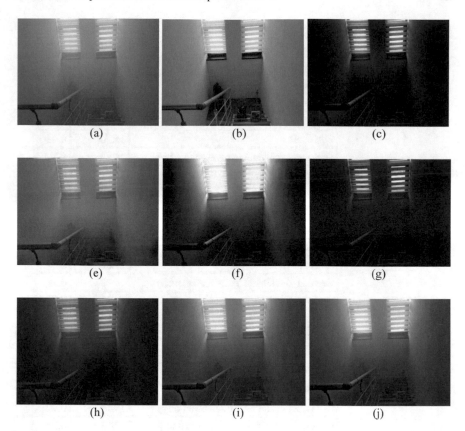

Fig. 4 Haze removal on the haze image "25_indoor_hazy.jpg" (i-haze) **a** Haze image. **b** Ground-truth haze-free image. **c–j** Results of Dhara et al.'s, Qin et al.'s, Berman et al.'s, He et al.'s, Zhu et al.'s, the previous and currently proposed methods, respectively

Table 1 Quantitative results for comparative methods on PSNR and SSIM methods

Dataset	Method	PSNR	SSIM
I-HAZE	Dhara et al	13.43	0.64
	Qin et al	15.65	0.70
	Berman et al	15.81	0.75
	He et al	11.91	0.58
	Zhu et al	16.66	0.73
	Our (single-channel transmission map)	16.32	0.74
	Our (multi-channel transmission map)	16.54	0.74

(continued)

Table 1 (continued)

Dataset	Method	PSNR	SSIM
O-HAZE	Dhara et al	16.27	0.69
	Qin et al	14.67	0.60
	Berman et al	15.71	0.73
	He et al	15.11	0.66
	Zhu et al	16.50	0.66
	Our (single-channel transmission map)	15.89	0.7
	Our (multi-channel transmission map)	15.94	0.7
SOTS-INDOOR	Dhara et al	19.60	0.86
	Qin et al	29.58	0.97
	Berman et al	17.28	0.78
	He et al	16.56	0.80
	Zhu et al	19.05	0.81
	Our (single-channel transmission map)	16.48	0.78
	Our (multi-channel transmission map)	16.81	0.8
SOTS-OUTDOOR	Dhara et al	16.62	0.80
	Qin et al	19.48	0.84
	Berman et al	17.96	0.83
	He et al	14.40	0.75
	Zhu et al	22.05	0.89
	Our (single-channel transmission map)	17.14	0.79
	Our (multi-channel transmission map)	22.23	0.87

Acknowledgements The work is supported by the Russian Fund for Basic Research. Grant no: 20-07-00441 and Grant no: 20-07-00055. Oleg Seredin is supported by the Ministry of Science and Higher Education of the Russian Federation within the framework of the state task FEWG-2021-0012.

References

1. Wang B, Niu B, Zhao P, Xiong NN (2021) Review of single image defogging. Int J Sens Networks 35:111–120. https://doi.org/10.1504/IJSNET.2021.113630
2. Cozman F, Krotkov E (1997) Depth from scattering. In: Proceedings of the IEEE computer society conference on computer vision and pattern recognition. IEEE, pp 801–806
3. He K, Sun J, Tang X (2011) Single image haze removal using dark channel prior. IEEE Trans Pattern Anal Mach Intell 33:2341–2353. https://doi.org/10.1109/TPAMI.2010.168
4. Berman D, Treibitz T, Avidan S (2017) Air-light estimation using haze-lines. 2017 IEEE international conference on computational photography ICCP 2017—Proceedings. https://doi.org/10.1109/ICCPHOT.2017.7951489

5. Zhu MZ, He BW, Zhang LW (2017) Atmospheric light estimation in hazy images based on color-plane model. Comput Vis Image Underst 165:33–42. https://doi.org/10.1016/j.cviu.2017.09.005
6. He J, Zhang C, Yang R, Zhu K (2016) Convex optimization for fast image dehazing. In: Proceedings of the international conference on image processing ICIP 2016-Augus, pp 2246–2250. https://doi.org/10.1109/ICIP.2016.7532758
7. Li Y, Tan RT, Brown MS (2015) Nighttime haze removal with glow and multiple light colors. In: Proceedings of the IEEE international conference on computer vision 2015 Inter, pp 226–234. https://doi.org/10.1109/ICCV.2015.34
8. Ancuti C, Ancuti CO, De Vleeschouwer C, Bovik AC (2020) Day and night-time dehazing by local airlight estimation. IEEE Trans Image Process 29:6264–6275. https://doi.org/10.1109/TIP.2020.2988203
9. Cai B, Xu X, Jia K et al (2016) DehazeNet: an end-to-end system for single image haze removal. IEEE Trans Image Process 25:5187–5198. https://doi.org/10.1109/TIP.2016.2598681
10. Qin X, Wang Z, Bai Y et al (2019) FFA-Net: feature fusion attention network for single image dehazing. https://doi.org/10.1609/aaai.v34i07.6865
11. Zhang S, He F, Ren W (2020) NLDN: non-local dehazing network for dense haze removal. Neurocomputing 410:363–373. https://doi.org/10.1016/j.neucom.2020.06.041
12. Ancuti C, Ancuti CO, De Vleeschouwer C (2016) D-HAZY: a dataset to evaluate quantitatively dehazing algorithms. In: Proceedings—international conference on image processing, ICIP. IEEE Computer Society, pp 2226–2230
13. Li B, Ren W, Fu D et al (2019) Benchmarking single-image dehazing and beyond. IEEE Trans Image Process 28:492–505
14. Shi L-FF, Chen B-HH, Huang S-CC et al (2018) Removing haze particles from single image via exponential inference with support vector data description. IEEE Trans Multimed 20:2503–2512. https://doi.org/10.1109/TMM.2018.2807593
15. Filin A, Gracheva I, Kopylov A (2020) Haze removal method based on joint transmission map estimation and atmospheric-light extraction. In: The 4th international conference on future networks and distributed systems (ICFNDS), pp 1–6
16. Dvoenko SDD, Kopylov AVV, Mottl VVV et al (2004) The problem of pattern recognition in arrays of interconnected objects. Statement of the recognition problem and basic assumptions. Autom Remote Control 65:127–141. https://doi.org/10.1023/B:AURC.0000011696.31008.5a
17. Ancuti C, Ancuti CO, Timofte R, De Vleeschouwer C (2018) I-HAZE: a dehazing benchmark with real hazy and haze-free indoor images. Lecture notes in computer science (including subseries lecture notes in artificial intelligence and lecture notes in bioinformatics). Springer, Cham, pp 620–631
18. Kuschmieder H (1924) Theorie der horizontalen Sichtweite. Beitr Phys freie Atmos 12:33–55
19. Gracheva I, Kopylov A (2017) Image processing algorithms with structure transferring properties on the basis of gamma-normal model. In: Communications in computer and information science. In Printing, pp 257–268
20. Singer S, Nelder J (2009) Nelder-mead algorithm. Scholarpedia 4:2928
21. Chen B-H, Huang S-C, Ye JH (2015) Hazy image restoration by bi-histogram modification. ACM Trans Intell Syst Technol 6:1–17
22. Ancuti CO, Ancuti C, Timofte R, De Vleeschouwer C (2018) O-haze: a dehazing benchmark with real hazy and haze-free outdoor images. In: Proceedings of the IEEE conference on computer vision and pattern recognition workshops, pp 754–762
23. Dhara SK, Roy M, Sen D, Biswas PK (2020) Color cast dependent image dehazing via adaptive airlight refinement and non-linear color balancing. IEEE Trans Circuits Syst Video Technol 8215:1–1. https://doi.org/10.1109/tcsvt.2020.3007850
24. Berman D, Avidan S, others (2016) Non-local image dehazing. In: Proceedings of the IEEE conference on computer vision and pattern recognition, pp 1674–1682
25. Zhu Q, Mai J, Shao L (2015) A fast single image haze removal algorithm using color attenuation prior. IEEE Trans image Process 24:3522–3533

Style Transfer Method Based on Deep Learning for Chinese Ink Painting

Yunyun Jiang and Hefei Wang

Abstract Image style transfer technology learn different drawing styles by making full use of the abstract and learning ability of the neural network, and then automatically generates similar style images from the target image. However, traditional style migration methods are not fully effective in learning the details of Chinese ink painting styles. In this paper, we propose a new style transfer method for Chinese ink painting, which uses the VGG19 model to extract the features of the input style image and the input content image, merge the extracted features to generate the content image with the input style. And we also design two regular items according to the characteristics generated by the Chinese ink painting style in the loss function. A series of experiment results demonstrate that our method can generate Chinese ink-and-ink-style images better.

Keywords Style transfer of Chinese ink painting · Deep learning · TV loss

1 Introduction

Before the invention of photography, paintings played an important role in spreading knowledge and civilization. However, with the rapid development of technology, humans no longer need paintings to record the history of civilization, but paintings still occupy an important position in modern society. We believe that artistic style is the biggest difference between painting and photo. Through these artistic styles, readers can appreciate the writer's mood when painting and feel the culture of the era in which the paintings are located. These are hard to be replaced by photographs. There are many styles in the field of painting, ranging from famous oil painting styles, abstract styles, classical styles, cubism styles, etc. to niche illustration styles.

Y. Jiang (✉) · H. Wang
School of Computer and Information Technology, Liaoning Normal University, Dalian, China
e-mail: jyy2903887735@163.com

H. Wang
e-mail: ww751584151@163.com

© The Author(s), under exclusive license to Springer Nature Switzerland AG 2022   473
N. H. T. Dang et al. (eds.), *Artificial Intelligence in Data and Big Data Processing*,
Lecture Notes on Data Engineering and Communications Technologies 124,
https://doi.org/10.1007/978-3-030-97610-1_37

It takes artists to spend a lot of time exploring to break through the existing artistic style and create a new and very personal style that belongs to themselves. However, due to the need for famous style paintings in the advertising and film industry, many artists waste a lot of time imitating the existing art style. In the face of this situation, researchers in many fields, including graphics, deep learning, etc., are trying to find a way to enable computers to produce images with any existing art style, which can reduce the time and economic cost of manual drawing. Then artists have more energy to devote themselves to the creation of artistic style.

In order to make the computer realize the technology of automatic drawing of artistic style images, the computer must first understand what the style of the picture is. In 1962, after Julesz et al. [1] pioneered that texture can be described by establishing a model based on statistical characteristics, a large number of researchers devoted themselves to the field of image texture generation in the 1970s and 1990s.

Based on the theory of Julesz et al., some scholars have successfully generated some texture through mathematical models and formulas [2, 3]. However, the establishment of the model is Time-Consuming and Labor-Intensive, and the texture generation effect is not good. Similar to the field of texture generation, the migration technology at this time also performs specific analysis based on images of a specific style, artificially establishes a dedicated mathematical model for the specific style and processes the image to be migrated to match the mathematical model established by the image. Since each transfer of a different style requires a different mathematical model, there is no exact name for the transfer of styles at that time. Researchers are working on their own, dealing with portrait styles and oil painting styles. Due to this limitation, image style transfer technology has stagnated. However, due to the birth of the game industry, the graphics card (GPU) has greatly improved the computing power of computers and promoted the rapid development of the field of deep learning. In such a large environment, the object recognition technology based on convolutional neural networks is becoming mature and the network model VGG19 for object recognition was proposed in the 2014 ILSVRC competition. In 2015, Gatys et al. [4, 5] creatively proposed the introduction of neural networks in style transfer. Based on the VGG19 model, machine learning styles of paintings can be trained to automatically generate images that match the input art images. Gatys et al. [4, 5] The results have made a qualitative leap in style transfer technology and on this basis, a large number of improved methods have been derived. However, because this method requires multiple iterations and back-propagation for optimization, it has high memory requirements and cannot guarantee generation speed when processing large images. In 2016, Ulyanov et al. [6] proposed a migration method based on training a feedforward generative network model. Compared with the method of Gatys et al., this method increased the generation speed by an order of magnitude and reduced memory consumption. The disadvantage is that although this method improves the speed of image migration, the network is restricted and cannot be arbitrarily applied to a new style [7]. In 2017, Huang and Belongie [8] achieved Real-Time arbitrary style conversion for the first time through the new AdaIN layer without reducing the speed of the feedforward model, but there were also limitations on the poor extraction effect of complex styles. In 2019, Chen et al.

[9] proposed a technique of style transfer based on the segmentation of the image foreground and background using a soft mask, which solved the previous limitation of processing the entire image. However, the processed foreground object is based on characters. It cannot meet the needs of segmentation for different prospects. In 2021, The ResNet network based on the traditional cyclic consistency network has the problem of gradient disappearance, which prevents the generator from continuing to learn. Lan Tian et al. [10] proposed the introduction of DenseNet's cyclic consistency confrontation network, which solved the problem of long training time for style transfer networks and avoided. The situation where the gradient disappears. In order to improve the performance of the generator, an attention mechanism is introduced, which improves the restoration of the output image color and the supplement of details, and improves the effect of image style transfer to a certain extent, but involves more parameters. In 2021, Li Chao et al. [11] proposed an image Multi-Style fusion algorithm based on deep learning. This algorithm can effectively increase the richness of image colors and solve the problem of color overflow. However, this method has the disadvantages that it consumes a long time, and the processing effect is closely related to the difference between the selected content image and the style image.

As one of the most distinctive Chinese paintings, ink painting has the characteristics of ethereal and hazy, subtle colors and rich artistic conception. Although the previously proposed techniques for style transfer are almost mature, the processing effect for ink paintings is not full enough. Taking the results of Gatys et al. as an example, this method is suitable for dealing with abstract and exaggerated works of art similar to Van Gogh's composition 'Star Moon Night', while migrating the ink painting style with ethereal artistic conception, the result did not meet our expectations.

With full consideration of traditional Chinese ink painting, we propose an improved style transfer method based on deep Learning for Chinese ink painting. Our contributions are as follows, we design a new loss function that added two regular terms, i.e. TV_loss and Smooth_loss. The new loss function can make our model can take more account of the image characteristics of Chinese ink painting. So our method can generate more texture of Close-Range objects and the blurring of the distant view, which truly conforms to the characteristics of ink painting.

The organizational structure of the paper is as follows, the first part introduces the development process of style transfer technology. The second part describes the methods used by the predecessors to transfer the style of ink painting images. In the third part, we combine the Above-Mentioned methods with the subtle and elegant characteristics of ink painting itself. The experimental method is improved, and two regular terms are added to realize image smoothing. The fourth part shows the results of the experiment and indicates the size of each parameter set in this experiment. The fifth part summarizes the experiment. Finally, the references of this article are listed.

2   Style Transfer Methods Based on Deep Learning

Image style transfer refers to inputting features and outputting generated pictures corresponding to such features. Before this article, Gatys et al. [5] proposed a method of using the VGG model to achieve image style transfer. They used the intermediate features extracted by the convolutional layer of the VGG model to generate content images corresponding to this feature. The VGG model can use the previous convolutional layer to extract the features of the input image, and use the following fully connected layer to convert the "features" of the image into category probabilities. Take the VGG19 model as an example. The model contains 16 convolutional layers and 3 fully connected layers. Different convolutional layers extract different features. The shallow convolutional layer can extract simple feature information such as color and texture in the image, while the deep convolutional layer will extract more complex High-Level feature information.

First, use noise to process the content image and use the processed noise image as the first generated result image (Fig. 1).

Use the VGG19 model to extract the feature performance of the input style image in the specified STYLE_LAYERS and the feature performance of the input content image in the specified CONTENT_LAYERS (Fig. 2).

Taking the mean square error between the generated image and the extracted content image as the loss, the automatic gradient descent algorithm is used to update the generated image to reduce the loss. After 5000 iterations, a result image that is similar to the style feature of the input style image is finally generated (Fig. 3).

2.1   *Loss Function*

The loss function is used to indicate the gap between the actual generated image and the expected generated image, including content loss and style loss. The loss function is usually used to calculate the loss to measure the effectiveness of the model. In order

Fig. 1 Initialize the generated image

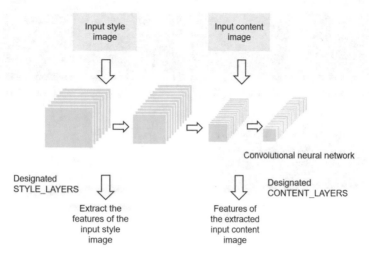

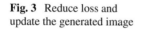

Fig. 2 Designated the convolutional layer to extract features

Fig. 3 Reduce loss and update the generated image

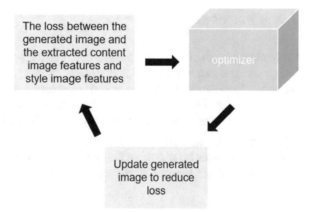

to ensure that the generated image is closer to the characteristics of the style image while retaining the original image content, the loss value needs to be continuously reduced. As shown in the formula in Fig. 4, c represents the input content image, s represents the input style image, and x represents the generated output image. The content loss and style loss are obtained by calculating the content difference between c and x and the style difference between a and x. α and β are the weight parameters of content loss and style loss, respectively.

$$Loss_{total}(\vec{c}, \vec{s}, \vec{x}) = \alpha Loss_{content}(\vec{c}, \vec{x}) + \beta Loss_{style}(\vec{s}, \vec{x}) \tag{1}$$

| Content image1 | Content image2 | Content image3 | Content image4 |

Style image

Fig. 4 Input style image and input content image

2.2 Content Loss

Content loss is used to measure the gap between the generated image and the content image and the mean square error is usually used to measure the similarity of two pictures. So we can use the mean square error between the input content image and the generated image to represent the content loss. We assume that the original image is G and the generated image is X, X_{ij}^{layer} is the feature representation of the layer generated image and G_{ij}^{layer} is the feature representation of the original content image of the Layer layer (where I represents the ith channel of the convolution and j represents the jth location of the convolution.) Content loss can be expressed as follows:

$$Loss_{content}(\vec{G}, \vec{X}, layer) = \frac{1}{2} \sum_{i,j} (X_{ij}^{layer} - G_{ij}^{layer})^2 \qquad (2)$$

The smaller the content loss, the higher the content similarity between the two pictures. Therefore, the gradient descent algorithm is used to continuously update the generated image and reduce the content loss, so as to achieve the goal that the generated image has a high degree of similarity with the content image.

2.3 Style Loss

The style loss is expressed by the mean square error of the Gram matrix between the input style image and the generated image. The Gram matrix calculates the correlation of feature planes captured across multiple layers. For pictures with similar styles, the corresponding Gram matrix has an approximate value.

The Gram matrix needs to flatten all the values of each channel, and then calculate it by multiplying the flattened value and its transpose. The so-called flattening is to ignore the spatial information in the image, which can be achieved by flattening the image (that is, multiplying the width and height of the image to tile the image into a One-Dimensional vector). Suppose that in the layer-th convolutional layer, the number of channels of the convolution feature is C, and the product of the width and height of the convolution is N, then the ith row and jth column of the Gram matrix corresponding to the convolutional feature of this layer is defined as:

$$Gram_{ij}^{layer} = \sum_k F_{i_k}^{layer} F_{j_k}^{layer} \tag{3}$$

$Gram_{ij}^{layer}$ meets the restrictions of $1 \leq i \leq C$, $1 \leq j \leq C$ and $1 \leq k \leq N$, Gramis actually the Gram matrix of the vector group F_1^{layer}, F_2^{layer} … F_i^{layer} … F_C^{layer}, and $F_i^{layer} = (F_{i_1}^{layer}, F_{i_2}^{layer} \ldots F_{i_k}^{layer} \ldots F_{i_N}^{layer})$.

Assuming that the input style image is A, the generated image is X, the Gram matrix of the input style image in the layer is A^{layer}, and the Gram matrix corresponding to the generated image in this layer is G^{layer}, then the style loss L_{style} can be expressed as

$$L_{style} = \frac{1}{4N_{layer}^2 M_{layer}^2} \sum_{ij} (X_{ij}^{layer} - A_{ij}^{layer})^2 \tag{4}$$

The denominator $4N_{layer}^2 M_{layer}^2$ is to normalize the style loss. The purpose is to prevent the magnitude of the style loss from being too large compared to the content loss. N is the product of the image width and height, and M is the number of channels of the image in the convolutional layer.

In practical applications, Multi-Layer style loss is often used instead of One-Layer style loss. Multi-Layer style loss is a weighted accumulation of Single-Layer style loss, namely

$$Loss_{style}(\vec{A}, \vec{X}) = \sum_{layer=0}^{L} \omega_{layer} L_{style_{layer}} \tag{5}$$

3 Proposed Methods

In order to highlight the obscurity and elegance of the ink painting, the generated image can be smoothed. We can ensure the smoothness of the image by increasing the regular term, controlling the total variation, and reducing the difference between adjacent pixels of the generated image. We use two regular terms TV_loss and Smooth_loss to achieve the purpose of smoothing the image. Note that because this experiment is the loss calculated on the feature domain obtained by convolution of the image, it should also be calculated in the feature domain when calculating the loss of TV_loss and Smooth_loss.

3.1 TV_loss

The total variation of TV_loss with order is defined as follows:

$$\Re_{V^\beta}(f) = \int_{\Omega} (\frac{\partial f}{\partial x}(x, y)^2 + \frac{\partial f}{\partial y}(x, y)^2)^{\frac{\beta}{2}} \tag{6}$$

However, when performing image processing operations, because the pixels of the image are discrete, the Above-Mentioned TV_loss total variation expression needs to be transformed from the integral in the continuous domain to the summation in the discrete domain of pixels. The expression is as follows:

$$\Re_{V^\beta}(x) = \sum_{i,j} ((x_{i,j-1} - x_{i,j})^2 + (x_{i+1,j} - x_{i,j})^2)^{\frac{\beta}{2}} \tag{7}$$

We multiply TV_loss by the weight (set TV_weight=2e−3) and count it into the total loss. By limiting the total variation, we can reduce the difference between adjacent pixels and achieve the effect of image smoothing.

$$L_{TV\_loss} = TV\_weight \times \Re_{V^\beta}(x) \tag{8}$$

3.2 Smooth_loss

In order to further achieve the purpose of image smoothing, we define a Smooth_loss function [12] to cooperate with the TV_loss regular term. Smooth_loss also achieves image smoothing by reducing the difference between adjacent pixels. Unlike the previously used TV_loss, the Smooth_loss is calculated by calculating the total variation of the input image:

$$I(x) = \sum \|\nabla x\|^2 \tag{9}$$

The total variation of the generated image:

$$O(x) = \sum \|\nabla x\| \tag{10}$$

We use the total variation difference between the input image and the generated image to represent the loss value returned by Smooth_loss

$$L_{\text{smooth}}(x) = \sum_{i=1}^{L} \omega_i O(x)(\omega_i = e^{I(x)}) \tag{11}$$

3.3 Total Loss

After adding two regularizations to smooth the image, the total loss expression is updated to:

$$\begin{aligned} Loss_{\text{total}}(\vec{c}, \vec{s}, \vec{x}) &= \partial(Loss_{content}(\vec{c}, \vec{x}) + \lambda L_{TV\_loss}(\vec{x}) \\ &\quad + \mu L_{smooth}(\vec{c}, \vec{x})) + \beta Loss_{style}(\vec{s}, \vec{x}) \end{aligned} \tag{12}$$

4 Results

This experiment uses multiple images of different input content to verify our improved method. By comparing the original model generated map and the generated map with the addition of TV_loss regular term and Smooth_loss regular term, it is concluded that the generated image obtained by adding TV_loss and Smooth_loss model processing is more elegant and hazy, which is the closest to the style of real ink painting. This experiment also compares the generated images with the generated images of other models and analyzes the differences between different models.

This experiment uses contentimage1.jpg, contentimage2.jpg, contentimage3.jpg, contentimage4.jpg as our input content image, and style.jpg as our style image. All the pictures mentioned above are from the Baidu website.[1]

The device properties are as follows: Device processor used: Intel(R) Xeon(R) CPU E3-1505 M v6 @3.00 GHz 3.00 GHz, Machine with RAM: 32 GB (31.8 GB available), System type: 64-bit operating system, based on ×64 processor, Software used: PyCharm 2019.3.3 × 64. The experiment is based on TensorFlow.

Set the number of iterations to 5000, the noise coefficient of 0.7, the width of the output image is 800, and the height of 600, the weight of the total variation regularization of TV_loss is set to 2e−3, and the weight of Smooth_loss is 0.6.

---

[1] http://image.baidu.com/

The effect of generating the original model and adding regular items TV_loss and Smooth_loss are shown below. According to the following four sets of comparisons, we can observe that although the general style of the original Model-Generated image has been formed, the texture of the foreground mountain is too disordered, and the background is not elegant enough. The effect of the image after adding TV_loss and Smooth_loss is significantly improved, such as The first set of effects shows that the texture of the foreground mountains is no longer too sharp, the background connection is also more coherent, and the whole is more blurred and hazy, which effectively improves the deficiencies of the original model (Fig. 5).

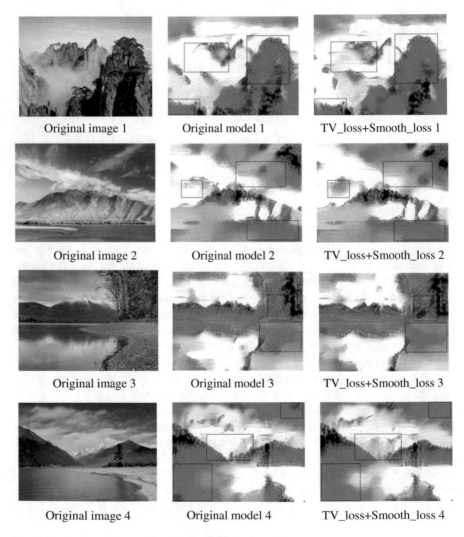

Original image 1   Original model 1   TV_loss+Smooth_loss 1

Original image 2   Original model 2   TV_loss+Smooth_loss 2

Original image 3   Original model 3   TV_loss+Smooth_loss 3

Original image 4   Original model 4   TV_loss+Smooth_loss 4

Fig. 5 Comparison of processing results of different models

5 Conclusion

5.1 Summary of Research Work

Image style migration is an important research direction in the field of computer image processing. This paper focuses on the optimization method of Chinese ink painting style migration based on VGG19. And we have proposed an optimized scheme. Aiming at the elegant and hazy characteristics of ink painting, we have introduced two regular terms: TV_loss and Smooth_loss. By limiting the total variation to reduce the difference between adjacent pixels of the generated image, the smoothness of the image is realized, and the texture is weakened, the background is coherent, and the overall hazy feeling is realized.

5.2 Future Expectations

Although the scheme proposed in this paper has achieved good experimental results in the task of style transfer of Chinese ink painting, there is still some work that needs to be optimized and perfected by us. But these issues will become the main research goals of our future work.

First of all, this article uses the VGG19 convolutional neural network to achieve image style transfer, and has achieved good experimental results, but it takes a lot of time in deep learning. Whether we can achieve the style conversion of high-quality traditional Chinese ink painting more efficiently and quickly is the main direction of our future work.

Secondly, the selected content image and style image have an important influence on the result of style transfer. Therefore, efficient selection of reasonable input image data and reducing the model's dependence on input data are also the work content that we need to continue to improve in the future.

Finally, the choice of hyperparameters is also very important. How to choose the most suitable hyperparameters for this experiment is also our work goal.

References

1. Julesz B (1962) Visual pattern discrimination. IRE Trans Inform Theory 8(2):84–92
2. Tyleek R, Sára R (2013) Spatial pattern templates for recognition of objects with regular structure. In: Gcpr
3. Portilla J, Simoncelli E (2000) A parametric texture model based on joint statistics of complex wavelet coefficients. Int J Comput Vis 40(1)
4. Gatys LA, Ecker AS, Bethge M (2015) Texture synthesis using convolutional neural networks. MIT Press
5. Gatys L, Ecker A, Bethge M (2016) A neural algorithm of artistic style. J Vis 16(12)

6. Ulyanov D, Lebedev V, Vedaldi A, Lempitsky V (2016) Texture networks: feed-forward synthesis of textures and stylized images
7. Li YJ, Fang C, Yang JM, Wang ZW, Lu X, Yang MH (2017) Diversified texture synthesis with feed-forward networks. In: 2017 IEEE conference on computer vision and pattern recognition (CVPR), pp 266–274. https://doi.org/10.1109/CVPR.2017.36
8. Huang X, Belongie S (2017) Arbitrary style transfer in real-time with adaptive instance normalization. IEEE
9. Chen C (2019) Research on image style transfer method based on foreground and background separation. North China University of Technology, China
10. Lan T, Xin YL, Yin XF, Liu WM, Jiang XY (2021) Unsupervised image style transfer based on generating adversarial network. Comput Eng Sci 43(10):1789–1795
11. Li C, Shi J, Jiang L (2021) Image multi-style transfer algorithm in deep learning. Softw Guide 20(07):171–176
12. Liu RS, Ma L, Zhang JA, Fan X, Luo ZX (2020) Retinex-inspired unrolling with cooperative prior architecture search for low-light image enhancement

Robust Median-Ternary Pattern for Traffic Light Detection

Thong Duc Trinh, Dinh Khanh Nguyen Diep, and Vinh Dinh Nguyen

Abstract Traffic light detection and classification still be challenging problems in driving assistance systems due to various unknown factors appearing along the driving roads, especially under night conditions. This research, therefore, introduces an efficient local feature that helps to make original feature input to be more robust under various driving conditions by investigating the benefits of ternary-encoding and global information. The proposed feature input is then transferred to the deep learning-based YOLO to detect and classify traffic light status. Experimental results show that the proposed method obtained stable performance with a detection rate of 73.18% and a classification rate of 92.22% under night conditions using the CCD dataset.

Keywords Local binary feature · Traffic light detection and classification · Hostile driving conditions

1 Introduction

A robust autonomous driving car often consists of several main features, such as obstacle detection, classification and segmentation, lane detection, traffic light, and traffic sign detection and classification. Currently, many giant companies, such as Google, Tesla, and Hyundai, have investigated much money to develop their own self-driving car. However, at this moment, there is no self-driving car that reaches

Supported by FPT University, Can Tho, Vietnam.

T. D. Trinh · D. K. N. Diep · V. D. Nguyen (✉)
FPT University Can Tho, Can Tho 9400, Vietnam
e-mail: vinhnd18@fpt.edu.vn
URL: https://cantho.fpt.edu.vn/

T. D. Trinh
e-mail: thongtdce130042@fpt.edu.vn

D. K. N. Diep
e-mail: dinhdnkce130209@fpt.edu.vn

level-5 because of failing to work under challenging driving conditions, such as night or snow. Recently, a self-driving Uber car killed a woman under a low-light condition at night because it cannot recognize a woman as a pedestrian [1].

This research, therefore, aims to investigate a new approach to improve the performance of traffic light detection and classification by introducing an efficient method that can encode more stable features under difficult driving conditions. Several algorithms have been introduced to produce a stable local feature under hard driving conditions (variation of illumination and unknown-noise conditions), such as local density encoding, support local pattern, local derivative patterns for various topics such as car detection, face detection, and texture classification. However, these existing local patterns still have a limitation to handle noise under real-driving conditions. This study, therefore, will introduce an efficient local feature method based on the median-ternary technique to produce a robust local feature for inputting to the deep learning model to detect and classify traffic lights. We verify the performance of the proposed system using the CCD dataset [2]. Experiment results showed that the proposed method provided good results under difficult conditions, such as at night with a detection rate of 73.18% and a classification rate of 92.22%.

The rest of this research paper is structured as follows. A survey of existing local pattern and object detection methods is conducted and discussed in Sect. 2. Section 3 describes the proposed method for producing an efficient feature for traffic light detection and classification. We verify the performance of the proposed system by conducting experiments on various datasets in Sect. 4. Finally, we conclude the research in Sect. 5.

2   Related Works

In recent years, many big companies and research institutes have been invested much money to develop robust driving assistance under various driving conditions. However, there are many challenging factors that affect the final accuracy of existing driving assistance systems, such as dark, illumination change, heavy rainy, night, and snowy. To solve the problems of reducing accuracy under difficult driving conditions, several researchers study the benefit of local patterns as processing time to filter noise in the input image as listed in Tables 1 and 2. Existing local patterns yield a robust feature by encoding the relationship of neighbor pixels within a local region. However, when the noise occurs, the existing local pattern might not handle well because of missing the global information. Motivated by the benefits of local ternary pattern [3] to handle noise, therefore, we proposed an efficient method by integrating the local feature with the global information along with the ternary information.

Recently, deep learning has been successfully applied in many computer vision, such as medical image processing, object detection, and classification, and behavior predictions [2]. Deep learning can be classified into two main classes: single-stage-based method and two-stage-based method. The single-stage-based methods, such as YOLO and SSD, achieve very fast processing time but their accuracy is lower than

Table 1 Survey on existing object detection for various computer vision applications

Author	Algorithm	Descriptions	Other	Year
[4]	Optical flow	The motion of vehicle detection, classification, and tracking algorithms in real time video surveillance		2019
[5]	Traditional method (artificial features, classifier) and Deep learning (DBN, RNN, CNN)	Combination of traditional methods and deep learning methods to deal with pedestrian occlusion and identify problems to be solved in the future	Pedestrian detection algorithms work well when the congestion level is from 0 to 10%, if the congestion level exceeds 50% then pedestrian detection becomes more difficult	2020
[6]		Summarizing related works about traffic lights detection (color segmentation, shape properties, classifiers)		2015
[7]	Color-based, shape-based, and machine learning based methods	Summarizing the popular traffic sign detection methods and discuss the future direction	The data sets use: German Traffic Sign Recognition Benchmark (includes 51,839 German traffic signs in 43 classes), Swedish Traffic Signs (more than 20,000 images in 7 classes) and Belgium Traffic Signs (more than 17,000 images in more than 100 classes)	2021
[8]	Using K-mean and Yolo-v5 to create anchor box for training and testing to detec traffic light	Achieve the accuracy of 63.33% and processing time 0.14 ms per frame	Experiment was conducted on BDD100K data set	2021
[9]	Combine Yolo-v3 and augment techniques for recognizing traffic light	Achieve the accuracy of 88%	Evaluate on Philippines data-set	2021

Table 2 Survey on existing local pattern for various computer vision applications

Author	Algorithm	Descriptions	Year
Local binary pattern			
[10]	Median binary pattern (MBP)	Median binary pattern (MBP) technique for texture classification, which aims to determine the localized binary pattern by thresholding the pix-els against their median value throughout a three-dimensional region 3×3	2007
[11]	Adaptive local binary patterns (ALBP)	? the mean and standard deviation of the local absolute difference is extracted to improve the classification efficiency of LBP ?	2009
[12]	Completed local binary pattern (CLBP)	CLBP creates a technique to effectively express the missing information in the LBP style, allowing for improved texture classification	2010
[13]	Local binary pattern variance (LBPV)	Using the LBP variance (LBPV) technique for texture classification, which treats the variance of each point as a weight of code value, and then accumulating the histogram	2017
[14]	Bayesian local binary pattern (BLBP)	Bayesian LBP (BLBP) is proposed to consider the matter of image formation, which is frequently overlooked by LBP techniques, resulting in incorrect accuracy and sensitivity to brightness fluctuation and noise. The dominant LBP (DLBP) texture classification approach, which captures descriptive textural information by using the most commonly occurring patterns	2008
[15]	Dominant local binary pattern (DLBP)	In the dominant LBP technique (DLBP), the most commonly occurring patterns are used to describe texture	2009
[16]	Robust local binary pattern	The target of RLBP is to create a descriptor that is resistant to point noise (e.g., Gaussian, salt and pepper, or Rayleigh noise)	2013
Local ternary pattern			
[17]	Completed local ternary pattern (CLTP)	The local ternary pattern is more noise resistant than the LBP. Additionally, build the related full Local Ternary Pattern, which will help to improve and expand its discriminating property	2014
Local density pattern			
[18]	Local density encoding (LDE)	A local pattern that takes the greatest similarity of the intensity of all pixels in a local region into consideration	2013
[19]	Local density shape context (LDSC)	The goal of the LDSC algorithm is to find the optimal spatial correspondence from the images	2018
[20]	Local density-based abnormal detector (LDBAD)	The purpose of LDBAD to find local abnormal data objects, which are characterized through three proposed measurements: local distance, local density, and influenced outliers degree	2021

the two-stage-based method, such as Faster RCNN, and ResNet. The two-stage-based methods achieved high accuracy, however, their processing is slow because of the generation region proposal stage. The performance of both single-stage and two-stage-based methods is still not satisfied with a driving assistance system. This research, therefore, proposed a robust feature pattern to improve the accuracy of the traffic light detection and classification of the existing model under difficult environmental conditions.

3 Proposed Method

The proposed system consists of two main stages: first, the input image was inputted to the proposed feature model to enhance its robustness. After that, the robust feature is then inputted to the deep learning model YOLOV5 [21] for detecting and classifying traffic light status as shown in Fig. 1.

Many computer vision applications, such as face detection and recognizing, texture classification, image retrieval, and object detection, have investigated the benefits of the local binary pattern (LBP) to improve their performance under illumination changes or noise conditions. Without loss of the generality, the LBP [17] can be described as follows:

$$
L_{binary}^{M,B}(x_c, y_c) = \sum_{m=0}^{M} \alpha(P_m, P_c) \times 2^m
$$
$$
P_m = (x_m, y_m); \ P_c = (x_c, y_c) \tag{1}
$$
$$
\alpha(P_m, P_c) = \begin{cases} 1, \ I_{pixel}(P_m) > I_{pixel}(P_c) \\ 0, \ else \end{cases}
$$

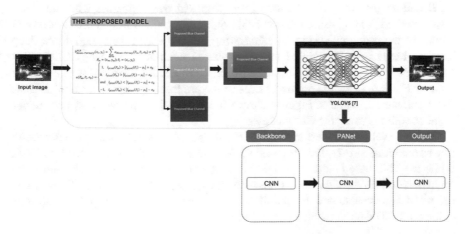

Fig. 1 The proposed system for traffic light detection and classification under night conditions

where P_c is the pixel intensity value of the center pixel, and P_m is the pixel intensity value of the m-th adjacent pixel. M is the number of adjacent pixels within the radius B. The LBP is computed by considering the relationship between the central pixel and adjacent pixels. However, the LBP fails to operate if the central pixel is affected by noise.

Therefore, this research introduces an efficient method to overcome the limitation of the LBP by providing a global-based feature to the LBP as follows:

$$
\begin{aligned}
P_{Mean-Ternary}^{M,B}(x_c, y_c) &= \sum_{m=0}^{M} \alpha_{Mean-Ternay}(P_m, P_c, \sigma_B) \times 2^m \\
P_m &= (x_m, y_m); \ P_c = (x_c, y_c) \\
\alpha(P_m, P_c, \sigma_B) &= \begin{cases} 1, \ I_{pixel}(P_m) > \left|I_{pixel}(P_c) - \mu_I\right| + \sigma_B \\ 0, \ I_{pixel}(P_m) > \left|I_{pixel}(P_c) - \mu_I\right| - \sigma_B \\ and \ I_{pixel}(P_m) < \left|I_{pixel}(P_c) - \mu_I\right| + \sigma_B \\ -1, \ I_{pixel}(P_m) < \left|I_{pixel}(P_c) - \mu_I\right| - \sigma_B \end{cases}
\end{aligned}
\tag{2}
$$

where σ_B is the standard variation of pixel intensity value within adjacent pixels of the central pixel. μ_I is the mean pixel intensity value of the entire input image. Thus, the proposed method is more robust than LBP because of the benefit of global information μ_I and local information σ_B.

4 Experimental Results

The proposed system was trained by using Google Collab [22] with CPU: Intel(R) Xeon(R) CPU @ 2.30 GHz (4 cores), GPU: Tesla P100-PCI-E-16GB, RAM 25.46 GB. We verified the performance of the proposed system by using the CCD dataset [2]. The performance of the proposed was compared with the Yolo-based method [21] by using the raw RGB input. Figure 2 described the experiment results of the proposed system by integrating the proposed robust model and deep learning-based Yolo. We applied the augmentation technique [21] to the CCD dataset to increase input size for training. The proposed method was tested by using 1262 images. Figure 3 show the training status of the proposed system in term of the loss functions after 15 epochs. Figure 4 shows the result of the proposed system under night conditions using the CCD dataset.

We also evaluate the performance of the proposed system by using the detection rate as mentioned in [2]. The proposed system obtained the detection rate of 73.18%, while the Yolo-based method [21] obtained the detection rate of 32.11% under the night condition under night conditions using the CCD dataset as shown in Table 3. The proposed system obtained the classification rate of 92.22% under night conditions using the CCD dataset as shown in Table 3.

Fig. 2 Experimental results of the proposed system under normal driving conditions using the traffic light dataset [2]. Experimental results shown that the proposed method successfully detect and classification traffic light status under normal driving conditions

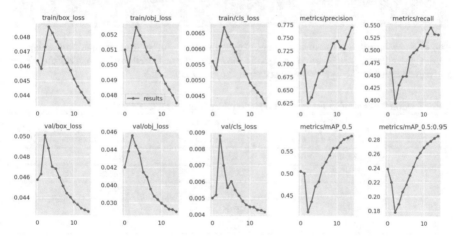

Fig. 3 Loss function of the proposed system after 15 epochs

The results stated that the proposed system obtained a promising performance under night conditions. However, the proposed still has a limitation of computation time of the prepossessing step by using the proposed pattern-based method. Therefore, we plan to implement the proposed method on a graphic processing unit (GPU) to accelerate the processing time of the proposed method.

Fig. 4 Experimental results of the proposed system under normal driving conditions using the traffic light datataset [2]. Experimental results shown that the proposed method successfully detect and classification traffic light status under night driving conditions

Table 3 Comparison of YOLOv5s and our model by using the CCD dataset

	YOLOv5s	Our model	
IoU ratio	Detection rates (%)	Detection rates (%)	Classification rates (%)
0.5	21.60	64.60	83.08
0.45	27.10	69.41	88.29
0.4	32.11	73.18	92.22
0.35	36.68	76.55	96.31
0.3	40.71	78.49	98.96

5 Conclusion

Traffic light detection and classification still be an open issue in developing a robust driving assistance system under various driving conditions. This research, therefore, develops a new local pattern that can help to improve the performance of the proposed system under difficult driving conditions, such as at night. We verified the performance of our system by using the CCD dataset. The results stated that the proposed system obtained a promising performance under night conditions.

However, the proposed still has a limitation of computation time of the prepossessing step by using the proposed pattern-based method. Therefore, we plan to implement the proposed method on a graphic processing unit (GPU) to accelerate the processing time of the proposed method.

References

1. NBCnews Homepage (2019). https://www.nbcnews.com/tech/tech-news/self-driving-uber-car-hit-killed-woman-did-not-recognize-n1079281
2. Nguyen VD, Tran DT, Byun JY, Jeon JW (2019) Real-time vehicle detection using an effective region proposal-based depth and 3-channel pattern. IEEE Trans Intell Transp Syst 20(10):3634–3646
3. Mahmoodi K, Ketabdari MJ, Vaghefi M (2014) Completed local ternary pattern for rotation invariant texture classification. Pattern Anal Appl
4. Jamiya S, Esther Rani P (2019) A survey on vehicle detection and tracking algorithms in real time video surveillance. Int J Sci Technol Res 8:2266–2276
5. Ning C, Menglu L, Hao Y, Xueping S, Yunhong L (2020) Survey of pedestrian detection with occlusion. Complex Intell Syst 7:577–587
6. Diaz M, Cerri P, Pirlo G, Ferrer M, Impedovo D (2015) A survey on traffic light detection. In: New trends in image analysis and processing—ICIAP 2015 workshops, pp 1–8
7. Ellahyani A, Eljaafari I, Charfi S (2021) Traffic sign detection for intelligent transportation systems: a survey. In: E3S web of conferences, vol 229, pp 2–10
8. Yan S, Lui X, Qian W, Chen Q (2021) An end to end traffic light detection algorithm based on deep learning. In: International conference on security, pattern analysis, and cybernetics, pp 370–373
9. Acoba AG, De Los Trinos MIP, Franco Cunanan C, Guerrero ND, Casuat CD (2021) A deep neural inferencing approach of assistive Philippine traffic light recognition: an augmented transfer learning approach. In: 2021 International conference on computational intelligence and knowledge economy (ICCIKE), pp 307–310
10. Hafiane A (2007) Median binary pattern for textures classification. In: Lecture notes in computer science, vol 4633, pp 387–398
11. Shen H (2009) Adaptive local binary patterns for 3d face recognition. In: 2009 Chinese conference on pattern recognition, pp 1–4
12. Guo Z (2010) A completed modeling of local binary pattern operator for texture classification. IEEE Trans Image Process 19(6):1657–1663
13. Mahmood T, Irtaza A, Mehmood Z, Mahmood MT (2017) Copy-move forgery detection through stationary wavelets and local binary pattern variance for forensic analysis in digital images. Forensic Sci Int 279:8–21
14. He C (2008) A Bayesian local binary pattern texture descriptor. In: 2008 19th International conference on pattern recognition, pp 1–4
15. Liao S (2009) Dominant local binary patterns for texture classification. IEEE Trans Image Process 18(5):1107–1118
16. Chen J (2013) RLBP: robust local binary pattern. In: Research gate, proceedings of the British machine vision conference, pp 122.1–122.10
17. Rassem TH, Khoo BE (2014) Completed local ternary pattern for rotation invariant texture classification. Sci World J 2014
18. Vinh ND (2014) Local density encoding for robust stereo matching. IEEE Trans Circ Syst Video Technol 24(12):2049–2062
19. Zhou Y, Cui P, Wang Y, Wang L, Yang S, Zhou X (2018) A local density shape context algorithm for point pattern matching in three dimensional space. Tehnicki vjesnik Technical Gazette 25(3)

20. Mahmoodi K, Ketabdari MJ, Vaghefi M (2021) Proposing a new local density estimation outlier detection algorithm: an empirical case study on flow pattern experiments. Pattern Anal Appl
21. Redmon J, Divvala S, Girshick R, Farhadi A (2016) You only look once: unified, real-time object detection. In: IEEE Conference on computer vision and pattern recognition, pp 779–788
22. Google Colab Homepage (2021). https://colab.research.google.com/

Stereo Matching Algorithm Based on Census Transform and Segment Tree Cost Aggregation

Madiha Zahari, Rostam Affendi Hamzah, Nurulfajar Abd Manap, and Haidi Ibrahim

Abstract Stereo matching is a complicated process due to the uneven distribution of textures, brightness and exposure to the image pair. Aiming to improve the accuracy of the stereo matching algorithm and increase the quality of the disparity maps, various stereo matching algorithms had been proposed recently. This paper proposed the stereo matching algorithm based on Census Transform at matching cost computation, followed by the non-local Segment Tree cost aggregation. WTA strategy is used in cost optimization and a weighted median filter is used in the final stage of the stereo matching process. The experimental results show that the combination of CT and ST is working effectively in terms of the absolute error and is comparable to the combination of CT and GF.

Keywords Stereo matching · Disparity map · Census transform · Segment tree · Non-local

M. Zahari (✉) · N. A. Manap
Faculty of Electronic and Computer Engineering, Universiti Teknikal Malaysia Melaka, Melaka, Malaysia
e-mail: madiha@utem.edu.my

N. A. Manap
e-mail: nurulfajar@utem.edu.my

R. A. Hamzah
Faculty of Electrical and Electronic Engineering Technology, Universiti Teknikal Malaysia Melaka, Melaka, Malaysia
e-mail: rostamaffendi@utem.edu.my

H. Ibrahim
School of Electrical & Electronic Engineering, Engineering Campus, Universiti Sains Malaysia, Pulau Pinang, Malaysia
e-mail: haidi@usm.my

© The Author(s), under exclusive license to Springer Nature Switzerland AG 2022   495
N. H. T. Dang et al. (eds.), *Artificial Intelligence in Data and Big Data Processing*,
Lecture Notes on Data Engineering and Communications Technologies 124,
https://doi.org/10.1007/978-3-030-97610-1_39

1 Introduction

Stereo matching technology recently become an important research area in the field of computer vision proportional to the increasing usage of three-dimensional reconstruction, virtual reality, unmanned driving, automated industrial monitoring, and vision-based objects [1–3]. The aims of providing accurate depth information of the images are the important criteria in stereo matching. Depth information from stereo images is acquired commonly by using stereo matching algorithm where the disparity value between left and right images is calculated. The process of determining the depth value of the stereo images are depending on how the stereo matching algorithm is designed to calculate the disparity value. The process can be categorized into two broad categories, namely the global method and the local method [4]. The global method calculates the disparity by optimizing a general energy function of all pixels within an image. Although these methods are highly accurate, they have a high degree of computing complexity [5]. On the other hand, the local method makes the stereo matching process less accurate than the global method. However, due to the lower complexity of the calculation volume, local methods are commonly employed in most stereo coupling systems.

Stereo matching aims to determine the disparity value denotes by d of a point at the position (x, y) in the left images corresponding to the point of $(x\text{-}d, y)$ in the right images [6]. Image rectification is applied to captured images as a pretreatment algorithm, to ensure that disparity can be easily achieved [7]. Scharstein and Szeliski [8] introduced the taxonomy of stereo matching in four steps that begin with the calculation of the matching cost, followed by cost aggregation, optimization of disparity and improvement of disparity. One of the challenging issues in the stereo matching algorithm is to find the corresponding point accurately in less texture region and in the brightness difference region [6]. In both regions, the possibility of the error matching process is high because the difficulty of finding the corresponding point between left image and right image. A lot of research in the process of developing a stereo correspondence algorithm has been done to find an accurate depth value, however there is still no perfect solution to this issue.

Various approaches were used to calculate matching costs. The most commonly used method recently is the nonparametric local transformation introduced by Zabih et al. [9] that includes Census Transform (CT) and Rank Transform (RT). Both methods are more robust to radiometric distortion compared with the previous method such as the sum of absolute difference (SAD), the sum of squared difference (SSD), and normalized cross-correlation (NCC) which are very sensitive to the amplitude distortion which leads to low accuracy [10]. Based on the evaluation of the similarity measures between CT and other matching costs, CT achieved the highest overall performance in the radiometric distortion environment either in simulation or real-time [11]. Due to this factor, researchers began to use CT to calculate the corresponding costs. Work done by Lee et al. [12] used three moded CT as the matching cost computation which resulted in a more accurate disparity map in the stereo matching algorithm. The combination between CT and other matching cost methods such as the combination of CT with gradient, CT with SAD also gained attention produced more accurate disparity maps [13–15]. The modified and improved CT also become popular method in matching cost computation which produced better improvement in disparity accuracy [16–18].

The second stage of stereo matching is the cost aggregation process, which has a direct impact on the accuracy and efficiency of the global coupling algorithm. In this stage, the most common local method usually adopted cost aggregation based on filtering where the process of filtering the cost volume occurred. The simple method used basic low pass filter such as Box Filter (BF) or Gaussian Filter However, this method unable to filter out the noise and produced fatten egde disparity map which lead to inaccuracy of depth map. Edge preserving filter such as Guided Filter (GF) and Bilateral Filter (BF) produces better results while keeping the fine edges [19]. GF gains more popularity compared to BF since it is more efficient and produces better results. Work done by Hosni et al. [20] showed that GF is more robust to be applied and produce better disparity map. Yang [21] proposed a nonlocal cost aggregation based on the minimum spanning tree (MST). Different from the common window-based method, the nonlocal method performs cost aggregation based on the overall pixels in the images. Mei et al. [22] proposed a segment tree (ST) structure to perform the nonlocal aggregation strategy and reported that aggregation using ST outperforms the performance of the local methods. Taking the advantages of the proposed cost aggregation by Mei and considering the advantages of CT from previous study. This research proposed a new local stereo matching algorithm which used CT as a matching cost computation method and ST as cost aggregation which aimed to reduce the stereo matching error and improve the accuracy of the disparity map. As in almost all studies, winner-take-all is used to calculate the initial disparity map and subject the initial disparity map to the weighted median filter for final disparity map. The performance of the proposed stereo matching algorithm is analyzed and compared with other matching method.

2 Proposed Stereo Matching Algorithm

This paper proposed method where matching cost computation is based on nonparametric local transform using Census Transform (CT). The cost aggregation is implemented by using then nonlocal Segment Tree (ST) aggregation. Then, the optimization step adopts a winner-take-all (WTA) approach which the well-known methods used for determining the initial disparity map [23]. The WTA strategy select the minimum aggregated corresponding value for each valid pixel. At this point, certain invalid or undesirable pixels always occur at occlusion and less textured areas. These undesired pixels will be detected by the Left-Right Consistency (LR) process [19]. Next, the filling process is carried out to replace invalid pixels by a valid minimum pixel value. Finally, the disparity enhancement step is implemented the weighted median filter.

2.1 Matching Cost Computation

Census Transform (CT) is the mapping process of the neighbouring surrounding pixel to a bit string which can represent the intensity value of the neighbouring pixel [9]. The transformations are based on Eq. (1):

$$CT(p) = \otimes_{q \in w_{CT}} cen(p, q) \tag{1}$$

where $\otimes$ refer to a bit wise catenation, p and q denote the target pixel and neighboring pixels respectively. The w_{CT} is the support window and cen(p, q) represents the binary function with the conditions as given by Eq. (2):

$$cen(p, q) = \begin{cases} 1, & I(p) \geq I(q) \\ 0, & otherwise \end{cases} \tag{2}$$

where I(p) and I(q) are the target pixel and neighboring pixels values respectively. By using the census based stereo matching, the matching cost is calculated by calculating the distance of the transformed pixel which stored in disparity space image [9]. The Hamming Distance (HD) is used to calculate the cost between two bit strings which is given by Eq. (3):

$$CT'(p, d) = HD(CT_l(p) - CT_r(p - d)) \tag{3}$$

where CT_l is the left bit string and CT_r is the right bit string. The left image is the reference image and the right image shifts horizontally from right to left. The initial matching cost was calculated based on Eq. (4):

$$M_c(p, d) = CT'(p, d) \tag{4}$$

2.2 Segment Tree for Cost Aggregation

In this work, the initial matching cost was aggregated using non local graph based segment tree cost aggregation method adopted from previous work done by Mei et al. [22]. In this paper, the non local cost aggregation is computed based on the following equation:

$$CA(p, d) = \sum_{q \in i} M_c(p, d) S(p, q) \tag{5}$$

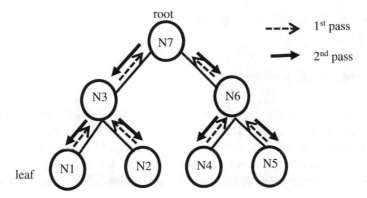

Fig. 1 The path for cost aggregation start from leaf to root and return from root to leaf

where q covers every pixel in the image I and S(p, q) denotes the weighted function which defined as follows:

$$S(p, q) = \exp\left(-\frac{D(p, q)}{\sigma}\right) \tag{6}$$

where $D(p, q)$ represent the distance of sum of edge weight along the path. The aggregation process is illustrated in Fig. 1 where the agrregation point starts from the leaf node to the root node for the first pass while the aggregation point start racing from root to leaf for the second pass.

The cost value at pixel p, is updated once all its children have been visited. The final cost aggregation of the inital matching cost at pixel p, with its parent $Pr(p)$ are based on the segment tree proposed by [22] defined as Eq. (7):

$$CA(p, d) = S(Pr(p), p) * CA(Pr(p), p) + (1 - S^2(Pr(p), p)) * CA^{\bar{\uparrow}}(p, d) \tag{7}$$

where $CA^{\bar{\uparrow}}(p, d)$ denotes the intermidiate aggregated costs, and set of cost of the children of pixel p.

2.3 Disparity Optimization

After completing the matching aggregation, Winner-Takes-All (WTA) strategy was implemented at this step in order to select the minimum matching cost as the initial disparity value [24]. The initial disparity, d_i is define as Eq. (8):

$$d_i = \arg min_{d \in R} CA(p, d) \tag{8}$$

where R is all the possible disparity value.

2.4 Disparity Refinement

The refinement of disparity consists of several steps, starting with the processing of the invalid disparity detection, then filling in the invalid disparity value and filtering using a weighted median filter. Invalid disparity values are determined by the inconsistency check process on the left disparity map relative to the right disparity map. The invalid disparity values are determined based on Eq. (9):

$$d_{LR}(p) - d_{RL}(p - d_{LR}(p)) \leq \tau_{LR} \tag{9}$$

where d_{LR} and d_{RL} refer to the left reference and the right reference of disparity maps. The τ_{LR} denotes the threshold value outlier detection at targeted position, p which set to 0. After the invalid disparity values are determined, the process of fill-in the invalid values take placed. In this process, the pixel value at invalid point is replaced with pixels values by refering to the left images. The invalid disparity values is replaced with a nearest valid disparity value where the location of the valid values must be at the same scan line or at the starting scan line as shown by Eq. (10):

$$d(p) \begin{cases} d(p - i), & d(p - i) \leq d(p + j) \\ d(p + j), & otherwise \end{cases} \tag{10}$$

where d(p) is a disparity value at the location of p, d(p − i) represents the location of the first valid disparity on the right side. After completing the fill-in process, the weighted median filter from [11] is adopted to refine the disparity map. Let D denote the disparity map obtained after normal-based plane fitting. The final disparity map,cD^F is refined as follows:

$$D^F(p) = \underset{q \in \Omega}{med}\{d(p)\} \tag{11}$$

3 Experimental Results and Discussion

The experiment was carried out to evaluate and determine the performance of the proposed algorithm from various perspectives. The results were evaluated based on the accuracy and the quality of the disparity map by using Middlebury Dataset 2014 which consists 15 training images [25]. Table 1 summarized the overall results of the absolute error for all pixels (all) and non occluded (nonocc) region based on three different aggregation. The results show that the quality of the disparity images using

Table 1 The comparative results of average error in the non-occluded region (nonocc) and all regions (all) while using CT as matching cost and three different cost aggregation

Method	Step 1 + GF		Step 1 + NL		Proposed method	
	%		%		%	
	nonocc	all	nonocc	all	nonocc	all
Adiron	2.59	3.59	3.88	4.63	3.31	3.79
ArtL	4.79	10.1	4.83	8.51	4.68	8.15
Jadepl	12.9	32.1	15.6	35.4	13.5	32
Motor	3.29	5.9	3.3	5.91	3.31	5.86
MotorE	3.18	5.67	3.38	5.9	3.41	5.98
Piano	4.34	5.4	5.09	5.99	4.33	5.24
PianoL	10.5	11.1	9.74	10.1	10	10.8
Pipes	6.64	12.1	6.48	11.8	5.97	11
Playrm	5.33	7.72	5.57	9.08	5.22	8.92
Playt	41.2	41.5	39.7	40	41.1	41.2
PlaytP	13	15.9	11.6	13.7	12.5	14.7
Recyc	3.16	3.55	3.92	4.18	3.17	3.45
Shelvs	8.64	8.99	6.53	6.99	7.13	7.56
Teddy	2.13	3.63	2.24	3.16	2.2	3.41
Vintage	10.8	12.1	17.8	18.5	18.4	18.9
Average	**7.53**	**11**	**8**	**11.3**	**7.78**	**11**

the ST as the cost aggregation method is comparable with the GF which commonly used in stereo matching as an edge preserving filter and outperform the basic non local method. Figure 2 shows the enlarge images of the experimental results of the Adirondack image based on three different cost aggregation. The results of the Adirondack show that the disparity images produce by using ST cost aggregation have better fine edge compared to the GF and NL. The results shows that the proposed algorithm is performed better in the pair of image with difference illumination such as ArtL and PianoL. The absolute error at all image regions for proposed algorithm also lower compared to the other two methods. It shows that the proposed methods outperform the quality of GF which commonly as edge preserving filter, although the overall accuracy GF still slightly higher.

Figure 3 shows the 15 images of experimental result of the final disparity map based on the proposed results. Table 2 shows the quantitative evaluation of the proposed algorithm compared with other framework. The overall results shows that the average absolute error of the proposed results are better and comparable with other method. The results of the proposed algorithm outperform other framework, especially at for image with different brightness and different exposure. The results from Fig. 3 can conclude that the algorithm proposed is robust upon processing different type of input image.

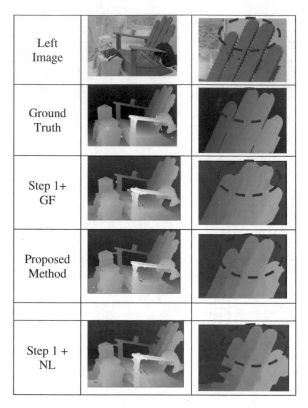

Fig. 2 The disparity map of Adirondack image based on three different cost aggregation

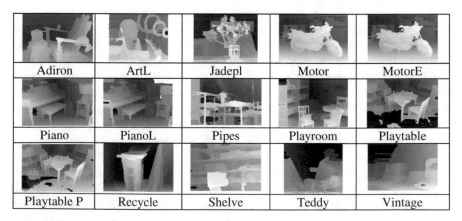

Fig. 3 The disparity map results of the Middlebury training dataset

Table 2 The comparison of average error in the non-occluded region (*nonocc*) and all region (*all*) between proposed algorithm and other framework

Method	SGMBP [26]		ADSR_GIF [27]		ReS2tac [28]		ADSM [29]		Proposed	
	%		%		%		%		%	
	all	nonocc	all	nonocc	all	nonocc	all	nonocc	all	nonocc
Adiron	6.5	3.87	6.4	4.84	5.39	3.48	14.3	13.3	3.79	3.31
ArtL	9.33	4.96	9	4.62	20.4	7.6	10.6	6.1	8.15	4.68
Jadepl	56.8	29.3	26.1	16.1	41.3	20.1	34.1	15	32	13.5
Motor	5.04	3.45	8.11	4.58	7.94	3.86	6	3.67	5.86	3.31
MotorE	5.43	3.89	11.4	7.72	7.88	3.89	8	5.67	5.98	3.41
Piano	4.77	3.82	6.15	5.2	10.4	5.97	7.37	7.08	5.24	4.33
PianoL	14.8	14.4	34	34.4	16.3	13.1	20.4	20.6	10.8	10
Pipes	7.85	3.94	14.9	7.53	17.9	8.82	12.1	6.57	11	5.97
Playrm	7.62	5.09	10.5	5.05	25.1	9.48	16.9	13.2	8.92	5.22
Playt	10.6	9.74	16.7	13	14.9	10.6	25.5	23.1	41.2	41.1
PlaytP	3.78	2.7	10	5.67	8.27	4.71	5.84	3.55	14.7	12.5
Recyc	3.19	2.91	4.2	3.37	4.34	3.19	5.83	5.76	3.45	3.17
Shelvs	5	4.64	9.97	9.49	16.1	14.7	17.2	17.2	7.56	7.13
Teddy	3.35	1.8	3.35	2.15	6.67	2.72	4.11	3.05	3.41	2.2
Vintage	30	26.1	10.9	9.64	15.2	12.3	11.1	10.1	18.9	18.4
Average	**11.2**	**7.25**	**11.3**	**7.81**	**13.9**	**7.55**	**12.3**	**8.95**	**11**	**7.78**

Figure 4 shows the comparison of the disparity map between the proposed methods and other method using the stereo image of Australia and Bicycle2 from Middlebury test dataset as input image. The experimental results show that the proposed method able to preserve the edge of the image better than the methods that established in the framework.

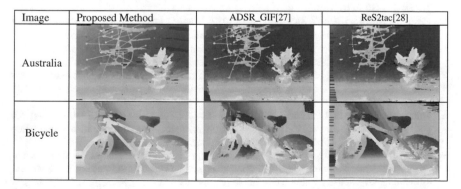

Image	Proposed Method	ADSR_GIF[27]	ReS2tac[28]
Australia			
Bicycle			

Fig. 4 Example of disparity map produced using proposed algorithm and using another method

4 Conclusion

In this paper, we proposed a stereo matching algorithm using CT at matching cost computation and the cost volume is aggregated by non-local cost aggregation method using ST, WTA at optimization step and WM filtering at the post processing step. The experiment results show that the proposed algorithm able to reduced the error and improve the accuracy of the final disparity map. Based on the experiment perform using Middlebury training dataset, preliminary results shows that the proposed algorithm able to reduce the average absolute error in the less texture image and at different brightness of the images while preserving the edge of the image. The experimental results show that the combination of CT and ST is working effectively in terms of the absolute error and is comparable with the combination CT and GF. In future, the proposed algorithm will be tested using more indoor and outdoor dataset.

Acknowledgements This work was supported by the Universiti Teknikal Malaysia Melaka.

References

1. Dixit M, Srimathi C, Doss R, Loke S, Saleemdurai MA (2020) Smart parking with computer vision and IoT technology. In: 2020 43rd international conference on telecommunications signal processing. TSP 2020, pp 170–174. https://doi.org/10.1109/TSP49548.2020.9163467
2. Zeng K, Cao G (2021) Application of VR technology in museum narrative design with computer vision models. In: Proceedings—5th international conference on computing methodologies and communication (ICCMC 2021), pp 913–916. https://doi.org/10.1109/ICCMC51019.2021.9418483
3. Li Y, Zhang Y (2020) Application research of computer vision technology in automation. In: Proceedings of the 2020 international conference on big data in management (CIBDA 2020), pp 374–377. https://doi.org/10.1109/CIBDA50819.2020.00090
4. Hou Y, Liu C, An B, Liu Y (2022) Stereo matching algorithm based on improved census transform and texture filtering. Optik (Stuttg) 249:168186 (October 2021). https://doi.org/10.1016/j.ijleo.2021.168186
5. Zhang K, Li J, Li Y, Hu W, Sun L, Yang S (2012) Binary stereo matching. In: Proceedings of the 21st international conference on pattern recognition (ICPR2012), pp 356–359
6. Dong Q, Feng J (2019) Outlier detection and disparity refinement in stereo matching. J Vis Commun Image Represent 60:380–390. https://doi.org/10.1016/j.jvcir.2019.03.007
7. Kang YS, Ho YS (2011) An efficient image rectification method for parallel multi-camera arrangement. IEEE Trans Consum Electron 57(3):1041–1048. https://doi.org/10.1109/TCE.2011.6018853
8. Scharstein D, Szeliski R, Zabih R (2001) A taxonomy and evaluation of dense two-frame stereo correspondence algorithms. In: Proceedings IEEE workshop on stereo and multi-baseline vision (SMBV 2001), pp 131–140. https://doi.org/10.1109/SMBV.2001.988771
9. Zabih R, Woodfill J (1994) Non-parametric local transforms for computing visual correspondence. In: Lecture Notes in Computer Science book series (including subseries Lecture notes in artificial intelligence and lecture notes in bioinformatics), vol 801 LNCS, pp 151–158. https://doi.org/10.1007/bfb0028345
10. Chai Y, Cao X (2018) Stereo matching algorithm based on joint matching cost and adaptive window. In: Proceedings of 2018 IEEE 3rd advanced information technology, electronic and

automation control conference (IAEAC 2018), pp 442–446. https://doi.org/10.1109/IAEAC.2018.8577495

11. Hirschmuller H, Scharstein D (2009) Evaluation of stereo matching costs on images with radiometric differences. IEEE Trans Pattern Anal Mach Intell 31(9):1582–1599. https://doi.org/10.1109/TPAMI.2008.221

12. Lee Z, Juang J, Nguyen TQ (2013) Local disparity estimation with three-moded cross census and advanced support weight. IEEE Trans Multimed 15(8):1855–1864. https://doi.org/10.1109/TMM.2013.2270456

13. Zhu C (2019) Hierarchical guided-image-filtering for efficient stereo matching

14. Wang W, Yan J, Xu N, Wang Y, Hsu F (2015) Real-time high-quality stereo vision system in FPGA. IEEE Trans Circuits Syst Video Technol 25(10):1696–1708. https://doi.org/10.1109/TCSVT.2015.2397196

15. Lee S, Lee JH, Lim J, Suh IH (2015) Robust stereo matching using adaptive random walk with restart algorithm. Image Vis Comput 37:1–11. https://doi.org/10.1016/j.imavis.2015.01.003

16. Rukundo O (2020) 4 × 4 census transform. In: 5th international conference on signal and image processing (ICSIP 2020), pp 381–385. https://doi.org/10.1109/ICSIP49896.2020.9339425

17. Luan X, Yu F, Zhou H, Li X, Song D, Wu B (2012) Illumination-robust area-based stereo matching with improved census transform. In: Proceedings of 2012 international conference on measurement, information and control (MIC 2012), vol 1, pp 194–197. https://doi.org/10.1109/MIC.2012.6273254

18. Zhu S, Yan L (2017) Local stereo matching algorithm with efficient matching cost and adaptive guided image filter. Vis Comput 33(9):1087–1102. https://doi.org/10.1007/s00371-016-1264-6

19. Yang Q, Ji P, Li D, Yao S, Zhang M (2014) Fast stereo matching using adaptive guided filtering. Image Vis Comput 32(3):202–211. https://doi.org/10.1016/j.imavis.2014.01.001

20. Hosni A, Rhemann C, Bleyer M, Rother C, Gelautz M (2013) Fast cost-volume filtering for visual correspondence and beyond. IEEE Trans Pattern Anal Mach Intell 35(2):504–511. https://doi.org/10.1109/TPAMI.2012.156

21. Yang Q (2012) A non-local cost aggregation method for stereo matching. In: IEEE conference on computer vision and pattern recognition, pp 1402–1409.https://doi.org/10.1109/CVPR.2012.6247827

22. Mei X, Sun X, Dong W, Wang H, Zhang X (2013) Segment-tree based cost aggregation for stereo matching. In: IEEE conference on computer vision and pattern recognition, pp 313–320. https://doi.org/10.1109/CVPR.2013.47

23. Zhu S, Wang Z, Zhang X, Li Y (2016) Edge-preserving guided filtering based cost aggregation for stereo matching. J Vis Commun Image Represent 39:107–119. https://doi.org/10.1016/j.jvcir.2016.05.012

24. Scharstein D, Szeliski R (2002) A taxonomy and evaluation of dense two-frame stereo correspondence algorithms. Int J Comput Vis. https://doi.org/10.1023/A:1014573219977

25. Scharstein D et al (2014) High-resolution stereo datasets with subpixel-accurate ground truth BT—pattern recognition, pp 31–42

26. Hu Y, Zhen W, Scherer S (2020) Deep-learning assisted high-resolution binocular stereo depth reconstruction. In: Proceedings—IEEE international conference on robotics and automation, pp 8637–8643. https://doi.org/10.1109/ICRA40945.2020.9196655

27. Kong L, Zhu J, Ying S (2020) Stereo matching based on guidance image and adaptive support region. Guangxue Xuebao/Acta Opt Sin 40(9):915001. https://doi.org/10.3788/AOS202040.0915001

28. Ruf B, Mohrs J, Weinmann M, Hinz S, Beyerer J (2021) ReS2tAC—UAV-borne real-time SGM stereo optimized for embedded ARM and CUDA devices. Sensors 21(11). https://doi.org/10.3390/s21113938

29. Ma N, Men Y, Men C, Li X (2016) Accurate dense stereo matching based on image segmentation using an adaptive multi-cost approach. Symmetry 8(12). https://doi.org/10.3390/sym8120159

Multiple Views and Categories Condition GAN for High Resolution Image

Huong-Giang Doan (iD)

Abstract GAN network has been become a novel augmented data with many variants. Our proposed method is inspired by the CGAN model, which could generate
images of various categories from latent representation and class controlling vector.
However, the CGAN model could not generate images with different viewpoints. In
addition, the synthetic images are of low-resolution ones. In this paper, a novel GAN
architecture, named High Resolution Categories Condition GAN (HC_GAN), is
proposed to improve the quality of synthetic images. Nevertheless, we also propose
a new method to generate images on different classes and viewpoints at a higher
resolution than input images, which is named High resolution Multiple Views and
Categories condition GAN ($HMVC\_GAN$). In this model, viewpoint controlling
vector and classes controlling vector are contemporaneously integrated into the high
resolution GAN model. Experiments are conducted on various hand gesture datasets,
in which hand images are complex hand shapes, appearances are highly diversified
on different viewpoints. Results show that our proposed method would obtain com-
petent quality not only in different categories but also in different directions.

Keywords Generative adversarial networks · Multiple views · Multiple
categories · High resolution GAN

1 Introduction

Hand gesture recognition [1–4] has been an attractive field in computer vision because
of its applications, such as: Human-Machine Interaction, sign language, entertainment, home appliances, and so on. The hand is a non-rigid and mobile object, whose
scales, shapes, and appearances are dramatically varied. Therefore, it is difficult to
capture a large enough dataset that can cover an entirety of all possible appearances
of hand poses in the real world. As a result, it is challenging to deploy a robust hand

H.-G. Doan (✉)
Control and Automation Faculty, Electric Power University, Hanoi, Vietnam
e-mail: giangdth@epu.edu.vn

© The Author(s), under exclusive license to Springer Nature Switzerland AG 2022
N. H. T. Dang et al. (eds.), *Artificial Intelligence in Data and Big Data Processing*,
Lecture Notes on Data Engineering and Communications Technologies 124,
https://doi.org/10.1007/978-3-030-97610-1_40

507

gesture recognition system. Besides, Hand skin color may be varied under different lighting conditions. Moreover, it is certain that data can be altered when it is captured at different instances. It has always been complicated to have enough data to produce effective data-driven models. Especially, deep learning systems require a large dataset to generalize. Thus, many augmentation methods have been proposed [5–7]. They can be divided into two main directions: (1) Traditional approaches with basic image manipulations, such as: rotation, flipping, color shifting, noise addition, and so on. These approaches have been proven to be not efficient enough because they are not capable of generating unfamiliar samples that can be completely discriminated from original data; (2) Deep learning-based approaches with GAN models [8–10]. These methods have become more favorable in augmentation because a well-trained model can freely generate unprecedented samples from the latent space with a certain level of control based on the random noise vectors. These GAN approaches have been a learning algorithm that allows machine to learn by itself.

However, recent GAN methods are not suitable for our hand gesture recognition system because of several problems: (1) Earliest models [8, 10] only work on small-scale images (e.g. 28 × 28 pixels or 32 × 32 pixels) and or simple samples like hand-written numbers. These models are not suitable for hand gesture of complex shapes, details, and appearances; (2) Some networks [9, 11] tried to improve image resolution (up to 256 × 256 pixels) but they simply translates from image to image without being able to create images conditioned on action classes; (3) Other GAN models could generate images for multi-class but they have bad quality or small resolution as [10]. As a consequence, in this study, we firstly propose a GAN architecture to improve the quality and resolution of synthetic hand images. We then consider and resolve how to create hand images on both various categories and on different camera viewpoints.

The remainder of this paper is organized as follows: In Sect. 2 we briefly survey recent works related to existing GAN approaches. Section 3 explains our proposed method. Experimental results and discussions are analyzed in Sect. 4. Finally, Sect. 5 concludes the paper and proposes research directions for future work.

2 Related Work

The authors in [8] proposed a Generative Adversarial Nets framework—GANs for estimating generative models via an adversarial process. The output images had resolution of 28 × 28 pixels. This model did not consist of any condition addition. Both Generator and Discriminator models used traditional feed-forward neural network architectures. Ideal data generation of authors in [8] has been a premise for later variants of GAN networks that can be divided following the type of output, such as image (static gesture/posture) or video (dynamic gesture/action). In this study, we focus on GAN networks that create static images, especially for hand postures. Recent GAN models [9, 11–14] replaced traditional feed-forward neural networks by convolution neural networks. DCGAN [12] (Deep Convolutional GAN) used convolution and transposed convolution layers in the discriminator and generator, respectively. The model outputs images of 32×32 pixels as in CIFAR-10 dataset. This dataset consists of 60,000 RGB images 32×32 pixels with 10 classes, each class consists of 6000 images. LSGAN [15] (Least Squares GAN) adopted the least squares loss function for the discriminator. One pitfall is that these models can only generate chaotic data not followed by given category information. Moreover, they worked on so small input images and output images. Datasets were used to evaluate quite simple (e.g. handwritten digit).

Other methods tried to translate, map or improve the resolution of images, such as [9, 11, 13, 14, 16]. CycGAN [14] (Cycle GAN) tried to learn the mapping between an input image and an output image using a training set of aligned image pairs. The authors in [11] proposed pix2pix GAN network that mapped through a general-purpose solution for image-to-image translation problems. Output images of DCGAN, startGAN, and LSGAN model are of small resolution at 32×32 pixels. CycGAN and pix2pix GAN networks worked on higher resolutions of 256×256 pixels and 128×128 pixels, respectively. Images generated by SRGAN [9] (Super-Resolution GAN) and ESRGAN [13] (Enhancement and Segmentation GAN) are both of 256×256 pixels. Moreover, they could not create images for a known certain class. VCGAN [17] (Video Colorization GAN) proposed a video colorization using end-to-end learning. Interpretive Representation Learning by InfoGAN [18] (Information Maximizing GAN) learned disentangled representations in a completely unsupervised manner. DRAGAN [19] (Deep Regret Analytic GAN) enabled faster training that results in improved stability with fewer mode collapses and led to generator networks with better modeling performance across a variety of architectures and objective functions.

Some conditional GAN models [10, 20, 21] could generate images of a certain category but not of a certain viewpoint. CGAN [10] (Conditional GAN) integrated a condition class to both the generator and discriminator, which generated MNIST digits conditioned on class labels at 28×28 pixels. The output images were simple, small, and has no viewpoint definition. ACGAN (Auxiliary Classifier GAN) similarly outputs images at 28×28 pixels. DualGAN [21] enabled image translators to be trained from two sets of unlabeled images from two domains. DiscoGAN [20] (Discover Cross-Domain Relations GAN) learned to discover relations between different domains that transfer style from one to another while preserving the key attributes such as orientation and face identity.

Data generation problems have been a necessity in order to augment data for deep learning training. In addition, hand gesture recognition still has to face many challenges because due to the non-rigidity nature of hand shapes combined with the complexity of details of hand appearances; and these are further perplexed by background, illuminations, etc. While datasets were captured under the limitation of both number and diversity of appearances. Existing GAN methods focused on digit or face images, with specific control of age or gender, and so on. For that reason, we proposed a new method to synthesize images conditioned on both specified class and deterministic viewpoint. Experiments are conducted on two multi-view and multi-category hand posture datasets.

In this work, a CGAN model is deployed for creating complex and detailed hand postures. Next, a new model is proposed to generate high-quality static hand images (128 × 128 pixels) on both multi-class and multi-view prior conditions. Our contributions are explained in Sect. 3. Finally, experiments are implemented on hand gesture datasets and results are presented in detail in Sect. 4.

3 Proposed Method

The CGANs model is an extension of the GANs model (Fig. 1a) that both Generator and Discriminator networks have additional conditioning as input cues. This extra controlling factor could be the class of the current image or any other properties. In [10], the authors used an auditioning input as category of data that is illustrated in Fig. 1b. In this research, we proposed a new framework that could generate data in different classes and views as shown in Fig. 1c. These problems are presented in details in Sects. 3.1 and 3.2:

3.1 Summary of Condition GAN

The GANs network (as illustrated in Fig. 1a) consists of two adversarial parts: a Generator (G) and a Discriminator (D). The content space Z was modelled by a

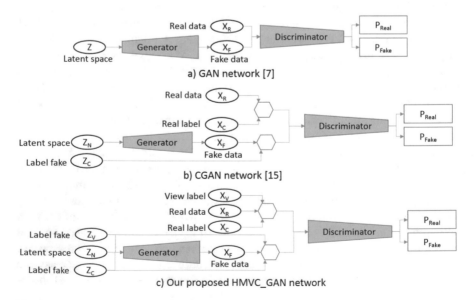

Fig. 1 The proposed framework for generative adversarial network

Gaussian Distribution $Z \sim P_z \equiv N(Z|0, I_{d_z})$ with I_{d_z} is the identity matrix of size $(d_z \times d_z)$. The G model builds a function to map from noise distribution $P_z(z)$ of input data space Z. The output G(z) is a fake image that is fed as one input of D model. The other input of the D model is a real data sample X_R. In [8], D network must be trained to balance between G(z) space and X_R space as presented in the following Eq. (1):

$$\min_G \max_D F_i(G, D) = E_{x_R \sim p_{x_R}(x_R)}[\log D(x_R)] + E_{z \sim p_z(z)}[\log(1 - D(G(z)))] \quad (1)$$

In [10], the authors used a controlling class vector (Fig. 1b) that was integrated into the GAN model to create synthetic images on multi-class. The vector consists of two optimal elements for image generation and category confinement as presented in the following Eq. (2):

$$\min_G \max_D F_c(G, D) = E_{x_R^c \sim p_{x_R^c}(x_R)^c}[\log D(x_R^c|y)] + E_{z^c \sim p_{z^c}(z^c)}[\log(1 - D(G(z^c|y)))] \quad (2)$$

In [10], Z_N is a noise vector, Z_c is an **embedding vector** that was used to control classes as inputs of Generator model. In addition, Z_c was also utilized as an input of Discriminator model of G(z) that was promoted to output 0. For other terms of D network, X_R is a real image and X_c is its class ground truth, which is used to encourage Discriminator model to output 1. Both z^c and x_R^c are concatenated by the above two elements, respectively, as presented in the following Eq. (3):

$$z^c = [Z_N, Z_c]^T; x_R^c = [X_R, X_c]^T \quad (3)$$

Both Generator and Discriminator networks of CGAN model in [10] were designed to be implemented with small images of 28×28 pixels of MNIST dataset. The CGAN model was suitable for this dataset because of its simplicity. Consequently, it is not suitable for complex objects such as face, hand, and so on. In this work, we trained the CGAN model [10] on MICAHandPose dataset [2] with 7 hand posture classes, whose synthetic images after training in 600 epochs are illustrated in Fig. 3a. The size of both input and output images is 28×28 pixels. It is apparent that generated hand images are blurred and there are not enough cues to present hand gestures. In this research, we firstly propose a new architecture that can generate larger images with better quality. In addition, we also propose a new GAN architecture that could generate images on multi-view and multi-categories as presented in detail in the next Sect. 3.2.

3.2 Proposed HMVC_GAN Network

(a) HC_GAN architecture

In this part, we are stimulated by [22], whose proposal could create images of size 128×128 pixels on a certain class. This resolution can cover the dimensional of our hand posture scene. The method required to fix embedding vector label for noise as in [10], and the length of this vector was not concerned. Furthermore, this model was not efficient in our work because of its unstable and divergent nature. Therefore, we propose a HC_GAN model to generate larger images that resolutions of both input and output images are 128×128 pixels. The detailed architecture of the HC_GAN model is presented in the left part of Table 1. In which, M denotes the number of classes in training dataset. In the discriminator part, we added noise elements that sigma scale was chosen at 0.1. This work aims to reduce over-fitting problem of the network and increase the accuracy: The HC_GAN network is then trained on multiple categories-labeled hand gesture datasets. Qualitative evaluations of this model are presented in detail in Sect. 4. Furthermore, this architecture is also utilized to design a multi-view and multi-category GAN in the next Sect. 3.2 (b).

(b) **View-point controlling with HMVC_GAN model**

In this study, a new HMVC_GAN is proposed to generate images not only on various categories but also on different views. We use the HC_GAN network to generate high resolution images (128×128 pixels) on multi-class. Our architecture is detailed in the right part of the Table 1. Where M is class number and L is viewpoint number in the dataset.

In this proposed network, viewpoint controlling was also integrated into the model as in the following Eq. (4). While content sub-space Z_N is noise vector that is also modelled by Gaussian distribution, similar to Z_N in Eq. (3). In this research, we combined Z_N with Z_c and Z_v as in Eq. (5). Which are two **one hot vectors** corresponding to categories and viewpoints, respectively. These **one hot vectors** are different from the **embedding vector** in [10] and [22]. Given z_F^{vc} and x_R^{vc}, two flows of discriminator are simultaneously implemented by value function as shown in Eq. (5):

$$\min_G \max_D V_{vc}(G, D) = E_{x_R^{vc} \sim p_{x_R^{vc}}(x_R^{vc})}[\log D(x_R^{vc}|y)] + E_{z_F^{vc} \sim p_{z_F^{vc}}(z_F^{vc})}[\log(1 - D(G(z_F^{vc}|y)))]$$

(4)

where

$$z_F^{vc} = [Z_N, Z_c, Z_v]^T; x_R^{vc} = [X_R, X_c, X_v]^T$$

(5)

Quantitative evaluation of our proposed multi-view and multi-class generation method will be presented in detail in the next Sect. 4.2.

3.3   GAN Deployment

In this study, two categories are considered: (1) Single view with CGAN [10], HC_GAN models, and (2) Multi-view with HMVC_GAN model. The two hand gesture datasets are utilized, including MICAHandPose and EPUHandPose. These datasets are labeled with both views and classes of hand postures. More specific

Table 1 Configuration of HC_GAN and HMVC_GAN

Generator

Layer	Architecture of HC_GAN	Output	Architecture of HMVC_GAN	Output
Input	Concatenation($Z_N(1\times100)$, $Z_C(1\times M)$)	$(100+M)\times1$	Concatenation($Z_N(1\times100)$, $Z_C(1\times M)$, $Z_V(1\times L)$)	$(100+M+L)\times1$
1	Linear ((100 + M)×1)	[(512+1)*4*4]×1	Linear ((100 + M + L)×1)	[(512+2)*4*4]×1
2	Reshape (8208×1)	513×4×4	Reshape(8224×1)	514×4×4
3	2DConvTrans-(N(513), K(4,4), S(2,2),P(1,1)), BN, ReLU	512×8×8	2DConvTrans-(N(514), K(4,4), S(2,2),P(1,1)), BN, ReLU	512×8×8
4	2DConvTrans-(N(512), K(4,4), S(2,2),P(1,1)), BN, ReLU	256×16×16	2DConvTrans-(N(512), K(4,4), S(2,2),P(1,1)), BN, ReLU	256×16×16
5	2DConvTrans-(N(256), K(4,4), S(2,2),P(1,1)), BN, ReLU	128×32×32	2DConvTrans-(N(256), K(4,4), S(2,2),P(1,1)), BN, ReLU	128×32×32
6	2DConvTrans-(N(128), K(4,4), S(2,2),P(1,1)), BN, ReLU	64×64×64	2DConvTrans-(N(128), K(4,4), S(2,2),P(1,1)), BN, ReLU	64×64×64
7	2DConvTrans-(N(64), K(4,4), S(2,2),P(1,1)), BN, ReLU	3×128×128	2DConvTrans-(N(64), K(4,4), S(2,2),P(1,1)), BN, ReLU	3×128×128
8	Tanh	3×128×128	Tanh	3×128×128

Discriminator

Layer	Architecture of HC_GAN	Output	Architecture of HMVC_GAN	Output
Input	Data (X(128×128), C(1×M)) (Real data (X_R, X_C)); Fake data (X_F, Z_C))	2×(3×128×128)	Data(X(128×128), C(1×M), V(1×L)) Real data (X_R, X_C, X_V); Fake data (X_F, Z_C, Z_V)	3×(3×128×128)
1	Concatenation (2×(3×128×128))	6×128×128	Concatenation (3×(3×128×128))	9×128×128
2	Noise, 2DConv-(N(6), K(4,4), S(2,2),P(1,1)), LReLU	64×64×64	Noise, 2DConv-(N(9), K(4,4), S(2,2),P(1,1)), LeakyReLU	64×64×64
3	Noise, 2DConv-(N(64), K(4,4), S(2,2),P(1,1)), BN, LReLU	128×32×32	Noise, 2DConv-(N(64), K(4,4), S(2,2),P(1,1)), BN, LReLU	128×32×32
4	Noise 2DConv-(N(128), K(4,4), S(2,2),P(1,1)), BN, LReLU	128×16×16	Noise 2DConv-(N(128), K(4,4), S(2,2),P(1,1)), BN, LReLU	128×16×16
5	Noise, 2DConv-(N(128), K(4,4), S(2,2),P(1,1)), BN, LReLU	256×8×8	Noise, 2DConv-(N(128), K(4,4), S(2,2),P(1,1)), BN, LReLU	256×8×8
6	Noise, 2DConv-(N(256), K(4,4), S(2,2),P(1,1)), BN, LReLU	512×4×4	Noise, 2DConv-(N(256), K(4,4), S(2,2),P(1,1)), BN, LReLU	512×4×4
7	Noise, 2DConv-(N(512), K(4,4), S(1,1)), BN, LReLU	1×1x1	Noise, 2DConv-(N(512), K(4,4), S(1,1)), BN, LReLU	1×1×1
8	Flatten	1	Flatten	1
9	Dropout, sigmoid		Dropout, sigmoid	

Table 2 Deployment of condition GAN models

Parameters	CGAN [10]	HC_GAN	HMVC_GAN
Generate image	28×28 pixels	128×128 pixels	128×128 pixels
Batch size	64	64	64
Number of epochs	600	600	330
Learning rate	$2 * 10^{-4}$	$2 * 10^{-4}$	$2 * 10^{-4}$
Loss function	Mean square error—MSE	Binary cross entropy—BCE	Binary cross entropy—BCE
Optimizer function	Adam	Adam	Adam

details of the two datasets are presented in Sect. 4. The parameters for training the Generators and Discriminators of three GAN models are presented in Table 2: Results of CGAN [10] model are utilized to compare with the results of our proposed HC_GAN model. FID values of these models are calculated and reported in Sect. 4.1.

3.4 FID Score

In this paper, Fréchet Inception Distance (FID) [23] is used to compare the performance of the generative component of GAN models. FID score measures the similarity between real dataset and synthetic dataset. It has been shown to correlate well with human judgement of visual quality and is most often used to evaluate the quality of samples of Generative Adversarial Networks. FID score stands for Fréchet distance [24] between two Gaussians fitted to feature representations of the Inception network.

In the original work, it is noted that FID score is computed for every 5 generations. Feature vectors are obtained at a hidden layer of InceptionNet-V3 [25] and size of these output vectors (F_R and F_Z) are (1×2048). The GAN model generates 2000 synthetic images for a class of a view. They are then utilized to compute FID scores.

(a) **FID^c score for a class condition GAN architecture:** Given real dataset D_R (containing real images denoted by X_R^i) and synthetic dataset D_Z (containing synthetic images denoted by X_F^i) on ith class ($i = 0 \div M$) that are inputs of InceptionNet-V3 model [25], output sets are taken at FC layer with F_R^i and F_Z^i, respectively. Then, the mean (μ^i) and covariance (C^i) of each feature vector set on the ith class ($i = 0 \div M$) are obtained which are presented by F_R^i ($\mu_{F_R}^i$, $C_{F_R}^i$) and $F_Z^i(\mu_{F_Z}^i, C_{F_Z}^i)$. Then, the difference of two Gaussians are measured by the Fréchet distance as in the following Eq. (6):

$$FID^i(F_R^i, F_Z^i) = ||\mu_{F_R}^i - \mu_{F_Z}^i||_2^2 + Tr[C_{F_R}^i + C_{F_Z}^i - 2(C_{F_R}^i C_{F_Z}^i)^{1/2}] \quad (6)$$

Firstly, $FID^i(F_R, F_Z)$ scores are computed on ith class ($i = 0 \div M$) of a dataset (as in Eq. (6)). Then, FID^c score is then computed by averaging FID^i of all classes as presented in the following Eq. (7):

$$FID^c(F_R, F_Z) = \frac{\sum_{i=1}^{M} FID^i(F_R^i, F_Z^i)}{M} \tag{7}$$

(b) **FID^{vc} score for a view and class condition GAN architecture:** In addition, $FID^{vc}(F_R, F_Z)$ score is used to evaluate our proposed view and class conditioned GAN network (HMVC_GAN model). Firstly, $FID_j^i(F_{Rj}^i, F_{Zj}^i)$ of the ith class ($i = 1 \div M$) and jth view ($j = 1 \div L$) is calculated as the following Eq. 8. Given the real dataset D_{Rj} (consists of real images X_{Rj}^i) and synthetic dataset D_{Zj} (consists of synthetic images X_{Fj}^i) on ith class of jth view that are inputs of InceptionNet-V3 model [25], FC layer is given feature vectors as F_{Rj}^i and F_{Zj}^i for real set and fake set, respectively. Then, mean (μ_j^i) and covariance (C_j^i) of each feature vector set on the ith class ($i = 0 \div M$) and j view ($j = 0 \div L$) are obtained and are denoted by $F_{Rj}^i(\mu_{F_{Rj}}^i, C_{F_{Rj}}^i)$ and $F_{Zj}^i(\mu_{F_{Zj}}^i, C_{F_{Zj}}^i)$. Then, the difference of two Gaussian distributions in ith class of jth view between real data and fake data is measured by the Fréchet distance as the following Eq. (8):

$$FID_j^i(F_{Rj}^i, F_{Zj}^i) = ||\mu_{F_{Rj}}^i - \mu_{F_{Zj}}^i||_2^2 + Tr[C_{F_{Rj}}^i + C_{F_{Zj}}^i - 2(C_{F_{Rj}}^i C_{F_{Zj}}^i)^{1/2}] \tag{8}$$

The FID_j^c scores are then calculated for all classes of a jth view where ($j = 1 \div L$), as presented in the following Eq. (9):

$$FID_j^c(F_R, F_Z) = \frac{\sum_{i=1}^{M} FID_j^i(F_{Rj}^i, F_{Zj}^i)}{M} \tag{9}$$

The $FID^{vc}(F_R, F_Z)$ score of dataset with L views is finally calculated as in the following Eq. (10). This score is average of all FIDs of all classes from all views.

$$FID^{vc}(F_R, F_Z) = \frac{\sum_{j=1}^{L} FID_j^c(F_R, F_Z)}{L} \tag{10}$$

The FID^{vc} score will be utilized to investigate in detail in the next Sect. 4.2.

4 Experimental Result

In this section, the experimental results present the evaluation on: (1) The impact of the generated images with higher resolution on multi-class; (2) The efficiency of our proposed method on both multiple views and multiple classes on various datasets. The code is implemented in Python language, under Pytorch deep learning

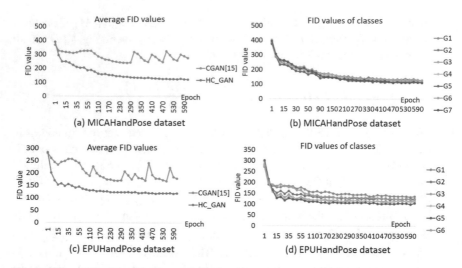

Fig. 2 FID results of the CGAN model and the HC_GAN model at various epochs

framework, and execute on NVIDIA GTX 1080 Ti GPU. Experimental protocols are explained thoroughly in the previous Sect. 3.4.

In this study, our experiments are investigated on multiple views and multiple classes. The first dataset is MICAHandPose [26], which has seven classes ($G_1, \ldots G_7$), and the second dataset is EPUHandPose [27], which has six classes ($G_1, \ldots G_6$). Both datasets are manual transferred with four directions: View 1 (0^0), View 2 (90^0)— original view, View 3 (180^0), View 4 (270^0).

4.1 Efficient of the HC_GAN Model

In this section, FID score is investigated for both the CGAN method [10] and the HC_GAN method, evaluated after every interval of 5 epochs from epoch 1 to epoch 600. Then, these values are averaged as in Fig. 2. This experiment is implemented in the 90^0 direction (corresponding to View 2 of both MICAHandPose and EPUHand-Pose). Have a glance at Fig. 2a and c, it is clear that FID values of these methods are dramatically reduced until epoch 200 on both datasets. Moreover, the FID values of the CGAN [10] method are extremely fluctuated from epoch 310–600, around 300 with MICAHandPose and 200 with EPUHandPose respectively; while those of the HC_GAN method are stable at about 120 for all. Especially, the FID values of the CGAN approach exceed those of the HC_GAN method by an order of magnitude. These results show that generated images of the HC_GAN model are more similar to real images of the training set than synthetic images of the CGAN method. In other words, the HC_GAN model converges better than the CGAN model, reaching its stable state from epoch 400 onwards.

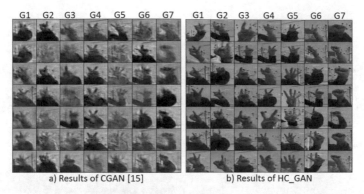

a) Results of CGAN [15]   b) Results of HC_GAN

Fig. 3 Results of the CGAN model and the HC_GAN model

Figure 2b shows FID values on seven classes of the HC_GAN model trained on MICAHandPose dataset. Figure 2d presents FID values on six classes of EPUHand-Pose dataset. It is apparent that FID values also converge on all classes. In addition, Fig. 3 illustrate some synthetic images of the CGAN and the HC_GAN models at 600 epochs. The HC_GAN architecture (Fig. 3b) performs better than the CGAN architecture (Fig. 3a) on all seven classes. This figure once again indicates that the HC_GAN architecture obtains images of better quality whilst output resolution (128×128 pixels) is also higher than that of the CGAN model (28×28 pixels).

Consequently, the HC_GAN architecture will be utilized in the HMVC_GAN model as presented in the previous Sect. 3.2 (b). Qualitative evaluations of this model will be presented in the following Sect. 4.2.

4.2   Evaluations of the HMVC_GAN Model

In this section, our model is evaluated on MICAHandPose and EPUHandPose datasets. The HMVC_GAN model is trained on the four aforementioned view-points. Figure 4 illustrates the FID values of the HMVC_GAN models are trained on the two datasets, obtained at different epochs. Figure 4b, c present the FID results of EPUHandPose and MICAHandPose, respectively. These results show that the HMVC_GAN model converges at approximately epoch 120 on all views. The averages of FID values of EPUHandPose are far smaller the these of MICAHandPose as illustrated in Fig. 4a. This result could be begotten by the fact that the number of classes of EPUHandPose (6 classes) is less than that of MICAHandPose (7 classes). Figure 5a, b illustrate the generated images of the HMVC_GAN model, which are trained on EPUHandPose and MICAHandPose after 300 epochs, respectively. It is obvious that our proposed method could exactly and clearly generate hand images on both a certain view and a determinant category. Each row illustrates the synthetic

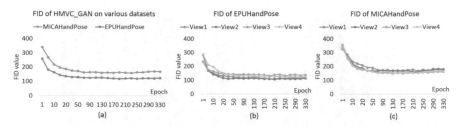

Fig. 4 FID results of our HMVC_GAN model trained on various datasets

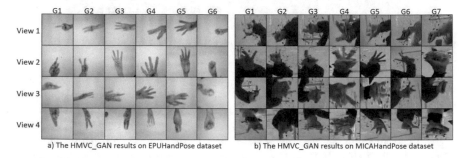

a) The HMVC_GAN results on EPUHandPose dataset   b) The HMVC_GAN results on MICAHandPose dataset

Fig. 5 Results of HMVC_GAN on various datasets after 300 epochs

images of a class and each column denotes that these images are from the same viewpoint. While Fig. 5a illustrates four views and six classes and Fig. 5b shows four views and seven categories.

5 Conclusion and Discussion

This paper introduced a novel HMVC_GAN architecture that can create synthetic images conditioned on a certain view and class. Experiments were conducted on two hand gesture datasets. The results indicated that our proposed method is not only can perform better than SOTA multi-class method but can also generate hand posture at a certain viewpoint. In future works, we will experiment on larger multi-viewpoint, multi-class hand gesture datasets and other datasets such as face, body, and so on. Moreover, we will evaluate the efficiency of synthetic data on improving recognition accuracy of *hand_in_wild* gesture system.

References

1. Al-Hammadi M, Muhammad G, Abdul W, Alsulaiman M, Bencherif MA, Mekhtiche MA (2020) Hand gesture recognition for sign language using 3D CNN. IEEE Access 8:79491–79509
2. Doan HG, Nguyen VT, Vu H, Tran TH (2016) A combination of user-guide scheme and kernel descriptor on RGB-d data for robust and realtime hand posture recognition. Eng Appl Artif Intell 49(C):103–113
3. Pisharady P, Vadakkepat P, Loh A (2013) Attention based detection and recognition of hand postures against complex backgrounds. Comp. Vis. 101:403–419
4. Ruffieux S, Lalanne D, Mugellini E, Abou Khaled O (2014) A survey of datasets for human gesture recognition. In: Kurosu M (ed) HCI'14, pp 337–348
5. Salama ES, El-Khoribi RA, Shoman E, Shalaby MA (2018) EEG-based emotion recognition using 3d convolutional neural networks. ACSA J 9(8)
6. Shijie J, Ping W, Peiyi J, Siping H (2017) Research on data augmentation for image classification based on convolution neural networks. In: CAC
7. Shorten C, Khoshgoftaar T (2019) A survey on image data augmentation for deep learning. Journal of Big Data 6:1–48
8. Goodfellow IJ, Pouget-Abadie J, Mirza M, Xu B, Warde-Farley D, Ozair S, Courville A, Bengio Y (2014) Generative adversarial nets. In: Proceedings of the NIPS, pp 2672–2680
9. Ledig C, Theis L, Huszar F, Caballero J, Cunningham A, Acosta A, Aitken A, Tejani A, Totz J, Wang Z, Shi W (2017) Photo-realistic single image super-resolution using a generative adversarial network. In: CVPR, 07 2017, pp 105–114
10. Mirza M, Osindero S, Conditional generative adversarial nets. CoRR arXiv:1411.1784
11. Isola P, Zhu JY, Zhou T, Efros AA (2017) Image-to-image translation with conditional adversarial networks. CVPR, pp 5967–5976
12. Radford A, Metz L, Chintala S (2016) Unsupervised representation learning with deep convolutional generative adversarial networks. In: ICLR'16
13. Wang X, Yu K et al (2018) ESRGAN: enhanced super-resolution generative adversarial networks. ECCV WS
14. Zhu JY, Park T, Isola P, Efros A (2017) Unpaired image-to-image translation using cycle-consistent adversarial networks. In: ICCV'17, pp 2242–2251
15. Mao X, Li Q, Xie H, Lau RYK, Wang Z (2016) Multi-class generative adversarial networks with the L2 loss function. CoRR arXiv:1611.04076
16. Jolicoeur-Martineau A (2019) The relativistic discriminator: a key element missing from standard GAN. ArXiv arXiv:1807.00734
17. Zhao Y, Po L et al (2021) VCGAN: video colorization with hybrid generative adversarial network. ArXvi arXiv:2104.12357
18. Chen X, Duan Y, Houthooft R, Schulman J, Sutskever I, Abbeel P (2016) InfoGAN: interpretable representation learning by information maximizing generative adversarial nets. In: Proceedings of NIPS'16, pp 2180–2188
19. Kodali N, Abernethy JD, Hays J, Kira Z (2017) How to train your DRAGAN. CoRR arXiv:1705.07215
20. Kim T, Cha M, Kim H, Lee JK, Kim J (2017) Learning to discover cross-domain relations with generative adversarial networks. In: ICML'17, pp 1857–1865
21. Yi Z, Zhang H et al (2017) DualGAN: unsupervised dual learning for image-to-image translation. In: ICCV'17
22. Sharma A (2021) Conditional GAN (CGAN) in pytorch and tensorflow
23. Heusel M, Ramsauer H, Unterthiner T, Nessler B, Klambauer G, Hochreiter S (2017) GANs trained by a two time-scale update rule converge to a nash equilibrium. In: Proceedings of the NIPS, pp 6629–6640
24. Dowson D, Landau B (1982) The Fréchet distance between multivariate normal distributions. J Multivariate Anal 12(3):450–455

25. Szegedy C, Vanhoucke V, Ioffe S, Shlens J, Wojna Z (2016) Rethinking the inception architecture for computer vision. In: CVPR, pp 2818–2826
26. Micahandpose dataset (2021). https://zenodo.org/record/5767154
27. Epuhandpose dataset (2021). https://zenodo.org/record/5767091

An Algorithm for Color Correction in Marine Biological Images Based on U-Net Network

Ruizi Wang and Zhan Jiao

Abstract With the rapid development of the economy, marine resources have attracted much attention. Marine biodiversity observation is especially important to work. However, due to the complex underwater environment, the captured marine biology has a color distortion and blur, which greatly hinders the marine biological observation work. The clean underwater image is taken at a precise measurement distance without the influence of light wavelength attenuation and scattering of impurities in the water, and all uncertain reasons. At present, it is still challenging to get clean and high-quality underwater images in complex marine environmental conditions through hardware equipment and external conditions. Not only is there huge difficulty and instability in operation, but also a huge loss in cost. Acquiring poor-quality underwater images has led to limited marine biological recognition technology, and indirectly affected the development of marine biological diversity research. Therefore, this paper proposes an algorithm for correcting the color in marine biological images based on the U-Net network. The first step is to build a deep convolutional neural network combined with the U-Net network, and make the network continuously learn the color difference between the input image and the output image; the second step is to use the image structure similarity algorithm to make the output image and real image undergo iterative learning and gradually approach the real image. After many experiments, it is proved that the algorithm proposed in this paper has a good effect on correcting the color of marine biological images.

Keywords Marine biological · Neural networks · Color correction

R. Wang (✉) · Z. Jiao
Liaoning Vocational College of Light Industry, Dalian, China

Z. Jiao
e-mail: 18940836609@163.com

© The Author(s), under exclusive license to Springer Nature Switzerland AG 2022   521
N. H. T. Dang et al. (eds.), *Artificial Intelligence in Data and Big Data Processing*,
Lecture Notes on Data Engineering and Communications Technologies 124,
https://doi.org/10.1007/978-3-030-97610-1_41

1   Research Background

Marine Biological is an important part of marine research, so it is particularly important to get clear images of marine biology images. Traditional machines require extremely strict external environments to capture clear images of marine biology images. However, the underwater environment is complex and it is difficult to get clear biology images with ordinary shooting methods. Therefore, Color correction algorithms are used to improve marine biology images. This technology has very important value (Fig. 1).

Under normal circumstances, we divide underwater biological image enhancement technologies into two categories, one is the traditional marine biological image algorithm, and the other is the algorithm for enhancing the marine biological image through neural networks. Traditional marine biological algorithms can be subdivided into biological image restoration and biological image enhancement. The biological image restoration algorithm mainly focuses on the detail restoration of biological images, while biological image enhancement focuses more on the adjustment of the overall color of underwater images.

Fig. 1 Marine biological images with deviated colors

1.1 Traditional Marine Biological Image Algorithm

The marine biological image restoration algorithm is represented by Liu [1] and others based on K-SVD dictionary learning is used to keep the structure and texture information of the original image. Han [2] and others propose a deep supervised residual dense network, which is used to better learn the mapping relationship between clear in-air images and synthetic underwater degraded images. Guzin [3] proposed two different approaches, global and local contrast enhancement techniques, to get better visual quality while enhancing image contrasts on underwater images. Wang [4] and others proposed Underwater Image Enhancement Convolution Neural Network using 2 Color Spaces that efficiently and effectively integrate both RGB Color Space and HSV Color Space in one single CNN. Dong [5] proposed the image super-resolution reconstruction algorithm SRCNN, To improve the visual appearance of the image, Li [6] proposed an adaptive algorithm for effective underwater image enhancement using a randomly wired neural network and synergistic evolution.

Marine biological images color correct the blue-green color, improve the color brightness, and enhance the visual effect. Galdran [7] proposed a red channel method to restore underwater images. This method restored the shortwave-related colors, thereby restoring lost information and improving contrast. Garcia [8] proposed homomorphic filtering is used to solve the problems of underwater fogging and blue-green color. It uses the gray-scale transformation principle of pixel blocks and its transformation frequency filtering method to improve the quality of the picture. Ancuti [9] proposed an underwater image enhancement algorithm based on fusion. Through the multi-scale fusion technology, the light and shadow generated when the image is output is avoided, the problems of unclear underwater images, poor contrast, and color differences are solved. Liu [10] and others propose an integrated approach, the Akkaynak-Treibitz model embedded physical model guides for network learning, and the generative adversarial network is adopted for coefficients estimation. Hui [11] proposed an underwater image enhancement algorithm based on histogram equalization. The algorithm makes a single unbalanced histogram into an evenly distributed histogram to enhance the image contrast, thereby achieving the effect of image optimization.

Traditional algorithms are ineffective for the underwater marine biological image. The marine biological image keeps blue and blurred. Therefore, the neural network image enhancement algorithm appeared.

1.2 The Neural Networks Image Enhance Algorithm

As the image processing technology based on deep learning continues to sink to specific domain tasks, traditional end-to-end tasks cannot further explore the unique characteristics of data in various domain tasks. The way to use specific domain knowledge and models to construct deep learning tasks became important. An end-to-end learning method, proposed by Li [12] and others, uses a large number of training image data sets and some underwater depth data as an end-to-end input, which can roughly learn the depth estimation of underwater scenes. In this way, the image characteristics are restored and the visual effect is enhanced. Li [13] proposed Water Net for optimization and Li [12] used an end-to-end learning method to optimize. Liu [14] and others proposed an underwater image enhancement algorithm based on a deep residual network framework, which improves the underwater image quality by introducing a super-resolution reconstruction model. However, this method has poor feature extraction ability, resulting in the information cannot be accurately restored during upsampling, so the image enhancement capability is limited. Yan [15] and others proposed a traditional neural network underwater image restoration method. Firstly, a degraded underwater image is generated according to the underwater image imaging model. Secondly, a degraded underwater image and real water are established. Download the mapping relationship of the image, and finally, train the neural network to learn the image features through this mapping relationship. In this way, the noise is reduced, the definition is enhanced, and the underwater image is restored. Because the network is small and only has a single mapping learning method, Therefore, the enhancement capability of the algorithm is weak.

2 Methodology

Given the existing algorithms, the ability of color correction is not strong, the overall marine biological images are not bright, and the details are blurred. This paper proposes an algorithm for correcting the color of marine biological images based on the U-Net network, which is used more thoroughly to correct the color in marine biological images.

2.1 Network Architecture of Our Algorithm

As we all know, the typical feature of U-net is jump connection. We will also use jump connections to create models. First, input a blurred image of marine biological, and after the image enters the network, the image is reduced in size through convolution and down-sampling. The purpose of this step is to extract the shallow features of the image. Next, in the operation of convolution and upsampling, the purpose of this step is to extract deep features. Finally, all the acquired features are fused, and then the enhanced image is output. The model structure of this article is shown in Figs. 2 and 3.

The network is composed of 5 encoders and 5 decoders. The output of each encoder is connected to the corresponding decoder in a hop. Each convolutional layer has a

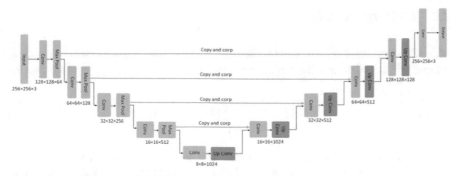

Fig. 2 Network architecture of our algorithm

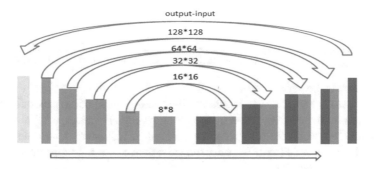

Fig. 3 Flow chart of the algorithm

3 * 3 * 2D convolution. The network execution process is: input a 256 * 256 * 3 size image, perform the first 128 * 128 convolution, and then go through the second 64 * 64 convolution. At the same time, the image feature obtained by the 128 * 128 convolution is jump-connected to the corresponding 128 * 128 convolution of the network. After 3 times of convolution, in the same way, it is up-sampled to become a 16 * 16 convolution. After 4 times of up-sampling, a 256 * 256 convolution is obtained. The obtained image and the input underwater image are learned again, Get the final optimized image.

Our network is fully convolutional. The ability to learn the color difference between the input image I(x) and the output image O was expressed as below:

$$C = O - I(x) \tag{1}$$

where C represents the difference between input and output pictures, O represents the output image, and I(x) represents the input marine biological images. Learning different features is a simple and effective way. The results of training in this way are more effective. The experiment proves that our algorithm can correct color and restore detailed information.

2.2 Evaluation Index

Mean Squared Error will be used to calculate the loss function, and marine biological images will be input into the trained network. The MSE calculation is expressed as:

$$LMSE = \frac{1}{m} \sum_{i=1}^{m} \|Oi - J(xi)\|^2 \qquad (2)$$

Here, it is agreed that m is the number of samples, and Oi is the value of the number i-th group generated in the model, $J(xi)$ is referred to as the value of the clear image in the number i-th group.

The algorithm incorporates the Structural Similarity Index loss value into the objective function to improve the similarity of the image structure.

$$SSIM(O, J(x)) = \frac{2\mu Y * \mu J(x) + C1}{\mu^2 Y + \mu^2 J(x) + C1} * \frac{2\sigma Y J(x) + C2}{\sigma^2 Y + \sigma^2 J(x) + C2} \qquad (3)$$

Y represents the generated biological image, $J(x)$ represents a clean underwater image, μY, $\mu J(x)$ respectively represents the average value of the generated biological image and the average value of the clean biological image, $\sigma^2 Y$ and $\sigma^2 J(x)$ respectively represents the variance of the generated biological image and the clean biological image. $\sigma Y J(x)$ represents the covariance, $C1$ $C2$ represents constants used to maintain stability.

The final loss function can be expressed as:

$$L = LMES + LSSIM \qquad (4)$$

Through the above method, the entire network is trained, and the obtained network model describes the mapping relationship between the color uncorrected image and the standard image.

PSNR (Peak Signal to Noise Ratio) is often used as a measurement method of signal reconstruction quality in image compression and other fields. It is often simply defined by the mean square error (MSE).

$$PSNR = 10 * \log_{10} \left(\frac{MAX_T^2}{MSE} \right) \qquad (5)$$

PSNR is Peak Signal to Noise Ratio, MSE means Mean Squared Error, T means clear picture, MAX_T^2 means the maximum possible pixel value of the picture.

3 Experimental Results

This algorithm uses more than 5000 real paired underwater environment data sets, and more than 500 underwater images to verify the experimental results, and has achieved good results.

3.1 The Comparison Pictures of Different Algorithms

The comparison of other algorithms is as follows.

As shown in Fig. 4, The Pix2Pix and ResNet make image details clearer, but can not enhance the subject information; the UWCNN method has no obvious effect on the image color correction; The FastGAN algorithm does not impose constraints on the generated images, resulting in loss of image information. The algorithm in this paper introduces a structural similarity loss function to constrain the characteristic structure information between the input image and the enhanced image, so that the generated image maintains the integrity of the main body information, and can effectively remove the color cast and improve the contrast. Therefore, the image generated by the model in this paper is closest to the visual effect of the real image.

3.2 The Comparison of Experimental Data

To verify the effectiveness of the algorithm in this paper, we compare it with Pix2Pix, Resnet, UWCNN, FastGan algorithms. To ensure the fairness of the experiment, the original network structure and parameters are used in all the algorithms, and the same training database and test base are used for comparison (Table 1).

4 Conclusion

This paper proposes an algorithm for correcting the color of marine biological images based on a U-net network. The algorithm uses input and output to learn residual color casts to improve the ability of the network. It aims to extract deep-level features of images. At the same time, the structural similarity is cited to improve the similarity of the image. Experiments show that the algorithm in this paper can effectively improve the clarity of marine biological images, correct the color, improve contrast, and enhance image visual effects, thereby achieving the purpose of improving image details, improving sharpness, contrast enhancement. Our work has brought convenience to the observation and identification of marine biodiversity, and further promoted the further development of the field of marine biodiversity

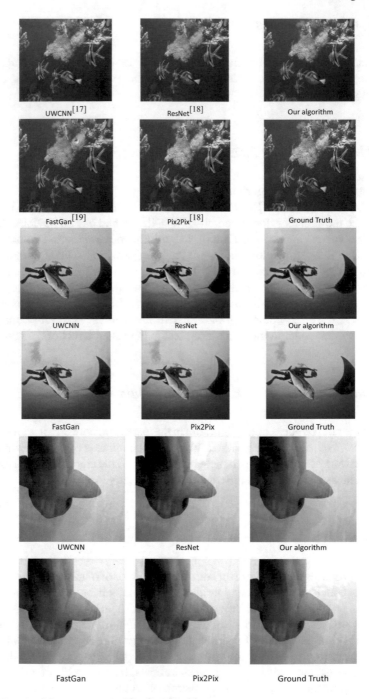

Fig. 4 Our algorithm compared with other algorithms

Table 1 Comparison of SSIM and PSNR for different algorithm

Model name	SSIM	PSNR
FastGan	19.89	0.89
UWCNN	21.35	0.91
Resnet	20.31	0.86
Pix2Pix	20.93	0.87
Our model	22.22	0.92

research, which has great research significance. But our algorithm needs a lot of data support to learn image features better, so when the program runs with a lot of data, its stability is poor. To make the algorithm more convenient, the model still needs to be analyzed and researched.

References

1. Liu X, Liu C, Liu X (2021) Underwater image enhancement with the low-rank nonnegative matrix factorization method. Int J Pattern Recogn Artif Intell 35(08)
2. Han Y, Huang L, Hong Z, Cao S, Zhang Y, Wang J (2021) Deep supervised residual dense network for underwater image enhancement. Sensors (Basel, Switzerland) 21(9)
3. Guzin U, Beste U (2021) Underwater image enhancement using contrast limited adaptive histogram equalization and layered difference representation. Multimedia Tools Appl (pre-publish)
4. Wang Y, Guo J, Gao H, Yue H (2021) UIEC^2-Net: CNN-based underwater image enhancement using two-color space. Signal Process Image Commun
5. Dong C, Loy CC, He K, Tang X (2016) Image super-resolution using deep convolutional networks. IEEE Trans Pattern Anal Mach Intell 38(2):295–230
6. Li Y, Chen R (2020) SE–RWNN: a synergistic evolution and randomly wired neural network-based model for adaptive underwater image enhancement. IET Image Process14(16)
7. Galdran A, Pardo D, Picón A et al (2015) Automatic red-channel underwater image restoration. J Vis Commun Image Represent 26:132–145
8. Garcia R, Nicosevici T, Cufi X (2003) On the way to solve lighting problems in underwater imaging. Oceans IEEE, San Diego
9. Ancuti C, Ancuti CO, Haber T et al (2012) Enhancing underwater images and videos by fusion. In: IEEE conference on computer vision and pattern recognition (Providence), vol 62, pp 81–88
10. Liu X, Gao Z, Chen BM (2020) IPMGAN: integrating physical model and generative adversarial network for underwater image enhancement. Neurocomputing (pre-publish)
11. Hui F (2007) A method improved on histogram equalization for image enhancement. Surving Mapping 30(004):152–154
12. Li J, Skinner KA, Eustice RM, Johnson-Roberson M (2018) WaterGAN: unsupervised generative network to enable real-time color correction of monocular underwater images. IEEE Robot Autom Lett 3(1):387–394
13. Li C, Guo C, Ren W, et al (2019) An underwater image enhancement benchmark dataset and beyond. IEEE Trans Image Process
14. Liu P, Wang G, Qi H et al (2019) Underwater image enhancement with a deep residual framework. IEEE Access. 99:1–1
15. Yan XU, Sun M (2018) Enhancing underwater image based on convolutional neural networks. J Jilin Univ 48(06):272–280

The Clustering Approach Using SOM and Picture Fuzzy Sets for Tracking Influenced COVID-19 Persons

H. V. Pham and Q. H. Nguyen

Abstract Recently, COVID-19 pandemic has increasingly affected the lives of the world population. Researchers have investigated various ways including conventional approach about the transmission routes of COVID-19 persons. However, it is difficult to track COVID-19 in real-time at anywhere. The paper has presented a novel approach using Self Organizing Maps with Picture Fuzzy Sets for tracking COVID-19 persons. The proposed clustering model has been grouped records of COVID-19 persons together with these rules in order to find similar features of COVID-19 person in large data sets. To confirm the effectiveness of this model, experimental results show that the proposed model has demonstrated by using SOM integrated with these rules for tracking COVID-19 persons.

Keywords Clustering approach · COVID-19 person · Identifying COVID-19 person · Deep learning in COVID-19 · Self-Organizing maps

1 Introduction

Recently, the pandemic of COVID-19 has widely spread in the world. Due to the unavailability of effective tracking person, monitoring social distancing is significant to strive against asymptomatic transmission. To do with self-protection and prevention of transmission of infection to another person spreading, many researchers have investigated to identify COVID 19 with its influenced people in simulations for decision making in identifying the person involved. Rajaraman et al. [1] have investigated deep learning model ensembles for detecting pulmonary manifestations of COVID-19 with chest X-rays. Apostolopoulos et al. [2] have investigated convolutional neural network architectures for medical image classification in a dataset of X-ray images in

H. V. Pham (✉)
Hanoi University of Science and Technology, Hanoi, Vietnam
e-mail: haipv@soict.hust.edu.vn

Q. H. Nguyen
University of Economics Ho Chi Minh City, Ho Chi Minh, Vietnam

© The Author(s), under exclusive license to Springer Nature Switzerland AG 2022
N. H. T. Dang et al. (eds.), *Artificial Intelligence in Data and Big Data Processing*,
Lecture Notes on Data Engineering and Communications Technologies 124,
https://doi.org/10.1007/978-3-030-97610-1_42

order to make the right decision of Covid-19 disease. Rahimzadeh and Attar [3] have conducted convolutional networks' model for classifying X-ray images into category classes consisting of normal, pneumonia, and COVID-19, using open-source datasets. Alom et al. [4] have focused on investigating a new method using Residual Recurrent Convolutional Neural Network combined with Transfer Learning (TL) model for detecting COVID-19 person. Punn et al. [5] have investigated to utilize deep learning models for analysis behaviors in prediction of the COVID-2019 in the world, developed by Johns Hopkins dashboard University. Patankar [6] has concerned LSTM-RNN for performance analysis, applied to deep learning-based models in order to find drugs for COVID-19. In addition, Ayyoubzadeh et al. [7] have applied to Linear regression integrated with long short-term memory (LSTM) in order to track COVID-19 person with its effectiveness. The studies have also concerned the all COVID-19 factors for previous incidence cases consisting of hand sanitizer, handwashing, and antiseptic topics. Bandyopadhyay and Dutta [8] have given the concepts of Long short-term memory (LSTM) combined with Gated Recurrent Unit (GRU) to predict COVID-19 data sets with its effectiveness. In related works [9], Huang et al. have investigated a new approach for forecasting Covid-19 with its influenced transitions in China using CNN model. Furthermore, Zeroual et al. [10] have indicated five deep learning methods to forecast the number of new COVID-19 cases including Recurrent Neural Network (RNN), Long short-term memory (LSTM), Bidirectional LSTM (BiLSTM). Shastri et al. [11] have investigated prediction for the future conditions of COVID-19. These studies have proposed deep learning techniques in decision making of COVID-19 cases in both India and USA. The other studies have investigated using Rule-based approaches [12–14] with context intelligence applied to solve various data sets under uncertain environment. Self-organizing maps (SOM) [15] model has investigated to deal with uncertain conditions for complex data sets. In the related works [16–19], these studies have concerned deep learning-based social distance for COVID-19 in order to track influenced COVID-19 people. However, these studies have not provided tracking influenced people with COVID-19 patients in large data sets with various fields.

The paper has presented a novel approach using Self Organizing Maps with Picture Fuzzy Sets for tracking COVID-19 persons. The proposed clustering model has been grouped records of COVID-19 persons together with these rules in order to find similar features of COVID-19 person in large data sets. The main contribution in this study is to find the right person among complex-relational groups, locations in real-time based on large datasets by clustering uncertain information quantified by Picture Fuzzy Sets of person in groups. To evaluate the effectiveness of this model, the proposed approach in simulations has demonstrated by using SOM integrated with these rules for tracking COVID-19 persons.

2 Research Background

In the context of the COVID-19 pandemic, the main attributes of a user (health status, visited places, meeting people, etc.) is used to collect information. Then, the four parameters of the Picture Fuzzy Set (PFS), such as agree, neutral, disagree, refusal, are applied to quantify the value of the attributes. This helps the authority agencies to easily detect and isolate places where have the confirmed cases of COVID-19 been there.

2.1 Picture Fuzzy Sets

A Picture Fuzzy Set [20] A on a universe X is an object of the form $A = \{(x, \mu_A(x), \eta_A(x), \nu_A(x)) | x \in X\}$, where $\mu_A(x) \in [0, 1]$ represents the "degree of positive membership of x in A", $\eta_A(x) \in [0, 1]$ is also called the "degree of neutral membership of x in A" and $\nu_A(x) \in [0, 1]$ represents the "degree of negative membership of x in A", and where μ_A η_A and ν_A satisfy the following condition:$(\forall x \in X)$ $(\mu_A(x) + \eta_A(x) + \nu_A(x) \leq 1)$ and $1 - (\mu_A(x) + \eta_A(x) + \nu_A(x))$, called the "degree of refusal membership" of x in A. Let $PFS(X)$ represent the set of all the PFSs on universe X.

2.2 Term and Definition

Suppose that a list of the new COVID-19 evident in Table 1 after gathering data from sensors as represented in a set of evident $E = \{E_1, E_2, E_3, \ldots, E_m\}$, based on

Table 1 List of the polices for evaluating

Criteria	Person	E_1	E_2	$\cdots$	E_m
C_1	P_1	$\{c_{11}, c_{12}, \ldots, c_{1s}\}$	$\{c_{11}, c_{12}, \ldots, c_{1s}\}$	$\cdots$	$\{c_{11}, c_{12}, \ldots, c_{1s}\}$
	$\cdots$	$\{c_{11}, c_{12}, \ldots, c_{1s}\}$	$\{c_{11}, c_{12}, \ldots, c_{1s}\}$	$\cdots$	$\{c_{11}, c_{12}, \ldots, c_{1s}\}$
	P_k	$\{c_{11}, c_{12}, \ldots, c_{1s}\}$	$\{c_{11}, c_{12}, \ldots, c_{1s}\}$	$\cdots$	$\{c_{11}, c_{12}, \ldots, c_{1s}\}$
C_2	P_1	$\{c_{21}, c_{22}, \ldots, c_{2t}\}$	$\{c_{21}, c_{22}, \ldots, c_{2t}\}$	$\cdots$	$\{c_{21}, c_{22}, \ldots, c_{2t}\}$
	$\cdots$	$c_{21}, c_{22}, \ldots, c_{2t}$	$c_{21}, c_{22}, \ldots, c_{2t}$	$\cdots$	$c_{21}, c_{22}, \ldots, c_{2t}$
	P_k	$c_{21}, c_{22}, \cdots, c_{2t}$	$c_{21}, c_{22}, \cdots, c_{2t}$	$\cdots$	$c_{21}, c_{22}, \ldots, c_{2t}$
$\cdots$	P_1	$\cdots$			
	$\cdots$				
	P_k				
C_n	P_1	$c_{m1}, c_{m2}, \ldots, c_{mv}$	$c_{m1}, c_{m2}, \ldots, c_{mv}$	$\cdots$	$c_{m1}, c_{m2}, \ldots, c_{mv}$
	$\cdots$	$c_{m1}, c_{m2}, \ldots, c_{mv}$	$c_{m1}, c_{m2}, \ldots, c_{mv}$	$\cdots$	$c_{m1}, c_{m2}, \ldots, c_{mv}$
	P_k	$c_{m1}, c_{m2}, \ldots, c_{mv}$	$c_{m1}, c_{m2}, \ldots, c_{mv}$	$\cdots$	$c_{m1}, c_{m2}, \ldots, c_{mv}$

a set of the category criteria $C = \{C_1, C_2, C_3, \ldots, C_n\}$. Each criteria set (C_i) has sub-criteria. The value of sub-criteria (c_{ij}) can be a digit or linguistic value. A set of the persons $P = \{P_1, P_2, P_3, \ldots, P_k\}$ will be checked among huge data sets.

2.3 Applied Picture Fuzzy Sets to Social Distance

Picture Fuzzy sets (PFS) represent as two kinds including directed and undirected graphs. Assume that let $G = (V, E)$ be a Picture Fuzzy Graph (PFG), where $V = \{v_1, v_2, \ldots, v_n\}$ represents a non-empty set of nodes or vertices, and $E = \begin{pmatrix} e_{11} & \cdots & e_{1n} \\ \vdots & \ddots & \vdots \\ e_{n1} & \cdots & e_{nn} \end{pmatrix}$ denotes a picture fuzzy relation on V. G is considered an undirected PFG if $e_{ij} = e_{ji}, \forall i, j = 1, 2, \ldots, n$. On the contrary, G is considered a directed PFG. For a directed PFG, some degree values on G such as the picture fuzzy in degree (d_I), picture fuzzy out degree (d_o), and picture fuzzy degree centrality (d) are expressed by in Eqs. (1), (2), and (3) respectively. They are given as follows.

$$d_I(v_i) = \sum_{j=1, j \neq i}^{n} e_{ij} \qquad (1)$$

$$d_o(v_i) = \sum_{j=1, j \neq i}^{n} e_{ji} \qquad (2)$$

$$d(v_i) = d_I(v_i) \oplus d_o(v_i) \qquad (3)$$

Assume that $\alpha = (\mu_\alpha, \eta_\alpha, \nu_\alpha)$ and $\beta = (\mu_\beta, \eta_\beta, \nu_\beta)$ are two PFSs. Then,

$$\alpha \wedge \beta = (\min\{\mu_\alpha, \mu_\beta\}, \max\{\eta_\alpha, \eta_\beta\}, \max\{\nu_\alpha, \nu_\beta\})$$
$$\alpha \vee \beta = (\max\{\mu_\alpha, \mu_\beta\}, \min\{\eta_\alpha, \eta_\beta\}, \min\{\nu_\alpha, \nu_\beta\})$$
$$\alpha \oplus \beta = (\mu_\alpha + \mu_\beta - \mu_\alpha\mu_\beta, \eta_\alpha\eta_\beta, \nu_\alpha\nu_\beta)$$
$$\alpha \otimes \beta = (\mu_\alpha\mu_\beta, \eta_\alpha + \eta_\beta - \eta_\alpha\eta_\beta, \nu_\alpha + \nu_\beta - \nu_\alpha\nu_\beta)$$

The directed PFG can be applied in various sectors where health is a typical example. In the context of the COVID-19 human profiles, PFG can be used to collect information from the digital profiles and social networks. A simple example of directed PFG to digital profiles of tracking COVID-19 person is shown in Fig. 1 as follows.

Then, some algorithms are used to create a weighted directed relationship between persons or between person and location, as well as to calculate distance between

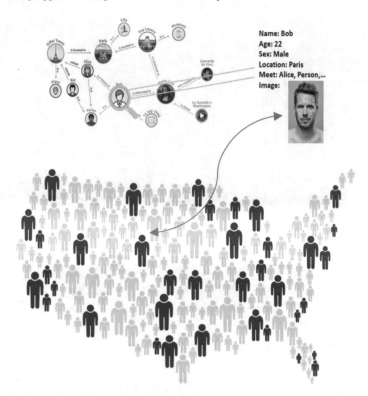

Fig. 1 A relevant example of directed PFS to COVID-19 profiles

PFGs. In conventional ways, COVID-19 epidemic is considered by social distance for each person with an individual profile (age, gender, nationality, location, phone, email, the presence of chronic diseases and etc.) [11, 21]. Figure 2 shows an example in features of the infection, from human-to-human transmission of the COVID-19 virus in the conventional management.

While people move around cities, all activities and features can be captured by data sensor, data with unstructured and structure data sets in real time. Figure 3 shows infected person providing much information with its profile.

In real-world problems, COVID-19 is depended on person behavior, moving demographic structure and mental values since COVID-19 is covered by countries from human-to-human transmission. Therefore, it is possible to capture data sources how to identify infected person influencing by cases of COVID-19. Sensor data is data normalization represented in fuzzy weight values [0, 1]. For getting these factor weights, we have applied a Sigmoid function for the data normalization and human profiles are described in Table 1 in details.

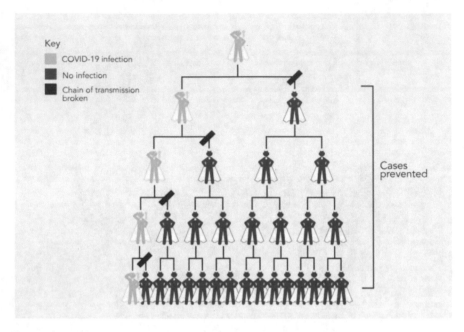

Fig. 2 Management of Human-to-human transmission of the virus [21]

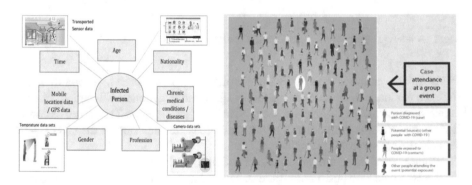

Fig. 3 The proposed model for infected person with real-time data collection

3 The Proposed Model

The proposed model shows applied SOM integrated with rule-based in order to cluster COVID-19 spread from human-to-human transmission, as shown in Fig. 4.

Inputs: All data sets are stored in the Data Cube with its multi-dimensional features, records of a human profile.

Outputs: SOM results show social distances by influenced people to indicate tracking person by cases of COVID-19.

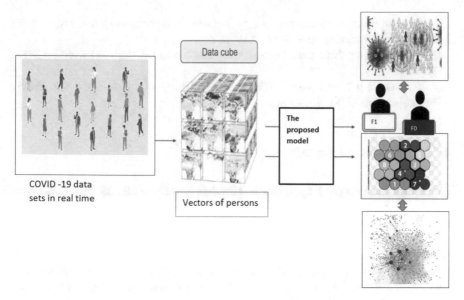

Fig. 4 The proposed model using SOM integrated with Rule-based for visualization

Step-1. Define human social distance. In data sets of the *Data Cube*, to calculate social distances among *human feactures* of pepople by SOM, the social distance $d_{P_i \rightarrow P_j}^S$ represent vectors $D_{P_i}^S$ and $D_{P_j}^S$ these attributes of persons P_i^S and P_j^S respectively in environment S as defined by Euclidean distance given by Eq. (4).

$$d_{P_i \rightarrow P_j}^S = ||D_{P_i}^S - D_{P_j}^S|| \qquad (4)$$

It is transformed the social distance using Eq. (3) $(v_i) = d_I(v_i) \oplus d_o(v_i)$

Step-2. Construct Data Cube for SOM Visualization. To construct a Data Cube called matrix $Cube_{n \times p}^S$ is to train by SOM, where n is the number of human vectors, p is the total number of dimensions.

Step-3. Calculate and update weights of social distances with Picture fuzzy rule-based.

Picture fuzzy rules represent the *j-th* factor f_j^S affected by $R_i^{f_j}$ to indicate human through sensors' data, as follows:

$$\text{IF } R_1^{f_j} \text{ AND } R_2^{f_j} R_m^{f_j} \text{ AND } == \text{"COVID-19" THEN}$$

Indicate F_j^S with marking points in the Cube database $\qquad (5)$

Step-4. Clustering SOM with human profile feactures

Calculate social human distance $d_{P_i \rightarrow P_j}^S$ is represented by a_{ij}^t and update weights to $Cube_{n \times p}^S$ by clustering with Picture fuzzy rule updated weights.

Recuduce the social distance $d^S_{P_i \to P_j}$ between two vectors $D^S_{P_i}$ and $D^S_{P_j}$ in order to find out clusterning groups on SOM map.

Step-5. Consider influnced human by social human distances based on SOM result

Repeat Step 3 and Step 5 with updated these weights to the Cube matrix toghether with SOM clustering.

4 Experimental Results

To confirm the proposed approach in simulations of COVID-19 cases, we have presented a scenario-based simulation to demonstrate how the proposed approach carried out the experiments. As considered in the sample datasets which represent data sources from: https://data.humdata.org/dataset/ecdc-covid-19, it has been extracted 500 records with random for anonymous human records from sensors collection of data sets, crowed with added with data collections. In order to evaluate the proposed system in simulation of COVID-19 for transmission human to human. The case study shows that SOM map was clustered groups of vectors in groups, as shown in Fig. 5.

In the case study, after the SOM training, a decision maker selected results on the map with several group clustering by SOM. The decision maker reduced by Steps 4

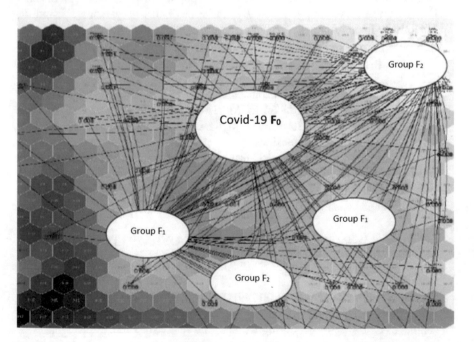

Fig. 5 The proposed model using SOM integrated with picture fuzzy rules for clustering

and 5 in the Sect. 3, the decision maker showed the influenced F_0 case in the location, as shown in SOM map of Fig. 6.

To evaluate our proposed system and establish a performance study we have investigated the system performs with respect to simulate results. The accuracy performance calculated by the actual correctness of the proposed model which were in the range 92–94% as shown in Fig. 7, based on around 24 h in random data sets. Experimental results in simulations show that the proposed model demonstrated significant was improvements in various locations, data segments.

The simulation results for our proposed model have been applied in real-world problems which resulted in improvements of influenced COVID-19 person in transmission human to human in simulations. The evaluation of the proposed model entails to support its practical application.

5   Concluding Observations

This paper has presented a new method for improvements in the identification of COVID-19 cased for influenced people in a large society. Our proposed approach uses SOM training integrated with a picture fuzzy rule-based approach to enhance of identification of F0 in case of COVID 19. Experimental results demonstrate that the proposed model provide enhanced detection accuracy with respect to COVID-19 person based on social human distance, as defined by SOM visualization.

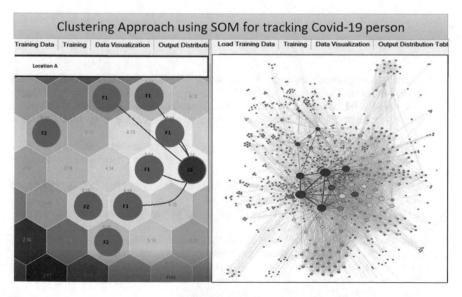

Fig. 6 SOM results for tracking persons in large data sets

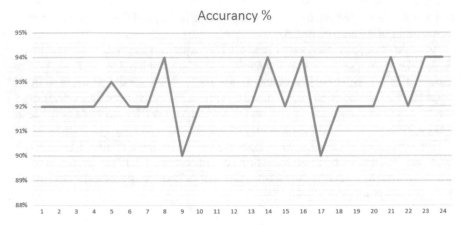

Fig. 7 Accuracy performance of the proposed model

In future research, while the proposed approach has demonstrated the principal challenge that forms the motivation in this study, we have identified a number of obvious extensions which consider investigation of tracking COVID-19 persons in real-time.

Acknowledgements This work was supported by the University of Economics Ho Chi Minh City (UEH), Vietnam under project CS-2021-51.

References

1. Rajaraman S, Siegelman J, Alderson PO, Folio LS, Folio LR, Antani SK (2020) Iteratively pruned deep learning ensembles for COVID-19 detection in chest X-rays. 8:115041–115050
2. Apostolopoulos ID, Mpesiana TA (2020) Covid-19: automatic detection from x-ray images utilizing transfer learning with convolutional neural networks. Phys Eng Sci Med 43(2):635–640
3. Rahimzadeh M, Attar A (2020) A modified deep convolutional neural network for detecting COVID-19 and pneumonia from chest X-ray images based on the concatenation of Xception and ResNet50V2. 19:100360
4. Alom MZ, Rahman M, Nasrin MS, Taha TM, Asari VK (2020) COVID_MTNet: COVID-19 detection with multi-task deep learning approaches
5. Punn NS, Sonbhadra SK, Agarwal SJM (2020) COVID-19 epidemic analysis using machine learning and deep learning algorithms
6. Patankar S (2020) Deep learning-based computational drug discovery to inhibit the RNA dependent RNA polymerase: application to SARS-CoV and COVID-19
7. Ayyoubzadeh SM, Ayyoubzadeh SM, Zahedi H, Ahmadi M, Kalhori SRN (2020) Predicting COVID-19 incidence through analysis of google trends data in Iran: data mining and deep learning pilot study. JMIR Publ Heal Surveill 6(2):e18828
8. Bandyopadhyay SK, Dutta SJM (2020) Machine learning approach for confirmation of covid-19 cases: positive, negative, death and release

9. Huang C-J, Chen Y-H, Ma Y, Kuo P-H (2020) Multiple-input deep convolutional neural network model for covid-19 forecasting in China
10. Zeroual A, Harrou F, Dairi A, Sun YJC (2020) Deep learning methods for forecasting COVID-19 time-series data: a comparative study. Solitons, Fractals 140:110121
11. Shastri S, Singh K, Kumar S, Kour P, Mansotra VJC (2020) Time series forecasting of Covid-19 using deep learning models: India-USA comparative case study. Solitons, Fractals 140:110227
12. Long CK et al (2021) A big data framework for E-Government in Industry 4.0. Open Comput Sci 11(1):461–479
13. Hai PV, Moore P et al (2014) Context matching with reasoning and decision support using hedge algebra with Kansei evaluation. In: SoICT '14: proceedings of the fifth symposium on information and communication technology, pp 202–210. https://doi.org/10.1145/2676585.2676598
14. Pham HV, Nguyen QH (2021) Intelligent IoT monitoring system using rule-based for decision supports in fired forest images. International conference on industrial networks and intelligent systems, pp 367–378
15. Pham HV, Cao T, Nakaoka I, Cooper EW, Kamei K (2011) A proposal of hybrid kansei-som model for stock market investment. Proceeding of the 1st international workshop on aware computing, pp 638–643
16. Magoo R, Singh H, Jindal N, Hooda N, Rana PS (2021) Deep learning-based bird eye view social distancing monitoring using surveillance video for curbing the COVID-19 spread. Neural Comput Appl 1–8. https://doi.org/10.1007/s00521-021-06201-5
17. Ahmed I, Ahmad M, Rodrigues JJPC, Jeon G, Din S (2021) A deep learning-based social distance monitoring framework for COVID-19. Sustain Cities Soc 65:102571. https://doi.org/10.1016/j.scs.2020.102571
18. Ahmed I, Ahmad M, Jeon G (2021) Social distance monitoring framework using deep learning architecture to control infection transmission of COVID-19 pandemic. Sustain Cities Soc 69:102777. https://doi.org/10.1016/j.scs.2021.102777
19. Rahim A, Maqbool A, Rana T (2021) Monitoring social distancing under various low light conditions with deep learning and a single motionless time of flight camera. PLoS ONE 16(2):e0247440. https://doi.org/10.1371/journal.pone.0247440
20. Lan LTH et al (2020) A new complex fuzzy inference system with fuzzy knowledge graph and extensions in decision making. IEEE Access 8:164899–164921
21. Zhang L, Li H, Lee W-J, Liao H (2021) COVID-19 and energy: influence mechanisms and research methodologies. Sustain Prod Consumption 27:2134–2152
22. Jiang G, Wang C, Song L et al (2021) Aerosol transmission, an indispensable route of COVID-19 spread: case study of a department-store cluster. Front Environ Sci Eng 15:46. https://doi.org/10.1007/s11783-021-1386-6

Sigmoid Local Pattern for Robust Car and Pedestrian Detection

Loc Duc Quan, Phuc Hoang Tran, Huy Quang Duy Nguyen, and Vinh Dinh Nguyen

Abstract In this paper, we propose a solution that can solve the problems of detecting cars and pedestrians under extremely difficult conditions, providing higher object detection and recognition rate. By analyzing the existing encoding-base feature method, we found that they do not consider the relationship between each channel when encoding the local region. The encoder was used to extract the information in the local region by considering the feature for each channel, separately. The fusion feature is combined by only considering the linear relationship between RGB input features and local features. This combination is not good enough because it is difficult to estimate the linear parameters during the running time. Therefore, it is necessary to introduce an efficient feature-based method that can help to resolve the current limitations of the existing methods. After analyzing the performance of the existing local pattern, we found that we can encode the local region by tolerating the relationship between the center features and neighbor features using an idea of a sigmoid activation function in the deep learning model. The proposed sigmoid-based method achieved the accuracy of 78% under extremely difficult conditions while the existing yolov5-based method achieved 65% under extremely difficult conditions.

Keywords Vehicle detection · Local pattern · Deep learning

Supported by FPT University, Cantho, Vietnam.

L. D. Quan · P. H. Tran · H. Q. D. Nguyen · V. D. Nguyen (✉)
FPT University, Can Tho 94000, Vietnam
e-mail: vinhnd18@fpt.edu.vn
URL: https://cantho.fpt.edu.vn/

L. D. Quan
e-mail: locqdce140037@fpt.edu.vn

P. H. Tran
e-mail: phucthce140628@fpt.edu.vn

H. Q. D. Nguyen
e-mail: huyndqce140623@fpt.edu.vn

© The Author(s), under exclusive license to Springer Nature Switzerland AG 2022   543
N. H. T. Dang et al. (eds.), *Artificial Intelligence in Data and Big Data Processing*,
Lecture Notes on Data Engineering and Communications Technologies 124,
https://doi.org/10.1007/978-3-030-97610-1_43

1   Introduction

According to safe driving, report [1], vehicle accident is one of the most accidents that often occurs along the road due to careless driving. We can save much money and save a human life if we can develop a good technique that can help the autonomous driving assistance system (ADAS) to detect vehicles under varied driving conditions on the road. In traditional computer vision applications, the researchers try to develop an algorithm to detect vehicles by using the RGB input images, or depth images [2]. Recently, many researchers have tried to propose to develop robust vehicle detection by using a local pattern as an input feature along with deep learning model [3]. These proposed techniques stated that they obtained very good performance under various weather conditions. The existing method-based local pattern still has several limitations:

1. They do not consider the relationship between each channel when encoding the local region. The encoder was used to extract the information in the local region by considering the feature for each channel, separately.
2. The fusion feature is combined by only considering the linear relationship between RGB input features and local features. This combination is not good enough because it is difficult to estimate the linear parameters during the running time.

Therefore, it is necessary to introduce an efficient feature-based method that can help to resolve the current limitations of the existing methods. After analyzing the performance of the existing local pattern, we found that we can encode the local region by tolerating the relationship between the center features and neighbor features using an idea of an activation function in the deep learning model. In addition, we proposed a new approach to combine the features of RGB and our proposed-based activation function feature without using any predefined threshold to determine how much feature is contributed to the final model.

We verified our system by using various popular international datasets and also used free and public traffic cameras in Vietnam. The results show that our system obtains very stable results under various road conditions. We obtain the detection rate of 78% while the yolov5 [4] achieve the detection rate 65% under difficult driving conditions using the CCD dataset. To review the advantages and disadvantages of existing vehicle detection and pedestrian detection-based method, we conduct a brief review in Sect. 2. From the limitation of existing methods, we introduce a robust and efficient method to extract and combine the features in Sect. 3. The performance of the proposed method is verified by using various dataset in Sect. 4. Finally, we point out the research conclusion and current limitations in Sect. 5.

2   Related Works

Car and pedestrian detection problems have been studied for many years. The key idea is to localize and classify obstacles as a car or pedestrians from the input images. Obstacle detection and classification algorithms were first developed by considering their own feature. This approach is often called a feature-based approach for detecting obstacles such as Haar-like features or histogram of gradient (HOG). However, the accuracy of shallow learning based on these traditional features is not good enough for industrial applications. From the limitation of the existing shallow-based learning method, recently, several deep learning methods have been studied by using various approaches such as single-stage or two-stage-based methods as listed in Table 1. With the help of a huge amount of data, the accuracy of existing deep learning-based models was increased significantly. However, the performance of existing these methods still depends on the input data.

3 Proposed Method

This section first briefly review the benefit and limitation of the traditional LBPs. Motivation by the current weakness of traditional LBP. Second, we suggest a new method to overcome the limitation of LBPs by developing a new feature constrain using the activation functions.

3.1 Strength and Weakness of LBPs

Local binary pattern (LBP) was introduced to improve the performance of texture classification by Ojala et al. in 1996 [13]. The idea is very simple is that they aim to encode the relation and extract local information by comparing the value of the center and neighbor pixel as follows:

$$L\left(x_{center}, y_{center}\right) = \sum_{i=0}^{N} \beta\left(x_{center}, y_{center}, x_i, y_i\right)$$

$$\beta\left(x_{center}, y_{center}, x_i, y_{iu}\right) = \begin{cases} 1 \ if \ I(x_{center}, y_{center}) > I(x_i, y_i) \\ 0 \ else \end{cases} \tag{1}$$

where $L\left(x_{center}, y_{center}\right)$ is the local encoding result at the center pixel in the input image. $\beta\left(x_{center}, y_{center}, x_i, y_{iu}\right)$ is the threshold function to evaluate the relationship between center pixel and neighbor pixels. N is the number of neighbor pixel. $I(x_i, y_i)$ is the value of the pixel at the position x_i, y_i.

According to the idea of local binary pattern, we realized that it is successfully encoded the information of sub-region in the image under illumination changes. Nguyen et al. have tried to investigate the benefits of LBP and LTP to improve the performance of obstacle detection [3]. However, they just investigated the benefit of local patterns by assuming that both LBP and LTP contribute equally to the final output feature. In addition, the local pattern feature is computed separated on each feature channel. There is no relationship between each channel when computing local

Table 1 Survey on existing car and pedestrian detection system

Survey			
Authors	Algorithms	Descriptions	Others
Pedestrian detection			
Jeon et al. [5]	Triangular pattern, T-CNN	Combining raw RGB image with novel local pattern called triangular pattern	Improve detection precision in rough conditions
Harshitha and Manikandan [6]	HOG & SVM	Real-time pedestrian detection system for autonomous vehicles	High precision, with prefect precision for nearby pedestrians
Su et al. [7]	AdaBoost, SVM, NCNN	Edge computing, integrate multiple networks on embedded vehicle	Optimize convolution operations by using greater amount of output channels, reduce its number at cost of reduced data
Lahmyed and Ansari [8]	HOG & SVM	Project clusters to extract and classify ROIs with LIDAR data	
Car detection			
Hsieh et al. [9]	GTOD	Input frame to GTOD api to make decision realtime, GTOD api trained by Coco dataset	
Sushmitha et al. [10]	SVM	Multiple car detection, recognition and tracking of approaches	Video is converted into frames in segmentation then followed by preprocessing and tracking of multiple cars
Aziz et al. [11]	HOG and KLT tracker	Object detection process are carried out using the HOG for detecting car. Object tracking process is performed using KLT	
Younis and Bastaki [12]	Fog removal algorithms	Dark channel prior (DCP) algorithm was introduced to remove fog. First, they compute the transmission map. This map is then updated by using the adaptive filter and edge-persevering filter. Finally, the fog was removed by using the new transmission map	DCP introduce an algorithm to eliminate the fog and increase the contrast of input features. Its processing time is slow and require huge memory size for storing parameters
YOLOv5 [4]	Increase the accuracy and performance of previous YOLO-based method	Introduce new HEAD architecture, and reduce training parameters	Surpass the performance of previous yolo versions
The proposed method	Sigmoid-based method for car and pedestrian detection	Fusing the benefits of SOTA deep learning model and sigmoid function for improving the accuracy of existing systems	Provide stable performance under difficult road conditions

patterns. Therefore, we suggest a new approach to improve the current limitation of traditional local patterns as mentioned in the next section.

3.2 The Proposed Sigmoid-Based Local Pattern

To overcome the limitation of existing local pattern, we introduce a new approach by integrating the sigmoid function to compute the values of the output pattern $S_{sig}^{\gamma}\left(x_{center}^{\gamma}, y_{center}^{\gamma}\right)$ at each image channel $\Omega = \{blue, green, red\}$ as follows:

$$S_{sig}^{\gamma}\left(x_{center}^{\gamma}, y_{center}^{\gamma}\right) = \sum_{i=0}^{N} F_{sig}\left(x_{center}^{\gamma}, y_{center}^{\gamma}, x_i^{\gamma}, y_i^{\gamma}, \lambda_{sig}\left(x_{center}, y_{center}, x_i, y_i\right)\right)$$

$$F_{sig}\left(x_{center}, y_{center}, x_i, y_i, \lambda_{sig}\right) = \begin{cases} 1 \; if \; I\left(x_{center}, y_{center}\right) > I(x_i, y_i) + \lambda_{sig} \\ 0 \; f \; I\left(x_{center}, y_{center}\right) \leq I(x_i, y_i) + \lambda_{sig} \end{cases}$$

$$\lambda_{sig}\left(x_{center}, y_{center}, x_i, y_i\right) = \zeta\left(\frac{1}{3}\sum_{\gamma \in \Omega}\frac{I(x_{center}^{\gamma}, y_{center}^{\gamma})+I(x_i^{\gamma}, y_i^{\gamma})}{2}\right)$$

$$\zeta\left(I(x, y)\right) = \frac{e^{I(x,y)}}{1+e^{I(x,y)}}$$

$$(2)$$

where $F_{sig}\left(x_{center}, y_{center}, x_i, y_i, \lambda_{sig}\right)$ is the function is used to encode the local information with the help the proposed threshold-based Sigmoid function $\lambda_{sig}\left(x_{center}, y_{center}, x_i, y_i\right)$. The proposed method successfully captures and encodes the local region as shown in Fig. 1. Our proposed algorithm produced robust features in comparison to the existing LBP's result as shown in Fig. 1. Our proposed sigmoid-based function provides more stable and clear features than those of existing LBP as described in Fig. 1. After that, the new feature $I_{sig}^{\gamma}(x_{center}^{\gamma}, y_{center}^{\gamma})$ is computed as follows:

$$I_{sig}^{\gamma}(x_{center}^{\gamma}, y_{center}^{\gamma}) = \kappa_{\gamma} \times I(x_{center}^{\gamma}, y_{center}^{\gamma})$$

$$\kappa_{\gamma} = \frac{\zeta\left(I(x_{center}^{\gamma}, y_{center}^{\gamma})\right) \times \zeta\left(S_{sig}^{\gamma}(x_{center}^{\gamma}, y_{center}^{\gamma})\right)}{\zeta\left(I(x_{center}^{\gamma}, y_{center}^{\gamma})\right) + \zeta\left(S_{sig}^{\gamma}(x_{center}^{\gamma}, y_{center}^{\gamma})\right)}$$

$$(3)$$

Thus, we proposed a new approach to combine the features of RGB by using the sigmoid activation function feature without using any predefined threshold to determine how much feature is contributed to the final model. The output feature $I_{sig}^{\gamma}(x_{center}^{\gamma}, y_{center}^{\gamma})$ at each image channel $\Omega = \{blue, green, red\}$ is then input to the Yolo-based deep learning model for training and testing.

Fig. 1 Experimental results of the proposed sigmoid local pattern in comparison to traditional LBP feature. **a** is the input image. **b** is the result of the LBP. **c** is the result of our proposed method using sigmoid function. **d** is the final feature combination results of our method

4   Experimental Results

To verify the accuracy of the proposed system, we used a deep learning-based YOLO algorithm [4] to integrate our proposed feature for training and testing. First, the sigmoid local patterns were computed to provide sigmoid features. The sigmoid features were then input into the deep learning-based YOLO for training. Various popular datasets were used to evaluate the performance of the proposed sigmoid-based feature, such as the CCD dataset [3] and public online traffic camera at Ho Chi Minh City, Vietnam. Google colab was used to train and validate the performance of the proposed system with 2.30 GHz CPU (4 cores), P100 Tesla GPU 16 GB.

The performance of the proposed system was verified by using the detection rate metric as follows:

$$\zeta_{rate} = \frac{\sum_N \sum_{I \in N, i \in I} IoU_i\,(P_i, G_i) \geq 0.5}{\sum_N \sum_{I \in N} G_i} \tag{4}$$

where IoU_i is intersection over union ratio that is used to computer the overlap region between the predicted output P_i and ground truth G_i.

Figure 2 indicates the value of the loss and train function of the proposed sigmoid-based method after running 300 epochs. We used 255,942 images for training, 178,844 images for validation and 25,549 images for testing by using CCD dataset and online traffic images at Ho Chi Minh city, Vietnam. The input image was resized to 640×640 before inputting to the training network. We reused the pre-trained COCO model initializing our input weights. We setup YOLO [4] with three anchor levels for multiple detection as follows: [10,13, 16,30, 33,23], [30,61, 62,45, 59,119], and [116,90, 156,198, 373,326]. The system was trained in 45 h.

The training precision of the proposed sigmoid-based system is approximate to 82% under hostile driving conditions after 300 training epochs. Experimental results (Fig. 3) show that the proposed sigmoid-based method achieved high accuracy of pedestrian detection rate under night conditions using the CCD dataset. Experimental results (Fig. 4) show that the proposed sigmoid-based method also achieved high accuracy of car detection rate under night conditions using the CCD dataset. We also verify the performance of the proposed system with the state-of-the-art yolo-based method [4] as described in Fig. 5. The first row presents the car detection results of the Yolov5 under a dark driving road. The second row presents the car detection results of the sigmoid-based function under the dark driving roads. We found that the proposed sigmoid-based method got better performance than the Yolov5-based method in terms of both confident score and number of detected objects. The proposed sigmoid-based

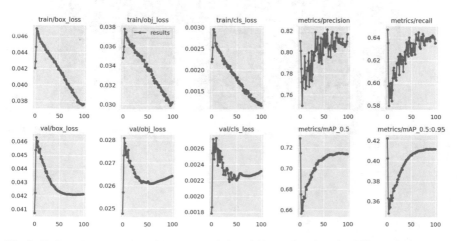

Fig. 2 Train and loss function of the proposed sigmoid-based function after 300 epochs

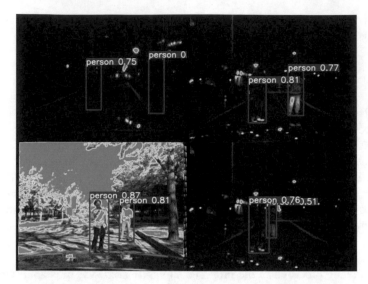

Fig. 3 The pedestrian detection results of the sigmoid function-based on YOLO deep learning model using the CCD dataset

Fig. 4 The car detection results of the sigmoid fucntion-based on YOLO deep learning model using the CCD dataset

method achieved the accuracy of 78% under extremely difficult conditions while the existing yolov5-based method [4] achieved 65% under extremely difficult conditions.

Figure 6 describes the detected results of the proposed sigmoid-based method and yolov5 [4] using the CCD dataset. The input conditions are many challenges

Fig. 5 The car detection results of the sigmoid fucntion-based on YOLO deep learning model and the Yolov5-based [4] using the CCD dataset. First row presents the car detection results of the Yolov5 under dark driving road. Second row presents the car detection results of the sigmoid-based function under dark driving road

Fig. 6 The car detection results of the sigmoid fucntion-based on YOLO deep learning model and the Yolov5-based [4] using the online traffic camera at HCM city, Vietnam. First row presents the car detection results of the Yolov5-based method [4]. Second row presents the car detection results of the proposed sigmoid-based method

that make the results of the proposed method, sometimes, confuse between car and motorbike shown in Fig. 6. However, we realized that the sigmoid-based function still provides more stable car ad pedestrian detection when compared with yolov5 [4].

5 Conclusion

To investigate a method that can help to improve the performance of the existing obstacle detection and classification. We found that we can encode the local region by tolerating the relationship between the center features and neighbor features using an idea of an activation function in the deep learning model. In addition, we proposed a new approach to combine the features of RGB and our proposed-based activation function feature without using any predefined threshold to determine how much feature is contributed to the final model. By verifying the performance of the sigmoid-based feature, we realized that the performance of our model is better than other compared methods under abnormal driving conditions. The processing time of the sigmoid-based feature is still high because of running a local encoding. Therefore, we plan to find a solution to speed up the processing time of the sigmoid-based feature.

References

1. Sivaraman S, Trivedi MM (2013) Looking at vehicles on the road: a survey of vision-based vehicle detection, tracking, and behavior analysis. IEEE Trans Intell Transp Syst 14(4):1773–1795
2. Ward IR, Laga H, Bennamoun M (2019) RGB-D image-based object detection: from traditional methods to deep learning techniques. In: Advances in computer vision and pattern recognition, pp 169–201
3. Nguyen VD, Tran DT, Byun JY, Jeon JW (2019) Real-time vehicle detection using an effective region proposal-based depth and 3-channel pattern. IEEE Trans Intell Transp Syst 20(10):3634–3646
4. YOLOv5. https://github.com/ultralytics/yolov5. Last accessed Nov 2021
5. Jeon H, Nguyen VD, Jeon JW (2019) Pedestrian detection based on deep learning. In: 45th Annual conference of the IEEE industrial electronics society, pp 144–151
6. Harshitha R, Manikandan J (2017) Design of a real-time pedestrian detection system for autonomous vehicles. In: IEEE Region 10 symposium (TENSYMP), pp 1–4
7. Su CL, Lai WC, Li CT (2021) Pedestrian detection system with edge computing integration on embedded vehicle. In: International conference on artificial intelligence in information and communication (ICAIIC)
8. Lahmyed R, Ansari ME (2016) Multisensors-based pedestrian detection system. In: IEEE/ACS 13th International conference of computer systems and applications (AICCSA), pp 1–4
9. Hsieh CH, Lin DC, Wang CJ, Chen ZT, Liaw JJ (2019) Real-time car detection and driving safety alarm system with Google tensorflow object detection API. In: International conference on machine learning and cybernetics (ICMLC), pp 1–4
10. Sushmitha S, Satheesh N, Kanchana V (2020) Multiple car detection, recognition and tracking in traffic. In: International conference for emerging technology (INCET), pp 1–5

11. Aziz MVG, Hindersah H, Prihatmanto AS (2017) Implementation of vehicle detection algorithm for self-driving car on toll road Cipularang using Python language. In: International conference on electric vehicular technology (ICEVT), pp 149–153
12. Younis R, Bastaki N (2017) Accelerated Fog removal from real images for car detection. In: IEEE-GCC Conference and exhibition (GCCCE), pp 1–6
13. Ojala T, Pietikäinen M, Harwood D (2016) A comparative study of texture measures with classification based on feature distributions. Pattern Recogn 19(3):51–59

Dynamically Adaptive Switching Based Median Mean Filter for Removal of High Density Salt and Pepper Noise

Shreyansh Soni, Dhananjay Raina, Jeeya Prakash, Bharat Garg, and Rana Pratap Yadav

Abstract This paper proposes a novel dynamic adaptive switching based median mean filter (DASBMMF) to remove salt and pepper noise (SAP) of high noise density (ND). The algorithm detects and flags noisy pixels in a separate array and replaces the corrupt pixel with the median of neighbouring noise free pixels. Further, if all the pixels present in the sliding selection window are corrupted, the size of the window is increased up to the threshold window size, to accommodate un-corrupted pixels. Lastly, values of pre-processed pixels are also taken into consideration since they give a better result in place of a corrupt pixel. Comparative analysis reveals that the proposed scheme has improved results in Peak Signal to Noise Ratio (PSNR) and Image Enhancement Factor (IEF) without any additional computational exertion (CPU Time) on the system, due to the simplicity of the algorithm and hence giving faster results.

Keywords Median filter · Noise density · PSNR · IEF · CPU time · Salt-and-pepper noise · Image processing

S. Soni · D. Raina · J. Prakash · B. Garg (✉) · R. P. Yadav
Thapar Institute of Engineering and Technology, Patiala, India
e-mail: bharat.garg@thapar.edu

S. Soni
e-mail: ssoni_be18@thapar.edu

D. Raina
e-mail: draina_be18@thapar.edu

J. Prakash
e-mail: jprakash_be18@thapar.edu

R. P. Yadav
e-mail: rpyadav@thapar.edu

1 Introduction

Pictures are regularly undermined by noises during transmission. Gaussian Noise is a statistical noise having a probability density function equal to normal distribution, also known as Gaussian Distribution. The image intensity in magnetic resonance magnitude images in the presence of noise is shown to be governed by a Rician distribution. Low signal intensities (SNR < 2) are therefore biased due to the noise. Impulse noise could either be fixed valued (Salt and Pepper) or random valued. The noisy pixels assume values between maximum and minimum grey level and the aim while removing such noisy pixels is to preserve the quality of the image. The pre-existing methods of filtration that we have studied and taken into consideration, do solve the purpose of restoring the image however are either too lengthy and complex or suffer on the front of accuracy.

The Weighted Median Filter (WMF) [20] and the Centre-Weighted Median Filter (CWMF) [12] were proposed as solution for further develop the middle filter by giving more weight to some chose pixels in the filtering window. Albeit these two filters can protect a larger number of subtleties than the middle filter, they are as yet executed consistently across the picture disregarding if the current pixel is without noise. An Adaptive Median Filter (AMF) [10] proposed in is acceptable at low and medium noise thickness levels. At higher noise densities, the quantity of substitutions of adulterated pixel increments extensively; expanding window size will give better noise expulsion execution; nonetheless, the first pixel esteems and supplanted middle pixel esteems are less corresponded. As a result, the edges are spread significantly. The Adaptive Centre Weighted Median Filter (ACWMF) [9] proposed in is utilized to eliminate high thickness motivation noise. It requires streamlined edges for both salt and pepper and irregular esteemed motivation noise types. However Progressive Switching Median Filter (PSMF) [18] performs efficiently, it is tedious and computationally difficult to implement.

A noise adaptive soft switching median (NASM) [1] channel is recommended, where a three-level various levelled delicate exchanging noise location measure is utilized. A novel robust estimation-based filter (REBF) [1] and The Modified Decision Based Unsymmetric Trimmed Median Filter (MDBUTMF) [1] have been proposed as of late to manage high density noise. The function of the robust estimation based filter (REBF) [17] channel is to identify the anomaly pixels and re-establish the first worth utilizing robust estimation. The Modified Decision Based Unsymmetric Trimmed Median Filter (MDBUTMF) [8] replaces the corrupt pixel by managed middle worth when other pixel esteems, 0's and 255's are available in the chosen window and when all the pixel esteems are 0's and 255's then the noise pixel is supplanted by mean worth of the relative multitude of components present in the chosen window.

Recently, other filtering techniques to denoise images at high noise density are presented in [4, 5]. An Adaptive Min-Max value-based Weighted Median (AMMWM) filter [3] computes highly correlated groups of noise-free pixels using minimum and maximum values of the current filter window to restore the processed Noisy Pixel.

A new filtering approach, Significance Driven Inverse Distance Weighted (SDIDW) filter [6] estimates the inverse distance weighted value using a minimum number of nearest original pixels. The weight of the original pixels is assigned based on the squared distance from the noisy pixel. However, at very high noise density, the output filtered image is reconstructed with poor visual clarity. The Adaptive Weighted Min-Mid-Max Value-based (AWM3F) [14] computes minimum, maximum, and middle values along with their weights based on the number of NFPs, which are closest to these values. Hybrid Decision-based filtering (HBDF) [7] de-noises images using a combination of linear, non-linear, and probabilistic techniques. In addition, to improve the quality, a universal add-on was presented which uses edge detection and smoothening techniques to brush out fine details from filtered images.

Our proposed technique has tried to create a balance between efficiency and accuracy while removing impulse noise from images. Our domain of work and image selection primarily consists of Magnetic Resonance Imaging (MRI) images, specifically those of brain cancer. Extraordinary information and experience on radiology are needed for exact growth location in clinical imaging. As indicated by a review, the occurrence of growths in the focal sensory system in India goes from around 5–10 cases for every 100,000 populaces. At the point when commotion is available in pictures of clinical diagnostics, it might prompt flawed examination to such an extent that tumorous growth cells might go undetected. This is the essential driver concerning why cerebrum malignant growth isn't perceived in the body until the third stage after which they metastasize and travel to different parts of the body through the circulatory system or the lymph framework which makes the malignancy nearly serious. Subsequently, a more productive and solid design of commotion evacuation in clinical diagnostics should be created which works on the precision of the outcomes as well as gets the biomedical hardware framework.

The paper is organized as follows: Sect. 2 presents proposed algorithm with its flow chart whereas Sect. 3 shows an illustration of denoising using proposed algorithm. Section 4 presents simulation results and analysis on various benchmark images with varying ND. Finally in Sect. 5, conclusions are drawn.

2 Proposed Filter Algorithm

The pseduo code of the proposed algorithm is shown in Algorithm 1 whereas the flowchart of the proposed algorithm is shown in Fig. 1. The proposed algorithm works in two stages. The first stage is the detection of noise in the image. We generate a flag mask over the image in which if the middle pixel (X_{ij}) of the sliding window is 0 or 255 (max or impulse value) then it is considered to be noisy and Flag $f_{i,j} = 0$. If it's not noisy ($f_{i,j} = 1$) then we move to the next pixel. The image is padded beforehand so that the windows fit perfectly throughout the image.

The second stage is to apply the proposed filter algorithm to the noisy pixel. The initial window size taken in our algorithm is a 3×3 window. If the number of surrounding healthy pixels is greater than 0 than we take the median of all such

Algorithm 1 Proposed filter (*Img*, *OutImg*)

Input *Img*: Input Image
Output *OutImg*: Output Image
for each pixel $X_{i,j}$ of *Img* **do**
  Set flag ($f_{i,j}$) value to 1, if non-noisy else to 0.
  if $f_{i,j} == 1$ **then**
    $P_{i,j} \leftarrow X_{i,j}$;
  else
    for n from W_{min} to W_{max} ▷ $W_{min} = 3$, $W_{max} = 7$
    if (non-noisy pixel > 0 in $w_{n \times n}$) **then**
      $P_{i,j} \leftarrow$ median(non-noisy pixel);
    else
      n=n+2;
    end if
    if n==7 && non-noisy pixel = 0 in $w_{n \times n}$ **then**
      if (i==pad+1 || i==row+pad && j>pad+1) **then**
        $P_{i,j} \leftarrow P_{i,j-1}$;
      else if (j==pad+1 || j==col+pad && i>pad+1) **then**
        $P_{i,j} \leftarrow P_{i-1,j}$;
      else if (i>pad+1 && j>pad+1 && j<col+pad && i<row+pad) **then**
        $P_{i,j} \leftarrow mean()$;
      end if
    end if
  end if
  $OutImg(i, j) = \leftarrow P_{i,j}$;
end for
Return *OutImg*;

pixels. If there are no noise-free pixels in the window, we then proceed to increase the size of the window. The size is increased from 3×3 to 5×5, and then the steps are repeated until we reach the 7×7 window. Upon reaching the threshold window size (7×7 here) instead of continuing to increase the window size as done in DAMF [13], we take the help of the previously processed pixels. If the pixel in question (X_{ij}) is noisy and in the last row or column of the image, we take the just previously processed available pixel and replace the noisy pixel with it. If the pixel is in the first or last row ($i = 1$ or $i = N$) we replace it with the processed pixel to its immediate left. And if the pixel is first or last column ($j = 1$ or $j = N$) we replace it with the processed pixel immediately above it. If the noisy pixel does not lie in the corner then we take a 3×3 window to take the mean of the pixels processed before pixel X_{ij} for denoising.

3 Illustration of Proposed Algorithm

To demonstrate the working of the proposed algorithm we take a 5×5 section of the Lena image corrupted by Salt and pepper noise of density 90% as shown in Fig. 2a.

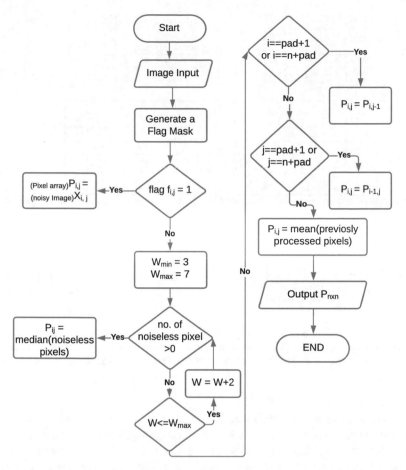

Fig. 1 Flowchart of the proposed filter

The noisy pixels can be identified by the maximum values they have (0 and 255). First, we create a flag for all the noisy pixels as shown in Fig. 2b, this will later help us in identifying the noisy pixels and then denoising them. We also pad our image to fit the sliding window for every pixel as shown in Fig. 2c. Now we put a 3×3 window over the noisy pixel and look for any healthy neighboring pixel. Since there are no healthy pixels in the 3×3 window, we increase the window size to 5×5. When increased to 5×5 we have 2 healthy pixels with values 118 and 120. Discarding the noisy pixels, we take the median of the healthy pixels: 119 as shown in (Fig. 2d). We then save this denoised value of the pixel in another Matrix and then continue the same methodology for the next pixel by shifting the window in the noisy image matrix. Figure 2 Illustrating methodology at threshold window condition.

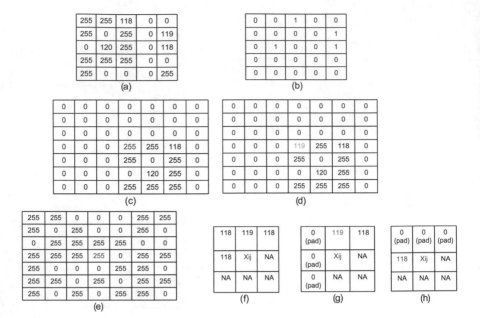

Fig. 2 **a** A noisy matrix **b** flagging noisy pixels **c** saving denoised value of the pixels in another matrix **d** methodology at threshold window condition **e** taking mean of pre processed pixels **f** X_{ij} is unknown pixel **g** Pi (segments) in case of first/last column **h** Pi (segments) in case of first/last row

For Non-corner regions, the pixel at the center of the 7×7 window is noisy, but there are no healthy pixels in the window in the image segment (Fig. 2e). Hence taking the median of the neighboring pixels will result in noise only. Therefore we take the help of the Pixel segment, i.e. the matrix we are filling up as keep on denoising the image by applying medians. In the Pixel segment, the pixels prior to the X_{ij} pixel are already processed. We take a 3×3 window and then take the mean of all such pre-processed pixels. We take the mean of all the available pixels (118, 118, 118, 119 = 118.25) and we then replace the noisy pixel (X_{ij}) with this value in Pixel Segment as shown in Fig. 2f.

Corner Conditions, if a noisy pixel is in the last or first row/column, we have to keep in mind that the 3×3 window section of the said pixel will contain padding pixels too from the Pixel Segment. Which will present us with an incorrect value of the noisy pixel. To deal with this problem we have some special conditions in our program. In such cases we don't have enough pre-processes signals to compute their average or median. Therefore we replace the X_{ij} (noisy pixel) with the immediate top pixel (119) in condition 1 (as shown in Fig. 2g) and with the immediate left pixel (118) in condition 2 as shown in Fig. 2h.

4 Simulation Results and Analysis

The efficiency of the proposed algorithm is analyzed over the existing techniques which are: AWSM [11], DAMF [13], FSBMMF [16], RSIF [15], SMF[19], TVWA [3], PSMF [18], BPDF [2]. We used different benchmark images as well medical images to simulate our results in MATLAB. The images were subjected to 10–90%. The quality metrics such as PSNR and IEF are extracted. We also take into consideration the complexity of our algorithm with the CPU time it takes to deliver the result.

The mathematical expression of PSNR is given by Eq. 1.

$$PSNR = 10 \log_{10}(MAX^2/MSE) \tag{1}$$

where MSE in Eq. 2 is mean square error and MAX is maximum pixel value of the image. In case of 8-bit grayscale images, the value of MAX is 255.

$$MSE = \frac{1}{M*N} \sum_{i=1}^{M} \sum_{j=1}^{N} (x_{i,j} - y_{i,j})^2 \tag{2}$$

Table 1 PSNR, IEF and computation time of proposed and existing SAP removal methods using Lena gray benchmark image with fluctuating noise density

Metrics	ND (%)	AWSM [11]	DAMF [13]	FSBMMF [16]	RSIF [15]	SMMF [19]	TVWA [3]	PSMF [18]	BPDF [2]	Prop.
PSNR (dB)	10	42.09	42.93	41.07	42.09	35.30	42.54	18.59	39.57	43.01
	30	35.95	36.8	34.35	24.64	30.20	37.20	13.37	32.92	36.81
	50	32.47	33.29	30.83	30.81	27.56	33.83	10.76	28.18	33.32
	70	29.32	30.17	28.26	27.41	26.05	30.57	8.38	22.81	30.33
	90	24.74	25.98	24.17	22.06	24.06	26.57	6.34	10.78	26.12
	Avg	**32.91**	**33.83**	**31.73**	**29.40**	**28.63**	**34.14**	**11.48**	**26.85**	**33.91**
IEF	10	464.67	558.67	360.62	461.38	97.22	512.15	2.05	257.32	573.69
	30	337.32	410.41	235.09	250.52	89.91	449.42	1.99	168.31	411.42
	50	252.55	306.68	172.65	172.44	81.15	344.04	1.70	94.57	306.64
	70	170.55	208.24	134.03	110.53	80.41	228.31	1.37	38.18	215.03
	90	76.52	101.55	67.30	41.19	65.51	116.83	1.10	3.07	104.91
	Avg	**260.32**	**317.11**	**193.38**	**207.21**	**82.84**	**330.15**	**1.64**	**112.29**	**322.33**
CPU time (s)	10	1.94	0.54	0.15	3.64	0.34	0.43	0.12	0.51	0.12
	30	4.18	0.86	0.27	9.43	0.67	1.21	0.12	1.34	0.35
	50	6.42	1.02	0.35	15.52	0.86	1.54	0.12	2.12	0.54
	70	8.31	1.32	0.47	21.84	1.12	1.91	0.13	2.85	0.69
	90	9.4	1.86	0.55	27.36	1.58	4.02	0.13	3.43	1.19
	Avg	**6.05**	**1.12**	**0.35**	**15.55**	**0.91**	**1.82**	**0.12**	**2.05**	**0.57**

The parameters M and N represent dimensions of image whereas x and y denotes input and output images, respectively. The mathematical of Image Enhancement Factor (IEF) is given below

$$IEF = \frac{MSE(Original, Noisy)}{MSE(Original, Restored)} \tag{3}$$

Both quantitative and qualitative analysis is done based on the extracted quality metrics and reconstructed images respectively. For benchmark images, we have taken two grayscale images (Lena gray and Doors and Windows). The following section represents the average quality metrics and run-time of proposed and existing algorithms on the benchmark images.

The results are presented in Tables 1 and 2 with the PSNR and IEF values for both the benchmark images. The results clearly indicate that the proposed algorithm shows better average PSNR values than most of the pre-existing techniques even at high noise densities. In Grayscale images Figs. 3 and 5 while we have comparable results with the TVWA [3] filter within the margin of error, the results were achieved in much lesser time with our algorithm. Finally, for the brain image results shown in Table 3, we see a complete dominance in performance by our algorithm. Our algorithm outperformed all the other algorithms in PSNR values and IEF values. While

Table 2 PSNR, IEF and computation time of proposed and existing SAP removal methods using Doors and Windows with fluctuating noise density

Metrics	ND (%)	AWSM [11]	DAMF [13]	FSBMMF [16]	RSIF [15]	SWMF [19]	TVWA [3]	PSMF [18]	BPDF [2]	Prop.
PSNR (dB)	10	30.77	22.67	31.07	29.71	28.10	30.64	19.55	29.97	32.93
	30	27.17	22.38	26.47	25.62	24.18	27.72	14.6	25.87	27.65
	50	24.45	22.01	23.75	23.35	22.07	25.53	11.62	22.63	25.05
	70	22.15	21.45	21.02	21.55	20.65	23.20	8.99	19.49	22.89
	90	19.56	20.02	19.81	19.38	19.29	20.47	6.85	11.2	19.92
	Avg	24.82	21.70	24.42	23.92	22.85	25.51	12.32	21.83	25.68
IEF	10	32.92	5.04	35.63	26.32	18.03	33.82	2.52	28.07	35.24
	30	43.22	14.53	37.41	30.97	30.97	21.93	2.41	32.28	48.92
	50	39.26	22.17	33.23	30.30	22.51	49.91	2.02	25.67	44.73
	70	32.67	27.38	29.71	28.03	22.76	40.84	1.55	17.43	38.04
	90	22.14	25.35	24.21	21.83	21.21	28.30	1.21	3.31	24.69
	Avg	34.04	18.89	32.03	27.49	23.09	34.96	1.94	21.35	38.32
CPU time (s)	10	4.33	1.10	0.23	6.27	0.68	0.54	0.17	0.96	0.29
	30	9.37	1.57	0.56	18.14	1.39	0.96	0.18	2.09	0.81
	50	14.05	2.05	0.88	29.79	2.01	1.67	0.18	3.07	1.32
	70	18.40	2.57	1.17	42.54	2.49	2.01	0.18	4.31	1.84
	90	19.54	3.8	1.49	53.36	3.22	4.27	0.18	5.32	2.96
	Avg	13.13	2.21	0.86	30.02	1.95	1.89	0.18	3.15	1.44

Table 3 PSNR, IEF and computation time of proposed and existing SAP removal methods using brain image with fluctuating noise density

Metrics	ND (%)	AWSM [11]	DAMF [13]	FSBMMF [16]	RSIF [15]	SWMF [19]	TVWA [3]	PSMF [18]	BPDF [2]	Prop.
PSNR (dB)	10	38.27	29.37	43.15	36.01	29.41	37,76	35.51	40.93	44.05
	30	30.56	24.79	37.44	31.59	23.90	32.38	24.02	34.12	39.32
	50	27.44	22.38	34.11	28.39	21.58	28.78	15.58	27.41	36.09
	70	24.37	20.72	31.35	24.88	19.79	25.71	10.42	22.31	33.18
	90	20.23	18.65	28.45	22.91	17.94	23.08	6.65	15.56	29.42
	Avg	**27.74**	**23.18**	**34.9**	**28.75**	**22.52**	**29.54**	**18.43**	**28.06**	**36.41**
IEF	10	309.95	40.85	942.22	187.86	40.20	273.38	161.26	556.9	1190.90
	30	157.12	41.28	768.86	196.11	34.12	229.20	34.45	355.01	1172.70
	50	128.14	39.58	583.38	158.78	33.10	173.40	8.37	127.77	928.46
	70	87.89	37.96	438.45	99.41	30.55	120.35	3.44	54.83	673.15
	90	43.73	30.57	288.83	80.73	25.54	84.21	1.91	14.98	358.33
	Avg	**145.46**	**38.04**	**604.34**	**144.57**	**32.70**	**176.10**	**41.86**	**221.89**	**864.70**
CPU time (s)	10	0.58	0.39	0.08	1.72	0.25	0.57	0.11	0.27	0.07
	30	1.19	0.51	0.13	2.58	0.33	0.68	0.10	0.61	0.14
	50	1.71	0.42	0.16	4.06	0.47	0.56	0.10	0.82	0.17
	70	2.41	0.65	0.21	5.63	0.44	0.92	0.10	1.13	0.22
	90	2.55	0.66	0.15	7.08	0.63	1.51	0.10	1.35	0.39
	Avg	**1.68**	**0.52**	**0.14**	**4.21**	**0.42**	**3.07**	**0.10**	**0.83**	**0.19**

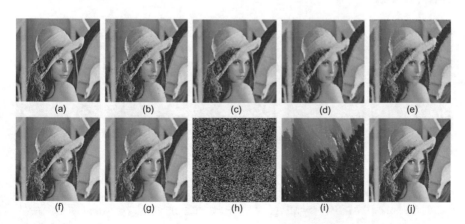

Fig. 3 Lena gray Image with 90% noise density filtered using: **a** original image, **b** AWSMF, **c** DAMF, **d** FSBMMF, **e** RSIF, **f** SMMF, **g** TVWA, **h** PSMF, **i** BPDF and **j** proposed filter

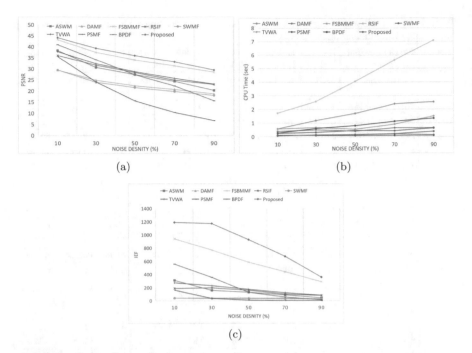

Fig. 4 PSNR, computational time and IEF of brain image using different filters for varying noise density (10–90%)

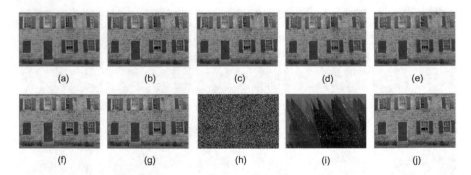

Fig. 5 Doors and Windows with 90% noise density filtered using: **a** original image, **b** AWSMF, **c** DAMF, **d** FSBMMF, **e** RSIF, **f** SMMF, **g** TVWA, **h** PSMF, **i** BPDF and **j** proposed filter

it took more time than FSBMMF [16] and PSMF [18] the output results were much better than what these filters offered. Figure 4 shows the graphical representation of the results obtained in the quantitative analysis of the brain image, which reaffirms the quality of our filter over the existing methods.

The simulation shows that our filter outperforms almost all the existent filters and provides better yet faster results for each benchmark images Figs. 3 and 5 as well as for medical images as shown in Fig. 6. On average the proposed algorithm provides

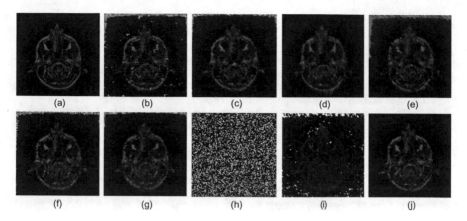

Fig. 6 Brain image with 90% noise density filtered using: **a** original image, **b** AWSMF, **c** DAMF, **d** FSBMMF, **e** RSIF, **f** SMMF, **g** TVWA, **h** PSMF, **i** BPDF and **j** proposed filter

3.5 dB and 14.09 higher $PSNR$ and IEF, respectively. And the results are 2.92 s faster than the existing filters, for grayscale images. The higher values of $PSNR$ and IEF tells us that the reconstruction of the image is of high quality. For the medical images Fig. 6, the results indicate that the proposed algorithm provides 9.83 dB higher PSNR and takes 1.18 s less than other algorithms. Figures 3 and 5 shows the quality of the restored image using a proposed filter is superior to the images restored using existing filters. Finally, the medical image as shown in Fig. 6 provides a better restoration results for tumor detection.

5 Conclusion

This paper presents a novel SAP noise removal algorithm that very efficiently restores the noisy pixel either by the use of an Adaptive median window or by using pre-processed pixels. The calculation recognizes and stores noisy pixels in a different array and replaces the corrupted pixel with the median of adjoining free pixels. Further, on the off chance that every one of the pixels present in the choice window are noisy, the size of the window is expanded up to it's threshold limit, to oblige healthy pixels. Ultimately, values of pre-processed pixels are taken into consideration. The presented algorithm has been evaluated over different benchmark images with varying noise density (0–90%). The quality metrics indicate the proposed algorithm provides a 9.83 dB higher PSNR value on average and 1.18 s faster than all the other algorithms for medical images. In the grayscale Lena Image, the algorithm provides 5.28 dB and 134.21 improvements in the PSNR and IEF respectively on average with the results being 2.92 s faster than other filters on average. In the Walls and Windows image, our novel technique has 3.50 dB better PSNR, 14.09 better IEF, and 5.23 s faster results.

References

1. Eng HL, Ma KK (2001) Noise adaptive soft-switching median filter. IEEE Trans Image Process 10(2):242–251
2. Erkan U, Gökrem L (2018) A new method based on pixel density in salt and pepper noise removal. Turk J Electr Eng Comput Sci 26(1):162–171
3. Garg B (2020) An adaptive minimum-maximum value-based weighted median filter for removing high density salt and pepper noise in medical images. Int J Ad Hoc Ubiquitous Comput 35(2):84–95
4. Garg B (2020) Restoration of highly salt-and-pepper-noise-corrupted images using novel adaptive trimmed median filter. Signal Image Video Process 14:1555–1563
5. Garg B, Arya K (2020) Four stage median-average filter for healing high density salt and pepper noise corrupted images. Multimedia Tools Appl 79(43):32305–32329
6. Garg B, Rana PS, Rathor VS (2021) Significance driven inverse distance weighted filter to restore impulsive noise corrupted x-ray image. J Ambient Intell Humanized Comput 1–12
7. Ghumaan RS, Sohi PJS, Sharma N, Garg B (2021) A novel hybrid decision-based filter and universal edge-based logical smoothing add-on to remove impulsive noise. Turk J Electr Eng Comput Sci 29(4):1944–1963
8. Jebamalar Leavline E, Gnana Singh D, Antony A (2013) Enhanced modified decision based unsymmetric trimmed median filter for salt and pepper noise removal. Int J Imaging Rob 11(3):46–56
9. Lin TC (2007) A new adaptive center weighted median filter for suppressing impulsive noise in images. Inf Sci 177(4):1073–1087
10. Meher SK, Singhawat B (2014) An improved recursive and adaptive median filter for high density impulse noise. AEU-Int J Electron Commun 68(12):1173–1179
11. Nair MS, Mol PA (2013) Direction based adaptive weighted switching median filter for removing high density impulse noise. Comput Electr Eng 39(2):663–689
12. Oliinyk V, Ieremeiev O, Djurović I (2019) Center weighted median filter application to time delay estimation in non-gaussian noise environment. In: 2019 IEEE 2nd Ukraine conference on electrical and computer engineering (UKRCON). IEEE, pp 985–989
13. Patel P, Majhi B, Jena B, Tripathy C (2012) Dynamic adaptive median filter (DAMF) for removal of high density impulse noise. Int J Image Graph Signal Process 4(11):53
14. Sharma N, Sohi PJS, Garg B (2021) An adaptive weighted min-mid-max value based filter for eliminating high density impulsive noise. Wirel Pers Commun 1–18
15. Veerakumar T, Esakkirajan S, Vennila I (2014) Recursive cubic spline interpolation filter approach for the removal of high density salt-and-pepper noise. Signal Image Video Process 8(1):159–168
16. Vijaykumar V, Mari GS, Ebenezer D (2014) Fast switching based median-mean filter for high density salt and pepper noise removal. AEU-Int J Electron Commun 68(12):1145–1155
17. Vijaykumar V, Vanathi P, Kanagasabapathy P, Ebenezer D (2008) High density impulse noise removal using robust estimation based filter. IAENG Int J Comput Sci 35(3)
18. Wang Z, Zhang D (1999) Progressive switching median filter for the removal of impulse noise from highly corrupted images. IEEE Trans Circ Syst II Analog Digital Signal Process 46(1):78–80
19. Zhang C, Wang K (2015) A switching median-mean filter for removal of high-density impulse noise from digital images. Optik 126(9–10):956–961
20. Zhang P, Li F (2014) A new adaptive weighted mean filter for removing salt-and-pepper noise. IEEE Signal Process Lett 21(10):1280–1283

Machine Learning Based Tomato Detection—A Practical and Low Cost Approach

Le Ngoc Quoc, Huy Q. Tran, Chuong Nguyen Thien, Ly Anh Do, and Nguyen Thinh Phu

Abstract In this paper, we focused on tomato detection using machine learning algorithms. Tomato data were divided into 2 forms: mature green and red ripe. The detection process was performed through three machine learning algorithms including stochastic gradient descent, random forest, and support vector machine. All images of tomato fruit were collected, processed, and analyzed by ourselves based on popular tomato varieties in Vietnam. Experimental results on Raspberry PI 4 showed that the machine learning algorithms provide impressive estimation results with an accuracy of 80% and more when applying background removal.

Keywords Tomato detection · Machine learning · Raspberry PI

1 Introduction

Tomato is a popular fruit not only in Vietnam but also in many parts of the world. Tomatoes contain a great source of vitamin C, potassium, folate, and vitamin K. Therefore, tomatoes are very good at reducing the risk of cardiovascular disease and cancer [1, 2]. Automating the process of growing, harvesting, and processing tomatoes is always an important topic that has attracted the attention of researchers. Process automation improves labor productivity, quality, and efficiency of the entire production process from cultivation to processing. In such a process, it is essential to identify the tomato fruit to monitor the quality and yield of the product. During the growing process, we can apply modern image processing techniques to detect the current health of tomatoes, count numbers, and predict yields. During the harvesting

L. N. Quoc · H. Q. Tran (✉) · L. A. Do · N. T. Phu
Falculty of Engineering and Technology, Nguyen Tat Thanh University, Ho Chi Minh City 70000, Vietnam
e-mail: tqhuy@ntt.edu.vn

C. N. Thien
Nha Trang University, Nha Trang, Vietnam
e-mail: chuongnt@ntu.edu.vn

© The Author(s), under exclusive license to Springer Nature Switzerland AG 2022   567
N. H. T. Dang et al. (eds.), *Artificial Intelligence in Data and Big Data Processing*,
Lecture Notes on Data Engineering and Communications Technologies 124,
https://doi.org/10.1007/978-3-030-97610-1_45

process, automatic robot models can recognize and select suitable tomatoes by them-selves. A similar detection operation is performed when tomatoes are processed in high-quality food factories. The detection and removal of low-quality tomatoes contribute significantly to improving the actual quality of the processed product. Therefore, studies using computer vision for fruits detection and mass estimation are always a potential topic [3, 4].

Recently, many research projects related to tomato fruit recognition using computer vision (CV) and artificial intelligence (AI), such as tomato detection [5–10], tomato disease degree [11–15], tomato processing defect [16] have been done with a lot of promising results.

Outstanding achievements in AI-assisted CV technology have helped research related to tomato identification achieve many positive results. In [5, 6], the authors used an improved version of YOLOv3 to identify tomatoes under many complex conditions with very high accuracy. To solve the complex scenes and performance constraints in embedded platforms, the authors in [7] proposed an improved version of YOLOv3-tiny. By applying this solution, their tomato recognition model proved the f1-score of 91.92% compared to approximately 80% when adopting popular YOLYv3-tiny. In addition to tomato fruit detection, ripeness has also been an inter-esting topic. Chunhua Hu et al. [8] applied Faster R-CNN and an intuitionistic fuzzy set to automatically detect the single ripe tomato and to help improve the harvesting capability of agriculture robots. In [9], Yo-Ping Huang et al. utilized the fuzzy mask R-CNN model to identify tomato ripeness. One of the main highlights of this research is that their solution could achieve an accuracy of 98%. The authors in [10] also applied mask R-CNN and a RealSense camera to detect tomatoes in a real green-house. Besides identifying tomato fruit and its degree of ripeness, the detection of tomato disease is also a major problem. This work was done through images of tomato leaves [11–13] or images of the tomato itself [14, 15]. Most of the above studies applied deep learning as a useful tool to increase the estimation efficiency of the proposed models. However, the application of these models to practical systems requires hardware platforms with advanced configuration and high cost.

In this paper, we apply different machine learning (ML) methods, such as stochastic gradient descent (SGD), random forest (RF), and support vector machine (SVM), to recognize tomato fruit. All data has been collected and processed by members of our research team. The training phase was executed on a personal computer. A trained model was embedded on the Raspberry PI 4 (RPI4), and the entire estimation process was performed on this very cheap and powerful mini computer. To improve the estimation results after training by three different ML algorithms, we conducted many different image processing techniques. This process is described in detail in the following sections.

2 Materials and Methodology

2.1 Testbed Setting and Data Collection

To collect the entire images of tomato fruit, the testbed is set as shown in Fig. 1. The experiments were executed on a personal computer (Windows 10 64 bit, Intel Core i3-9100F, 3.6 GHz, 8 GB Ram), a RPI4 (model 4 B+ , 4 GB Ram), and a Logitech HD C270 webcam. We prepare 2000 images of positive data (Fig. 2) and 2000 images of negative data (Fig. 3) under the support of the Logitech webcam. Each image has

Fig. 1 Experimental testbed

Tomato_Pos0395   Tomato_Pos0396   Tomato_Pos2676   Tomato_Pos2677

Tomato_Pos0400   Tomato_Pos0401   Tomato_Pos2681   Tomato_Pos2682

Fig. 2 Positive traning data

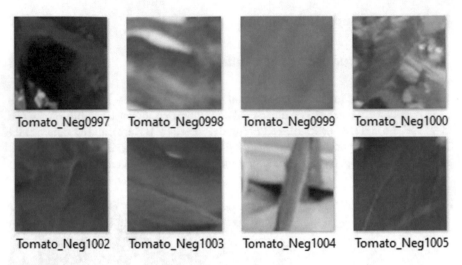

Fig. 3 Negative traning data

a resolution of 100×100 pixels.

2.2 Methodology

In this section, the whole process of processing, training, testing, and estimation was implemented as shown in Fig. 4. All data collected in Sect. 2.1 were first divided into

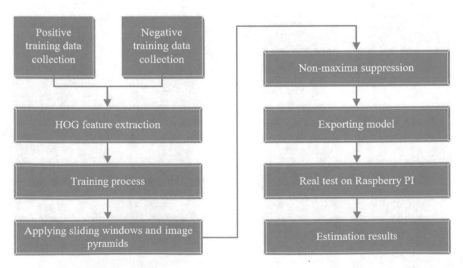

Fig. 4 Experimental flow chart

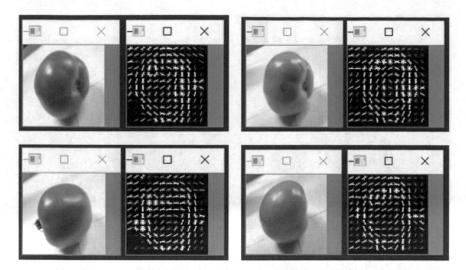

Fig. 5 Feature extraction

2 groups (i.e., training data (80%) and testing data (20%)). We then extracted the histogram of oriented gradients (HOG) from these samples (Fig. 5). Next, the training process was executed based on the above data and three different ML algorithms (i.e., SGD, RF, and SVM) [17, 18]. After training, we continued to apply sliding windows and image pyramids to locate and detect tomatoes in the image at various scales and locations as shown in the left sub-figures of Fig. 6. In order to remove overlapping bounding boxes and select the strongest one, we applied non-maxima suppression to calculate the overlap ratio between bounding boxes. The obtained results in the right sub-figures of Fig. 6 demonstrated great capability of this method. After that, the trained model was exported and embedded into RPI4. Finally, we utilized this trained model on RPI4 and Logitech webcam to detect the presence of real tomatoes and classify them into two classes including mature green tomatoes and red ripe tomatoes. To perform this classification task, we calculate the mean pixel value at four positions, which is 25 pixels far away from the bounding box center. This value was then compared to a thresholding value to assign the current object either mature green or red ripe.

In the next step, we removed the background of the entire training data as shown in Fig. 7. The main purpose of this procedure is to eliminate the noise around the main object. This method helped to improve the quality of the sample after the process of HOG extraction. Therefore, the final estimation results were also enhanced.

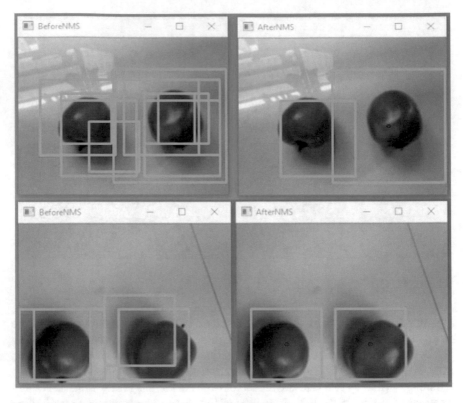

Fig. 6 Estimated results before and after non-maximum suppression

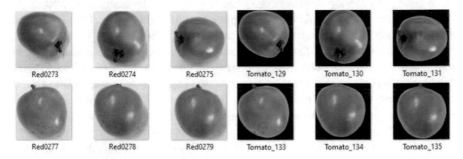

Fig. 7 Traning data: before and after background removal

3   Results

To prove the performance of the suggested solution, we analyzed the precision, recall, and f1-score after the training and testing phases. It is worth noting that the performance before and after background removal was different as presented in

Table 1 Experimental results before background removal

Algorithms	Precision	Recall	F1-score
SGD	0.74	0.64	0.72
RF	0.65	0.81	0.72
SVM	0.67	0.38	0.48

Table 2 Experimental results after background removal

Algorithms	Precision	Recall	F1-score
SGD	0.90	0.85	0.89
RF	1.00	0.70	0.82
SVM	0.94	0.85	0.89

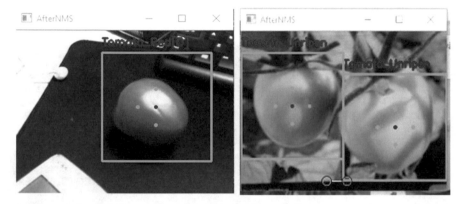

Fig. 8 Final estimation

Tables 1 and 2. All three algorithms showed an impressive achievement in all aspects. The obtained results proved that the improvement in the quality of HOG extraction played an important role in the final estimation step.

After tomato detection, we located 4 specific locations around the bounding box center as illustrated in Fig. 8. From these locations, we computed the mean pixel value and assigned the label for every testing image. This process helped to classify the estimated tomato into two groups including mature green and red ripe. In this case, we obtained an accuracy of more than 80% when using 30 random images to evaluate the performance of the trained model on RPI4.

4 Conclusion

In summary, all the ML algorithms used in this paper demonstrated their ability to detect tomatoes after the trained model is embedded in RPI4. During the training phase, all ML algorithms provided a high estimation accuracy of 80%. However, the SVM algorithm showed its superiority over the rest of the algorithms when the model after training was tested thoroughly. Obviously, the application of these algorithms on a low-cost minicomputer like the RPI4 has provided an ideal opportunity for the implementation of AI and computer vision technology with a low cost and an acceptable level of efficiency for small and medium farms in Vietnam and other developing countries.

References

1. Malik MH, Zhang T, Li H, Zhang M, Shabbir S, Saeed A (2018) Mature tomato fruit detection algorithm based on improved HSV and watershed algorithm. IFAC-PapersOnLine 51:431–436
2. Tomatoes 101: nutrition facts and health benefit. Available online: https://www.healthline.com/nutrition/foods/tomatoes. Accessed on 25 Nov 2021
3. Sa I, Ge Z, Dayoub F, Upcroft B, Perez T, McCool C (2016) Deepfruits: a fruit detection system using deep neural networks. Sens 16(8):1222. https://doi.org/10.3390/s16081222
4. Lee J, Nazki H, Baek J, Hong Y, Lee M (2020) Artificial intelligence approach for tomato detection and mass estimation in precision agriculture. Sustain 12(21):9138. https://doi.org/10.3390/su12219138
5. Lawal MO (2021) Tomato detection based on modified YOLOv3 framework. Sci Rep 11:1447. https://doi.org/10.1038/s41598-021-81216-5
6. Liu G, Nouaze JC, Touko Mbouembe PL, Kim JH (2020) YOLO-tomato: a robust algorithm for tomato detection based on YOLOv3. Sens 20(7):2145. https://doi.org/10.3390/s20072145
7. Xu Z, Jia R, Liu Y, Zhao C, Sun H (2020) Fast method of detecting tomatoes in a complex scene for picking robots. IEEE Access 8:55289–55299. https://doi.org/10.1109/ACCESS.2020.2981823
8. Hu C, Liu X, Pan Z, Li P (2019) Automatic detection of single ripe tomato on plant combining faster R-CNN and intuitionistic fuzzy set. IEEE Access 7:154683–154696. https://doi.org/10.1109/ACCESS.2019.2949343
9. Huang Y-P, Wang T-H, Basanta H (2020) Using fuzzy mask R-CNN model to automatically identify tomato ripeness. IEEE Access 8:207672–207682. https://doi.org/10.1109/ACCESS.2020.3038184
10. Manya A, Hubert F, Schadeck FF, Dick L, Marcel M, Nanne F, Gerrit P, Ron W (2020) Tomato fruit detection and counting in greenhouses using deep learning. Frontiers Plant Sci 11:1759. https://doi.org/10.3389/fpls.2020.571299
11. Shijie J, Peiyi J, Siping H, Haibo S (2017) Automatic detection of tomato diseases and pests based on leaf images. 2017 Chinese Automation Congress (CAC), pp 2537–2510. https://doi.org/10.1109/CAC.2017.8243388
12. Zhang Y, Song C, Zhang D (2020) Deep learning-based object detection improvement for tomato disease. IEEE Access 8:56607–56614. https://doi.org/10.1109/ACCESS.2020.2982456
13. Tian Y, Zheng P, Shi R (2016) The detection system for greenhouse tomato disease degree based on android platform. 2016 3rd International Conference on Information Science and Control Engineering (ICISCE), pp 706–710. https://doi.org/10.1109/ICISCE.2016.156
14. Wang Q, Qi F, Sun M, Qu J, Xue J (2019) Identification of tomato disease types and detection of infected areas based on deep convolutional neural networks and object detection techniques. https://doi.org/10.1155/2019/9142753
15. Wang Q, Qi F (2019) Tomato diseases recognition based on faster RCNN. In: 2019 10th International Conference on Information Technology in Medicine and Education (ITME), pp 772–776. https://doi.org/10.1109/ITME.2019.00176

16. Shi X, Wu X (2019) Tomato processing defect detection using deep learning. In: 2019 2nd World Conference on Mechanical Engineering and Intelligent Manufacturing (WCMEIM), pp 728–732. https://doi.org/10.1109/WCMEIM48965.2019.00153
17. Pedregosa F, Varoquaux G, Gramfort A, Michel V, Thirion B, Grisel O, Blondel M, Prettenhofer P, Weiss R, Vanderplas J et al (2011) Scikit-learn: machine learning in python. J Mach Learn Res 12:2825–2830
18. Bradski G (2000) The OpenCV library. Dr. Dobb's J Softw Tools

Intelligent Systems and Networking

Designing Water Environment Monitoring Equipment for Aquaculture in Vietnam

Phat Nguyen Huu⬤, Quang Tran Minh, and Quang Tran Minh

Abstract Nowadays, the application for aquaculture has become popular with many benefits. In several specific cases, the productivity and quality of shrimp can be mentioned such as temperature, pH, dissolved oxygen (DO) concentration, and salinity of culture water. In the field of shrimp farming, the water environment is extremely important that determines the quality of output. Therefore, the farmers need to regularly check parameters such as temperature, pH, DO, or including salinity to ensure the best environment. In the article, we design a water parameter meter that will help people quickly check the parameters of water for suitable solutions. These parameters are temperature, pH, salinity, and DO concentration. Temperature not only affects the habitat of shrimp but also pH and DO concentration. These parameters are important for shrimps that help them develop. Our device can measure the parameters with accuracy 98% for one minute and send them to the server or mobile. The results show that the device is suitable for real environments.

Keywords Big data · Monitoring system · pH meter · Shrimp farming · Signal processing

P. N. Huu (✉) · Q. T. Minh
Hanoi University of Science and Technology (HUST), Hanoi, Vietnam
e-mail: phat.nguyenhuu@hust.edu.vn

Q. T. Minh
e-mail: quang.tm182744@sis.hust.edu.vn

Q. T. Minh
Faculty of Computer Science and Engineering, Ho Chi Minh City University of Technology (HCMUT), 268 Ly Thuong Kiet, Dist. 10, Ho Chi Minh City, Vietnam
e-mail: quangtran@hcmut.edu.vn

Vietnam National University Ho Chi Minh City (VNU-HCM), Linh Trung Ward, Thu Duc District, Ho Chi Minh City, Vietnam

N. H. T. Dang et al. (eds.), *Artificial Intelligence in Data and Big Data Processing*,
Lecture Notes on Data Engineering and Communications Technologies 124,
https://doi.org/10.1007/978-3-030-97610-1_46

1 Introduction

Today, technology applying for aquaculture becomes popular with many benefits. In the field of shrimp farming, the environment is extremely important that determines the quality of output. Several factors directly affect the productivity, yield, and quality of shrimp that can be mentioned such as temperature, water pH, dissolved oxygen (DO) concentration, and salinity of the water. In practical application, water parameter meters are developed with many types of sensors. The farmers need to regularly check the parameters to ensure that shrimps have the best environment. In this article, we design a water parameter meter that will help people quickly check the water parameters for suitable solutions. Besides, temperature not only affects the shrimp habitat but also pH and DO concentration. These parameters make sure the shrimp live in a safe environment. The results of measuring DO concentration will help people determine its concentration since they can use a reasonable and cost-effective oxygen aerator. Normally, routine water measurement jobs require a linkage machine to be installed to measure continuously over a period of time. The installation also wastes time and it is necessary to check the system after the fixing period. Therefore, it has several limitations when installing metering systems for a long time.

In the paper, we can design these parameters of water quickly. We use rechargeable batteries that make compact and highly portable.

The rest of the paper is presented as follows. In Sect. 2, we will present related work. In Sects. 3 and 4, we present and evaluate the effectiveness of the proposed model, respectively. Finally, we give a conclusion in Sect. 5.

2 Related Work

Today, there are many companies to focus on designing sensors to measure the water environment [2–6, 8–12, 15, 16, 18]. Horiba company [5, 11] designs pH and DO meters with high accuracy. Their products are easy to use. However, it is difficult to use because of its high cost. Hach [3, 16] also has similar products. They include one original sensor and a rugged meter. The rugged meter is efficient and flexible for laboratory water. It is a simple operation, reliable, and has high accuracy. However, the cost of products is very high and they are often used in laboratories or factories.

Sensorex [12, 15] also designs pH and ORP sensors. User-programmable sensor ensure that their system always obtain accurate data. It is also compatible with wide range. HANNA [4, 5] offers an affordable and basic benchtop pH/mV meter that features calibrate with one or two points automatically with five pre-programmed buffers. All data are automatically compensated for temperature variations from the thermistor probe providing from pH or mV. The probe provides a highly accurate $0.5\,°C$ for precise temperature compensation. Lutron [9] offers a pH meter that can display mV value and can be compensated for temperature automatically. The meter has a low battery and is calibrated at one or both points. They can apply for many applications such as water conditioning, aquariums, beverage, fish hatcheries, food processing, laboratory, and industry.

A product of EXTEXT instruments [2, 8] is the DO700 (9-in-1) kit that measures DO concentration/saturation, pH, mV, temperature, conductivity, TDS, salinity, and resistivity. It is measured salinity automatically and manual barometric pressure compensation for DO measurements. In addition, it has automatically calibrated 4, 7, and 10 pH with three-point for better accuracy. It consists of a DO probe, pH/mV/temp electrode, polymer conductivity cell, pH and conductivity calibration solutions, batteries, and hard carrying case.

DO meter of Milwaukee [10, 18] is portable with automatic calibration and temperature compensation (ATC) specifically designed for spot sampling applications. DO measurements can be displayed in parts per million (ppm = mg/L) or percent saturation. The temperature is indicated in Celsius from 0 to 50 °C with 0.1 resolution. The meter compensates salinity and altitude automatically after manual input. Calibration is very simple and fast. The low battery indicator is suitable for applications such as aquaculture, wastewater, environment, and education. In our products, the meter can measure DO, pH, salinity, and temperature. The device can compensate for temperature automatically with a battery charging circuit. Besides, the price is not too expensive that is suitable for farmers.

3 Developing Product

The meter measures four parameters, namely pH, temperate, salinity, and DO concentration. We use sensor E201-C to measure pH and DO 1200 to measure DO concentration.

3.1 Temperature Sensor

The diagram and schematic of measuring temperature circuits are shown in Figs. 1 and 2. NTC 15k is a rheostat. Therefore, you need to install an additional 15k resistor in parallel to measure the value of rheostat as shown in Fig. 2.

In Fig. 2, we choose a resistor with a value of 15k Ω. The analog voltage comes in from Con2 in Fig. 2. It is put into the OPA 2277. The ADC then receives a digital

Fig. 1 Diagram of the circuit for measuring temperature

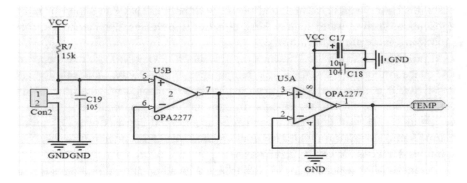

Fig. 2 Schematic of the circuit for measuring temperature

voltage based on the voltage division formula that we obtain the resistance value of the temperature sensor. Finally, we apply the Stein-Hart temperature equations to determine temperature.

The data output is defined as follow

$$V_{out} = V_{in} * \frac{R_t}{R_t + R}, \tag{1}$$

$$R_t = R * \frac{V_{in}}{V_{out}} - 1. \tag{2}$$

Temperature is calculated by Stein-Hart [13] as

$$T = \frac{1}{A + B * \ln(R_t) + C * (\ln(R_t))^3}, \tag{3}$$

where T is temperature, R_t is NTC rheostat. Other parameters are defined as

$A = 1.009249522 \times 10^{-3}$,
$B = 2.378405444 \times 10^{-4}$,
$C = 2.019202697 \times 10^{-7}$.

3.2 pH Sensor

The pH or potential ion [H+] is the concentration of H ions in an acidic solution dissolved in water. pH indicates the acidic or basic nature of an electrolyte solution. The pH scale is divided into two ranges:

Solution with pH between 0 and 7 is an acidic solution.
Solution with pH of 7–14 is a basic solution.

pH = 7 indicates that the solution is neutral (pure water).

To measure pH, the sensor works on Galvanic electromotive force. An electrode immersed in an electrolyte will produce an electromotive force according to Nernst as [14]

$$E = \frac{RT}{nF} Ln\,[H+],$$ (4)

$$E = \frac{2.303RT}{nF} Log\,[H+],$$ (5)

where E is Galvanic electromotive force and R is Bolzman constant. $T = 273 + t$ is absolute temperature and n is the valence of H ion. F is Faraday constant and [H+] is concentration of H+ ions.

Since the sensor needs to work continuously in the water environment, it needs to be durable and the glass electrode PH sensor is one of the best choices. The glass electrode consists of a very thin-walled (0.02 mm) glass bulb with AgCl solution and a silver electrode. The glass electrode has similar properties as the H electrode. Although the glass electrode is fragile, it works well and stably. The biggest disadvantage is that the resistance of the electrode is very large (108–109). Therefore, it is required that the measuring device has an input resistance Rv = 1011–1012.

3.3 DO Sensor

DO is the parameter of water that requires for respiration of aquatic organisms (fish, amphibians, aquatic animals, insects, etc.). It is usually produced by dissolution from the atmosphere or photosynthesis of algae. The oxygen concentration of water is expressed in mg/l or percent saturation based on temperature.

The formulas for calculating mass of saturated oxygen in water (Weiss Equation) at zero salinity is calculated as follows

$$DO_o = 1.42905 exp\,(2.00907 + 3.22014T_s + 4.05010T_s^2 \\ + 4.94457T_s^3 - 0.256847T_s^4 + 3.88767T_s^5)$$ (6)

where T_s is a scaled temperature defined as:

$$T_s = \ln\left(\frac{298.15 - t}{273.15 + t}\right).$$ (7)

DO concentration is in mg/L and t is the water temperature in degree Celsius.

The structure of galvanic DO tip is as shown in Fig. 3. The basic molecule of the probe is an ion selection membrane that allows only oxygen ions (O_2) to penetrate. The measuring chamber is an alkaline electrolyte (e.g. KOH). The anode electrode

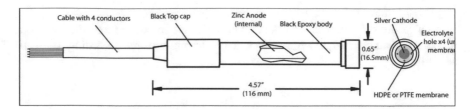

Fig. 3 Structure of DO1200 sensor [1, 7, 17]

is made of lead (Pb) and the cathode is a silver rod. The cathode electrode is placed close to the oxygen permeable membrane. Oxygen is dissolved in test water and permeable through Teflon (or polyethylene) permeable membrane.

Oxygen reaches the cathode and the oxidation process occurs as follows:

$$O_2 + 2H_2O + 4e^- = 4OH, \tag{8}$$

and at anot, we have

$$2Pb_2 \rightarrow Pb_2 + 4e^-. \tag{9}$$

Charges flowing through the electrodes induce an electric current and reduce all oxygen-permeable through the selectively permeable (IS) membrane. The amount of oxygen that permeates the membrane depends on the DO concentration of test water. Therefore, the current flowing between two electrodes depends on the amount of DO of test water.

The oxygen near the cathode is usually considered to be zero because of polarization. Therefore, the amount of oxygen that permeates the membrane depends on the pressure exerted on the membrane. The current DO probe is partly dependent on atmospheric pressure.

The sensor used is DO1200 that is a galvanic probe. Specifications of sensor includes

1. Temperature range is from 0 to 50 °C.
2. Response time is 5 min to reach 95% of official value.

Output voltage when oxygen is saturated from 24 to 42 mV. Structure of DO1200 is shown in Fig. 3.

The measuring circuit is capable of measuring for two modes, namely pH and DO as shown in Figs. 4 and 5. Both of these sensors use a BNC connector. The

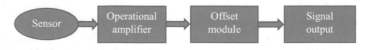

Fig. 4 Diagram of the pH and DO measuring circuit

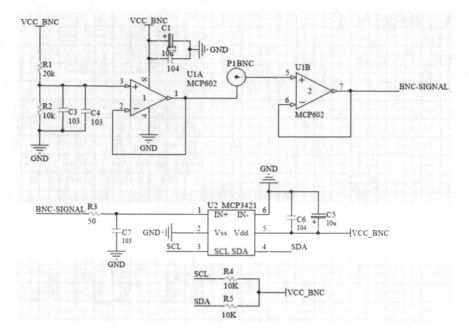

Fig. 5 Schematic of pH and DO measuring circuit

analog voltage of pH or DO goes through MCP-602. They then run through MCP-3421 with a resolution of 18 bits. The MCP3421 performs at rates of 240 samples per second. Since MCP-3421 uses an inter-integrated circuit (I2C) to communicate with two signal lines, namely serial clock line (SCL) and serial data line (SDA). Therefore, the signal after being performed ADC has transmitted I2C and ATmega-328 microcontroller directly. We use MCP-3421 since its resolution of bits is large, high rates of samples, simple I2C transmission protocol, and high accuracy.

4 Result

4.1 Setup

To perform the system, we use as shown in Fig. 6. Input obtained from measuring temperature and pH or DO sensor is analog voltage. It is then performed ADC by MCP-3421. The voltage signal received from the temperature sensor is used to determine the thermistor. We will calculate the temperature at the time of measurement. Finally, we display measurement results on an LCD screen-based on measuring values.

Fig. 6 Diagram of testing
system

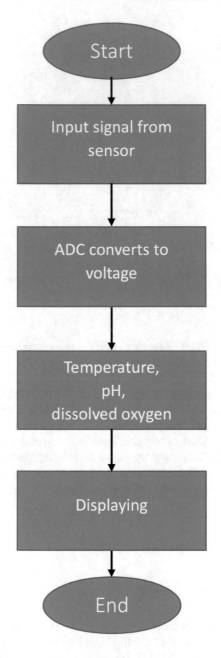

We perform for two cases as follows.

Case 1: Salinity does not affect by DO result.

Case 2: The measuring temperature is affected by pH and DO result directly. Therefore, the temperature result is not accurate since it will lead to deviation of
pH and DO in several cases.

4.2 Result

We compare with actual temperatures at 0, 15 and 25 °C obtained as shown in Table 1.

The difference between the actual temperature and test temperature is quite small.
It can be seen that the maximum absolute error is 0.8 °C. This shows the accuracy in
our measurements and sensor accuracy.

Conducting pH measurement of solutions with pH is 4.00, 6.86, 7.60 and 9.18,
we obtain the following results as shown in Table 2.

Our pH calibrations are 4, 6.86, and 9.12. The difference between actual and test
pH is very small. The measurement results have the largest absolute error of 9%. It
can be seen that the accuracy of the sensor is good.

The result is the result of our measurements in a shrimp pond in Nam Dinh city as
shown in Figs. 7, 8 and 9. The results are at two different dates, taken from different
periods. The split time is a minute or hour. In Fig. 7, we can see that the temperature
and pH parameters are stored after every 10 min. The temperature and pH change
from 1 to 1.5 °C and 0.5 to units, respectively. In Fig. 7, the temperature and pH
parameters are stored after every one hour. We see that the temperature and pH
also change from 1 to 2 °C and 0.5 to 1 unit, respectively. We find that the value of
temperature and pH is still at a safe level that is suitable for the habitat of shrimp. The
ideal pH for shrimp is from 7.5 to 8.5. Therefore, our results are safe and suitable

Table 1 Results of measured temperature

No.	Real temperature (°C)	Measured temperature (°C)	Accuracy (%)
1	0	0.8	–
2	15	15.1	99.33
3	25	25.3	98.80

Table 2 Results of measured pH

No.	Real pH	Measured pH	Accuracy (%)
1	4.00	4.08	98.00
2	6.86	6.90	99.41
3	7.60	7.51	98.81
4	9.18	9.2	99.78

588   P. N. Huu et al.

```
*LOG1.txt - Notepad
File  Edit  Format  View  Help
Nhiet do: 26.28        pH: 8.87
10:36:4       8/4/21
Nhiet do: 28.16        pH: 8.74
10:36:18      8/4/21
Nhiet do: 28.25        pH: 8.61
10:37:20      8/4/21
Nhiet do: 27.89        pH: 8.61
10:39:9       8/4/21
Nhiet do: 27.80        pH: 8.34
10:40:32      8/4/21
Nhiet do: 28.16        pH: 8.21
10:41:34      8/4/21
Nhiet do: 27.80        pH: 8.48
10:43:24      8/4/21
Nhiet do: 27.71        pH: 8.87
10:46:38      8/4/21
Nhiet do: 27.80        pH: 8.87
```
(a)

```
*LOG1.txt - Notepad
File  Edit  Format  View  Help
Nhiet do: 23.47        pH: 8.08
2:50:40       9/4/21
Nhiet do: 22.25        pH: 8.21
3:10:3        9/4/21
Nhiet do: 22.25        pH: 8.08
3:35:27       9/4/21
Nhiet do: 22.60        pH: 8.08
6:37:3        9/4/21
Nhiet do: 21.82        pH: 8.08
8:35:11       9/4/21
Nhiet do: 21.30        pH: 8.08
10:24:24      9/4/21
Nhiet do: 21.99        pH: 8.08
11:8:43       9/4/21
Nhiet do: 22.86        pH: 8.08
12:8:9        9/4/21
Nhiet do: 21.99        pH: 8.21
```
(b)

Fig. 7 Result of measuring the parameters for **a** minute and **b** hour

Fig. 8 Result of measuring for real environment

for shrimp habitat. Differences in pH can cause several external influences such as time, temperature, a season of the year, and different organisms.

5 Conclusion

In the paper, we have designed a device that can measure the water parameters at a low cost. We get the same result as in Fig. 8. The results are quite accurate compared to the test results in Tables 1 and 2.

Fig. 9 Performing experimental measurements for shrimp culture

In the future, we need to design and improve the system to be able to measure more parameters and make the product compact and easy to use.

Acknowledgment Work by Quang Tran Minh (HCMUT) acknowledges the support of time and facilities from Ho Chi Minh City University of Technology (HCMUT), VNU-HCM.

References

1. DO1200, DO1200TC (2021) Laboratory dissolved oxygen sensor-DO1200. Accessed 20 Oct 2021. https://www.isweek.com/product/laboratory-dissolved-oxygen-sensor-do1200_917.html
2. Extech (2021) Extech DO700: portable dissolved oxygen meter. Accessed 5 Oct 2021. http://www.extech.com/products/DO700
3. Hach (2021) HQ30D portable dissolved oxygen meter, laboratory kit for water quality, with luminescent DO sensor, 1 m cable. Accessed 5 Oct 2021. https://www.hach.com
4. Hana (2021) Portable pH meter for food and dairy. Accessed 5 Oct 2021. https://www.hannainst.com/hi99161-haccp-ph-meter-for-food-and-dairy.html
5. Hanna M (1995) An intelligent architecture for the simulation of product quality in a machining centre. In: IEE Colloquium on manufacturing simulation, pp 3/1–3/9. https://doi.org/10.1049/ic:19951023

6. Hsu L, Selvaganapathy P, Brash J, Fang Q, Xu CQ, Deen M, Chen H (2014) Development of a low-cost hemin-based dissolved oxygen sensor with anti-biofouling coating for water monitoring. Sens J IEEE 14:3400–3407. https://doi.org/10.1109/JSEN.2014.2332513
7. Jia F, Kacira M, Ogden KL (2015) Multi-wavelength based optical density sensor for autonomous monitoring of microalgae. Sensors 15(9):22234–22248. https://doi.org/10.3390/s150922234
8. Luo J, Dziubla T, Eitel R (2017) A low temperature co-fired ceramic based microfluidic Clark-type oxygen sensor for real-time oxygen sensing. Sens Actuators B Chem 240:392–397. https://doi.org/10.1016/j.snb.2016.08.180
9. LUTRON (2021) Mi605 dissolved oxygen/temperature professional portable meter. Accessed 5 Oct 2021. http://www.us.milwaukeeinstruments.com/component/content/article/35-martini-portable-meters/76-products-g-martini-portabel-meters-g-mi605
10. Milwaukee (2021) pH meter LUTRON PH-230SD. Accessed 5 Oct 2021. https://www.lutroninstruments.eu/ph-redox-orp-meters/ph-meter-lutron-ph-230sd/
11. Nomura D (2003) New challenge to pH measurement what will come next to the glass electrode? Feature Article 66–71
12. Sensorex (2021) TX105 pH/ORP loop powered 4–20 mA transmitter. Accessed 5 Oct 2021. https://sensorex.com/product/tx105-transmitter/
13. Steinhart JS, Hart SR (1968) Calibration curves for thermistors. Deep Sea Res Oceanogr Abstracts 15(4):497–503. https://doi.org/10.1016/0011-7471(68)90057-0
14. Wahl D (2005) A short history of electrochemistry—part 1. Galvanotechnik 96:1600–1610+vii
15. Wang Y, Rajib S, Collins C, Grieve B (2018) Low-cost turbidity sensor for low-power wireless monitoring of fresh-water courses. IEEE Sens J 18(11):4689–4696. https://doi.org/10.1109/JSEN.2018.2826778
16. Wei Y, Jiao Y, An D, Li D, Li W, Wei Q (2019) Review of dissolved oxygen detection technology: from laboratory analysis to online intelligent detection. Sensors (Basel, Switzerland) 19
17. Wei Y, Jiao Y, An D, Li D, Li W, Wei Q (2019) Review of dissolved oxygen detection technology: from laboratory analysis to online intelligent detection. Sensors 19(18). https://doi.org/10.3390/s19183995
18. Wei Y, Li W, An D, Li D, Jiao Y, Wei Q (2019) Equipment and intelligent control system in aquaponics: a review. IEEE Access 7:169306–169326. https://doi.org/10.1109/ACCESS.2019.2953491

Reliability Evaluation of Active Distribution Network Based on Improved Sequential Monte Carlo Algorithm

Zhuang Guanqun, Zhang Yifu, Zhang Yu, Hao Qiankun, Han Wenqi, and Zhang Haifeng

Abstract Reliability evaluation of active distribution network for distributed generation (DG) is a hot topic. Due to the uncertainty of DG, the radial network structure of the traditional distribution network has changed, and the system control strategy has become more complicated. Aiming at the problem between the calculation accuracy and speed of the Sequential Monte Carlo method (SMC), this paper propose an improved SMC method. Firstly, a structural hierarchical model has be added to the topology of the active distribution network, which can simulate the grid operation indicators with a tree structure; Secondly, this paper adopts the sampling method based on state transition, only sampling at the simulation point of state transition, which can improve the sampling efficiency to a certain extent; Finally, compared with traditional SMC method, the calculation time of our method is reduced by nearly half, and the error rate is reduced by more than 3 times, which proves the effectiveness of the algorithm used in this paper.

Keywords Active distribution network · Reliability evaluation · SMC · Structural hierarchical model

Z. Guanqun · Z. Yifu · Z. Yu · H. Wenqi · Z. Haifeng
Electric Power Research Institute, State Grid Jilin Electric Power Co.Ltd, Changchun, China
e-mail: zhuangguanqun2021@163.com

Z. Yifu
e-mail: zyifu2021@163.com

Z. Yu
e-mail: zhangyu1202111@163.com

H. Wenqi
e-mail: hanwq1111@163.com

Z. Haifeng
e-mail: zhanghf1107@163.com

H. Qiankun (✉)
Northeast Electric Power University, Jilin, China
e-mail: 20172762@neepu.edu.cn

© The Author(s), under exclusive license to Springer Nature Switzerland AG 2022   591
N. H. T. Dang et al. (eds.), *Artificial Intelligence in Data and Big Data Processing*,
Lecture Notes on Data Engineering and Communications Technologies 124,
https://doi.org/10.1007/978-3-030-97610-1_47

1   Introduction

With the development of the world economy, the research on the reliability of active distribution networks for distributed generation (DG) is the current research hotspot [1]. The traditional distribution network presents a radial structure in actual operation, and the addition of DG makes its structure more complicated. Due to the instability of DG (such as photovoltaic and wind) in the active distribution network, the grid control strategy becomes more complicated [2]. The active distribution network connects the power generation system, the distribution system and the users, which is a very important role in the entire power system [3]. According to the relevant statistics of the State Grid Corporation of China (SGCC), the vast majority of faults are caused by the distribution network in all power grid faults. Therefore, providing safe, high-quality, and reliable electrical energy through the active power distribution system is significance [4].

Generally, the SMC algorithm is commonly used in reliability evaluation, but there are two significant problems: the one is how to evaluate the failure rate after the introduction of DG; the other is SMC method faces the problems of low efficiency and long simulation time. This paper proposes an improved sequential Monte Carlo simulation algorithm to solve the above problems. The innovation of this paper is embodied in the structural hierarchical division of the original distribution network topology using the structural hierarchical model. As the same time, State transition sampling is used to o solve the problem of low sampling efficiency.

2   Related Works

Until now, many scholars have made a lot of research results on the reliability of traditional distribution networks [5, 6]. However, there are relatively few reliability assessments in active distribution networks [7–10]. According to the different models, reliability evaluation methods can be divided into two categories: one is the evaluation method based on deterministic indicators, and the other is the evaluation method based on statistical probability indicators [11, 12].

The deterministic index evaluation method uses a deterministic technology. If a component fails and does not affect the entire system, then this is reliable system [11]. In the implementation of the specific evaluation strategy, it is necessary to conduct a deterministic evaluation based on the failure rate and repair rate of each component in the distribution network. The advantage of the deterministic index method is that the mathematical model used to calculate accurate results for a single network structure of the distribution network system, but the randomness of network component failures cannot be analyzed well [13]. Commonly used deterministic index evaluation methods include Failure Mode and Effect Analysis (FMEA) [14], Minimum Path Method [15], Minimum Cut Set Method [16], Network Equivalence Method

[17] and Fault Traversal Method [18]. FMEA evaluates the reliability of the distribution network by traversing possible faults and generating a failure mode impact table [14]. The minimum path method analyzes the connectivity of all components of the distribution network and determines the minimum path at each load point [15]. The minimum cut set method uses the load minimum search method to transform the reliability analysis process of the distribution network into a set of load point fault events [16]. The network equivalence method uses an equivalent device to replace part of the distribution network, which has a certain effect on both simple and complex distribution networks [17]. The fault traversal method is to traverse all possible faults of the distribution network to obtain system reliability [18].

The evaluation method based on statistical probability indicators uses random sampling to form a set of faults in the distribution network, and achieves reliability measurement through statistical analysis of the probability of failure. The typical method is Monte Carlo simulation [19]. The sampling times of Monte Carlo simulation method is easier to deal with random events in actual operation, but the calculation accuracy of the simulation method is inversely proportional to the calculation efficiency. Monte Carlo simulation method can be divided into SMC [20] and Non-Sequential Monte Carlo (NSMC) [21]. Compared with NSMC, SMC can accurately simulate the operation process of distribution network components and load points, and evaluate the reliability of the distribution network accurately. NSMC does not consider the actual operating sequence of the system, and only evaluates the risk indicators of system operation by extracting a large number of sample data. In literature [22], the SMC was used to evaluate and analyze the reliability of the distribution network with distributed power generation, and achieved good results. In literature [21], the analytical method and variance reduction technology are introduced into the NSMC to reduce the time required for a single sampling simulation, and the equal sampling method is used to ensure the calculation accuracy. At the same time, the number of sampling is reduced, which improves the efficiency of power grid risk reliability assessment to a certain extent.

3 Traditional SMC Method

SMC is a mathematical term published in 1993, which can be used to evaluate the reliability of complex systems. Through set the distribution network components and load point network structure, SMC can counts the probability distribution of the occurrence and repair rate, and realizes the reliability index evaluation of the distribution network system. With the continuous increase of the number of simulations, the accuracy of the overall reliability results will continue to improve until the steady-state output of the reliability index is obtained.

In a distribution network system, there are N components $\{x_1, x_2, \ldots, x_N\}$. For each of x_i, the probability of failure is a function about time. The commonly used failure curve is bathtub curve. In bathtub curve, the entire cycle for component x_i is

Fig. 1 Typical component
failure curve

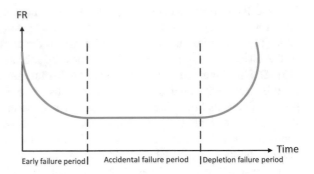

divided into three phases: early failure period, accidental failure period and depletion failure period [20]. The bathtub curve is shown in Fig. 1.

During the accidental failure period, the failure rate of component x_i can be replaced by a constant value. Therefore, the probability failure $F_{x_i}(t)$ and the probability of repair $G_{x_i}(t)$ to component x_i at time t are shown in formulas (1) and (2):

$$F_{x_i}(t) = \lambda_{x_i} e^{-\lambda_{x_i} t} \tag{1}$$

$$G_{x_i}(t) = \mu_{x_i} e^{-\mu_{x_i} t} \tag{2}$$

Among them, λ_{x_i} represents the failure rate of the component x_i, and μ_{x_i} represents the repair rate of the component x_i.

According to formulas (1) and (2), the continuous working time TTF_{x_i} and fault repair time TTR_{x_i} of component x_i are shown in formulas (3) and (4) respectively:

$$TTF_{x_i} = -\frac{1}{\lambda_{x_i}} \log(\delta_{x_i}) \tag{3}$$

$$TTR_{x_i} = -\frac{1}{\mu_{x_i}} \log(\xi_{x_i}) \tag{4}$$

Among them, δ_{x_i} and ξ_{x_i} are random numbers in the interval (0, 1) respectively.

Using SMC to sample the above random quantities, the operating state sequence of each component is shown in Fig. 2.

In Fig. 2, state 0 represents a fault state of the component, and state 1 represents the normal working state of the component.

The general process of using SMC to calculate the reliability of the distribution network is shown in Fig. 3.

Distribution network system Initialization. Enter the original configuration parameter data of the distribution network system (including network structure, device reliability parameters, etc.), end conditions (maximum number of simulations, calculation

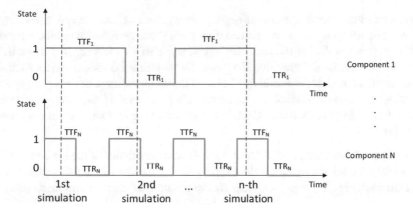

Fig. 2 Continuous working and repair time sequence of components

Fig. 3 SMC flow of the
distribution network

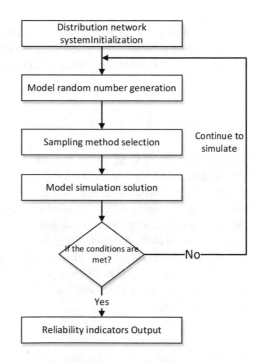

accuracy, etc.), reliability indicators, etc., to construct an applicable random process
probability model to make the problem The solution is consistent with the statistical
characteristics of the random variables in the model, that is, the parameters in the
established probability model are the solutions to the reliability problem.

Model random number generation. Set the number of simulations according to the set reliability requirements, and continuously generate random numbers for reliability evaluation during the simulation process. This step is very important in the Monte Carlo method. Generally, the better the randomness of the random number generation, the higher the accuracy of the reliability assessment of the entire system. Usually, a uniformly distributed random number is generated first, and then a random number that obeys a special distribution is generated, and a simulation experiment is performed.

Sampling method selection. According to the characteristics of the random process probability model, design and select appropriate sampling methods to sample each random variable. Various sampling methods such as direct sampling, stratified sampling, and correlation sampling can be used.

Model simulation solution. According to the random probability model established in the first step, simulate and simulate to find the random solution of the problem. In the simulation, it is usually assumed that the probability density function of the reliability parameters of the components in the network satisfies a certain distribution. Through many random simulation experiments, the indicators of all components are calculated, and the reliability indicators are calculated on this basis.

4   Improved SMC Method

Although the SMC is accurate, but there are two significant problems. First, the introduction of DG has led to major changes in the operating characteristics of the traditional distribution network, which has a greater impact on system reliability assessment. The main difference after joining the DG is that after the system fails, the DG can continue to operate in islands to continue supplying power to important loads. Then one of the problems encountered here is how to determine the failure rate of the islands. Second, the traditional SMC sampling method generally uses fixed timing sampling, which has low sampling efficiency and long simulation time. This paper proposes an improved sequential Monte Carlo simulation algorithm to solve the above problems.

4.1   Simulation Based on Structural Layered Model

In order to solve the problem of the reliability assessment of the distribution network caused by the introduction of DG, we use the structural hierarchical model to divide the topology of the original distribution network. Assume that a distribution network system with N components $\{x_1, x_2, \ldots, x_N\}$ contains a total of M types of DG sources ($M \leq N$), denoted as $\{DG_1, DG_2, \ldots, DG_M\}$, for any DG_i, it contains K_i components. This article divides the N components into two parts, one is the components

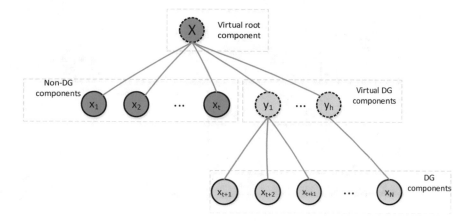

Fig. 4 The hierarchical structure of the components

that do not contain the DG source, and the other is the components that are contained in the DG source. In order to construct the hierarchical relationship, for each DG source DG_i, virtual encapsulate it into a large component, denoted as y_i, and then construct the hierarchical relationship as shown in Fig. 4.

After the transformation as shown in Fig. 4, a tree structure composed of real and virtual components is formed. The solid line nodes in the tree represent the components that actually exist in the distribution network system, and the dashed lines represent the virtual structured components. The reliability of the entire system can be calculated upwards from the leaf nodes of the tree one by one. For a node with multiple sub-nodes, it is necessary to calculate the reliability index for all of its sub-nodes, and then take the minimum reliability value. The reliability index of the virtual root device calculated in this way can be used as the reliability evaluation value of the entire distribution network system. In addition, when calculating continuous working time and fault repair time, it is necessary to calculate from the leaf nodes layer by layer.

4.2 Sampling Method Based on State Transition

In order to solve the problem of low sampling efficiency and long simulation time of the traditional SMC method, this paper proposes a sampling method based on state transition. The sampling time point t is completely taken as the time at the state transition for simulation. This sampling method can better reflect the whole As the state of the system changes, the convergence speed of the reliability assessment will be accelerated. The new simulated sampling for the situation in Fig. 5 is shown.

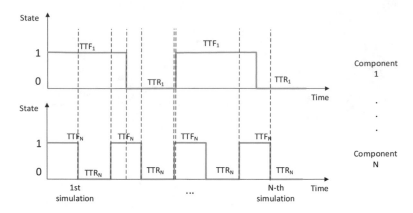

Fig. 5 New continuous working and repair time sequence of components

5   Experiments and Results

In this paper, We select the IEEE-RBTS6 under main feeder line F4, which add 2 DG access sources for simulation analysis to verify the performance of the method proposed. The system contains 23 load points, represented by a distribution transformer, 21 load switches, 4 circuit breakers, and the DG access source contacts the 17th and 20th feeder sections. In literature [8, 23] has given specific network topology and load data.

The reliability index selects LOLP (loss of load probability) and EENS (expected energy not supplied). LOLP is closely related to the reliability of the power generation and transmission synthesis system. The larger the LOLP, the more likely it is for the power generation company to implement market power; the smaller the LOLP, it indicates that the power generation market is close to a perfectly competitive market. EENS refers to the expected number of reductions in the power demand of the load due to the shortage of power generation capacity in the system in a given period of time.

It can be seen from Table 1 that the improved algorithm in this paper can meet the calculation accuracy requirements (Table 2).

Table 1 System reliability index calculation results

Index item	Traditional SMC method	Our method
LOLP	0.0945	0.0873
EENS	321,120.58	320,324.15

Table 2 The performance comparison with different methods

Methods	Time	Errs
Traditional SMC method	950	0.098
Our method	487	0.026

In the experiment, a total of 10,000 samples were taken, and the calculation time and calculation error of the two methods were compared. Under the same number of samples of the courseware, the algorithm in this paper achieved better accuracy and saved more time.

6 Conclusion

Aiming at the problem between the calculation accuracy and speed of the SMC, this paper propose an improved SMC method. The innovation of this paper is embodied in the structural hierarchical division of the original distribution network topology using the structural hierarchical model. As the same time, State transition sampling is used to o solve the problem of low sampling efficiency. In this paper, we only consider the impact of a single fault on the distribution network. Future work is to consider the impact of multiple faults on the distribution network.

Acknowledgements Special thanks are given to State Grid science and technology project [Title: Research and System Development of Power Quality Monitoring and Power Supply Reliability Evaluation of Active Distribution Network of Jilin Electric Power Research Institute Co. Ltd. Grant Number: JLDKYGSWWFW202106012], which offer grants to sponsor this project.

References

1. Chai WK, Wang N, Katsaros KV et al (2015) An information-centric communication infrastructure for real-time state estimation of active distribution networks. IEEE Trans Smart Grid 2134–2146
2. An Y, Liu D, Chen F et al (2019) Risk analysis of cyber physical distribution network operation considering cyber attack. Power Syst Technol 2345–2352
3. Wang S, Liang D, Ge L et al (2016) Analytical FRTU deployment approach for reliability improvement of integrated cyber-physical distribution systems. IET Gener Transm Distrib 2631–2639
4. Xu D, Wang Y, Zhang Y, Wu K, Zhao C (2020) Reliability assessment of active distribution network considering cyber failure in uncertain environment. Autom Electr Power Syst 134–142
5. Li B, Zhu J, Li P et al (2015) Island partition of distribution network with unreliable distributed generators. Autom Electr Power Syst 59–65
6. Zhou L, Sheng W, Liu W, Ma Z (2020) An optimal expansion planning of electric distribution network incorporating health index and non-network solutions. CSEE J Power Energ Syst 681–692
7. Jiang Z, Liu J, Xiang Y (2017) Discussion on key technologies of reliability evaluation of distribution network information physics system. Electr Power Autom Equip 30–42
8. Ding J, Ma C, Chen X, Fu B (2020) Reliability evaluation of active distribution network considering "Supply-Grid-Load". Power Syst Clean Energ 18–26
9. Zhao C, Qian H, Wang D, Qiu J, Jiang P, Lu Y (2020) Reliability evaluation of active distribution network considering load response. Electr Saf Technol 5–8
10. Karimi H, Niknam T, Dehghani M, Ghiasi M et al (2021) Automated distribution networks reliability optimization in the presence of DG units considering probability customer interruption: a practical case study. IEEE Access 98490–98505

11. Li J, Meng A, Liu X, Wang P (2019) Research of power grid reliability evaluation scheme based on sequential Monte Carlo algorithm. Chin J Electron Dev 82–87
12. Ge H, Asgarpoor S (2011) Parallel Monte Carlo simulation for reliability and cost evaluation of equipment and systems. Electr Power Syst Res 347–356
13. Nemeth CW, Rackliffe GB, Legro JR (2010) Ampacities for cables in trays with firestops. IEEE Power Eng Rev 55–56
14. Liu J, Luo H (2021) FMEA meter and standardization assessment method of active intelligent distribution network. Power Capacitor React Power Compensation 150–158
15. Yu M, Gang W, Yang L, Mu L, Shenwei D (2021) Reliability evaluation of DC distribution network considering islanding source-load uncertainty. Trans China Electrotech Soc 1–14
16. Ding Q (2020) Reliability analysis of multi structure DC distribution network. Electr Eng Mater 35–36
17. Li Y, Wang Q (2018) Construction of network equivalent model based on reliability evaluation of distribution network. Telecom Power Technol 70–71
18. Xu Z, Zhou J (2005) An improved fault traversal algorithm for complex distribution system reliability evaluation. Power Syst Technol 64–67
19. Phuc TK, Thang VT, Binh PH et al (2015) A computing tool for composite power system reliability evaluation based on monte carlo simulation and parallel processing. Lecture Notes Electr Eng 279–285
20. Song Y, Wang K, Zhu Y, Liu Y, Yuan Q (2020) Summary of sequential Monte Carlo simulation method for reliability assessment of distribution network. Electr Energ Manag Technol 90–99
21. Luo Q, Xie X (2020) Risk assessment of power network dispatching based on non-sequential Monte Carlo. J Shen yang Univ Technol 138–142
22. Khodaei A, Shahidehpourm, Kamalinia S (2010) Transmission switching in expansion planning. IEEE Trans Power Syst 1722–1733
23. Huang Z, Jin P, Wang J et al (2019) Fault recovery strategy of the distribution network with dg based onhierarchical partition. Power Syst Clean Energ 6–16

A Multi-condition WiFi Fingerprinting Dataset for Indoor Positioning

Ninh Duong-Bao, Jing He, Trung Vu-Thanh, Luong Nguyen Thi, Le Do Thi, and Khanh Nguyen-Huu

Abstract In recent years, indoor positioning is getting more attention, and WiFi fingerprinting is one promising method to track the position of a person. Until now, there is a lack of public datasets that can be used to compare fairly among different positioning algorithms. In this paper, a multi-condition WiFi fingerprinting dataset is introduced and can be accessed freely for further researches. The raw data were collected over four months in an office room that has a total of 205 reference points. Moreover, this dataset focuses on the different environmental conditions such as the density of people, the subject direction, the period in a day, etc. Various conditions were set up not only in the offline phase but also in the online phase. To support the understanding of the dataset, some materials and software are also provided.

Keywords WiFi · Fingerprinting · RSS · Dataset · Indoor positioning

N. Duong-Bao · J. He · T. Vu-Thanh
College of Computer Science and Electronic Engineering, Hunan University, Changsha, China
e-mail: duongbaoninh@hnu.edu.cn

J. He
e-mail: jhe@hnu.edu.cn

T. Vu-Thanh
e-mail: vuthanhtrung@vnu.edu.vn

N. Duong-Bao
Faculty of Mathematics and Informatics, Dalat University, Lam Dong, Viet Nam

T. Vu-Thanh
VNU School of Interdisciplinary, Vietnam National University, Ha Noi, Viet Nam

L. N. Thi
Faculty of Information Technology, Dalat University, Lam Dong, Viet Nam
e-mail: luongnt@dlu.edu.vn

L. Do Thi · K. Nguyen-Huu (✉)
Department of Electronics and Telecommunications, Dalat University, Lam Dong, Viet Nam
e-mail: khanhnh@dlu.edu.vn

L. Do Thi
e-mail: ledt@dlu.edu.vn

© The Author(s), under exclusive license to Springer Nature Switzerland AG 2022
N. H. T. Dang et al. (eds.), *Artificial Intelligence in Data and Big Data Processing*,
Lecture Notes on Data Engineering and Communications Technologies 124,
https://doi.org/10.1007/978-3-030-97610-1_48

1 Introduction

Location-Based Services (LBS) has become more popular and its market is growing firmly over the past few years. Positioning is one of the main components of LBS with its wide range of applications to determine the position of a user in different environments such as indoors and outdoors. The outdoor positioning is covered by the global navigation satellite system (GNSS) where it achieves good and reliable positioning information. However, the GNSS cannot provide good results in indoor environments due to the loss of its signal caused by the blocking or reflecting by walls and roofs. Thus, there is a need to develop the positioning systems for indoors separately.

It is a fact that these days people spend 90% of their time per day living and working indoors, especially in urban areas. This emphasizes the importance of indoor positioning investigation. In the field of indoor positioning, the position of a person can be determined by different technologies such as radio frequency identification (RFID) [1], ultrawideband (UWB) [2], Bluetooth [3], etc. Using WiFi, i.e. WiFi fingerprinting [4–7], is one of the most common technologies. This technique includes two main phases: the offline and the online phases. In the former phase, the radio map of the area where the user works is built. In the latter phase, the user position can be tracked by using the WiFi received signal strength (RSS) values from all available access points (AP). This technique has several advantages compared to others. First, it is an infrastructure-less approach since it utilizes the deployed APs which are already set up in many buildings and we do not need any additional equipment. Second, it can be implemented easily on smart devices such as smartphones which are indispensable in our daily lives. However, there are two major problems with using the WiFi fingerprinting technique. The first problem is that it takes a lot of time to collect the RSS signals at each reference point (RP) during the offline phase. Thus, the number of chosen RPs is proportional to the collecting time. The second problem is that the RSS values are affected by many factors such as the density of people, the density of electrical devices, or the user direction.

Currently, there is still a small number of public WiFi fingerprinting datasets that can be used to compare different positioning algorithms fairly. In this paper, we introduce a multi-condition WiFi fingerprinting dataset to the research community so that different research groups can apply their new methods and compare the positioning performance together. The dataset was created with the consideration of different environmental conditions not only in the offline phase but also in the online phase. This aims to help investigate the environmental factors that can affect the performance of WiFi-based positioning. The main contributions of this work are listed as follows:

- A WiFi fingerprinting dataset is proposed. The dataset is composed of a set of 205 RPs with RSS values collected from five APs by a single subject in four months. For each RP, the RSS values were collected 100 times, which means 20,500 scanning times to create the radio map. In the online phase, the RSS values were collected at each RP twice. Thus, there are 20,910 scanning times in total to create this dataset.

- Different conditions are considered in the offline phase. The subject collected the data in various conditions such as the different number of people in the area, different directions, or different timelines in a day, etc. These conditions create noise data that might be interesting for the researchers to handle.
- Different conditions are also considered in the online phase. The changing conditions are applied not only in the offline phase but also in the online phase. This makes it a more challenging environment to test the performance of the positioning algorithms.
- Some materials and software are also attached to the dataset to provide a better understanding of the dataset.

The rest of the paper is organized as follows: Sect. 2 presents the related work. Section 3 describes the dataset in detail. The data collection methodology is proposed in Sect. 4. Section 5 describes examples to use the proposed database. In the end, the discussion and conclusion are given in Sects. 6 and 7.

2 Related Works

Since positioning is one important key of LBS, there are many kinds of researches conducted to solve the tracking problems. However, the number of public datasets about this topic is not many. Using wireless technologies, some datasets were created and published. Alhomayani and Mahoor [8] introduced their dataset to track the user position outdoors. They collected different types of data not only wireless types such as WiFi, Bluetooth, cellular strengths but also sensor types such as accelerometer, gyroscope, and magnetometer, etc. A total of 122 RPs were set up at four different areas inside the university campus.

Comparing the outdoors and indoors datasets, the number of the dataset collected indoors is bigger. These indoor databases can be divided into two approaches. The first one is the large-scale approach where the wireless signals were collected in big areas such as multi-floor in a building and multi-building [9–11]. In [12], the author collected the TV, FM, and cellular signals during one year and he also mentioned the conditions during the collection phase such as the number of people in the area as well as the weather. The two university buildings with the area of $100 \text{ m}^2 \times 50 \text{ m}^2$ and $80 \text{ m}^2 \times 80 \text{ m}^2$ were used for data collection. With the same target to provide the dataset in big areas, Torres-Sospedra et al. [9] proposed one of the biggest public datasets which were conducted by more than 20 subjects with multiple devices. The data were obtained from three buildings with multi-floors. The total covered area was $108,703 \text{ m}^2$ with approximately 20,000 RPs. To investigate long-term data collection in large-scale areas, Mendoza-Silva et al. [10] and Ashraf et al. [11] introduced their datasets which were collected in 1.5 years and five years. The former one used a single smartphone and the data were collected by one user while the latter one created their dataset using different smartphones as well as different users. In addition, different conditions were also considered such as the mobility of the users and the orientations of the smartphones in [11].

Some datasets were also proposed to work with small-scale areas. In [12], the author not only covered the large-scale areas as discussed above but he also collected the raw wireless data in a small apartment ($14 \text{ m}^2 \times 7 \text{ m}^2$). PerfLoc [13] was another dataset that considered the small size area such as an office room, even though they also covered the bigger areas such as two industrial warehouses. Using four Android smartphones, three data types were collected including the WiFi, cellular, and sensor, respectively. Our dataset also focuses on a small size area (i.e. an office room) but we attempted to increase the number of RPs to cover the whole room as well as we collected the data in four months with different environmental conditions not only in offline phase but also in online phase.

3 Dataset Description

The proposed dataset includes three Excel format (*.xlsx) files that store the WiFi data. The files are divided into two folders named "Offline" and "Online". The former folder has one file that has RSS values collected from 205 RPs to build the radio map. The latter one has two files that could be used to test the positioning algorithms in two cases where the environmental conditions are different.

Figure 1 shows the contents of the file named "Training_Data" in the "Offline" folder. The recorded file includes eight columns as follows:

- "ID_RP" is the ID of the RP and is numbered from 0 to 204 to show a total number of 205 RPs. For each RP, the subject scanned for the RSS values 100 times, thus, there are 100 rows for each RP.

	ID_RP	X (m)	Y (m)	AP1 (dBm)	AP2 (dBm)	AP3 (dBm)	AP4 (dBm)	AP5 (dBm)
1								
2	RP0	0.0	0.5	-60	-48	-65	-45	-53
3	RP0	0.0	0.5	-58	-49	-65	-43	-54
4	RP0	0.0	0.5	-60	-48	-65	-45	-54
5	RP0	0.0	0.5	-59	-48	-64	-51	-54
6	RP0	0.0	0.5	-58	-50	-65	-47	-52
7	RP0	0.0	0.5	-58	-48	-62	-47	-53
8	RP0	0.0	0.5	-57	-49	-59	-48	-54
9	RP0	0.0	0.5	-59	-49	-63	-48	-52
10	RP0	0.0	0.5	-58	-49	-63	-48	-54
11	RP0	0.0	0.5	-59	-49	-64	-48	-54

Fig. 1 The recorded WiFi data file of the "Offline" folder

ID_RP	AP1 (dBm)	AP2 (dBm)	AP3 (dBm)	AP4 (dBm)	AP5 (dBm)
RP0	-57	-43	-66	-46	-51
RP1	-57	-45	-64	-44	-55
RP2	-52	-49	-67	-45	-49
RP3	-52	-49	-64	-51	-60
RP4	-41	-45	-61	-51	-47
RP5	-38	-53	-64	-48	-50
RP6	-39	-41	-66	-57	-56
RP7	-52	-55	-65	-50	-46
RP8	-58	-46	-66	-46	-47
RP9	-65	-50	-63	-49	-52
RP10	-62	-52	-63	-49	-45
RP11	-68	-49	-63	-44	-47

Fig. 2 The recorded WiFi data file of the "Online" folder—case 1

- "X" and "Y" are the created X and Y coordinates of each RP that the subject stood on and made a new RSS scanning. The distance between two RPs is 0.5 m. The (X, Y) coordinate is (0.0, 0.5) at the first RP (RP0) and is (8.5, 6.0) at the last RP (RP204).
- "AP1" to "AP5" describe the RSS values collected at each RP from five APs. Each row shows the RSS values after each scanning time.

Figure 2 describes the contents of two files named "Case 1" and "Case 2" in the "Online" folder. The recorded file includes six columns with two main information: "ID_RP" and "AP1" to "AP5". The difference is that in the online phase, the subject collected the WiFi data at each RP once for each case. Thus, there are only 205 rows in total which correspond to 205 scanning times.

4 Data Collection Methodology

The WiFi data were collected in an office room that had an area of 9.0 m^2 × 6.5 m^2 in four months (from November to December of 2018 and from March to April of 2019). Since it was a working office, the environment was complicated where it consisted of furniture (e.g. tables, desks, or shelves), electrical devices (e.g. laptops and desktops), and amount of people (e.g. 10 people). Five APs and 205.

RPs were set up in this office. The coordinates of the APs and RPs are shown in Fig. 3. The five APs were fixed at different positions with heights of 1.1–1.6 m from the ground and they could be seen from any position in the room (i.e. line of sight). The 2D coordinates of five APs were (2.5, 0.0), (4.75, 3.25), (9.0, 5.5), (9.0,

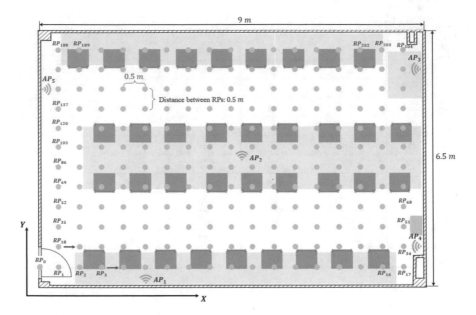

Fig. 3 Distribution of five APs and 205 RPs in the room

1.0), and (0.0, 5.0), respectively. The 205 RPs were created in a grid spaced 0.5 m apart. The coordinates of the RPs were constructed as follows. The first RP (RP0) was chosen at the entrance door and its coordinate was (0.0; 0.5), the next RP (RP1) was at 0.5 m apart on the right side of RP0, and so on. There were 17 RPs in a row and 12 RPs in a column.

The WiFi data collection was conducted by a subject holding one Android smart-phone (Samsung Galaxy Note 3 SM-N900T) in front of his body. The subject stood on each RP and used the smartphone to scan the RSS values. To collect the data, a wireless local area network (WLAN) system was built. It consisted of one webserver which would wait for the RSS values sent from the Android smartphone and store them in a folder. Figure 4 shows the system used for data collection. The user interface of the Android application is shown in Fig. 5. The media access control addresses (MAC addresses) and names of five APs were assigned to the application. When the subject stood on a RP, he clicked the "REFRESH" button to start a new WiFi scan to collect new RSS values from the APs. Then he clicked the "UPLOAD" button to upload the values in the XML format to the webserver.

The performance of WiFi fingerprinting is affected by the changes of the surrounding environment such as the density of people, the electrical devices, or the subject direction, etc. Therefore, when collecting the WiFi data, the subject inten-tionally changed the environmental factors that can make the RSS values fluctuate much. In the offline phase, different conditions were set up as follows:

- The density of people working and moving in the room: the number of people was changed from a small group of 1–5 people to a bigger group of 16–20 people.

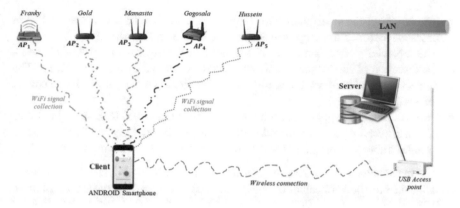

Fig. 4 Prototype of WLAN system for data collection

Fig. 5 Android application
interface

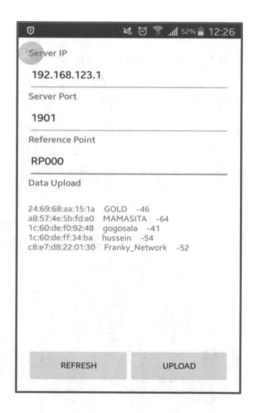

- The density of electrical devices: the number of laptops and desktops was changed from 1 to 24 devices.
- The temperature in the room: cool or warm.

608 N. Duong-Bao et al.

- The height of the smartphone from the ground: the height was 1.3 m or 1 m in cases we assumed the subject was standing or sitting, respectively.
- The subject direction: since the subject held the smartphone in front of his body, the direction of the subject is the same as the direction of the smartphone. The subject changed his direction in four main directions (i.e. 90°) at one RP to collect the data.
- Time: in the offline phase, at each RP, the subject made 100 RSS scanning times and each scanning was conducted in different periods not only in a day but also in months. Table 1 summarizes the different periods when collecting the WiFi data.

The mean values of five APs at each RP are shown in Fig. 6. The mean value is calculated from 100 RSS values of one AP at one RP. Figure 7 shows the distribution of RSS values of AP1 over 205 RPs. It can be seen in the figure that the RPs that are closer to the AP1 position would have stronger RSS values.

In the online phase, the subject used the smartphone to collect the RSS values at each RP once in a random direction. The WiFi data was collected in two cases

Table 1 The collected RSS values at one RP

Period	Morning (8:00 to 11:45)	Noon (12:00 to 15:50)	Afternoon (16:00 to 18:45)	Night (19:00 to 20:30)	Sum
11/2018	9	5	9	2	25
12/2018	8	5	9	3	25
03/2019	9	5	8	3	25
04/2019	9	5	9	2	25
Sum	35 (times)	20 (times)	35 (times)	10 (times)	**100 (times)**

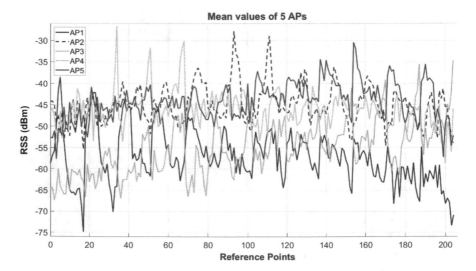

Fig. 6 Mean values of five APs at each RP

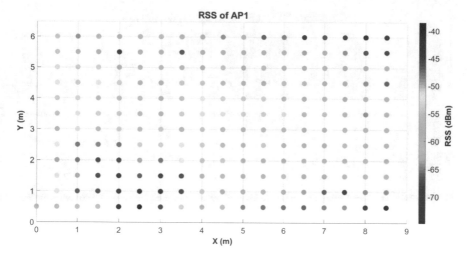

Fig. 7 Distribution of AP1 in the room

Table 2 Environmental conditions in case 1 and case 2

Conditions	Case 1	Case 2
Density of people	1–9	6–13
Density of electrical devices	11	20
Temperature	Cool	Warm
Height from the ground (m)	1.3	1.3
Subject direction	Random	Random

with different environmental conditions. The conditions are described in Table 2. Therefore, there are 410 sets of RSS values that would be used for testing cases.

5 Examples of Data Use

To prove the efficacy of the dataset, K-Nearest Neighbor (KNN) and Weighted KNN (WKNN) algorithms [14] are implemented to test the positioning performance in two cases. These two algorithms are common position matching algorithms to estimate the position of a user. The experiments are carried out using K = 3, 7, and 11. The positioning error is defined as the distance between the actual RP position where the subject was standing and the estimated position using KNN or WKNN. As aforementioned, in the online phase, the subject stood on each RP and used the smartphone to scan for new RSS values once. Thus, there are 410 test positions in two cases which consist of different environmental conditions. The positioning results using KNN and WKNN in two cases are figured out in Figs. 8 and 9 where (a) shows the results over 205 RPs and (b) shows the different results at some RPs (i.e.

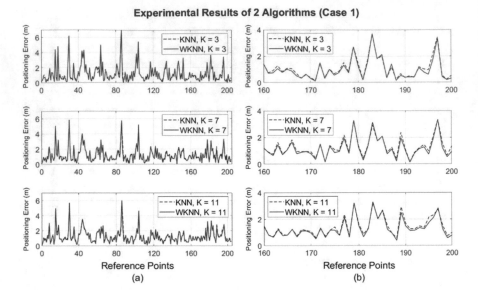

Fig. 8 Positioning results in case 1

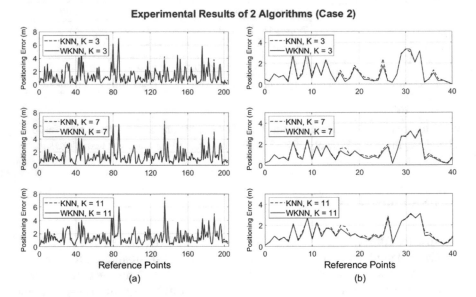

Fig. 9 Positioning results in case 2

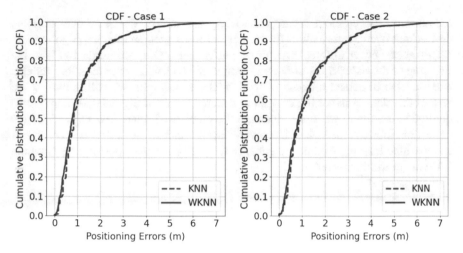

Fig. 10 Cumulative position error distributions for two test cases

from RPs 160 to 200 in case 1 and 0–40 in case 2). The results show that with the same chosen K value, the WKNN obtains slightly better positioning results than the KNN. The average errors of KNN with K = 3, 7, and 11 in case 1 are 1.22, 1.26, and 1.30 m, while the errors of WKNN are 1.17, 1.21, and 1.24 m. For case 2, the average positioning errors with different K values are 1.32 m, 1.45 m, and 1.52 m for KNN and 1.26 m, 1.38 m, and 1.45 m for WKNN, respectively. The results point out that using K = 3 gives the best positioning performance. Furthermore, the tracking result in case 2 is worse than in case 1 since the conditions in case 2 are more complicated than in case 1 (e.g. the increased number of people as well as electrical devices). Figure 10 shows the cumulative distribution function (CDF) of KNN and WKNN (K = 3) in two test cases. The results also show that the performance of WKNN is a bit better than KNN.

6 Discussion

To deal with the practical aspect of positioning, the large dataset for indoor positioning has become more popular. Here, the large dataset can be defined as the large amount of RPs to cover the vast area of a building including multi-floor areas. However, some problems can affect the performance of the WiFi fingerprinting technique. The first problem is the high-dimension data since the position of a person in real scenarios (e.g. big buildings or shopping malls) should be determined from hundreds of nearby WiFi APs. The second problem is the changes in environmental conditions can cause the fluctuation of WiFi RSS values which can degrade the positioning accuracy. So the question is how to deal with these problems? Big data analytic with different types of machine learning methods can provide good solutions to help solve this problem. It can process large amounts of WiFi RSS data, analyze, and extract useful information for further purposes such as creating a good radio map that can cover different environmental changes. Our current dataset is at the first step of the data acquisition. Even though the collection time took four months and we tried to collect a bunch of RSS values at 205 RPs, we still think that the size of this dataset is not big enough to become a part of big data. To improve its quality, in the future we aim to increase the number of smart devices used for the collection, increase the size of the covered area, increase the collection period, and use machine learning methods to enhance the performance of positioning using the WiFi fingerprinting technique.

7 Conclusion

In this paper, a multi-condition WiFi fingerprinting dataset is proposed and published freely to the research community. The dataset was created by a subject holding a smartphone in an office room. The raw data collection spanned over four months to build the radio map which included 205 reference points. Different environmental conditions such as the density of people, the density of electrical devices, the subject direction, or period in a day, were considered when collecting the WiFi data not only in the offline phase but also in the online phase. Even though the creation and collecting process of this dataset was laborious and time-consuming, it is worthy as this dataset can help widen the knowledge of positioning using the WiFi specifically in dynamic changes of environments. Moreover, by allowing this dataset to be accessed freely, researchers can apply different algorithms onto it, compare the results, and improve their methods if needed. Besides, reading materials and software are included with this dataset to provide information in detail on our collecting methodology.

In future work, we will try to increase the size of this dataset by collecting the WiFi data in bigger areas such as multi-floor buildings. Instead of using only one subject and one smartphone, we will use the crowdsourcing approach which includes multiple people and multiple smart devices. Moreover, in the online phase, we will apply different machine learning methods to enhance the performance of positioning using the WiFi fingerprinting technique.

8 Supplementary Materials

Supplementary material associated with this article can be found, in the online version, at https://github.com/luongnt1983/IPS.

Acknowledgements This work was supported in part by National Natural Science Foundation of China (NSFC) (61775054), and by National Natural Science Foundation of Hunan Province (grant no. 2020JJ4210).

References

1. Seco F, Jiménez AR (2018) Smartphone-based cooperative indoor localization with RFID technology. Sensors 18:266–289
2. Zampella F, Jiménez AR, Seco F (2013) Robust indoor positioning fusing PDR and RF technologies: the RFID and UWB case. In: International conference on indoor positioning and indoor navigation, pp 1–10
3. Zhuang Y, Yang J, Li Y, Qi L, El-Sheimy N (2016) Smartphone-based indoor localization with Bluetooth low energy beacons. Sensors 16:596–616
4. Deng Z-A, Wang G, Qin D, Na Z, Cui Y, Chen J (2016) Continuous indoor positioning fusing WiFi, smartphone sensors and landmarks. Sensors 16:1427–1447
5. Li Y, Zhuang Y, Zhang P, Lan H, Niu X, El-Sheimy N (2017) An improved inertial/WiFi/magnetic fusion structure for indoor navigation. Inform Fusion 34:101–119
6. Yu J, Na Z, Liu X, Deng Z (2019) WiFi/PDR-integrated indoor localization using unconstrained smartphones. EURASIP J Wirel Commun Netw 2019:41–54
7. Ninh DB, He J, Trung VT, Huy DP (2020) An effective random statistical method for indoor positioning system using WiFi fingerprinting. Futur Gener Comput Syst 109:238–248
8. Alhomayani F, Mahoor MH (2021) OutFin, a multi-device and multi-modal dataset for outdoor localization based on the fingerprinting approach. Sci Data 8:66–80
9. Torres-Sospedra J, Montoliu R, Martínez-Usó A, Avariento JP, Arnau TJ, Benedito-Bordonau M, Huerta J (2014) UJIIndoorLoc: a new multi-building and multi-floor database for WLAN fingerprint-based indoor localization problems. In: 2014 international conference on indoor positioning and indoor navigation, pp 261–270
10. Mendoza-Silva GM, Richter P, Torres-Sospedra J, Lohan ES, Huerta J (2018) Long-term WiFi Fingerprinting dataset for research on robust indoor positioning. Data 3:1–17
11. Ashraf I, Din S, Ali MU, Hur S, Zikria YB, Park Y (2021) MagWi: benchmark dataset for long term magnetic field and Wi-Fi data involving heterogeneous smartphones, multiple orientations, spatial diversity and multi-floor buildings. IEEE Access 9:77976–77996
12. Popleteev A (2017) AmbiLoc: a year-long dataset of FM, TV and GSM fingerprints for ambient indoor localization. In: 2017 international conference on indoor positioning and indoor navigation, pp 1–4
13. Moayeri N, Ergin MO, Lemic F, Handziski V, Wolisz A (2016) PerfLoc (Part 1): an extensive data repository for development of smartphone indoor localization apps. In: 2016 IEEE 27th annual international symposium on personal, indoor, and mobile radio communications, pp 1–7
14. Xingbin G, Zhiyi Q (2016) Optimization WIFI indoor positioning KNN algorithm location-based fingerprint. In: 2016 7th IEEE international conference on software engineering and service science (ICSESS), pp 135–137

Design and Experimental Evaluation of MAS-GiG Model for Crowd Evacuation Planning in Case of Fire

Dinh Thi Hong Huyen, Hoang Thi Thanh Ha, and Michel Occello

Abstract In this paper, we proposed the design and experimental evaluation of the Multiagent System-Group in Group (MAS-GiG) model for crowd evacuation planning in case of fire. We incorporated two MAS-GiG models for an application. Case studies were conducted in the arrivals and departures halls of stations. The evacuation guide strategy is implemented based on a pre-built plan. The evacuation guide is only done one way through the exits. The results are the number of total evacuation times and remaining occupants after evacuation. The smaller these two values are, the better the proposed model is. The experiment was performed at a departure hall that has a structure similar to the 1st floor, Danang international airport.

Keywords Emergency management · Evacuation planning · Crowd behavior · Evacuation guide · Emergency crowd evacuation

1 Introduction

Planning to guide the evacuation of crowds in emergencies has an important role in saving people. Many studies have been proposed and implemented. However, each study takes its perspective by not covering all of the issues of evacuation. Because emergencies often happen quickly and unexpectedly, individuals in the crowd are not prepared, so they have unexpected reactions and behaviors. The crowd became chaotic causing many dangers to the occupants.

D. T. H. Huyen (✉)
Quynhon University, Quynhon, Binhdinh, Vietnam
e-mail: dinhthihonghuyen@qnu.edu.vn

H. T. T. Ha
DaNang University of Economics, The University Of DaNang, DaNang, Vietnam
e-mail: ha.htt@due.edu.vn

M. Occello
Grenoble Alpes University, Grenoble, France
e-mail: michel.occello@lcis.grenoble-inp.fr

© The Author(s), under exclusive license to Springer Nature Switzerland AG 2022
N. H. T. Dang et al. (eds.), *Artificial Intelligence in Data and Big Data Processing*,
Lecture Notes on Data Engineering and Communications Technologies 124,
https://doi.org/10.1007/978-3-030-97610-1_49

615

There are some related studies such as the paper [1] proposed a group model based on social relationships. The study [2] proposed a multi-level multi-agent model MAS-GiG and applied it to navigate the crowd in case of fire. Modeling crowd behavior is in case of fire [3]. Evacuation planning is based on four ways of arranging exits in different directions [4]. Emergency evacuation planning is accomplished by providing the safest and nearest evacuation route [5]. Planning emergency crowd evacuation is based on the current structure of a local airport [6].

The evacuation planning algorithm is based on a route network model. The evacuees are grouped based on the location next to each other [7]. The authors presented a novel approach for interactive navigation and planning multiple agents in crowded scenes with moving obstacles. They used a precomputed roadmap that provides global connectivity for wayfinding and combines it with fast and localized navigation for each agent [8].

There are several advantages of the proposed approach as. Firstly, applying the MAS-GiG model to manage and guide the evacuation of crowds by each level. At each level, there are respective representatives and members. The representative manages and controls the members of the group. He represents the group to receive instructions from the higher level. Then he passes the instructions to the members; Secondly, the structure of the physical environment is described by a graph G = <V, E> . With V is the set of vertices (the intersection points from the paths), E is the set of edges (the segment connecting two intersecting points). Then we applied Dijkstra's algorithm to determine the shortest path from 1 vertex (one location) to 1 vertex (an exit) of graph G; Thirdly, we applied an integrated MAS-GiG model that guides the crowd move to follow the chosen route; Finally, the proposed approach has not occurred conflicts between agents/groups. In this approach, we focus on the application design and experiment evaluation to determine the appropriate time to make an evacuation, minimizing the total evacuation time and the number of people remaining in the evacuation.

According to [9] guidelines for emergency evacuation at airports in case of fire, we combine two MAS-GiG models for planning to guide the evacuation of crowds in case of fire. A model that represents the passenger system is called base-MAS-GiG. This model consists of an individual level and a group level. A model that represents the airport emergency management system is called manage-MAS-GiG. Depending on the application scope, the number of levels of the model for the application is determined. The actions/events are described according to the timeline model. They changed from a normal situation to an emergency. The main contribution of the paper includes combining two MAS-GiG models in planning to guide the evacuation of crowds in case of fire at a flight lounge; Designing the application, evaluating the experimental results to determine the advantages of the proposed method. We experimented with three methods and compared the results. Three methods include the proposed method, MAS-GiG [2], and the Auto method.

The next parts of the paper are presented as follows: Sect. 2 presents the proposed approach. Section 3 presents the experiment and the obtained results. Finally, Sect. 4 presents the discussion and conclusion.

2 The Proposed Approach of Planning for Crowd Evacuation in Case of Fire

2.1 The Framework

The proposed approach is an evacuation planning method for optimizing emergency management in the public crowd. Firstly, based on the spatial data we established the two-dimensional evacuation environment. Secondly, the scenarios are set up for evacuation simulation. Thirdly, the evacuation simulation was performed by three methods: automatic method, MAS-GiG method [2], and the proposed method. Finally, statistics and compare results. The framework includes some modules (see Fig. 1).

   Data acquisition. Data acquisition includes spatial data and occupant data. Spatial data includes flight lounges, exits, and corridors. The occupant data includes spatial distribution of the occupants in the simulated environment, walking speed, and physical traits. These are important factors in the emergency crowd evacuation experiment.

Fig. 1 The framework for the proposed approach

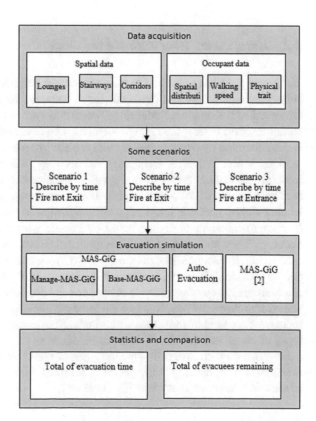

Scenarios. We tested three scenarios based on the available physical structure of each lounge. The scenarios represented actions/events by the timeline model [10]. Each scenario has a different fire location that represents locations with similar characteristics. Scenario 1, the fire location does not coincide with the exits, it represents the fire locations that are not in the exits in the lounge. Scenario 2, the fire location is an exit where few occupants know this exit. Scenario 3, the fire location is the entrance to the lounge, for which most of the lounge occupants know this exit.

Evacuation simulation. The evacuation simulation was tested for three methods on the same simulation environment and same occupant data. From there, we evaluate the proposed method based on its test results compared to the other two methods. The proposed method incorporates the manage-MAS-GiG model and base-MAS-GiG model. Auto evacuation does not use an evacuation guide model and a method using a MAS-GiG [2].

Statistics, compare results. We chose the total number of evacuation times and the total number of remaining occupants to compare the results. These two parameters are important in emergency evacuation situations. The lower the total number of evacuation times and the total number of remaining occupants, the better the evacuation method.

2.2   Emergency Evacuation Modeling

The paper [11] has described human movement behavior in an emergency. These behaviors include Learned Exit Knowledge behavior, Random Movement behavior, Observed Exit Knowledge behavior, Directed Movement behavior, Panicked Movement behavior, Follow the Herd Movement. The Learned Exit Knowledge behavior is demonstrated by people familiar with the environment. They usually are the staff working there. When an emergency occurs they will guide the evacuation. The Random Movement behavior is shown by people who don't know where the exit is. The Observed Exit Knowledge behavior is demonstrated by people who can locate exits by personal knowledge but are unsure to choose the exit. The Directed Movement behavior is demonstrated by people who are emergency management agencies. They have enough data, environmental data, and occupants to give proper evacuation instructions. The Follow the Herd Movement behavior is demonstrated by people who don't know where the exit is, they decide to follow the crowd. This case often causes the currents to move in the opposite direction. The Panicked Movement behavior is shown by people who are a great psychological influence. They decide to increase the speed of movement (running), often causing panic, congestion. These types of behaviors are shown in a crowd when an emergency occurs. If the occupants move on their own following the types of behavior described above without an evacuation plan, causing many deaths. Therefore, planning to guide the evacuation of crowds in an emergency is really necessary.

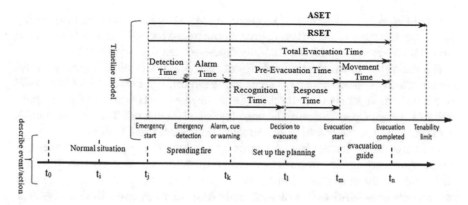

Fig. 2 Describe the process by timeline mode

Describe the process by timeline model. During the simulation, actions/events take place by time, the timelines are described corresponding to the timeline model [10, 12, 13] (see Fig. 2).

At the starting time t(0), there are already some passengers, they do something like sitting on a waiting chair, looking for a seat, going to a shop, going to a restaurant, going to the toilet, etc. Some passengers go into the lounge and do similar actions. The initial relationships are also determined, they travel alone or in groups, which flight they take, and the time of the flight. The staff in the emergency management agency are also in their place to carry out the task. At this time, the base-Mas-GiG and manage-MAS-GiG models have been applied. These actions/events continue to take place to time t(j). The fire appears at time t(j) and it is spread from t(j) to t(k) time. When receiving fire information, for individuals going individually, what will they do? Depending on their knowledge they will exhibit one of the behaviors described in the literature [11]. For individuals going in groups, what will they do? They will communicate with each other to make decisions for the group. For the airport emergency management department (AirportOperator), they will establish an evacuation guide plan and send this plan to the lounge representatives (Guide leader). In each lounge, the Guide leader sends this plan to the area representatives (Guides). In each area, the Guide sends this plan to the group representatives (Group leaders). These actions take place in the time interval t(k) to t(m). This is the Pre-evacuation time in the timeline model [13]. At time t(m), the Group leader guides the members in the group to follow the plan. At time t(n) the evacuation is complete.

The proposed model for the application. MAS (Multiagent system) is organized according to the AEIO approach [14]. A are agents, E is an environment, I are the interactions, and O is an organization. The MAS-GiG is a multi-level multi-agent model. The mechanism of formation of levels is carried out from the bottom up. A group at level 1 is formed from agents at level 0. The formation of the group at this level is based on social relationships such as family, couple, friends, colleagues. At level 2, a group is formed from the group representatives at level 1, that the distance between them is less than or equal to a constant r. The remaining levels are formed similar to level 2. The number of levels of the MAS-GiG is determined depending on the specific application. In this approach, the base-MAS-GiG model is the group model. It consists of two levels: the individual level and the group level.

According to the document [9], the airport emergency management agency always has a hierarchy system to manage areas in the airport. Apply the MAS-GiG model to this system. There are three levels of the manage-MAS-GiG model for the application. They are Operation level, Lounge level, and Area level. The representatives of the respective levels are AirportOperator, Guide leader, and Guide.

Each area in the lounge has occupants who belong to the passenger system. A Guide represents an area in a lounge and he manages the occupants in this area. He belongs to the emergency management system. This is a natural association of two models base-MAS-GiG and manage-MAS-GiG from two different systems. Base-MAS-GiG represents the passenger system, manage-MAS-GiG represents the emergency management system at the airport.

Evacuation plan. The evacuation plan includes two phases. A phase is based on the physical structure of the lounge to make a path graph G = <V, E> , with V being a set of vertices, E is a set of edges. We applied Dijkstra's algorithm to select the shortest evacuation route from a point to an exit (a vertex to a vertex). The second phase used a combined model of two MAS-GiGs to guide evacuation along the selected route (see Fig. 3).

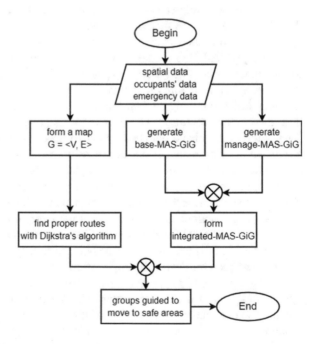

Fig. 3 The overview of planning to guide the evacuation

3 The Experiment

3.1 Design Framework for Application

Some modules in the framework for the application are described in Fig. 4.

JsonParser is a helper used to parse data from the scenario.json file, which contains the entire information about the environment, flight schedules, and the time and location of a fire occurring. **MASH platform** is a simulation program that is used to simulate agents' activities based on our specific purpose. In other words, it is considered as a programming library and used to simulate our scenario. **Presentation** is the view of the program. It reads data from the environment or agents' models and renders them into the view. **ProgressController** is a controller used to manage the entire progress of a certain scenario from the beginning to the ending. **CenterModel** is a central data model used to manage all objects of the program. To facilitate managing, we create two smaller data models **AirportMap** and **AgentManager**, which are components of CenterModel. The first one, AirportMap is used to handle data of static objects related to the environment such as paths, chairs, restaurants, shops, etc. Therefore, this model also handles tasks related to finding a route from a location to another. The second one, AgentManager is used to handle data of dynamic objects, which are agents such as passengers, guides, fires, airport operators, etc. **Initialization** is a module used to initialize the program. This process is through three steps. First, it is generating the vertices and edges of the graph based

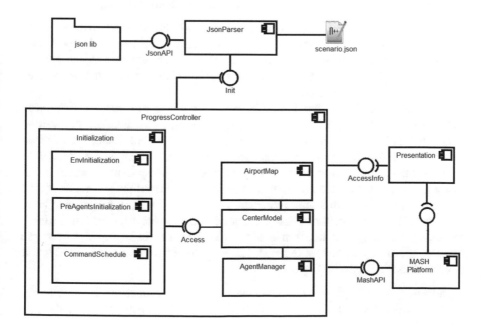

Fig. 4 The framework for an application

on the raw data read from the scenario file. This step is called **EnvInitalization**. Second is **PreAgentsInitialization**, which is based on the data of flight schedules to distribute the number of passengers into the lounge at random positions before the simulation really starts (t0). The last step is **CommandSchedule**, which creates events that will occur at a specific time of the simulation progress in the future (after t0). For example, when a fire event or a delay notification of a flight occurs as well as how many passengers arrive at the entrance of lounges. Additionally, the number of passengers is distributed into lounges at a specific time in steps two and three by adopting the IATA pattern [15].

3.2   Input Parameters for Simulation

We experiment with the proposed method, the MAS-GiG model [2], and the Automatic evacuation. All three methods are tested in the same environment, with the same parameters. The result is the total evacuation time and the total number of people remaining in the evacuation. The departure hall includes 4 lounges from lounge 4 to lounge 7. There are 8 areas and 3 exits in each lounge. Each lounge has its evacuation plan. The exit with the largest evacuation time is also the lounge evacuation time. Similarly, the evacuation time of the departure hall is calculated by the evacuation time of the lounge with the largest evacuation time. The number of occupants remaining after evacuation is equal to the total number of remaining occupants of all lounges. In this study, we experimented with three methods in lounge 5.

Simulation environment. The environment is a two-dimensional space with the same structure as the departure hall, first floor, Da Nang international airport [16]. From spatial data, we create detailed maps including environmental objects such as rows of seats, paths, shop, restaurant, toilet, massage chairs, walls, stairs, entrances entrance, exit, etc. (see Fig. 5).

We briefly describe the structure of lounge 5, the rest of the lounge are similarly distributed. In the lounge, there is an area for passengers to sit called the waiting area. An area for souvenirs is called a shop, an area is called a restaurant and an area is a restroom. Based on the actual lounge structure, we created a 2D map to simulate the experiment. In each area are numbered correspondingly such as (1): safe area, (2): boarding gate, (3): entrance to the lounge from the second floor, (4): exit to go down the ground floor (emergency exit of the lounge), (5): waiting area, (6): shop, (7): restaurant, (8): restroom, between rows of seats, is the path. In the three test scenarios, the ground floor entrance (4), the aircraft exit (2), and the concourse entrance (3) are used as three exits E1, E2, E3. Three fire positions are located at three positions F1, F2, F3 (Fig. 5).

Agents. The agents represent the occupants in the simulation space, each agent has a set of attributes, knowledge, roles, interactions, and relationships described [2].

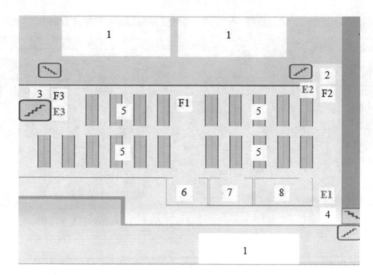

Fig. 5 The simulation environment for the lounge 5

Parameters. According to [17], the walking speed on horizontal roads is 1.12 m/s. The speed going down the stairs is 0.8 m/s. The speed of going up the stairs is 0.5 m/s. The size of the exit is 80 cm, the size of the emergency exit is 130 cm [18]. We choose a rate of 100% of the number of normal evacuees, they are likely to receive equally awake and healthy.

In the simulation program, evacuation time is a period from a fire occurring to the last agent moving to a safe area. The number of remaining occupants is counted based on the study [19], a person is stuck in about 3–5 min during the evacuation, he has died. This means that if an agent does not move for a period of 3 to 5 min, we assume he is dead. From there, the number of remaining occupants has added the total.

3.3 Experimental Result

We assume fire occurs in lounge 5 at 3 representative locations F1, F2, F3 corresponding to three scenarios 1, 2, 3. Experiments are carried out in this lounge. In the simulation, the blue dot represents non-group passengers, the yellow dot represents group passengers, the red dot represents Group leader, the black dot represents Guide and the gray dot represents Guide leader. The simulation interface is shown that fire appears at 3 positions F1, F2, F3 (see Fig. 6).

For each test time, the number of agents representing passengers is about 700–800 agents. To compare the experimental results between the three methods, we tested 3 scenarios corresponding to the three fire locations F1, F2, and F3. Each scenario for each method was tested 10 times then averaged the total evacuation time for each

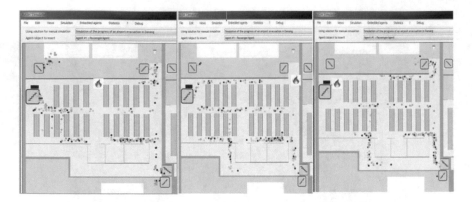

Fig. 6 The simulation interface is shown that fire appears at 3 positions F1, F2, F3

method for comparison (Fig. 7). There are some reasons we chose the test many times. Firstly, agents enter the lounge allocated to chairs randomly. Secondly, agents can do any activities and the period time for each activity differently. For example, an agent can choose to go shopping but another just wants to sit on a chair. Therefore, the number of passengers in each area of the lounge will differ between tests when a fire occurs, leading to different results.

The results show that the total evacuation time of the proposed method is lower than the MAS-GiG method [2] and the Auto method. The total evacuation time of the Auto method was higher than the two methods using MAS-GiG. The two methods using MAS-GiG, the total evacuation time for the 3 scenarios are equivalent. For the Auto method, the majority of occupants move to the entrance. In scenario 3, the entrance has fire, the occupants have to return to the remaining two exits. Therefore, the total evacuation time of this method for scenario 3 is higher than scenario 1 and scenario 2.

Similar to calculating the total evacuation time, we calculated the total number of remaining occupants after evacuation. The results show that the total number

Fig. 7 The total evacuation time for three scenarios

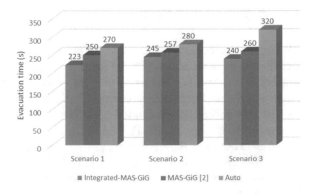

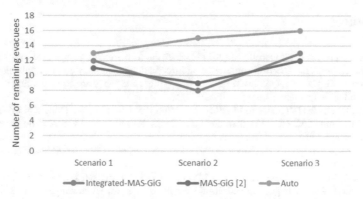

Fig. 8 The total remaining occupants for three scenarios

of remaining occupants in the two methods using the MAS-GiG model is not significantly different. This number is lower than the Auto method (Fig. 8).

4 Discussion and Conclusion

The study has achieved initial results from scenario design, application design, application of the MAS-GiG model in planning to guide evacuation of crowds when there is a fire at any location in the lounge. All three methods provide a reasonable total time, total number of evacuees.

The proposed method has advantages in emergency evacuation for an application with an environment similar to an airport. The evacuation plans are established and coordinated at each level. Each level has its evacuation plan and is managed and coordinated by a representative. This avoids overcrowding at the operator level, the total evacuation time is short, the number of deaths is small.

We compare the proposed method with the applied MAS-GiG method in [2]. At the area level, the proposed method has the advantage of combining two completely natural systems, the individuals and groups present in each area. The area's representative is Guide. He manages and controls the area. For the method in [2], at this level, a group is formed and a leader is selected. The leader at this level is Guide.

Evacuation results depend on spatial data and occupant data such as movement speed, distribution rate, size of occupants. Therefore, in the next study, we will test on many data samples, with the parameters changed for each scenario to choose the time to start the evacuation and reasonable parameters to bring the optimal results. We continue to research evacuation guidance planning on two-way flow and expand the range of applications.

References

1. Huyen DTH, Ha HTT, Occello M (2021) Detect the formation of groups in a crowd based on social relations. In: The 10th conference on information technology and its application
2. Huyen DTH, Ha HTT, Occello M (2021) Navigating emergency crowd evacuation using masgig model. Fundamental Appl Res (FAIR 2021)
3. Huyen DTH, Ha HTT, Occello M (2021) Mô Hình Hóa Hành Vi Đám Đông Khi Có Cháy Dựa Vào Mô Hình MAS-GIG. Hội thảo quốc gia lần thứ XXIV: Một số vấn đề chọn lọc của Công nghệ thông tin và truyền thông–Thái Nguyên
4. Kurdi HA, Al-Megren S, Althunyan R, Almulifi A (2018) Effect of exit placement on evacuation plans. Europ J Operat Res
5. Haris M, Mahmood I, Badar M, Alvi MSQ (2018) Modeling safest and optimal emergency evacuation plan for largescale pedestrians environments. In: Proceedings of the 2018 winter simulation conference
6. Chen J, Liu D, Namilae S, Lee S-A, Thropp JE (2019) Effects of exit doors and number of passengers on airport evacuation effeciency using agent based simulation. Int J Aviation Aeronautics Aerospace
7. Han L, Guo H, Zhang H, Kong Q, Zhang A, Gong C (2020) An efficient staged evacuation planning algorithm applied to multi-exit buildings. Int J Geo-Info
8. van den Berg J, Patil S, Sewall J, Manocha D, Lin M (2008) Interactive navigation of multiple agents in crowded environments. In: Conference: proceedings of the 2008 symposium on interactive 3D graphics, SI3D 2008, February 15–17, 2008, Redwood City, CA, USA
9. https://www.emergency-live.com/of-interest/emergency-in-airports-how-is-an-airport-building-evacuation-carried-on/
10. Van de Leur P (2005) Building evacuation. Rules and Reality. HERON 50:237–246
11. Jafer S, Lawler R (2016) Emergency crowd evacuation modeling and simulation framework with cellular discrete event systems. Trans Soc Model Simul Int 92(8):795–817
12. Lovreglio R, Kuligowski E, Gwynne S, Boyce K (2019) A pre-evacuation database for use in egress simulations. Fire Saf J 105:107–128
13. Ng C MY, Chow WK (2006) A brief review on the time line concept in evacuation. Int J Archit Sci 7(1):1–13
14. Demazeau Y (1995) From interactions to collective behavior in agent-based systems. In: European conference on cognitive science, Saint-Malo France
15. Han L, Guo H, Zhang H, Kong Q, Zhang A, Gong C (2020) An ecient staged evacuation planning algorithm applied to multi-exit buildings. Int J Geo-Info ISPRS 9:46. https://doi.org/10.3390/ijgi9010046
16. https://danangairportterminal.vn/vi/maps/
17. Michelle O'Neill (2005) Guide for evaluating the predictive capabilities of computer egress models. National Institute of Standards and Technology
18. Kuligowski ED, Peacock RD, Reneke PA et al (2015) Movement on stairs during building evacuations. NIST Technical Note. https://doi.org/10.6028/NIST.TN.1839
19. https://www.fireengineering.com/firefighting/survivability-profiling-how-long-can-victims-survive-in-a-fire-2/#gref

AI Powered Tomato Leaf Disease Detection and Diagnostic

Duc-Huy Pham, Thi-Thu-Hien Pham, Ngoc-Bich Le, and Thanh-Hai Le

Abstract In this study, the common diseases on tomato plants including Bacterial Spot, Yellow Leaf Curl, Late Blight along with healthy leaves were identified in the conditions of greenhouse environment and simultaneous leaves images. Image-based artificial intelligence (AI) detection and classification were used. Specifically, the YOLO recognition model was nominated and conducted. The processing procedure was divided into two phases. The first phase identifies and separates each tomato leaf on the plant and the second phase recognizes diseases on the tomato leaf surface that was isolated from the previous model. For the leaf recognition process, the majority of leaves were identified and separated for observable leaves. For the process of disease detection, the disease detection model achieved the best results when identifying Healthy leaves with an F1-score of 0.97, followed by Yellow Leaf Curl disease of 0.93. The feasibility of the proposed models was demonstrated by conducting the actual experiments with a two-class identification model including the disease class (Disease) and the non-disease (Healthy) class. The model results correctly identified 88% Disease and 86% Healthy. This moderate result is due to the inability to actively induce the desired disease in the tomato plant to create an adequate abundant dataset.

D.-H. Pham · T.-H. Le (✉)
Department of Mechatronics, Faculty of Mechanical Engineering, Ho Chi MinhCity University of Technology (HCMUT), Ho Chi Minh City, Vietnam
e-mail: lthai@hcmut.edu.vn

D.-H. Pham
e-mail: huy.phamduc2999@hcmut.edu.vn

T.-T.-H. Pham · N.-B. Le (✉)
School of Biomedical Engineering, International University, Ho Chi Minh City, Vietnam
e-mail: lnbich@hcmiu.edu.vn

T.-T.-H. Pham
e-mail: ptthien@hcmiu.edu.vn

D.-H. Pham · T.-T.-H. Pham · N.-B. Le · T.-H. Le
Vietnam National University Ho Chi Minh City, Linh Trung Ward, Thu Duc City, Ho Chi Minh City, Vietnam

© The Author(s), under exclusive license to Springer Nature Switzerland AG 2022 627
N. H. T. Dang et al. (eds.), *Artificial Intelligence in Data and Big Data Processing*,
Lecture Notes on Data Engineering and Communications Technologies 124,
https://doi.org/10.1007/978-3-030-97610-1_50

Keywords Digital image processing · Computer vision · AI in agriculture · Tomato leaf disease detection · YOLO

1 Introduction

Digital transformation in agriculture has been creating remarkable successes in many aspects of agriculture including the detection and diagnosis of many plant diseases. The growing demand for food along with increasingly strict requirements for the quality of agricultural products has prompted countries to shift their focus to automation in the agricultural sector. In agricultural problems, early detection of plant diseases is very important affecting quality and yield. The quality of fruits, vegetables, and rice is affected by plant diseases. The loss of yield and quality due to diseases will therefore have an economic impact. Therefore, fast and effective detection and early diagnosis of plant diseases are crucially essential.

The robust development and effectiveness of artificial intelligence (AI) in recent times promises to be a great tool for solving problems in agriculture, including applications for the diagnosis and detection of plant diseases. Mohanty et al. [1] are one of the first groups who applied deep learning to detect plant diseases with images. In their study, 14 different crops and more than 25 diseases were classified with 99.35% accuracy. Followed by a series of studies applying deep learning to disease detection for plants [2–5]. More recent, Kaur et al. [6] deployed machine learning models to improve the process of plant disease detection in early stages to improve grain security. In another attempt, Pooja et al. [7] utilized Convolutional Neural Networks (CNNs) techniques to detect diseased rice plant leaves. CNNs model was applied and trained to classify common rice diseases. The accuracy of 95% was attained in this attempt.

Tomato is one of the major plants in agriculture. Consequently, its diseases detection catches a lot of AI researchers. Specifically, in [8], Faster Region-based Convolutional Neural Network (Faster R-CNN), Region-based Fully Convolutional Network (R-FCN), and Single Shot Multibox Detector (SSD) were considered and evaluated. Nine different types of tomato diseases and pests were effectively recognized. Benoso et al. [9] applied K-NN, ANN, SVM, Random Forest, and Naïve Bayes to detect three kinds of disease including late blight, mosaic virus, and sheet by Septoria. The results demonstrate the best performance of Random Forest. Zhang et al. [10] proposed an improved faster RCNN to detect healthy tomato leaves and four diseases including powdery mildew, blight, leaf mold fungus, and ToMV. The proposed model worked very well especially in complex scenarios from a plant's surrounding area.

It is obvious in the above studies on detecting tomato leaf diseases, there are many types of diseases and each study focuses on certain diseases. On the other hand, the application environment also plays an important role in the performance of identity models. In this study, our aim is to discover and demonstrate a solution to effectively detect common tomato diseases including Bacterial Spot, Yellow Leaf Curl, Late Blight in a greenhouse environment and using the mobile robot's camera.

These requirements require the ability of the algorithm to work well in the natural environment and identify diseases on images with simultaneous leaves. In addition, light, fast, and efficient are also important requirements. In this study, we only selected 3 common diseases for the purpose of verifying the proposed solution. With satisfied results achieved, the extension to other diseases will be carried out in further studies.

2 Methodology

2.1 Tomato Leaf Diseases

Plant diseases, in general, are the consequences of bacteria, fungus, nematodes, viruses, pests, weeds, insects, photo plasma and other pathogens affects. Diseases' signs and symptoms comprise ooze, cottony mass, or a blank mass on the plant [6]. Regarding tomato leaf diseases, the common diseases on tomato leaves depicted in Fig. 1 include (1) Bacterial Spot: This disease is caused by Xanthomonas vesicatoria bacteria. Symptoms include many small spots. The spots may have a yellow halo. The leaf center is dry and often torn; (2) Early Blight: caused by Alternaria linariae

Bacterial Spot   Early Blight   Yellow Leaf Curl

Late Blight   Septoria Leaf Spot   Leaf Mold

Fig. 1 Tomato leaf diseases [11]

and A.solani fungi. Symptoms of the disease are brown spots appearing on mature leaves; (3) Late Blight: caused by the Phytophthora infestans fungus. Symptoms appear on young leaves in the form of black spots. Appearance of white mold at the edge of the diseased area on the underside of leaves; (4) Leaf damage: caused by the fungus Septoria lycopersici. Symptoms are tiny black spots. In case of severe disease, leaves turn yellow, die, and fall off the tree; (5) Tomato Yellow Leaf Curl: caused by yellow leaf virus. Symptoms in diseased plants are curled leaves, yellow edges (withered), leaves are smaller than normal; (6) Leaf Mold: occurs when the humidity inside the greenhouse is high. Symptoms of the disease are the appearance of yellow spots on the upper surface of the leaves. Scattered olive-green spores appear on the underside of diseased leaves.

2.2   Machine Learning Models for Leaf Diseases Detection

YOLO (You Only Look One) [12]: is a family of object recognition models that are quite popular today. In terms of accuracy, YOLO may not be the best algorithm, but it is the fastest in the class of object detection models.

Different from R-CNN [8], the model works by first dividing the input image into a grid of cells, where each cell is responsible for predicting bounding boxes and the confidence of containing objects inside. Then, a class probability map with the confidences is combined into a final set of bounding boxes and labels.

In this study, YOLOv5 was nominated due to its ability to process at high speed, does not require large data training resources, can be trained on Google Colab when personal computers are not qualified to perform.

2.3   Data Set

With the goal of identifying diseases on tomato leaves, we performed a two-step treatment process. First, identify the leaves on the tomato plant and separate each leaf into a single image. Then, from those pictures, the disease was identified on the leaves. To do this, trained two datasets were trained including: (1) The first model was trained for leaf recognition. The data set was collected from online sources, from diseased and non-diseased leaf data, from the referenced document [11], and from the self-taken data. Then label each with "leaf"; (2) The second sample was trained to recognize leaf diseases, the data set was taken from the reference source [11]. In this study, three diseases were identified as Bacterial Spots, Yellow Leaf Curl and Late Blight, labeled as shown in Fig. 2. From [11], the data set was selected a total of 1600 images with 400 images for each disease.

Fig. 2 Trained classes image labeling

2.4 Data Preprocessing and Feature Extraction

Data preprocessing

Preprocessing of the trained data is a very important step before it is applied to the training model. For the purpose of diversifying and enhancing the training data, making the object recognition faster and more accurate in the future, here the data is enhanced by: flipping the image vertically and horizontally, rotating the image 90°, and utilizing the Mosaic Data Enhancement algorithm of YOLO model (as shown in Fig. 3).

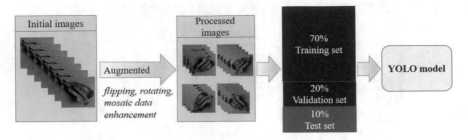

Fig. 3 Data processing steps

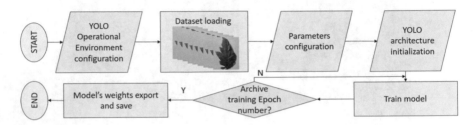

Fig. 4 Training steps

Data training steps

The object recognition training was conducted on Google Colaboratory with the steps as shown in Fig. 4. The training sample was divided into three parts: Train, Valid and Test with the ratio of 70, 20, and 10%, respectively.

3 Recognition Models

3.1 Tomato Leaves Recognition Model

Figure 5 is the result of the leaf recognition model after training with the following numbers of samples: (1) 1000 samples: The model is capable of identifying and zoning leaves, but the detection level and reliability are not high. The bounding box is wider compared to the leaves and there are many errors. Therefore, it is necessary to take more samples on the actual tree for training; (2) 1500 samples: The model recognized more leaves and with greater accuracy. However, the number of detected leaves is still not much, the leaves that are hidden or stacked are still not recognized or misidentified; (3) 2000 samples: The model detected the majority of leaves with visible surface.

| 1000 samples | 1500 samples | 2000 samples |

Fig. 5 Tomato leaves recognition results with different sample quantity

3.2 Diseased Leaf Detection Model

The disease recognition model was trained with the classification sample grouped into three diseased classes and one non-diseased class named Bacterial Spots, Yellow Leaf Curl, Late Blight and Healthy. A total of 160 samples, equally divided among four classes, (i.e. 40 samples for each class) were used for the trained model to evaluate its accuracy.

To assess the percentage of data that are correctly classified, how each particular disease is classified, which class is most correctly classified, and which class data is often misclassified into another class. a confusion matrix has been applied. Since the number of classes is four, the size of this matrix is 4 × 4. The confusion matrix of the model when classifying diseases with the above 160 test samples is shown in Fig. 6.

Figure 7 presents the results of the healthy leaf recognition model on the test samples. The model has correctly identified and correctly labeled 100% healthy leaf samples as shown by the confusion matrix. As shown in Fig. 7, it can be seen that 85% of the results have a confidence score above 0.9. Some samples have lower confidence scores due to a number of factors such as branches with more than one leaf (H36, H37, H38, and H40), leaf regrowth causing leaf margins to curl (H6 and H35).

Figure 8 shows the results of detecting Bacterial Spots Disease. The results showed that the model recognized and correctly labeled 100% of the test samples of Bacterial Spots, similar to the results of the confusion matrix. The proportion of samples with a confidence score over 0.9 accounts for 92.5%. For some leaf samples with complex

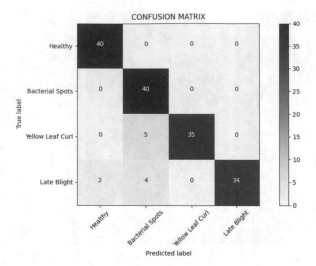

Fig. 6 Confusion matrix of 160 samples recognition model

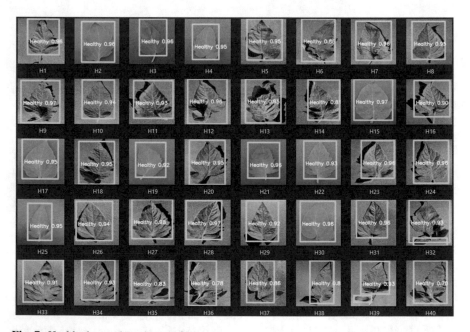

Fig. 7 Healthy leaves detecting results

deformation as shown in Bs16 and Bs40 samples, the confidence score is lower than 0.8 because the model has not been trained much on this data type.

For Yellow Leaf Curl Disease, based on the results in the confusion matrix and Fig. 9, the rate of correct identification and labeling was 87.5%. In which, five samples

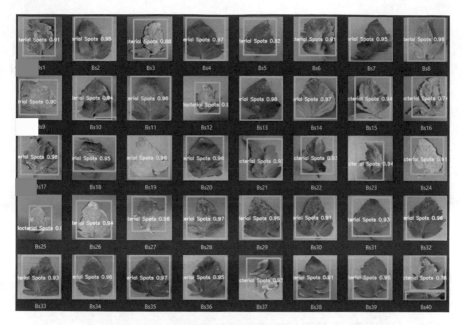

Fig. 8 Bacterial Spots diseased leaves detecting results

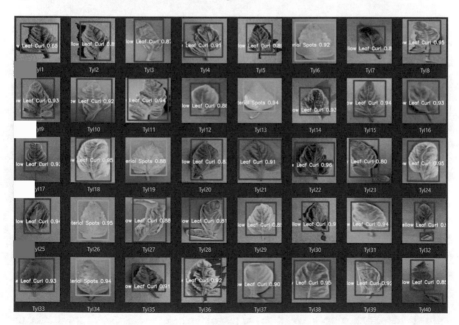

Fig. 9 Yelow leaf curl diseased leaves detecting results

Fig. 10 Late blight diseased leaves detecting results

including Tyl6, Tyl13, Tyl19, Tyl26, and Tyl34 were falsely detected as Bacterial
Spots disease. The reason for this misdetection is that the disease morphology on
these leaves is quite similar to that of Bacterial Spots. There are 66% of correctly
recognized samples with a confidence score of more than 0.9.

For Late Blight Disease, the model obtained 85% of correct identification and
labeling. In Fig. 10, four samples including Lb6, Lb11, Lb22 and Lb34 were wrongly
identified as Bacterial Spots and two samples Lb12 and Lb27 as Healthy. Because
this is a disease that causes leaves to be deformed in many different ways, many
results have confidence scores below 0.8 such as Lb2, Lb14, …, even less than 0.7
such as Lb9 and Lb32.

Table 1 shows the model evaluation parameters including Precision (P), Recall
(R), and F1-score (F1) when identifying the above four classes in which TP (True
Positive) is the correctly classified class, FP (False Positive) is the number of another
class erroneously classified as true class, and FN (False Negative) is the number
of true class which is incorrectly classified as another class. These parameters are
calculated based on the results of the confusion matrix in Fig. 6.

As shown in Table 1, the F1-score of the Healthy class reached the highest value
of 0.97 among the remaining three classes. This shows that the model has a better
ability to identify healthy leaves than the rest of the disease classes. Yellow Leaf Curl
disease is the disease with the best recognition model in the disease classes with an
F1-score of 0.93. Followed by Late Blight with an F1-score of 0.92. With the lowest
F1-score of 0.90, Bacterial Spots is the class that the model recognizes as the worst
out of the four classes. The reason for this result is that there are quite a few leaves

Table 1 Precision (P), Recall (R), and F1-score (F1)

Parameter	Equation	Healthy	Bacterial spots	Leaf curl	Late blight
Precision (P): is a measure of how many of the positive predictions made are correct (true positives)	$\frac{TP}{TP+FP}$ (1)	0.95	0.82	1	1
Recall (R): is a measure of how many of the positive cases the classifier correctly predicted, over all the positive cases in the data	$\frac{TP}{TP+FN}$ (2)	1	1	0.88	0.85
F1-score (F1): is a measure combining both precision and recall requiring both to have a higher value for the F1-score value to rise	$\frac{2\times P\times R}{P+R}$ (3)	0.97	0.90	0.93	0.92

that are mistakenly identified for this disease, causing the False Positive of this class to be high, leading to a low Precision value (0.82).

Since the importance of identifying each class is the same, the model's F1-score average was calculated utilizing Macro–Average F1-score:

$$Macro\ F_1 = \frac{1}{N}\sum_{i=1}^{N} F_{1i} = \frac{0.97 + 0.90 + 0.93 + 0.92}{4} = 0.93 \qquad (4)$$

4 Result and Discussion

Real tomato's diseased leaf detection experiment.

It is unmanageable to actively create the desired diseases on actual tomato plants. Therefore, we only evaluated the diseased leaves by training the recognition model according to the following real leaf samples as shown in Fig. 11.

The trained model was tested on a sample of 70 leaf images that were identified and segmented from the tomato plant images and evaluated based on a 2 × 2 confusion matrix (Fig. 12). Diseased leaves are called Positive, Healthy leaves are called Negative. Consequently, (1) True Positive (TP): The number of diseased leaves predicted to be diseased; (2) False Positive (FP): The number of healthy leaves predicted to be diseased; (3) True Negative (TN): The number of healthy leaves predicted to be healthy; (4) False Negative (FN): The number of diseased leaves that are predicted to be healthy.

Diseased leafs   Healthy leafs

Fig. 11 Leaf samples

Fig. 12 Confusion matrix of
70 samples test

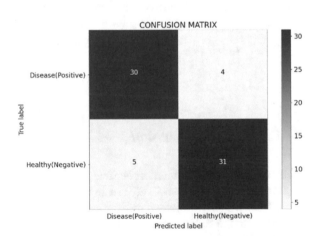

Because this is a classification problem with two data classes, the model has been evaluated through the parameters of True Positive Rate, False Negative Rate, False Positive Rate, and True Negative Rate. These parameters can be calculated based on the confusion matrix results in Fig. 12 as shown in Table 2.

The evaluation results in Table 2 show that the model correctly identified 88% of diseased leaves and 86% of healthy leaves. The FPR parameter is the false-positive rate, which is the percentage of healthy leaves that the model predicts to be diseased.

Table 2 Evaluated parameters of model's real disease detection

True Positive Rate: $TPR = \frac{TP}{TP+FN} = \frac{30}{30+4} = 0.88$	False Negative Rate: $FNR = 1 - TPR = 0.12$
False Positive Rate: $FPR = \frac{FP}{TN+FP} = \frac{5}{31+5} = 0.14$	True Negative Rate: $TNR = 1 - FPR = 0.86$

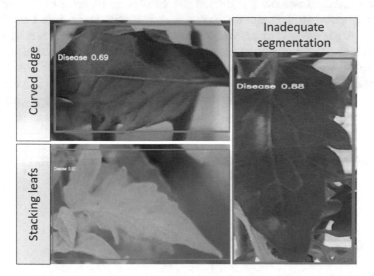

Fig. 13 Recognition errors

FPR has a fairly high value of 14%. However, for the problem of disease identification, this value is acceptable. This error arises due to the following factors (Fig. 13): (1) Healthy leaves often have a curved edge when growing because the leaves are quite large and long; (2) Leaves to identify the disease are overlapped with many other leaves; (3) The leaf recognition model cannot completely separate the leaves, causing the leaves to be defective. The FNR parameter is a false negative rate, which is the percentage of diseased leaves that the model predicts to be healthy. FNR value is slightly high with 12% because disease samples collected from actual trees are not abundant. Because we cannot actively create diseases for plants.

5 Conclusion

In this study, the YOLO recognition model was conducted for AI application in detecting diseases on tomato leaves. The processing is divided into two phases. The first phase used the model to identify and separate each tomato leaf on the plant and the second phase was the disease recognition model on the tomato leaf surface that was isolated from the previous model. For the leaf recognition process, the model was trained with 1600 different leaf samples. The majority of leaves were identified and separated for observable leaves. For the process of disease detection, experiments were performed on 160 samples evenly divided into four classes. The disease detection model achieved the best results when identifying Healthy leaves (i.e. Healthy class) with an F1-score of 0.97, Yellow Leaf Curl disease was the best recognized disease among diseased classes with an F1-score of 0.93, followed by Late Blight with 0.92 and the lowest was Bacterial Spots with 0.90. Finally, the actual experimental process with a two-class identification model including the disease class (Disease) and the non-disease (Healthy) class was conducted on 70 test samples. The model results correctly identified 88% Disease and 86% Healthy. The rate of mistakenly identifying healthy leaves as Disease is 14% and the rate of missing disease, mistaking Disease for Healthy is 12%. This moderate result is due to the inability to actively induce the desired disease in the tomato plant to create an adequate abundant dataset.

Acknowledgements This research is funded by Vietnam National University Ho Chi Minh City (VNU-HCM) under grant number DS2020-28–02.

References

1. Mohanty SP, Hughes DP, Salathé M (2016) Using deep learning for image-based plant disease detection. Front Plant Sci 7:1419
2. Ferentinos KP (2018) Deep learning models for plant disease detection and diagnosis. Comput Electron Agric 145:311–318
3. Sladojevic S, Arsenovic M, Anderla A, Culibrk D, Stefanovic D (2016) Deep neural networks based recognition of plant diseases by leaf image classification. Comput Intell Neurosci 3289801:11
4. Wang G, Sun Y, Wang J (2017) Automatic image-based plant disease severity estimation using deep learning. Comput Intell Neurosci 2917536:8
5. Ramcharan A, Baranowski K, McCloskey P, Ahmed B, Legg J, Hughes DP (2017) Deep learning for image-based cassava disease detection. Front Plant Sci 8:1852
6. Kaur H, Prashar D, Madhuri (2019) Applications of machine learning in plant disease detection. Think India J 22(17)
7. Pooja K, Madhuri S, Mahalaxmi IH, Mamatha S, Keerthi R (2020) Rice plant disease detection. Int J Res Appl Sci Eng Technol 8(VI). https://doi.org/10.22214/ijraset.2020.6033
8. Fuentes A, Yoon S, Kim S, Park D (2017) A robust deep-learning-based detector for real-time tomato plant diseases and pests recognition. Sensors 17(9):2022
9. Benoso BL, Perales JCM, Galicia JC (2020) Tomato disease detection employing pattern recognition. Int J Comput Opt 7(1):35–45
10. Zhang Y, Song C, Zhang D (2020) Deep learning-based object detection improvement for tomato disease. IEEE Access 8:56607–56614. https://doi.org/10.1109/ACCESS.2020.2982456
11. Huang M-L, Chang Y-H (2020) Dataset of tomato leaves. https://data.mendeley.com/datasets/ngdgg79rzb/1
12. Redmon J, Divvala S, Girshick R, Farhadi A (2016) You only look once: unified, real-time object detection. In: 2016 IEEE conference on computer vision and pattern recognition (CVPR), pp 779–788. https://doi.org/10.1109/CVPR.2016.91

Monitoring Employees Entering and Leaving the Office with Deep Learning Algorithms

Viet Tran Hoang, Khoi Tran Minh, Nghia Dang Hieu,
and Viet Nguyen Hoang

Abstract This study attempts to create a system to monitor employees entering and leaving the office using face recognition. In addition, the system also signals by LED when recognizing a staff who has clearance to enter or notifies those who do not in the area. Events of entering and leaving from staff are written into a log file for management purposes. The face-detection and image processing utilize Multi-task Cascaded Convolutional Network. Feature data is then extracted from the processed images by FaceNet, which is classified by the Support Vector Machine algorithm into a model. Information of employees and logs are saved in MySQL database, which is also used in a web application using Python and Django web framework.

Keywords Face · Face recognition · Feature extraction · Training · Learning (artificial intelligence) · Face detection · Face alignment · Detectors · Convolutional neural nets · Cascaded convolutional neural network · Databases · Cameras · Real-time systems · Facenet · Python · Support vector machine · Mysql · Django · Raspberry pi

1 Introduction

As a part of facial image processing applications, face recognition systems' significance as a research area are increasing recently. They are usually applied and preferred for people and security cameras in modern life. These systems can be

V. T. Hoang (✉) · K. T. Minh · N. D. Hieu · V. N. Hoang
Can Tho University, Can Tho, Vietnam
e-mail: thviet@ctu.edu.vn

K. T. Minh
e-mail: khoim3721005@gstudent.ctu.edu.vn

N. D. Hieu
e-mail: dhnghia@ctu.edu.vn

V. N. Hoang
e-mail: nhviet@ctu.edu.vn

© The Author(s), under exclusive license to Springer Nature Switzerland AG 2022   641
N. H. T. Dang et al. (eds.), *Artificial Intelligence in Data and Big Data Processing*,
Lecture Notes on Data Engineering and Communications Technologies 124,
https://doi.org/10.1007/978-3-030-97610-1_51

used for person verification, video surveillance, crime prevention, and other similar security activities.

Nowadays, with the non-stop growing quality of hardware like cameras, CPUs, or especially GPUs, the computing time for training process in machine learning, has reduced significantly. Face recognition, which is a case of machine learning, uses some algorithms to extract features from the input face, converting them into mathematical vectors to classify them.

Face recognition system is a complicated image-processing problem with many affecting factors like pose variations, angle, facial expression, illumination, occlusion, imaging conditions, and time delay (for recognition). It is a combination of face detection and recognition applications in image analysis. Detection techniques are used to find the position of the faces in a given image, while recognition techniques are used to classify said faces base on extracted facial feature components.

There are many methods to develop a face recognition application with many training algorithms and models. The applications related to this technology haven't been very popular in our city yet. Therefore, more attention should be spent on learning and practicing face recognition technology. Thus, in this work we developed a system utilizing the technology to apply in practice.

In the rest of the introduction we briefly review the related work and the main topics necessary to understand our system discussed in Sect. 2. Section 3 details the results of our system before drawing some conclusions and discussing further directions in Sect. 4.

1.1 Related Work

An attempt to tackle the task of checking attendance automatically has been proposed in [1], where the idea is to use face recognition technique, with Eigenface values, Principle Component Analysis (PCA), and Convolutional Neural Network (CNN) to implement an automated attendance management system for students of a class.

Multiple approaches have been proposed in order to overcome some of the critical challenges of face recognition under difficult conditions. A system for real-time video-based face recognition has been introduced in [2], which tackled three key challenges: computational complexity, in the wild recognition, and multi-person recognition.

Since face recognition in real-time has been a rapidly growing challenge, another work [3] has proposed the PCA facial recognition system, where the PCA is a statistical method under the broad heading of factor analysis, which aims to truncate the substantial quantity of data storage to the size of the feature space that is required to represent the data economically. However, differently from [1], this work's system has been built with OpenCV, Haar Cascade, Fisher Face, Local Binary Pattern Histogram (LBPH), and Python.

According to studies mentioned in work [4], deep learning approaches, specifically convolutional neural networks (CNNs), can achieve remarkable progresses in a variety of computer vision tasks, including face detection and alignment. Other proposed face detectors at the time, such as one that utilizes Haar-like features by Viola and Jones [5] to train cascaded classifiers, may degrade significantly in real-world application [6–8], while deformable part models (DPM) usually need high computational expense [9–11]. Yang et al. [12] train deep CNNs to obtain high response in face regions, but this approach can be time costly in practice due to its complex CNN structure.

To overcome these challenges, [4] has introduced a multi-task Cascaded Convolutional Network (MTCNN) based framework for joint face detection and alignment, which achieves superior accuracy across several challenging benchmarks while keeping real time performance. Inspired by the good performance of this framework in practice, we adopt the same approach with MTCNN for our image processing task.

After taking notes from the mentioned works and considering among many new algorithms as well as technologies, we introduce our own approach to develop a system in this work, using MTCNN and FaceNet. The major contributions of this paper are summarized as follows: (1) We propose a new face detection and recognition system utilizing MTCNN and FaceNet that will be used to monitor employees entering and leaving the office. (2) In addition to the face detection/recognition modules, we connect them with a database to store data, and a web application to present those data for users. (3) We also successfully send out a signal to an LED according to the system's recognition result. In practice, this signal can be used to announce verbally with text-to-speech technology, or automatically open any door that uses magnetic lock.

1.2 Convolutional Neural Network

A CNN is a Deep Learning algorithm that takes an input image, assigns parameters (weights and biases) to several aspects in it, and can recognize one from the rest. The pre-processing required in a CNN is lower than other classification algorithms. With enough training, CNNs have the ability to learn these characteristics.

The architecture of a CNN is analogous to that of the connectivity pattern of Neurons within the Human Brain and was inspired by biological processes in which the connectivity pattern between neurons imitates the organization of the animal visual cortex. Individual cortical neurons respond to stimuli only in a restricted region of the visual field called the Receptive Field. The receptive fields of various neurons partially overlap so that they cover the whole visual area [13].

Thanks to the reduction in the number of parameters involved and the reusability of weights, CNN's architecture performs a better fitting to the image dataset. In other words, the network can be trained to grasp the sophistication of the image better.

1.3 Multi-Task Cascaded Convolutional Networks

The MTCNN model consists of three separate convolutional networks: the P-Net, the R-Net, and the O-Net as shown in Fig. 1, where "MP" means "max pooling" and "Conv" means "convolution". The step size in convolution and pooling is 1 and 2, respectively.

For every image passed in, the network creates an image pyramid, to detect faces of all different sizes. In other words, it creates multiple different copies of the same image in different sizes to search for different sized faces within the image.

For each scaled copy, a 12×12 kernel will go through every part of the image, scanning for a face. It starts in the top left corner, a section of the image from (0,0) to (12,12). This portion of the image is passed to Proposal Network (P-Net), which returns the coordinates of a bounding box if it notices a face. Then, it will repeat that process after shifting the 12×12 kernel sideways (or downwards), and it continues doing that until it has gone through the entire image. How many pixels the kernel moves by every time is known as the stride.

Each kernel would be smaller relative to a large image, so it would be able to find smaller faces in the larger-scaled image. Similarly, the kernel would be bigger relative to a smaller image, so it would be able to find bigger faces in the smaller-scaled image.

With each of these 12×12 kernels, 3 convolutions are run through with 3×3 kernels. After every convolution layer, a Parametric Rectified Linear Unit (PReLU) layer is implemented (when you multiply every negative pixel with a certain number α, which is to be determined through training). In addition, a Maxpool layer is put in after the first PReLU layer (Maxpool takes out every other pixel, leaving only the largest one in the vicinity).

After the third convolution layer, the network splits into two layers. The activations from the third layer are passed to two separate convolution layers, and a Softmax layer after one of those convolution layers (Softmax assigns decimal probabilities to every result, which all add up to 1. In this case, it outputs two probabilities: the probability that there is a face in the area and the probability that there isn't a face).

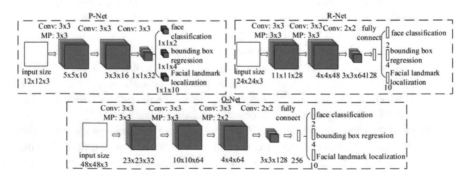

Fig. 1 MTCNN structure (*Source* [13])

Refinement Network (R-Net) has a similar structure, but with even more layers. It takes the P-Net bounding boxes as its inputs and refines its coordinates.

Similarly, R-Net splits into two layers in the end, giving out two outputs: the coordinates of the new, more accurate bounding boxes, as well as the machine's confidence level of each of these bounding boxes.

Finally, Output Network (O-Net) takes the R-Net bounding boxes as inputs and marks down the coordinates of facial landmarks.

O-Net splits into three layers in the end, giving out three different outputs: the coordinates of the bounding box, the coordinates of the five facial landmarks (locations of the eyes, nosc, and mouth), and the confidence level of each box.

1.4 FaceNet

FaceNet is a face recognition system introduced by Google in their paper [14] in June 2015. It is a system that, given a picture of a face, will extract high-quality features from the face and predict a 128-element vector representation these features, called a face embedding.

The model is a deep convolutional neural network trained via a triplet loss function that encourages vectors for the same identity to become more similar (smaller distance), whereas vectors for different identities are expected to become less similar (larger distance).The L2 distance (or Euclidian norm) between two faces embeddings directly corresponds to its similarity: faces of the same person have small distances and faces of distinct people have large distances, which is exactly like measuring the distance between two points in a line to know if they are close to each other. In an n-dimension Euclidean space, the L2 distance between two points can be expressed as

$$d_{L2}(x, y) = \sqrt{\sum_{i=1}^{n} (x_i - y_i)^2}$$

Once this space has been produced, tasks such as face recognition, verification, and clustering can be easily implemented using standard techniques with FaceNet embeddings as feature vectors. This method achieves high accuracy (99.65% on the LFW dataset) despite being old, some new researchers still use it, like MaskTheFace [15]. It is one of the best free facial recognition solutions in 2021 [16].

FaceNet system uses a deep convolutional network trained to directly optimize the embedding itself, instead of an intermediate bottleneck layer used by previous deep learning approaches, resulting in much higher efficiency. For implementation of FaceNet within this work, we use some code from [17], which uses Support Vector Machine (SVM) to classify the feature vectors.

1.5 Support Vector Machines

SVM is a class of machine learning algorithms, belongs to the area of supervised learning methods, which need labeled, known data to classify new unseen data. Using the idea of kernel substitution, it can deal with many tasks such as classification, regression, and novelty detection. Besides face recognition, SVM also has applications in handwritten characters recognition, text classification, bioinformatics, etc.

Its approach to classifying the data starts by trying to create a function that splits the data points into the corresponding labels with (a) the least possible amount of errors and (b) with the largest possible margin. This is so because larger empty spaces around the splitting function result in fewer errors because the labels are better distinguished from one another overall.

Figure 2 shows that a data set may be separable by multiple functions without any error. Therefore, the margin around a separation function is used as an additional parameter to evaluate the quality of the separation. In this case, the separation p_2 is the better one, since it distinguishes the two classes in a more precise manner.

Formally, SVM tries to find one or multiple optimal hyperplanes in an n-dimensional space. The first attempt in the process of splitting the data is to try to linearly separate the data into the corresponding labels. For example, for a task of predicting if a day is rainy or sunny uses a data set with n data points, where each data point consists of a label $y \in \{rainy, sunny\}$ and an attribute vector $\vec{x}$ containing the data values for the specific session. The SVM now tries to find a function that separates all the data points $(\vec{x}, y)$ with $y = rainy$ from all the data points $(\vec{x}, y)$ with $y = sunny$.

If the data is completely separable linearly, the separating function can be used to classify future events.

The data may not be well linearly separable or not linearly separable at all, which is often the case for real-world data. In the example given above, it can happen for example that two days have the exact same weather attributes such as temperature, humidity, etc. but it only rains on only one of them. This leads to inseparable data

Fig. 2 Visualization of an SVM

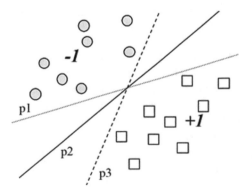

since the same attribute vector has different labels. This can be resolved by mapping the n-dimensional input data to a higher-dimensional space, where the data can be separated linearly.

In order to avoid overfitting in SVM, the data has to be preprocessed to identify noise and accept some misclassifications. Otherwise, the accuracy values of the SVM will be flawed and result in more erroneous classification for future events.

2 Work Details

In this section we present the design of our Monitoring employee entering and leaving the office system. As a first step, we need to identify what the users require from the system, which are the following tasks:

- Registration: new user can register to be recognized by the system in the future, this procedure must have the system taking photos of the user's face and letting the user input their personal information.
- Model training: the system processes and learns the face images from registered users to output a trained model that can recognize them.
- Face recognition and Door opening: the system can detect a person's face in front of the camera and recognize their identity, if they are a user who is authorized to enter, the system will automatically open the door and save the event to log. An announcement can be played by voice as an optional step.
- Web application: the system also provides a website for both users and admin to access, view information, and perform tasks from their own device, as long as it is connected to the same network as the server's.

With these requirements in mind, after discussing many possible approaches, we have decided to come up with the system design as Fig. 3 demonstrates. This system design consists of 4 main parts:

- **Training module.** This part handles new user registration, taking raw images, preprocessing them which outputs face images to train into the recognition model.
- **Recognition module.** Using the trained model, this part detects and recognizes faces from live-acquired images then performs accordingly
- **Database.** Storing personal data from users, as well as other necessary information for the Django web application to run.
- **Web application.** Providing users and admin a place to view their personal information and logs from the database, as well as performing some other tasks depending on permission.

A message sequence chart (MSC) is illustrated in Fig. 4 to show what messages will be exchanged between the four mentioned modules of the system.

In the rest of this section, we will go into detail about each part, explaining the components involved in them, how they operate together to reach their own goal, and the general workflow when the system is running.

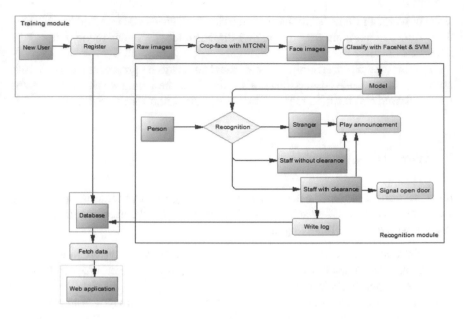

Fig. 3 System design

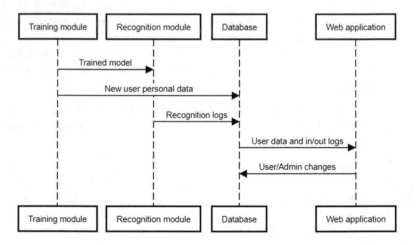

Fig. 4 System message sequence chart

2.1 Training Module

The training module, as shown in Fig. 5, is the first part of the system, which is usually installed in the administrator's office. Here new employees can register into the system by inputting their ID, name, and other required information, which are

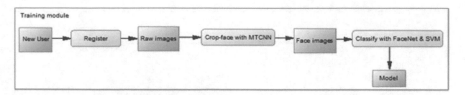

Fig. 5 Training module design

input into the database and used to create a class name for the classification model later.

The next step is to take raw face images from staff. By default, the system will take up to 50 images, one shot every 5 frames (by default), unless terminated prematurely by the administrator. During this process, the staff is advised to move their face around in front of the camera so it can take pictures from multiple angles. The 5 skipped frames by default is to give time for the staff to move their face, preventing multiple similar images. The more various face angles captured in the images, the better the data is. The images then will be saved into a created folder with a name generated from the staff's name and ID.

The administrator can repeat this registration step for multiple staff members. Until everyone is registered, the image preprocessing and model training can begin, using the technologies that are discussed in the previous chapter. This results in a model, which is uploaded to the server.

In the training process, however, with new data and a previously trained model, instead of having to retrain the entire dataset over again, a tweak was made to the original SVM to reuse the classified data in the model so it can reduce the training time for the new data. An option to retrain the entire model is still available.

2.2 Recognition Module

The module shown in Fig. 6 is installed on every door of the system. It consists of the following components: an LED, a camera, and a speaker, all connected to a Raspberry Pi, which has to connect to the same network as the server's.

When the system is running, the camera will open and look for faces in front of it. The live acquired images are constantly streamed to the server, where the face detection and recognition computations take place. This process uses the model which is the result of the training module mentioned in the previous section. Once at least a face is detected in the frame, if the person is a staff and has clearance to the area, a log is automatically written into the database with the staff's identity, current timestamp, in/out status, and door's identity.

At the same time, the log's data will also be written down into a local text file, which will be read by the API view, which returns the data in form of JSON. After the JSON data is returned, the local text file will be reset. The Raspberry Pi is

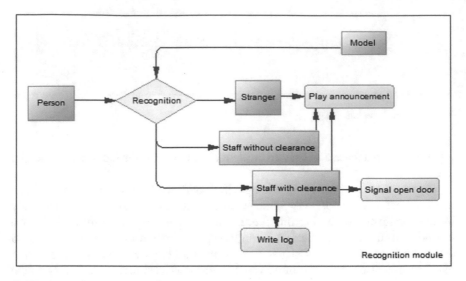

Fig. 6 Recognition module design

programmed to constantly access this API view, if it reads a log data, the speaker will announce their name and the LED will be lit if they have clearance to enter.

2.3 Database

In order to support running the system on multiple doors at once, besides the two tables Staff and Log to save data mentioned previously, a table Gate is also created to store every gate's name, IP address, and voice option.

In addition to that, several tables are also generated automatically by the Django framework to support its web application, which will be discussed later.

In total, our system has 13 tables and 12 relational references. MySQL is used to store all the data, and the database's design can be seen in Fig. 7.

2.4 Web Application

Once the previous three modules work, a requirement from the admin to access the database for management and administrative tasks has arisen, as well as staff users also want access the database to view their own entering/leaving logs and personal information. To solve this requirement, a web application is created so that admin and users can access from their device's browser directly without downloading any extra applications.

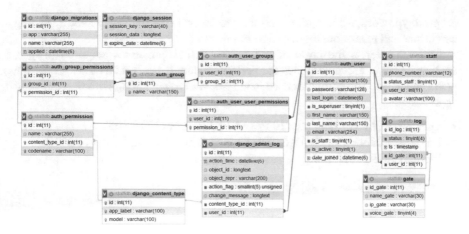

Fig. 7 System database diagram

When a staff registers into the system (the process in Sect. 3.1), an account of this web application is also created with their ID and a default password, which can be changed later after login. An admin account was already created when developing the application, extra admin accounts can be created if the need arises.

Because Django's authentication is used, the Staff table described in Fig. 6 has a one-to-one relationship with Django's generated Auth_user table. Its purpose is to extend Django's default User model so more related information like phone number, clearance, and avatar can be stored.

By fetching data from the database described in Sect. 2.3, the web application can render and display plenty of useful information as well as providing the user with some functions they can use.

For admin, this application allows them to manage all users and gates, including create, update, view, delete. They can also view and delete log records from everyone, but not editing them. New admin accounts can also be added.

For users, this application allows them to view and edit their personal information except for their ID, change their account password, and only view their own log records. They can also upload an avatar that will be displayed when they log in, the image will be saved in folder *media*.

3 Results

Following the design as discussed in Sect. 2, with the technology and knowledge introduced in Sect. 1, we have succeeded in developing and implementing a complete hardware and software system in Cantho University Software Center. The developed system has been tested for many live acquired images and the results are satisfactory for prototype work, with requirements that are set out in the starting of Sect. 2. Testing conditions and results are discussed in this section.

The computer is the brain of the system, which processes acquired images, analyzes images, and determines the person's name. It also acts as a server for the web application. The computer used in the test is a typical laptop with the following specifications:

- Operating System: Windows 10 Pro 64-bit (10.0, Build 18,363)
- Processor: Intel® Core™ i7-6500U CPU @ 2.50 GHz (4CPUs), ~ 2.6 GHz
- Memory: 8192 MB RAM

3.1 Training Module

The registration procedure starts by letting the user enter their name, id, and other information. This data is saved into the database and used to create a class name for the classification model later.

The next step is taking raw images. By default, the system takes 50 raw images in size 640 × 480, which are saved in folder *DataSet\FaceData\raw\ < fullname >_ < id >* with *< fullname >* and *< id >* taken from user input. For each staff, this step takes about 1 min at max, which can be quickened by reducing the number of frames it skips in between shots (as explained in Sect. 2.1).

This registration step can happen at any point of time and multiple times, even when the system is running live since it only uploads information from user to database and saves taken raw images in the server's folder.

Once all the staffs' raw images are taken, the admin can decide to process the images and train the model at any time, preferably when the system is not running.

Image processing is handled by MTCNN, it detects the face area and crops it, resizing to 160 × 160 and save in folder *DataSet\FaceData\processed\ < fullname >_ < id >* .

In order to demonstrate the effectiveness of our proposed system, we have conducted an experiment between two methods, i.e. Histogram of Oriented Gradients (HOG) and MTCNN for the image processing step. The simulation is performed on the same laptop, with specifications we have mentioned at the start of Sect. 3 above. The testing dataset for both methods consists of 25 20-s video clips, each contains only one different person facing the screen with changes in various poses and illuminations. The comparison result is illustrated in the following graphs in term of detection rate and time.

Each video frame from the dataset is proceeded into a single image, which is then passed through HOG and MTCNN for face-detecting task. If one method recognizes a face in the image, the corresponding frame will be marked as detected by said method. As shown in Fig. 8, MTCNN (orange line) performs the task slightly better overall with an average of 436 detected frames throughout 25 videos, comparing to HOG (blue line) with an average of only 395 detected frames. Having too.

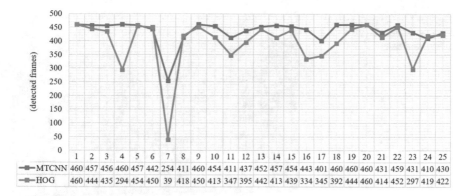

Fig. 8 Detection rate graph and data table

	1	2	3	4	5	6	7	8	9	10	11	12	13	14	15	16	17	18	19	20	21	22	23	24	25
MTCNN	460	457	456	460	457	442	254	411	460	454	411	437	452	457	454	443	401	460	460	460	431	459	431	410	430
HOG	460	444	435	294	454	450	39	418	450	413	347	395	442	413	439	334	345	392	444	460	414	452	297	419	422

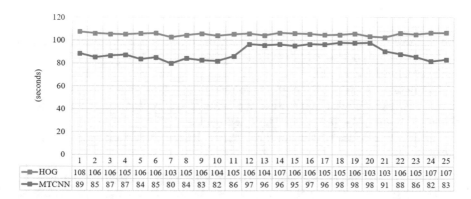

Fig. 9 Detection time graph and data table

	1	2	3	4	5	6	7	8	9	10	11	12	13	14	15	16	17	18	19	20	21	22	23	24	25
HOG	108	106	106	105	106	106	103	105	106	104	105	106	104	107	106	106	105	105	106	103	103	106	105	107	107
MTCNN	89	85	87	87	84	85	80	84	83	82	86	97	96	96	95	97	96	98	98	98	91	88	86	82	83

many lost frames in processing results in less data to train, which will negatively impact the quality of model after training.

In term of speed, in Fig. 9, MTCNN (orange line) also performs the face-detecting task faster in 89 s each video in average, whereas HOG (blue line) proceeds through the same dataset a bit slower in 105 s each video in average.

In conclusion, our experiment has demonstrated the effectiveness of MTCNN in our proposed system when comparing to another method like HOG. It performs not only more precisely but also faster as well. Its higher detection rate and speed ensure the image processing step provides good data (fewer lost frames) in a swift manner.

3.2 Recognition Module

The recognition procedure starts by loading the model and opening the camera(s). This process takes about 30 s to a minute, depending on the computer's CPU, GPU

Fig. 10 Live acquired image
from the webcam with a
recognized face

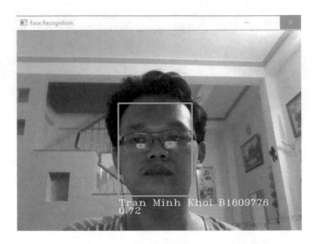

(unavailable in this testing computer), and memory. Once opened, the camera(s) will stream live acquired images back to the computer, which will be presented on screen in the separated window(s).

It only takes 2 s max for the system to recognize a face when it's on-screen, but it will only announce the result when the same face is recognized at least 5 times within 50 frames, in order to avoid false recognition.

Once the user's face is recognized like in Fig. 10, one of the following three cases happen:

- If they are staff and have clearance to enter, a log record is created with the user's ID, the gate they're seen at, current timestamp, and in/out status depending on the latest log record of the user in the database. Then, this new log record is then inserted into the database while the system writes its content down to a local file for the API view to read, which later will be accessed by the Raspberry Pi (as explained in Sect. 2.2)
- If they are staff but don't have clearance to enter, no log record is created while the system writes down the staff's ID and name to a local file for the API view.
- If they are strangers, no log record is created while the system writes down 'stranger' to the local file for the API view.

Depending on the three previous cases, the API view will read the content of the local file and the Raspberry Pi will act accordingly once accessing it:

- If they are staff and have clearance to enter, the Raspberry Pi will greet/say goodbye (according to the record's in/out status) to the staff's name and light up the LED light.
- If they are staff but don't have clearance to enter, the Raspberry Pi will greet the staff's name and tell them they are not allowed to enter the area.
- If they are strangers, the Raspberry Pi will greet them in general and tell them they are not allowed to enter the area.

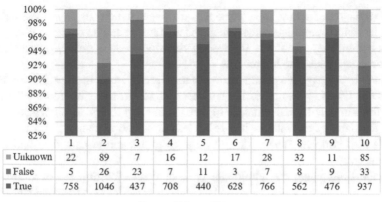

	1	2	3	4	5	6	7	8	9	10
■ Unknown	22	89	7	16	12	17	28	32	11	85
■ False	5	26	23	7	11	3	7	8	9	33
■ True	758	1046	437	708	440	628	766	562	476	937

■ True　■ False　■ Unknown

Fig. 11 Recognition accuracy

In order to evaluate our system's recognition accuracy, we have set up a testing simulation as follow. We collect face data of 10 different people, 50 images each to be trained by the system, this will fit the practical condition when implementing in real life environment, as you should not take too many images of each person, otherwise it can lead to high storage/time cost. After that, we show the system a large amount of images taken of the same 10 people so it can perform recognition task. As demonstrated in Fig. 11, the system reaches an average accuracy of 94% over 10 people, ranging from 89% up to 97% for each individual case.

3.3　Web Application

Our web application starts with a login form, and Fig. 12 shows the sitemap for the other pages one can access from there. For users, after logging in the web application, the first page user sees is their log records, with the timestamp sorted in descending order.

Fig. 12 Web application sitemap

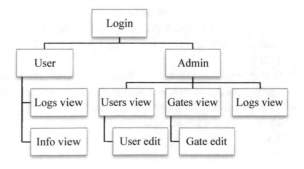

Then, they can access their personal info page, which displays the user's first name, last name, email, phone number, and avatar that can be changed if they want.

For admin, they can log in the web application to do a variety of management tasks on users, gates and log records.

On the Users page, the admin can view the list of current users with summarized details and filters. When clicking on a user, the admin can view more information about them in details, and edit each of them, including their account's password.

On the Gates page, since each Gate has less information than each User, its listing table is also simpler. When clicking on a gate, the admin can view and edit its information.

On the Logs page, the admin can view or delete (editing is not allowed) the available log records, and filter them base on many options on the right side.

4 Conclusions

The main theme of this study is to provide solutions to address some of the challenges that arise in developing and deploying a system to monitor employees entering and leaving the office using deep learning. In this section, we describe the main contributions and discuss some further research directions.

4.1 Summary of Contributions

After researching deep learning in general as well as face recognition in particular, referencing many other articles and examples that had tackled this problem before, we have managed to build a system to monitor employees entering and leaving the office. The system's benefits are described as following:

- With the currently complicated Covid-19 situation, practicing social distancing is crucial, with this system, employees can check attendance without directly interacting with other people or touching any surface.
- The system can take images continuously for multiple employees without interruption in between any two.
- The system can detect people's faces from a distance, ideally within three meters from the camera for better image quality, which will provide better recognition results.
- The system can perform tasks automatically like playing announcements accordingly, opening doors, and logging events without any further input.
- User can access their own log history from anywhere, as long as they are connected to the server's network, they can view when and where they entered or left the area.

- Admin can manage various information on the system, especially controlling the users' permission to enter the area and enabling a door's announcement.
- Admin can use the log information from the system to monitor staff's performance at work as a reference to evaluate their KPI.
- The logs that are saved in the database can be accessed at any point in the future if a situation requires them.

4.2 Future Directions

The system is designed, implemented, and tested. Test results show that the system has acceptable performance. On the other hand, the system has some future works for improvements and extended functionality.

Using another model implementation. Facenet using Keras instead of Tensorflow and Similarity learning supports training only new user data without retraining the entire dataset again or interrupting the system, which reduces the training time.

Extending to other field applications. This system can be applied to other applications like at parking lot where you can automatically check-in and out; or in the dormitory where you can monitor the students entering or leaving the area and some other management tasks.

Including body temperature check. By integrating this system with a thermal camera, it can be very useful in the current pandemic situation where it can check the person's body temperature in addition to their face before letting them enter.

Acknowledgements We would like to express our sincere gratitude to the College of Information and Communications Technology at Can Tho University for assisting us with techinical expertise, as well as Can Tho University Software Center for facilitating our work with equipment and testing environment. Without their support, carrying out this research would not have been possible.

References

1. Sawhney S, Kacker K, Jain S, Singh SN, Garg R (2019) Real-time smart attendance system using face recognition techniques. In: 2019 9th international conference on cloud computing, data science and engineering (Confluence), pp 522–525. https://doi.org/10.1109/CONFLU ENCE.2019.8776934
2. Arachchilage SW, Izquierdo E (2019) A framework for real-time face-recognition. In: 2019 IEEE visual communications and image processing (VCIP), pp 1–4. https://doi.org/10.1109/ VCIP47243.2019.8965805
3. Khan M, Chakraborty S, Astya R, Khepra S (2019) Face detection and recognition using openCV. In: 2019 international conference on computing, communication, and intelligent systems (ICCCIS), pp 116–119. https://doi.org/10.1109/ICCCIS48478.2019.8974493
4. Zhang K, Zhang Z, Li Z, Qiao Y (2016) Joint face detection and alignment using multitask cascaded convolutional networks. IEEE Signal Process Lett 23(10):1499–1503. https://doi.org/ 10.1109/LSP.2016.2603342,Oct

5. Viola P, Jones MJ (2004) Robust real-time face detection. Int J Comput Vision 57(2):137–154. https://doi.org/10.1023/B:VISI.0000013087.49260.fb
6. Yang B, Yan J, Lei Z, Li SZ (2014) Aggregate channel features for multi-view face detection. In: IEEE international joint conference on biometrics, pp 1–8. https://doi.org/10.1109/BTAS. 2014.6996284
7. Pham M, Gao Y, Hoang VD, Cham T (2010) Fast polygonal integration and its application in extending haar-like features to improve object detection. In: 2010 IEEE computer society conference on computer vision and pattern recognition, pp 942–949. https://doi.org/10.1109/ CVPR.2010.5540117
8. Zhu Q, Yeh M-C, Cheng K-T, Avidan S (2006) Fast human detection using a cascade of histograms of oriented gradients. In: 2006 IEEE computer society conference on computer vision and pattern recognition (CVPR'06), pp 1491–1498. https://doi.org/10.1109/CVPR.200 6.119
9. Mathias M, Benenson R, Pedersoli M, Van Gool L (2014) Face detection without bells and whistles. In: European conference on computer vision, Springer, Cham, pp 720–735. https:// doi.org/10.1007/978-3-319-10593-2_47
10. Yan J, Lei Z, Wen L, Li SZ (2014) The fastest deformable part model for object detection. In: 2014 IEEE conference on computer vision and pattern recognition, pp 2497–2504. https://doi. org/10.1109/CVPR.2014.320
11. Zhu X, Ramanan D (2012) Face detection, pose estimation, and landmark localization in the wild. In: 2012 IEEE conference on computer vision and pattern recognition, pp 2879–2886. https://doi.org/10.1109/CVPR.2012.6248014
12. Yang S, Luo P, Loy C, Tang X (2015) From facial parts responses to face detection: a deep learning approach. In: 2015 IEEE international conference on computer vision (ICCV), pp 3676–3684. https://doi.org/10.1109/ICCV.2015.419
13. Saha S (2018) A comprehensive guide to convolutional neural networks—the ELI5 way. 16 December 2018. [Online] Available: https://towardsdatascience.com/a-comprehensive-guide-to-convolutional-neural-networks-the-eli5-way-3bd2b1164a53. Accessed 2 January 2021
14. Schroff F, Kalenichenko D, Philbin J (2015) FaceNet: a unified embedding for face recognition and clustering. In: 2015 IEEE conference on computer vision and pattern recognition (CVPR), pp 815–823. https://doi.org/10.1109/CVPR.2015.7298682
15. Anwar A, Raychowdhury A (2020) Masked face recognition for secure authentication. arXiv preprint arXiv:2008.11104
16. Pospielov S (2021) What is the best facial recognition software to use in 2021?," 11 March 2021 [Online] Available: https://towardsdatascience.com/what-is-the-best-facial-recognition-software-to-use-in-2021-10f0fac51409. Accessed 1 December 2021
17. Sandberg D (2021) Face recognition using tensorflow," 17 April 2018. [Online]. Available: https://github.com/davidsandberg/facenet. Accessed 2 January 2021

Deep Learning for Rice Leaf Disease Detection in Smart Agriculture

Nguyen Thai-Nghe, Ngo Thanh Tri, and Nguyen Huu Hoa

Abstract Vietnam is a country that has the advantage in agriculture, especially rice. Rice is one of the primary food grains which provides sustenance to almost fifty percent of the world population and promotes a huge amount of employment. Hence, detection and prevention of diseases on rice are very important to help enhance rice production. This work proposes an approach for rice leaf disease detection on mobile devices using deep learning technique. Specifically, it proposes using EfficientNet which is a variant in deep learning networks for classification. This approach also utilizes the pre-trained model on imageNet for transfer learning. The proposed model can detect five types of images including three common types of diseases on rice leaf (e.g., brown spot, hispa, and leaf blast), healthy rice leaf, and other leaves. The model was trained on 1790 images and produced 95% validation accuracy. Finally, this model was converted to *tflite* format for running on mobiles or IoT devices. An Android application was built for rice leaf disease detection using the proposed model. When applying in practice on a mobile device, it took about 1.7 s for detecting and providing a treatment solution for a disease, thus this could be a useful solution to help the farmers.

Keywords Rice leaf disease detection · Smart agriculture · Deep learning · EffcientNet

N. Thai-Nghe (✉) · N. T. Tri · N. H. Hoa
Can Tho University, 3/2 street, Ninh Kieu District, Can Tho City, Vietnam
e-mail: ntnghe@cit.ctu.edu.vn

N. T. Tri
e-mail: trib1707000@student.ctu.edu.vn

N. H. Hoa
e-mail: nhhoa@ctu.edu.vn

1 Introduction

Rice is one of the world's most significant food. It is a staple meal for more than 4 billion people throughout the world, providing 27% of calories in low- and middle-income nations.[1] Vietnam has about 1.51 million hectares of rice producing about 11 rice tons.[2] Rice is one of the primary food grains which provides sustenance to almost fifty percent of the Vietnamese population and promotes huge amount of employment.

The climate changes and global warming have affected to agriculture field, especially the rice. Many rice diseases such as brown spot, hispa, and leaf blast cause reducing both quality and quantity of rice production and collection. For example, in Southeast Asia, the diseases cause more than 10% yield loss on average throughout entire lowland rice cultivation. A severely diseased field might lose up to 45% of its output.[3] Thus, how to apply advanced technologies to help the farmers is necessary.

This work proposes an approach for rice leaf disease detection on a mobile application using deep learning. Specifically, it proposes using EfficientNet [1], an approach in deep learning networks, for disease classification. This work also utilizes the pre-trained model on imageNet for transfer learning. The proposed model can detect five types of images including three common types of diseases on rice leaf (e.g., brown spot, hispa, and leaf blast), healthy rice leaf and other leaves. After building, training, and testing the model, it was converted to *tflite*[4] format for running on mobiles or IoT devices. Finally, an Android application was built for rice leaf disease detection using the trained model.

2 Related Works

There are several works in rice leaf detection using different techniques such as image segmentation, feature extraction, feature selection and machine or deep learning for classification [2]. In [3], the authors applied AlexNet [4] technique to detect the three prevalence rice leaf diseases termed as bacterial blight, brown spot as well as leaf smut and got about 95% accuracy due to adjusting an efficient technique and image augmentation.

The authors in [5] were developed ensemble model for classifying six types of rice diseases: leaf blast, false smut, neck blast, sheath blight, bacterial stripe disease, and brown spot. Results showed that overall accuracy of 91% was achieved. The smartphone app was developed based on the above model. Other works can be found in [2, 6–7].

---

[1] http://www.knowledgebank.irri.org.

[2] https://www.mard.gov.vn/.

[3] http://www.knowledgebank.irri.org/training/fact-sheets/pest-management/diseases/item/brown-spot.

[4] https://www.tensorflow.org/lite/guide/android.

In this work, the proposed model uses EfficientNet [1] for rice leaf disease classification. After building, training, and testing the model, it was deployed on mobile device applications.

3 Proposed Method

In this part, first, the datasets are collected and pre-processed. Then, the classification model was built based on the EfficientNet [1]. Next, this model was evaluated and compared with other baselines. Finally, an Android application was developed to deploy the classification model.

3.1 Datasets for Experiments

The datasets have been collected from both manual and public sources.[5] The raw data has several problems such as wrong labels, multiple diseases on a leaf. Thus, we have pre-processed normalized these data. Each of the images was resized to 300 × 300 pixels as input for deep learning model. After pre-processing, the dataset has 1790 files for five classes as presented in Fig. 1. These data include healthy leaf (346 files), hispa disease (429 files), brown spot disease (345 files), leaf blast disease (337 files), and others (333 files, e.g., grass, other kinds are not the rice leaf). We use 60% for training and 40% for testing as presented in Table 1.

3.2 Classification Model

The proposed model is based on EffcientNet B3 [1] as presented in Fig. 2. This is a variant of Convolutional Neural Networks which has 384 layers and 10,783,535 parameters. The input shape used for this network is 300 × 300 × 3 to fix with the input images pixels. The main building block of this network consists of MBConv [10] to which squeeze-and-excitation optimization [9] is added.

In this model, BN denotes for the Batch Normalization and Swish is an activation function that takes the form $f(x) = x * sigmoid (\beta x)$, where β is a learnable parameter. Swish exhibits one-sided boundedness at zero, smoothness, and non-monotonicity. Models and hyperparameters generated for ReLU were simply replaced with this Swish activation functions [8].

MBConv denotes for Mobile Inverted Residual Bottleneck. To develop even more efficient layer structures, MobileNetV2 [10] introduced the linear bottleneck and

---

[5] https://www.kaggle.com/minhhuy2810/rice-diseases-image-dataset.

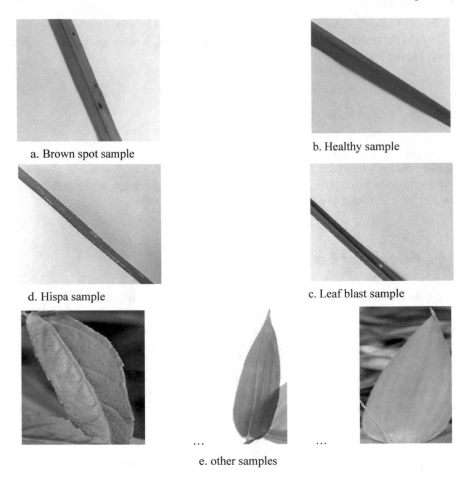

a. Brown spot sample

b. Healthy sample

d. Hispa sample

c. Leaf blast sample

...   ...

e. other samples

Fig. 1 Samples of images: three rice leaf diseases, healthy leaf and others

Table 1 The ratio of training/testing data

Class	#Training images	#Testing images
Brown spot	214	131
Healthy	201	145
Hispa	266	163
Leaf blast	204	133
Others	189	144

inverted residual structure. A 1 × 1 expansion convolution is followed by depth-wise convolutions and a 1 × 1 projection layer to define this structure. If and only if the input and output have the same number of channels, they are connected using a residual connection. The MBConv in EfficientNet [11] looks like MobilenetV3

Fig. 2 Architecture of the proposed model based on effcientNet B3 [1]

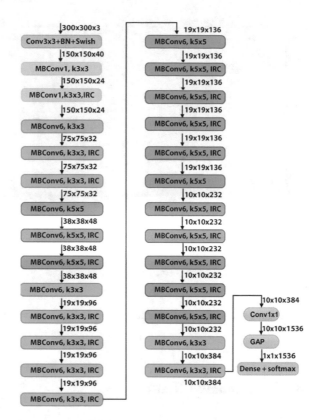

and is presented in Fig. 3, which is a combination of MobileNetV2 and SQUEEZE-AND-EXCITATION BLOCK [9] (SE block) to build the effective models. Layers are also upgraded with modified swish nonlinearities.

The IRC denotes Inverted Residual Connection. The Squeeze and Excitation (SE) is a block that aims to improve the quality of representations produced by a network by explicitly modeling the interdependencies between its convolutional features' channels. It has global information access and can re-calibrate filter responses in two

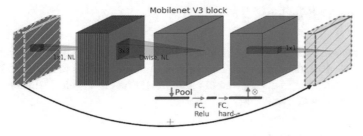

Fig. 3 MBConv architecture [11]

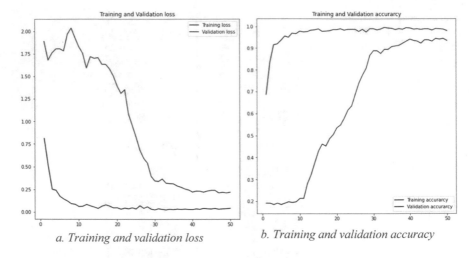

a. Training and validation loss   b. Training and validation accuracy

Fig. 4 Loss and accuracy during training and validation

processes, squeezing and excitation, before feeding them into the next transformation
[9].

4 Experimental Results

The model was trained on Google Colab using Keras library.[6] The model parameters
are set as (optimizer = 'Adam', learning_rate = 0.001, epochs = 50, batchsize =
32 and image_dims = (300, 300, 3)). After training the model, it was evaluated and
compared with other baselines. This model was then converted to *tflite*[7] format so
that it can be run on mobile or IoT devices. Finally, an Android application was built
for rice leaf disease detection using the pre-trained model. For evaluating the model,
this work uses the Precision, Recall, F-Measure and Accuracy.

5 Model Validation

Figure 4 presents the loss function and accuracy values during training and validation
phases. After 40 epochs, the model becomes stable and reaches more than 95%. The
model accuracy is high during training and validation phases and it does not suffer
from the problem of overfitting.

---

[6] https://keras.io/api/optimizers/adam/.

[7] https://www.tensorflow.org/lite/guide/android.

5.1 Results of Classification

The proposed method is compared with other baselines such as Support Vector Machines (SVM) and the Decision Tree to check whether the deep learning approach can give better results for developing applications in the next stage. The dataset is divided into 60% for training and 40% for testing. Figure 5 presents the classification results of the proposed model on the test set. Results show that the average precision of the model is higher than 95%.

Figures 6 and 7 show the classification results of the SVM and Decision Tree. These models were trained using the HOG features which are supported in the *sklearn*

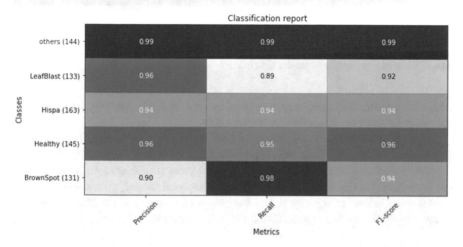

Fig. 5 Classification results of the efficientNet

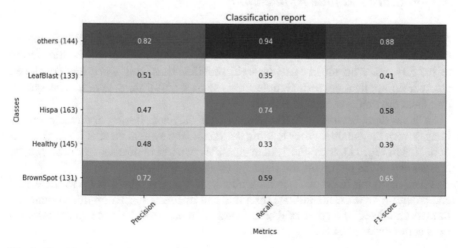

Fig. 6 Classification results of the SVM

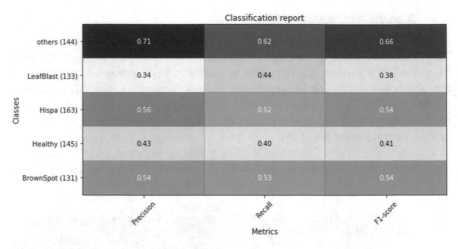

Fig. 7 Classification results of the decision tree

and *skimage* libraries. Results are not good as expected. The Healthy and LeafBlast classes are almost misclassified by both models.

Figure 8 presents the confusion matrix of the EffcientNet model. The values on the diagonal of this matrix are very high (more than 95% on average), this means that the model performance is really good. Figure 9 presents the comparisons between the three models. These results show that the proposed model, which uses the deep learning—EfficientNet approach, has the best performance. Thus, it is selected for integrating to the mobile application for rice leaf disease detection.

5.2 Building Mobile Application

After selecting the prediction model, we have converted the model to *tflite* format so that it can be run on mobiles and IoT devices. This application will be transferred to the farmers, who mostly use Android mobiles, thus, this work has developed an Android application with Google Firebase as a real-time database. The system interfaces are demonstrated in Fig. 10.

In this system, the users can load images from the photo library or directly capture images from the camera. After loading the image, the system checks it is a rice leaf or not (as in Fig. 11), then, the system classifies what kind of disease and shows the reasons as well as treatment solutions for this kind of disease.

When applying in practice on a mobile device, it took about 1.7 s for detecting and providing treatment solutions for a rice leaf image. Thus, using this system, the farmers can detect the types of diseases and get recommendations from the system for a better production plan.

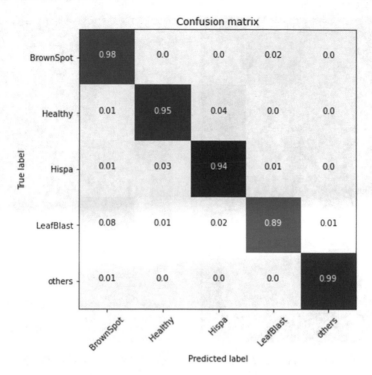

Fig. 8 Confusion matrix of the proposed model

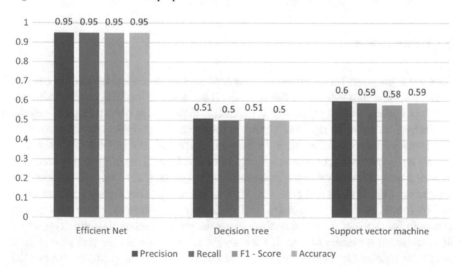

Fig. 9 Evaluation comparison between the models

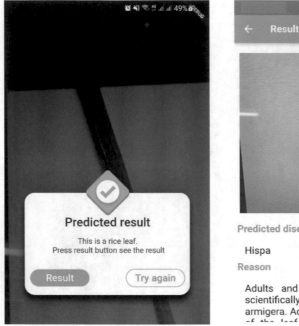
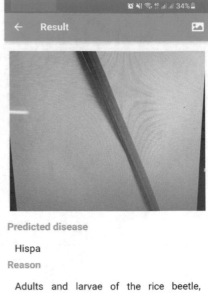

a. Predict from camera   b. Disease reason and way of treatment

Fig. 10 Snapshots of rice leaf disease detection using smart phones

6 Conclusion

This work proposes an approach for rice leaf disease detection on mobile devices
using deep learning technique. Specifically, it proposes using EfficientNet which is
a variant in deep learning networks for classification. This approach also utilizes the
pre-trained model on imageNet for transfer learning. The proposed model can detect
five types of images including three common types of diseases on rice leaves in
Vietnam (e.g., brown spot, hispa, and leaf blast diseases). Two other classes are also
included such as healthy rice leaf and other leaves (grass, plant, etc.). The model was
trained on 1790 images and produced 95% of accuracy. This model was converted
to *tflite* format for running on mobile or IoT devices. Finally, an Android application
was built for rice leaf disease detection using the pre-trained model. When applying
in practice on a mobile device, it took about 1.7 s for detecting and providing a
treatment solution for a disease, thus this could be a useful solution to help the
farmers.

This work continues to improve by collecting more data (both the numbers of
images for each disease and other kinds of rice diseases) to re-train the models as
well as to compare with other deep learning approaches. Moreover, the application
will be developed for multi-platforms so that it can be run on iOS devices.

Fig. 11 Checking whether an image is rice leaf

References

1. Tan M, Le QV (2019) EfficientNet: rethinking model scaling for convolutional neural networks. ArXiv, abs/1905.11946
2. Sethy PK, Barpanda NK, Rath AK, Behera SK (2020) Image processing techniques for diagnosing rice plant disease: a survey. Proc Comput Sci 167:516–530
3. Matin M, Khatun A, Moazzam M, Uddin M (2020) An efficient disease detection technique of rice leaf using alexNet. J Comput Commun 8:49–57. https://doi.org/10.4236/jcc.2020.812005
4. Krizhevsky A, Sutskever I, Hinton GE (2012) ImageNet classification with deep convolutional neural networks. Commun ACM 60:84–90
5. Deng R, Tao M, Xing H, Yang X, Liu C, Liao K, Qi L (2021) Automatic diagnosis of rice diseases using deep learning. Front Plant Sci 12:701038. https://doi.org/10.3389/fpls.2021.701038
6. Bera T, Das A, Sil J, Das AK (2019) A survey on rice plant disease identification using image processing and data mining techniques. In: Abraham A, Dutta P, Mandal J, Bhattacharya A, Dutta S (eds) Emerging technologies in data mining and information security. Advances in Intelligent Systems and Computing, vol 814. Springer, Singapore. https://doi.org/10.1007/978-981-13-1501-5_31
7. Hong Son N, Thai-Nghe N (2019) Deep learning for rice quality classification. In: 2019 international conference on advanced computing and applications (ACOMP), pp 92–96. https://doi.org/10.1109/ACOMP.2019.00021
8. Ramachandran P, Zoph B, Le QV (2017) Swish: a self-gated activation function. arXiv: Neural and Evolutionary Computing
9. Hu J, Shen L, Sun G (2018) Squeeze-and-excitation networks. IEEE/CVF Conf Comput Vision Pattern Recogn 2018:7132–7141. https://doi.org/10.1109/CVPR.2018.00745.2018

10. Sandler M, Howard AG, Zhu M, Zhmoginov A, Chen L (2018) MobileNetV2: inverted resid-
    uals and linear bottlenecks. In: 2018 IEEE/CVF conference on computer vision and pattern
    recognition, pp 4510–4520
11. Howard AG, Sandler M, Chu G, Chen L, Chen B, Tan M, Wang W, Zhu Y, Pang R, Vasudevan
    V, Le QV, Adam H (2019) Searching for mobileNetV3. In: 2019 IEEE/CVF international
    conference on computer vision (ICCV), pp 1314–1324
12. Alhichri HS, Alswayed AS, Bazi Y, Ammour N, Alajlan NA (2021) Classification of remote
    sensing images using efficientNet-B3 CNN model with attention. IEEE Access 9:14078–14094

Speech-Based Traffic Reporting: An Automated Data Collecting Approach for Intelligent Transportation Systems

Huy Nguyen G., Thanh Nguyen, Hieu Le Trung, and Quang Tran Minh

Abstract In this paper, we explore the possibility of an automatic data collecting feature for Intelligent Traffic Systems (ITS). Our proposed feature collects users' speech reports and uses deep learning approaches to analyze the unstructured data, then provides traffic information to the community, helping to increase users' awareness and better routing system. The basic procedure is to transform speech report to text form, we used the popular Conformer model for this automatic speech recognition (ASR) task, and then extract useful information from the output, this is the task for intent classification and slot filling, our choice is the state-of-the-art JointIDSF model for Vietnamese language. We trained the Conformer on the VIVOS dataset and gain promising result (25 and 25.5% WER on the train and test set, respectively). For the JointIDSF, we got 0.63 F1 score on the test set. We experimented this feature on the existing urban traffic estimation system deployed for real-world applications in Ho Chi Minh City, namely UTraffic, to confirm the effectiveness of the proposed solution.

Keywords ITS · Traffic data collecting · Deep learning · UTraffic · ASR · JointIDSF · Conformer

Huy Nguyen G. (✉) · T. Nguyen · H. L. Trung · Q. T. Minh
Faculty of Computer Science and Engineering, Ho Chi Minh City University of Technology (HCMUT), 268 Ly Thuong Kiet, Dist. 10, Ho Chi Minh City, Vietnam
e-mail: huy.nguyen.1209@hcmut.edu.vn

T. Nguyen
e-mail: thanh.nguyen2612@hcmut.edu.vn

H. L. Trung
e-mail: hieu.letrunghieu@hcmut.edu.vn

Q. T. Minh
e-mail: quangtran@hcmut.edu.vn

Huy Nguyen G. · T. Nguyen · H. L. Trung · Q. T. Minh
Vietnam National University Ho Chi Minh City (VNU-HCM), Linh Trung Ward, Thu Duc District, Ho Chi Minh City, Vietnam

© The Author(s), under exclusive license to Springer Nature Switzerland AG 2022
N. H. T. Dang et al. (eds.), *Artificial Intelligence in Data and Big Data Processing*,
Lecture Notes on Data Engineering and Communications Technologies 124,
https://doi.org/10.1007/978-3-030-97610-1_53

1   Introduction

One of the most challenging and annoying problems in urban city life is traffic congestion. Traffic congestion is not primarily a problem, its origin was actually a solution to our basic mobility problem. Basic operational requirements coming from different social systems, economy, education require people to go out and interact with each other at the same hours; this makes traffic conditions too complicated to be controlled without breaking our modern social life. Fortunately, traffic congestion at the same time is a motivation for researching fields to build adaptive solutions that progressively mitigate the issue. Many efforts have been made, like improving infrastructure, optimising traffic light management, enforcing laws, eco-friendly transportation, intelligent transportation systems, etc. Nevertheless, in the fast-changing world of technology nowadays, intelligent transportation systems (ITS) are becoming more prevalent thanks to their high efficiency, cost effectiveness and feasibility.

Within the last two decades, numerous ITS technologies have been developed, such as city-wide traffic regulation control, general transport information assistance, smart parking, etc. Information collected from diverse sources in the ITS for instance GPS, actuators, multimedia locators, internet community, generates unbounded data measurements from Trillion byte to Petabyte. As a consequence of the enormous volume and diversity of data, convention system frameworks are inadequate and do not fulfill the requirements for big data analytics. Shifting to data-centric solutions is now a debating point. Many of the big data mechanisms with artificial intelligence approaches, including text mining, data science, robotics, knowledge discovery and much more bring unforeseen innovations to the field. Nevertheless, data fed into such systems conventionally comes from fixed sensors, loop detectors, RFID, camera systems which are infeasible to deploy everywhere, especially in developing countries like Vietnam due to severely high deployment and maintenance costs. As a result of the convergence of mobile communications and intelligent terminal technology, the transportation system is provided with a new approach in alleviating the traffic congestion through mobile crowd-sourcing technique. This significantly empowers the current intelligent transportation system with enormous and modern data source to make good use of specialized big data solutions and novel system architectures.

The most popular traffic system today can arguably be named Google Maps [1]. Their navigation technology, the high geographical availability, efficient routing with shortest path all are based on historical data, statistical records, and human inputs. Another well-known one is Waze [2] doing very impressive work on traffic status monitoring and serving to end user, which undoubtedly takes advantage of the user driving information. In specific, Waze application allows users to directly report traffic condition, but still requires traditional human-operated software. Even though, such prominent applications have been extensively developed and proven to be a design pattern for efficient and scalable traffic system, yet their systems are not applicable in some cases due to expensive development cost or unsuitable for specialized regions. To name one similar system specialized for Ho Chi Minh City, an existing crowd-sourcing and traffic awareness system namely **UTraffic (BKTraffic)** [3], it constructs a way that encourages online users to share their traffic data, timely report about traffic status, and helps people monitor traffic conditions effortlessly. However the current system collects data via GPS, mobile sensors, or manual users' typing report form, which can be a real struggle for drivers with their handheld devices when driving [4]. With the eruption of artificial intelligent in general and natural language processing in particular, we are motivated to adapt and propose an appropriate approach to support drivers in contributing traffic data in a legitimate, effortless and convenient way. To achieve the objective, we introduce a modern approach that requests users' traffic reports in the form of voice records and with the aid of big data analysis techniques and state of the art natural processing language models, we implement a full-fledged analytic pipeline to collect–process–serve data from end-to-end and produce a real-time traffic condition monitoring service. The speech report consists of information about the **location** where the users want to report, the **reason** that is causing any trouble/accident on the road and the current **velocity** of the driver. These details can be used as an effective information for routing procedures that are based on velocity, or used as data for traffic predicting models. Our main contributions in this paper:

 (i) We devise a feature that allows users to send natural speech records demonstrating traffic congestion situations.
 (ii) Applies natural language processing models: speech-to-text recognition, intent classification and slot filling classification.
(iii) Real time data collecting and processing analytic pipeline.

The structure of the paper is as follows. In Sect. 2, we summarize related work that proposed our innovation in the way of approaching to this problem. The main part are discussed in Sect. 3, where we briefly introduce our existing systems, data collecting strategies and deeply focus on our two models. In Sect. 4, we carry on experiments and come up with the results and measurements of our proposing system. The final part of this paper is our future work to due with the challenges we encountered.

2   Related Works

A few intelligent traffic systems in foreign countries have been put into operation, such as the VICS [5] in Japan. Their system collects data from administrators and prefectural police headquarters, then handles them in an information center and broadcasts the information to users by FM and radio channels, or by infrared beacons installed on the streets. This method, however, demands a huge amount of human labor and budgets, therefore is not a candidate for implementation for countries like Vietnam. Other traditional methods of collecting velocity and traffic density by using loop detectors have been used for a long time and still have improvement [6]. Nonetheless, physical devices have trouble with critical weather conditions or when it comes to a huge volume of vehicles passing through, or even damages inside the devices. Their degree of coverage is also questionable because of the pricing level and the complexity of installation.

In Vietnam, there are also works that try to reduce traffic loads. The Voice of the People of Ho Chi Minh City (VOH)—95.6 MHz [7] or The Voice of Vietnam (VOV)—91 MHz [8] are the two radio stations that collect data from traffic users (drivers, walkers, etc.) and contributors to widespread the information back to the people through their radio channels. Still, these systems require a lot of human work, intended contributions and users' proactiveness to access the information. Their scope of coverage depends entirely on the number of contributors that are actively reporting traffic data thus are not easy to scale. Recently, researchers from Ho Chi Minh City University of Technology (HCMUT) proposed a system called "Smart BK Traffic" [9]. It works on a principle that collects data from the city buses' GPS and publishes the analyzed information on a website. It largely reduced the workload for human labor, but had a major drawback on routes that are not included in the bus system network.

The most popular traffic routing application worldwide can easily be named Google Maps. As a giant in the industry, they have explosive resources and technology to collect traffic data, from satellites to contributors and billions of GPS from users' devices. Surely we cannot have massive sources of data like Google, and cannot follow their approaches because of obvious reasons. Our existing intelligent traffic system, namely UTraffic [3], aiming to collect data from the community and relevant sources to give traffic warning, increase traffic users' safety and awareness, as well as to solve the heavy traffic load problem in Ho Chi Minh City with minimal expense and human effort. To our best knowledge, only a few navigation softwares allow users to report traffic conditions lively, and the most successful of them all is Waze [2]. Waze allows users to report multiple situations they encounter on the road. The mechanism is fairly simple, users just need to choose the kind of problem they want to report, choose the level of seriousness they feel about the problem, (e.g. for traffic jams, they can choose between Moderate, Heavy and Standstill) and take a picture to verify. With only a few steps needed, the application optimized users' experience. However, in our opinions, there are flaws that can be overcome. The fact that users can select the level of seriousness makes this important information become biased. Different personalities can have different viewpoints of whether a traffic jam should be considered "Moderate" or "Heavy". To address this, we aim to provide a system that can automatically infer information just by using the unstructured data users provided (voice record, video, or image). In this case, users only need to provide minimal data that won't take time to think, thus enhancing their experience as well as information accuracy.

With the rise of deep learning in particular and machine learning in general, we can have automatic systems that can collect traffic data from more friendly and common sources. Deep learning approaches to recognize speech have been around since many years ago. The most famous architecture for sequence data, like speech and text, is the sequence-to-sequence model, with remarkable researches like the LAS [10], Transformer [11] and recently, the Conformer [12]. As for natural language understanding (NLU), since the birth of the Transformer, massive language models have been proposed for many different tasks. One of which is the BERT pre-trained model [13]. For any language, BERT just needs a large enough text corpus to create a probabilistic language model, that can acts like a foundation for many artificial neural networks to build upon, and have a great sense of "understanding" the learnt language. Intent classification and slot filling are the two NLU tasks that were recently embedded with BERT to achieve novel results [14].

3 Utraffic Analyzer

3.1 System Overview, Architecture and Technology

The intelligent transport system is a complex dynamic system with various elements such as drivers, passengers, pedestrians, vehicles, roads, and some key components: positioning systems, communication channels and computing elements. This paper will deal with a comprehensive description of a part of the proposed system which focuses mainly on computing elements and data collection processes. Figure 1 shows a schematic overview of the described system. We have considered the MongoDB database service as the main storage for a variety of reasons such as fast query, security, horizontal scalability, etc. For the API server, we chose Apache and for the computing server, we used the resource of high performing computer (HPC) at Ho Chi Minh City University of Technology. The goal of this architecture is to extract the level of service (LOS) from two incoming sources: Form-based reports and speech reports. Noting that in our system **UTraffic**, the basic unit for any computation is **segment**: many segments are linked together to form a street, and lastly to form the whole map information. The system allows clients to send reports about traffic in two ways, by inputting data into a pre-designed form and by talking (speech). The form reports will be mapped directly to segment reports and stored in the database. With the speech reports, the data will be stored in the database first before being processed. The computing server is built to get the unprocessed speech reports from the database periodically and extract the list of segments, velocity and causes of traffic to map to the segment reports. After having the segment reports in database, the API server will process and display them as the level of service (LOS) to the client application and other service like routing service also takes this information into account for calculating the best path between two stations, which is modeled as a path finding with minimum traveling time problem in a graph.

3.2 Data Pipeline and System Workflow

To utilize the existing data warehouse, we use GridFS of MongoDB to store and manage audio records from users' reports. GridFS provides a powerful, scalable

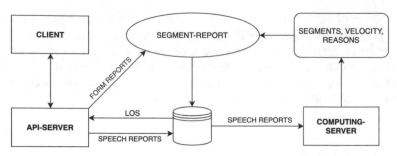

Fig. 1 System architecture

solution to store and manage a large number of unstructured data like video or audio files without file size limit. When a user requests to POST a speech report from client to API server, a FormData object is packed with a field of audio file in binary format (BLOB) and a field of segment IDs list to be reported explicitly. At the time the request is received at server, the following actions will be done sequentially:

1. Recording file part is extracted from the payload and a GridFS module immediately creates a new document into the SpeechRecord collection in the database, then a unique identifier is sent back to the server.
2. Secondly, server computes necessary data related to a traffic report including *report's source*, *report's creator*, *time period* when the report is made, informative timestamps (*processed_date*, *createdAt*, *updatedAt*), list of segment IDs reported, and *speech_record* ID indicating where to find the recording file in SpeechRecord collection. This data is used as an intermediate form, namely SpeechReport, to leverage the transformation process from speech report to segment report. Finally, existing services will seamlessly use the traffic status reporting information.

A SpeechReport document contains a field named *processed_date* with null value initially; it performs as a flag controlling whether a record has been transformed yet. At transformation stage, all speech reports and recording files that haven't been transformed into segment reports will be taken out and be fed into the pipeline sequentially. To mitigate the cost of processing speech reports, we schedule our analytic server to run these jobs in micro-batch style every 3 min. When a report is sent to the pipeline, it must be uniformly converted to waveform audio file format with the help of FFmpeg before feeding into deep learning models, the encoded audio stream is then mapped to Vietnamese text by an Automatic Speech Recognition model and finally a pre-trained Intent Classification and Slot Filling model will extract and output traffic related information. Output results are combined with data fields to produce segment reports just like normal traffic reports by form and are stored into SegmentReport collection in the end. Noting that a speech report may have many segments to be reported, which means that a single speech report can be mapped to many segment reports, and after processing the report *processed_date* will be set.

3.3   The Deep Learning Models

Automatic Speech Recognition (ASR) Since the last decade, deep neural networks have been the *de facto* choice for any ASR systems. Sequence to sequence structures with the inspiration from recurrent neural networks (RNNs) have been widely used for natural language processing tasks. However, RNN suffers from the well-known *bottleneck problem*, in which the model fails to capture long dependency information (in the case of ASR, a long duration speech). Moreover, the recurrence nature of RNN, the fact that they process sequence data one frame at a time, leads to a high time complexity in both training and inferring phase, which is no longer ideal for relatively real-time systems. The Transformer [11] addressed many problems of RNN by avoiding the recurrence processing mechanism for better time performance, and is based entirely on *self-attention* mechanism, which better helps in capturing long dependencies information. Since its arrival, Transformers have been inspiring many natural language processing system and made significant improvement in both performance and result. To enhance the power of the Transformer, Anmol Gulati et al. proposed a new architecture in the second quarter of 2020, called the Conformer [12]. Their work combines the characteristics of the Convolution Neural Network (CNN) and the Transformer, to make use of their ability to capture local and global dependencies feature information, respectively.

As we are proposing a new feature for the UTraffic system to report traffic conditions by speech, our ASR model expects the input utterance to have a long duration, and have many local information about location, reason for reporting and velocity. Example for such kind of report would be: "Hôm nay tại **tuyến đường Lý Thường Kiệt** xảy ra **tai nạn giao thông** nên làm tắc nghẽn đường, vận tốc di chuyển **10 km/h** (Today at **Ly Thuong Kiet street** occurred an **accident** causing traffic jam, moving velocity is about **10 km/h**). As we are highlighting the cluster keywords in a speech report, we find that the Conformer, which can capture both local and global dependencies information, will be a reasonable architecture.

One of the most challenging, and if not the hardest, issue for us is the data for training. So far Vietnamese has not have many large open-source speech datasets. Our attempt for collecting data is to gather speeches from different contributors, from our friends, families to the community around us. This method, however, is not promising because it will take a long time to have a large enough dataset and require a consistency effort from numerous labor. Currently, we are using the VIVOS dataset [15], which is a small and clean corpus, but if we want to furthure improve the system, bigger training data is required.

Intent Classification and Slot Filling After turning users' speech into the text form, we need to extract information from the transcript. Ideally, we would want to know what is the users' intention through their speech, are they giving a report or not, if the intention is not classified as a traffic report, we can discard it. If it truly is a traffic report, we want to know where in the text transcript is the location the users want to report, the reason why they report and the velocity of their vehicle at that moment. This is a well-known task in natural language understanding (NLU), called Intent Classification and Slot Filling. Both of them are classification problems, where the former aims to predict a user's intention through a document and the latter aims to extract certain parameters inside that document. For example, consider the above traffic report, we would want to label the whole sentence as **traffic report**, and the term tuyến đường Lý Thường Kiệt (Ly Thuong Kiet street) as **location, tai nạn giao thông** (traffic accident) as **reason** and **10 km/h** as velocity.

In recent years, state-of-the-art deep learning models have been growing around the publication of BERT [13]. This is a pre-trained language model that can be fine-tuned with additional output layers to perform different tasks in NLP. Qian Chen et al. proposed a joint model built on top of BERT to perform indent classification and slot filling [14], and got a novel result back in 2019. One year later, VINAI from Vietnam extended this work with additional attention layers, which they called the Joint-tIDSF [16]. Their work used a BERT version of Vietnamese, called PhoBERT [17], which was trained by VINAI themself using a massive Vietnamese text corpus. Joint-tIDSF is the best candidate for our system because it were built on a Vietnamese language model and up to now it gives the state-of-the-art result.

4 Experiments and Results

4.1 Deep Learning Models

Automatic Speech Recognition (ASR) We use the ESPNETv2 [18] toolkit for our ASR model. We trained a sequence-to-sequence architecture, where the encoder is a stack of eight Conformers and the decoder is a stack of eight Transformers, with CTC loss function. Each layer of Conformer or Transformer has the dimension of 512, where the former use the Relative Multiheaded Attention mechanism proposed in [19] and the latter just use normal Multiheaded Attention. The convolution layer has the kernel size of 7, padding size of 3 and stride 1. We trained this model on the VIVOS dataset [15] with 150 epochs, batch size of 30, and use warm-up learning rate. The Language model used for better decoding is a two layers sequence-to-sequence model using RNN with LSTM cell and tokenized at the character level. It was trained with 20 epochs and the batch size of 20. Our model obtained **25%** **WER** on the training set and **25.5% WER** on the test set. Up to now, we are satisfied with this result, given that VIVOS is a small corpus and our limitation in time to experience more architectures.

It has to be noted that VIVOS is a very *clean* dataset, meaning that it has minimum background noise and was recorded with modern equipments. It is hard to simply use the trained model in production, since we will expecting the users' recordings to be of lower quality. Some future works that we think might be useful is to preprocess the training set. This involves adding background noise to simulate a real-life envi-

Table 1 Example of a speech report label

Text	kẹt xe ở hầm thủ thiêm do có tai nạn xe di chuyển với vận tốc 8 ki lô mét trên giờ
Translated	Traffic congestion in thu thiem tunnel because of a car accident velocity 8 km per hour
Slot filling label	O O O B-location I-location I-location O O B-reason I-reason O O O O O B-velocity I-velocity I-velocity I-velocity I-velocity I-velocity

ronment, speed perturbation and pitch shifting to generalize the model with human voice and conditions.

Intent Classification and Slot Filling We collected and generated around one hundred speech reports to train the JointIDSF model. Our Intent Classification task currently have two labels, which are *Unknown (UNK)* and *Report* since we only expect only the intention to report from users. Unfortunately, because we only have one hundred texts to be considered as a report, we still cannot train the indent classification task properly, models for such task would require huge data corpus. For the slot filling part, we constructed eight labels, which are *UNK, O, B-location, I-location, B-velocity, I-velocity, B-reason* and *I-reason*. Token *UNK* and *O* are used for unknown labels and labels that are not significant, respectively. Token **-location, *-reason* and **-velocity* are used to identify the information about location, reason and velocity. Any token that start with a *B* implies that it should be the start of a sequence of terms, followed by tokens that start with an *I*. For example, the term **đường Lý Thường Kiệt** (Ly Thuong Kiet street) should be labeled as *B-location I-location I-location I-location*, the term **vận tốc khoảng hai mươi ki lô mét trên giờ** should be labeled as *O O O B-velocity I-velocity I-velocity I-velocity I-velocity I-velocity I-velocity*. An example is illustrated in Table 1.

We trained JointIDSF with 50 epochs and archive an **F1 score of 0.63** on the test set. The result is not satisfied, when looking at some validation results, we found some interesting scenarios. For example, a speech like **Đường Cộng Hòa là một tuyến đường lớn ở thành phố Hồ Chí Minh do đó thường xuyên xảy ra kẹt xe vận tốc di chuyển khoảng mười ki lô mét trên giờ** (Cong Hoa street is a major road in Ho Chi Minh City so it usually has traffic jam moving velocity of about ten kilometers). Our models labeled both **Đường Cộng Hòa** (Cong Hoa street) and **thành phố Hồ Chí Minh** (Ho Chi Minh City) as **location**, which does not satisfy our requirements since we only need to label the street, information about Ho Chi Minh City is redundant, this "misunderstanding" is the reason why the F1 score is not high. In technical terms, the model is doing quite well what it is supposed to do (labelling locations in a document), but to our specific requirements, we should not care about information of cities, rather only information about streets. One of the ways to counter this is to concretely create smaller labels for **location**, for example, **street** and **city**. However, more labels mean more training data, which presses the importance of collecting and generating data for our models.

Information about **location**, **reason** and **velocity** can be used for: (1) Any intelligent transportation systems that have routing procedures based on velocity, (2) awareness systems that show real-time traffic status, such as accidents, floods, level of traffic jams, etc. on the city map, (3) a social network or community-based application that informs or propagates a report from one user to the others to avoid unnecessary traffic jam, accidents, (4) predictive systems that require a huge amount of real-time data to estimate future traffic conditions or average time to travel on the road. There are a lot of possibilities with just four simple pieces of information. In the future, more details can be added to the report, such as the type of vehicle users are driving in order to estimate road density, the current weather as additional features for traffic predictive models, etc.

4.2   Analytic Service Experiment

(a) *Experimental environment*

The two models are pre-trained by Google Colab and stored in AWS S3 bucket; therefore, it can be loaded and used anywhere. In the end, we evaluate our system on a single machine with CPU 24 cores, 64 GB memory running on Ubuntu 20.04 LTS. Nevertheless, system is configured to process and measure performance of speech record audio file.

In terms of the dataset, we prepared 80 recording files of present residents of Ho Chi Minh City which are variant in length and accents. Each record file is wrapped and sent to server via reporting API as a single traffic report. After all requests have been processed and inserted to MongoDB data storage, our analytic server will starts running and measuring time of each internal processing phase.

(b) *Result*

Our system currently only takes a single stream of speech reports and handles them one-by-one. We ran the service on a pure CPU-based environment, no GPU was used for deep learning models to decode. We examine the processing time of each model in comparison to the duration of the speech. For a sample of 80 speeches handled one at a time, we have the statistic in Table 2. Intuitively, by looking at the standard deviation, we can observe that the ASR's inferring time varies a lot whereas the IDSF's seems to be more stable. To understand which one of those two is having more effect on the overall time, we plotted a correlation map Fig. 2. Surprisingly, among the two models, ASR contributes almost fully to the total time. It also has a strong correlation with the speech duration, or we can say the length of the input, whereas the IDSF model is relatively independent with the input characteristics (Fig. 3 summarizes the relationship among time duration). We think that the overall time to process one speech report is long and insufficient for an intelligent system in production (mean value is 31.85 s).

Table 2 Time performance statistic (seconds, rounded to 2 decimals)

	Duration (s)	ASR time (s)	IDSF time (s)	Overall time (s)
Mean	11.29	19.33	12.43	31.85
Std	3.66	10.67	0.52	10.77
25%	8.74	12.09	12.21	24.39
50%	10.57	16.31	12.53	29.05
75%	13.26	22.52	12.82	12.82

where
Duration: duration of the speech report
ASR time: decoding time of the ASR model
IDSF time: decoding time of the JointIDSF model
Overall time: total time to process one speech report $\approx$ ASR time + IDSF time

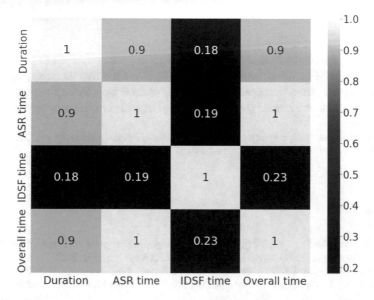

Fig. 2 Correlation map between metrics

This is mostly due to the large size of the ASR model and the CPU environment which we use for decoding. Likewise, processing one report at a time is not ideal, further implementation will require a batch-like processing procedure.

5 Conclusion and Future Works

In this paper, we have proposed an approach that leverages deep learning models to develop a traffic speech reporting feature for intelligent transport systems. The adoption of big data architectures and knowledge of deep learning help bringing

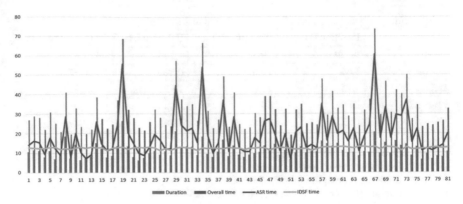

Fig. 3 Processing time of ASR and JointIDSF models on different in length records

useful and convenient services to users from ITS in particular and other application fields in general. For future work, we plan to improve the two models, migrate and implement the system comprehensively and continue conducting experiments on more data to further evaluate the feasibility and usability of the new approach.

Acknowledgements This research is funded by Vietnam National University Ho Chi Minh City (VNU-HCM) under grant number DS2021-20-01.

References

1. Google Maps Homepage (2021). https://www.google.com/maps/. Last accessed Oct 2021
2. Waze Homepage (2021). https://www.waze.com/. Last accessed Oct 2021
3. UTraffic Homepage (2021). https://bktraffic.com/home/. Last accessed Oct 2021
4. Mai-Tan H, Pham-Nguyen H-N, Long NX, Minh QT (2020) Mining urban traffic condition from crowd-sourced data. SN Comput Sci 1(4):1–16
5. VICS Homepage (2021). https://www.vics.or.jp/en/. Last accessed Oct 2021
6. Coifman B (2001) Improved velocity estimation using single loop detectors. Transp Res Part A Policy Pract 35(10):863–880. https://doi.org/10.1016/S0965-8564(00)00028-8
7. The Voice of the People of Ho Chi Minh City (VOH) Homepage (2021). https://voh.com.vn. Last accessed Oct 2021
8. The Voice of Vietnam (VOV) Homepage (2021). https://vovgiaothong.vn. Last accessed Oct 2021
9. Han T (2021) Tránh kẹt xe bằng hệ thống smart BK traffic. https://www.hcmut.edu.vn/vi/event/view/noi-san-bk/1579-tranh-ket-xe-bang-he-thong-smart-bk-traffic, SGGP Online (2014). Last accessed Oct 2021
10. Chan W, Jaitly N, Le QV, Vinyals O (2015) Listen, attend and spell: a neural network for large vocabulary conversational speech recognition. In: IEEE International conference on acoustics, speech and signal processing (ICASSP 2015)
11. Vaswani A, Shazeer N, Parmar N, Uszkoreit J, Jones L, Gomez AN, Kaiser Ł, Polosukhin I (2017) Attention is all you need. In: Advances in neural information processing systems, pp 5998–6008

12. Gulati A, Qin J, Chiu C-C, Parmar N, Zhang Y, Yu J, Han W, Wang S, Zhang Z, Wu Y, Pang R (2020) Conformer: convolution-augmented transformer for speech recognition. arXiv
13. Devlin J, Chang M-W, Lee K, Toutanova K (2018) BERT: pre-training of deep bidirectional transformers for language understanding. CoRR
14. Chen Q, Zhuo Z, Wang W (2019) BERT for joint intent classification and slot filling
15. AILAB University of Science HCMC (2021). https://ailab.hcmus.edu.vn/vivos. Last accessed Oct 2021
16. Dao MH, Truong TH, Nguyen DQ (2021) Intent detection and slot filling for Vietnamese. In: Proceedings of INTERSPEECH 2021
17. Nguyen DQ, Nguyen AT (2020) PhoBERT: pre-trained language models for Vietnamese. In: Findings of ACL: EMNLP 2020, pp 1037–1042
18. Watanabe S, Hori T, Karita S, Hayashi T, Nishitoba J, Unno Y, Enrique Yalta Soplin N, Heymann J, Wiesner M, Chen N, Renduchintala A, Ochiai T (2018) ESPnet: end-to-end speech processing toolkit. In: Proceedings of INTERSPEECH 2018, pp 2207–2211
19. Dai Z, Yang Z, Yang Y, Cohen WW, Carbonell J, Le QV, Salakhutdinov R (2019) Transformer-xl: attentive language models beyond a fixed-length context. In: Proceedings of the 57th annual meeting of the association for computational linguistics, pp 2978–2988

Implementing Software-Defined Networks in Heterogeneous 5G Communications to Provide Security and Intelligent Resource Management

Omid Mahdi Ebadati E. and Hamid Reza Ebadati E.

Abstract Software Defined Network (SDN) is a technology that facilitates novel design and management of mobile networks. The biggest challenge of SDN security is that it uses centralized control. Therefore, it is necessary to provide different methods security and resource management in fifth-generation mobile networks, which has an erratic and congested network architecture due to rapid increasing data growth. One of the solutions that can be effective in this area is the use of SDN-based networks, which need to be introduced and articulated by a programmable management platform for data traffic and security management. In this research, based on the fifth-generation mobile network, the network demand is analysed to propose a model to optimize network performance in SDN. The result shows that the average energy consumption of each node is dramatically reduced and as far as incrementing of nodes, the node data rate and energy consumption is decreasing, which shows the efficiency of the model.

Keywords Software-defined network · Heterogeneous · Security · 5G · Wireless networks · Mobile

1 Introduction

Software-defined Network is to design, construct, manage and maintain components of multiple computer networks with a software-based approach. The SDN general rule is to separate network control processes (route, access control, bandwidth, security, etc.) from data packet transfer processes. A server equipped with a controller monitors all control functions, and therefore, this technology has transformed the layered structure of the TCP/IP protocol, which is the unrivaled protocol in the

O. M. Ebadati E. (✉)
Department of Mathematics and Computer Science, Kharazmi University, Tehran, Iran
e-mail: ebadati@khu.ac.ir

H. R. Ebadati E.
Department of Computer Engineering, Islamic Azad University, Tehran, Iran

© The Author(s), under exclusive license to Springer Nature Switzerland AG 2022
N. H. T. Dang et al. (eds.), *Artificial Intelligence in Data and Big Data Processing*,
Lecture Notes on Data Engineering and Communications Technologies 124,
https://doi.org/10.1007/978-3-030-97610-1_54

685

present networks. Large network operator companies strategically and efficiently need to take advantage of the economic benefits and superior applications of this new technology, as well as forcing IT managers to invest and take research measures in this area [1].

In software-defined networks, the control part is separated from the sending or transmitting part of the data and is designed and implemented for a particular network section. These networks increase the intelligence and security of the network, and by use of virtualization tries to reduce the cost of the network. Despite the many positive points in these networks, due to different vector attacks [2, 3], such as sending invalid packages floods to large resources can threaten the security of the network and also face a new challenge in providing services.

The fifth generation of mobile network (5G) is now the latest generation of mobile cellular communication systems. Among the goals of the fifth-generation are to increase the speed of data transmission up to 20 Gigabit per second (one Gigabit per second per user), increase the capacity to serve network subscribers, reduce network latency and consume less power in network and user equipment (such as mobile phones), as well as support for more speed of moving users (such as users riding cars on highways or high-speed trains). In this generation, at least one million mobile broadband communications users per square kilometer should be able to connect to the network, and each user should be able to experience a minimum speed of 100 Mbps at the busiest [4]. Reduction of energy consumption is also a measure that has been considered in this generation and operators' transmitter, and receiver devices should be entered into the energy-saving mode in low hours and activated quickly, which is not mentioned in the fourth generation. As a result, the fifth generation, for example, allows people to communicate in streaming during peak hours of mobile consumption. Furthermore, its other operational goal is to improve support for machine-to-machine communications, i.e. IoT, at lower cost, lower battery consumption and less latency than the fourth generation.

Based on the above introduction, the goals and innovation of this research include:

- Review SDN on 5G networks
- Mapping the challenges of SDN, especially in 5G networks
- Simulation and implementation of SDN networks in a simulated 5G network
- Presenting a SDN architecture model for optimal use in 5G networks
- Implementation of simulated model and investigation of energy and routing and programming capability for model development in 5G.

2 Research Background

2.1 5G Networks and Its Development

Cambell et al. introduced and expressed various features, applications, challenges related to software-defined networks and tried to fully address SDN networks and the benefits of this network, such as architecture and protocols [5]. In 2014, Nunes et al. conducted a research to express a new approach of network that enables managers to use OpenFlow protocol, which is the first software intermediary between control and sending layers, enabling direct access and making changes in a multi-layer program of network equipment, such as switches, virtual and physically routers [6]. In another research, authors tried to reduce the time of packet transfer during virtual switch flows by providing a new method, which will increase network efficiency by affecting the response time and bandwidth used in the network [7]. In another research, authors review software-based data centers and providing energy-efficiency solutions using software-defined networks, while they defined the exact and complete implementation of software-based data centers and its technical and structural model to address the new approach of data centers [8]. In some research, authors focused on cognitive computing and artificial intelligence to achieve security in several application areas for Industrial Cyber-Physical Systems [9].

Yang et al. divided the specific challenges of SDN into seven general categories, including, unauthorized access, data leakage, data change, malicious programs, service bans, configuration issues and SDN security level, and examined each one separately [10]. Other researchers have investigated methods about potential standards and development directions and present recent research efforts in the field of fifth generation of mobile networks and addressing its challenges. They describe advantages and disadvantages of this generation of networks [11]. In 2019, another study by Angalaban and his colleagues introduced an LTE/Wi-Fi integrated architecture with the use of SDN in mobile sector and improved components that provide movement toward next-generation mobile networks [12]. In 2017, Jiang et al. proposed a Radio Resource Sharing Initiative (RRS) between different mobile infrastructures to provide free access to all radio resources available around the mobile device. This proposed RRS project that enabled and exploitable using the concept of SDN and virtualization of radio access resources [13]. Another research described the features and architecture of SDN as well as 5G networks and demonstrated the requirements and solutions to use this type of networks in 5G mobile communications [14]. In another study, Tuncer et al. introduced and articulate a software-defined security architecture for 5G based on SDN. They showed how security applications can be easily improved in the architecture through the interface [8].

One of the issues related to using of 5G networks for IoT while providing appropriate speed and quality compared to different generations of existing networks is the security and privacy issue of individuals in these networks [15]. Moreover, software-defined radio access networks (SD-RAN) introduced. Authors believe that software-defined design in RAN can be a key step to support network segmentation, RAN sharing, flexible and appropriate frequency spectrum management and other important and key feature in 5G networks [16].

2.2 Software-Defined Networks

SDNs increase the intelligence of networks and provide capabilities such as planning, scalability, flexibility, automation and implemented by transferring the data control section from switches and hardware routers to virtual network software layers and using a centralized software controller [6]. SDN enables to work and program on L2 by using the OpenFlow protocol. By adding SDN and OpenFlow to network equipment in addition to the possibility of using the tools provided in the device itself, researchers are able to use software techniques to provide programs for intelligent management of the entire network and will no longer be limited to programs offered by network equipment providers [17]. Since software-defined networks focus on separating the controller and data layer, the SDN architecture consists of three layers of infrastructure, control and application. The controller is located in the control layer, and it is the most important layer in this type of network architecture. In the following, different components of this network are expressed (Fig. 1).

Application Layer: From the perspective of this layer, the new network appears to be a series of logical and unified switches. Commercial applications located in this layer include, security applications, network virtualization applications, network monitoring, access control, etc.

Control Layer: Control level includes a set of software controllers that provide an integrated and comprehensive control through the programming interfaces of applications to monitor network behaviors. In this layer, there are two application programming interfaces that are responsible for the controller's communications with other controllers and layers, these two interfaces are:

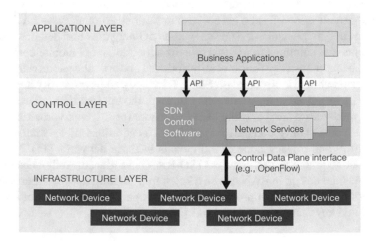

Fig. 1 Software-defined network structure [18]

- **Southbound**: The main task of this interface is to link the control layer with its bottom layer, i.e. the infrastructure layer, as well as the OpenFlow and Forces protocols, which are responsible for programming the network software.
- **Northbound**: This bound is responsible for linking the application layer and the control layer. In fact, it allows programs to interact with the control unit. With the software structure, it enables users to schedule the data control and guidance section, and the controller manages the data sending section by adding, modifying, or deleting the entries of the switch routing table [11].

Infrastructure Layer: This layer is located at the lowest level of architecture, known as the data level, and has physical and virtual switches. The role of virtual switches is connecting virtual servers with virtual network card and control traffic congestion. In general, this layer is responsible for managing and directing packages in a suitable way that path is determined by the controllers in the control layer [19].

SDN Protocols

- **Forces Protocol**: This protocol is provided by the IETF working group, which aims to separate the central control unit from data exchanges, and that is the conceptual definition of software-defined networks.
- **OpenFlow Protocol**: It allows direct access and changes to the network equipment program such as switches and routers, both physically and virtually. The lack of an open interface in the data for sending program has caused today's network equipment to be integrated, closed and software-based controller.

OpenFlow Architecture

The OpenFlow protocol architecture, includes a series of flow tables and an abstract layer that has been used to securely communicate between SDN layers. The flow table contains three information fields that are presented in Fig. 2.

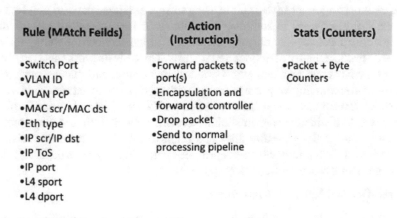

Fig. 2 OpenFlow protocol architecture [19]

Fig. 3 Controllers in
software-based networks
[19]

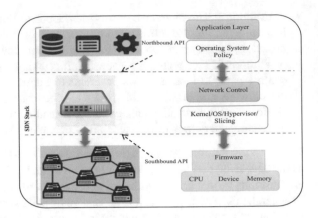

In an OpenFlow network, switches exist in both pure and hybrid forms. Pure switches do not have any of the old network's properties with in-switch control properties and rely entirely on separating controllers to decide whether to send packages. Hybrid switches, however, support OpenFlow protocol alongside traditional operations and protocol. Using the OpenFlow protocol, the controller can connect to the switch and add or change certain commands in the switch flow tables. SDN switches and controllers are communicated on a channel by the OpenFlow protocol. The details of parameters of SDN stack illustrated in Fig. 3.

Recently, two important types of controllers have been emerged:

- **Centralized controllers**; such as NOX, Maestro, Beacon, SNAC, Trema, Floodight.
- **Distributed controllers;** such as: DIFANE, Onix, HyperFlow.

Software-defined Networks Advantages

One of the most important advantages of using software-defined networks is programming, elasticity and trust. In addition, it enables the network operator to implement the protocols, rules and policies to enforce control of the services, including routing, engineered traffic, Quality of Service and security. Thus, the network meets the needs of the user, adapts and manages and configures in a centralized manner using open application programming interfaces standard without changes to network devices and data sectors. Barmpunakis et al. have investigated 5G networks and utilized different layers as well as distinctive parts in Slicing Network in bandwidth, and they noted the advantages of using SDN. Finally, this paper refers to cross layer control in software-defined network technology which can provide appropriate layers in the architecture of these networks based on frequency and proportional [1].

Software-Defined Network Challenges

One of the major challenges faced by network equipment designers and manufacturers to migrate to SDNs is the of massive network infrastructure as it includes a huge variety of equipment and protocols of the network. Another challenge of these networks is the separation of data and control sections, which if the allocation of resources of different programs is not done correctly, will cause interference in the decision making of the controller unit, and the efficiency of the whole network be affected. The biggest challenge for SDN is the fact that these networks use centralized controllers. If a SDN server is attacked or a hacker gets access to the available network controller, it can take over network traffic entirely. Increasing the use of multimedia and demand-driven services requires fundamental changes in network structures [20].

Virtualization

Virtualization or the use of virtual machines is one of the methods used in SDN. In this way, using virtual tools in the network, network scalability can be investigated more efficiently. A virtual network which its components are resources such as switches, routers and network equipment and can be easily programmed. It is easy to take a code from the user and applies it to the virtual network.

The comparison table of the most important background researches related to the present study is as follows (Table 1).

In this research, based on the studied literatures, and investigating the challenges and problems of SDN networks to implement in enterprise networks, we propose a more optimal model in terms of routing and energy consumption and try to solve other common problems in software-defined networks, due to impossibility of practical implementation of this model, it has been tried to simulate this model.

3 Research Methodology

Today, network's requirements are included, improving efficiency and establishing wider connections. Companies are increasingly forced to face security regulations and there is a growing demand to provide different connection environments for the user through different devices at different times.

By empire the network hardware, it is no longer compulsory to enforce policies on network hardware. Using a centralized software application that acts as a control layer allows networks to be virtualizes. This process is similar to server virtualization:

- **Server Virtualization**: The process of creating various virtual machines (VM) and differentiating them from physical servers.
- **Network virtualization**: The process of creating virtual networks that are separated from the physical components of the network.

A key aspect of SDN is the user interface, which provides for the network. SNMP API interface is more considered than other interfaces in SDN. The API allows programmers to interact with the network, obtain information about the status of the network, schedule the forwarding table, and can avoid existing hardware differences between platforms in order to easily implement their goals despite the heterogeneous environment.

Table 1 Comparison of important related literatures

Authors	Years	Problems	Features
Balaram et al. [18]	2017	Not checking 4G networks and its functions Failure to investigate the advantages and functions of SDN in 4th and 5th generation networks	1. Using OpenFlow Protocol in SDN 2. Review of Using OpenFlow in SDN
Yang et al. [10]	2019	Battery Consumption Not investigating the effectiveness of 5G networks Failure to check data architecture and data transfer rates	1. Segmentation of networks at the time of sharing 2. Stream review and reduction of energy consumption
Jiang et al. [13]	2017	Not checking effective protocols for optimizing SDN network's lack of the overview of data transfer rates and its effects on 5G networks	1. Investigating the different features of SDN networks 2. Explaining the main architecture in SDN 3. Proposing the use of SDN and the possibility of optimizing consumption and cost in 5G networks
Barmpounakis et al. [1]	2019	Not checking all the effective layers in the network Limitations in frequency presentation and optimization	1. Network bandwidth components in the frequency domain 2. Proposing the use of management layers to utilize appropriate frequency in SDN-based software networks in 5G generation
Anbalagan et al. [12]	2019	Restrictions on online game reviews Not checking the effectiveness of 5G network	1. Optimizing SDN network traffic 2. Check the quality of the service 3. Check the quality of the experience
Ebadati et al. [21]	2020	Limitations in practical simulation and implementation and the number of nodes used	1. Comparison of current network infrastructures and problems with cost reduction in SDN use 2. Energy Optimization 3. Routing Optimization 4. Volume of data transfer rate
Althobaiti et al. [9]	2021	Limitation on time management in view of the deep learning facts	1. Industrial Cyber-Physical Systems 2. Elect an optimal subset of features 3. Hyperparameter optimization

As the result, SDN is a framework that will allow network administrators to dynamically and automatically control and manage a large number of network devices, services, topology, traffic routes and package management policies using high-level languages and APIs.

4 Implementation

In this research, the 5G network in the form of software-defined networks is simulated since researchers have different opinions about the capabilities and characteristics of the 5G network. There is not any experience of real implementation in the past, so, we want to simulate this network and obtain relative results from this simulation. Matlab ver. 2016b software has been used in simulation of the proposed model, as in other previous researches used the same. For network and SDN functions, tools such as Mininet and IPref have also been used.

In the implementation of SDN, the followings are considered:

- **Virtualization**: Using network resources without worrying about where they are physically located.
- **Orchestration**: The ability to control and manage thousands of devices with one command.
- **Programmable**: Change performance during execution.
- **Dynamic upgrade capability**: Resizing and numbering.
- **Automating**: It includes automating, troubleshooting, implementing policies, reducing downtime, preparing/dividing resources, adding workloads, locations, devices and resources.
- **Visibility**: Ability to monitor resources and connections.
- **Efficiency**: Optimizing the efficiency of network devices by engineering traffic/bandwidth management, capacity optimization, load balancing, rapid error management.
- **Multi-tenancy**: Providing an architecture that a single instance of software (application) runs on a server and then provides services to several tenants. Tenants need complete control over addresses, topology, routing and security.
- **Service integration**: Load balancing, firewalls, intrusion identification system (IDS), demand-based supply and traffic route placement.

4.1 Database

In this research, a database for both local and public space is used. In this simulation, we use my_sql database and transfer all the required information through this database. The main reason for using this database is its availability at all levels, allowing us to benefit from it in any network of any size. In this way, we were able to run the network first in a client and then close it to a wider network and transfer it to a higher level, so we were able to resize the network and check all bugs and network problems.

4.2 Programming Language

Programming languages used in controllers are common programming languages such as Python, Java and C++, and in certain cases, Ruby and JavaScript, which make them quickly learn to use these controllers. In industry applications that require fast execution time, java can be a good option. In addition, Java also supports multi-threading. On the other hand, there are problems with multi-threading in Python. It should also be noted that there are limitations in memory management in C/C++.

4.3 Memory Consumption

The amount of memory consumption is one of the most important issues among controllers. The Beacon controller consumes less memory than other controllers, so it has a relatively better performance. The best operating system to simulate this research is Linux, because it has a very good stability and also requires few resources, thus providing higher speed and accuracy than other operating systems. In this research, Debian-based distribution is used, which is the best and lightest operating system for single-range computers such as Raspberry Pie. This distribution is very close to the Debian distribution and has a good support. Also, the number of software packages provided for it is adequate. Simulation performed in a system with Intel Octa Corei7 CPU and 8 GB RAM. In this simulation scenario, we selectively distributed 50 nodes with a range of 10 m in a space of 100*100 dimensions and finally analyzed the amount of energy consumption and data transfer rate in them.

In Fig. 4, the nodes' paths are designed in such a way that the lowest amount of energy is consumed by the sum of nodes in the grid. In the next figure (Fig. 5), we will examine the amount of energy consumption in the designed grid.

In Fig. 5, we can see the rate of data transfer, which, it turns out, the distance of nodes that has a little effect on the rate of data transfer. This indicates that the model that is used in this study is not only reducing energy consumption, nevertheless, has been able to take a high rate in data transfer as well.

Figure 6 shows the energy consumption of each node in the network. The figure shows as the energy is decreasing during the lifespan of the network, it reached to 0.1, which is a quite significant amount and shows that the proposed model has reduced energy consumption in SDN nodes.

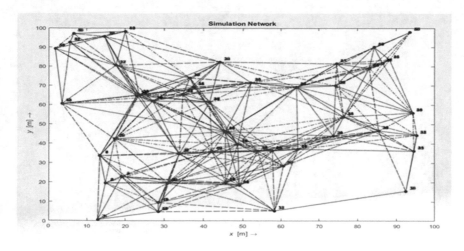

Fig. 4 Routes to send information packets in the software-centric network

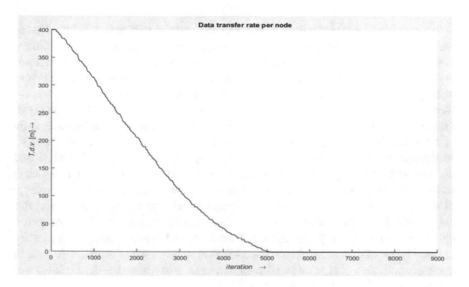

Fig. 5 Data transfer rate in the designed network

5 Conclusions and Future Work

In this paper, initially, Software-defined Networks and its applicant in mobiles networks have been investigated. The main aim is to examine and propose the best solution by using SDN in 5G mobile networks. Accordingly, we were able to simulate and analyze a SDN-based on 5G mobile network and obtain efficient results. In this research, we have tried to using different algorithms for allocating and managing

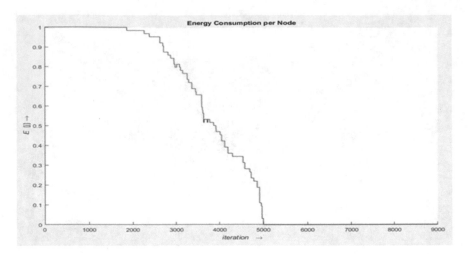

Fig. 6 Average energy consumed by each node

data sources; and we were able to find distinctive ways to reduce source loss. One of these algorithms was the use of intelligent time-based resource allocation algorithms to solve resource allocation problems and minimize resource loss. One of the features of software-defined networks is accuracy in routing, which helps us to have a very intelligent routing in the network. In addition to reducing traffic in the network, we were able to track packages, and increase security in these networks. In continue, through this routing, we are able to improve resource management and reduce waste of resources by reducing network consumption. In this study, we conducted two approaches to increase the security and efficiency of SDN network. We increased the tolerance of the network against error by using the OpenFlow algorithm which is based on load and energy distribution. Therefore, we are able to maintain the energy of the grid for a long time, which caused more nodes to survive and can transfer data. Using the OpenFlow protocol in the hierarchical structure of SDN can ensure optimal allocation of resources in the software-defined network along with security in these types of fast networks. It is possible to use software-defined networks in heterogeneous networks, such as 5G communications in a way that its efficiency and capabilities can be used in the best way.

Some of the limitations of this research are as follows:

• Failure to implement different security layers and check security issues that increase the rate of data transfer.
• Lack of access to large virtualization resources, such as powerful servers, for accurate and effective implementation and high data transfer rate.
• Implementation with a limited number of nodes, which needs a stronger hardware to investigate the average energy consumption and data transfer rate. However, due to the need for high density in 5G networks and optimization of data transfer rate, this will definitely have a low and limited effect in high volume.

Reduction of energy consumption and longevity is one of the most important research issues in the field of SDN networks due to severe limitation of resources. Operational SDN is the establishment of an effective topology control protocol to increase lifespan and reduces energy consumption. For the future work, topology and challenges in SDN are suggested to consider and design a sleep-based topology control algorithm. Furthermore, a comprehensive study of sleep-based topology control algorithms in SDN is recommended to identify the operational domain for future study purpose.

References

1. Barmpounakis S, Maroulis N, Papadakis M, Tsiatsios G, Soukaras D, Alonistioti N (2020) Network slicing—enabled RAN management for 5G: cross layer control based on SDN and SDR. Comput Netw 166:106987
2. Silva M, Teixeira P, Gomes C, Dias D, Luís M, Sargento S (2021) Exploring software defined networks for seamless handovers in vehicular networks. Veh Commun 31:100372
3. Valdovinos IA, Pérez-Díaz JA, Choo K-KR, Botero JF (2021) Emerging DDoS attack detection and mitigation strategies in software-defined networks: taxonomy, challenges and future directions. J Netw Comput Appl 187:103093
4. OMDIA. https://omdia.tech.informa.com/OM016694/IoT-in-Private-Cellular-Networks
5. Campbell AT, Katzela I, Miki K, Vicente J (1999) Open signaling for ATM, internet and mobile networks (OPENSIG'98). SIGCOMM Comput Commun Rev 29:97–108
6. Nunes BAA, Mendonca M, Nguyen X, Obraczka K, Turletti T (2014) A survey of software-defined networking: past, present, and future of programmable networks. IEEE Commun Surv Tutorials 16:1617–1634
7. Kim H, Feamster N (2013) Improving network management with software defined networking. IEEE Commun Mag 51:114–119
8. Tuncer D, Charalambides M, Clayman S, Pavlou G (2015) Adaptive resource management and control in software defined networks. IEEE Trans Netw Serv Manag 12:18–33
9. Althobaiti MM, Kumar KPM, Gupta D, Kumar S, Mansour RF (2021) An intelligent cognitive computing based intrusion detection for industrial cyber-physical systems. Measurement 186:110145
10. Yang Z, Cui Y, Li B, Liu Y, Xu Y (2019) Software-defined wide area network (SD-WAN): architecture, advances and opportunities. In: 2019 28th international conference on computer communication and networks (ICCCN), pp 1–9
11. Yu M, Wundsam A, Raju M (2014) NOSIX: a lightweight portability layer for the SDN OS. SIGCOMM Comput Commun Rev 44:28–35
12. Anbalagan S, Kumar D, Raja G, Balaji A (2019) SDN assisted Stackelberg game model for LTE-WiFi offloading in 5G networks. Dig Commun Netw 5:268–275
13. Jiang M, Xenakis D, Costanzo S, Passas N, Mahmoodi T (2017) Radio resource sharing as a service in 5G: a software-defined networking approach. Comput Commun 107:13–29
14. Network Computing. https://www.networkcomputing.com/cloud-infrastructure/role-iot-gateways-network
15. Sicari S, Rizzardi A, Coen-Porisini A (2020) 5G in the internet of things era: an overview on security and privacy challenges. Comput Netw 179:107345
16. Tuncer D, Charalambides M, Clayman S, Pavlou G (2015) On the placement of management and control functionality in software defined networks. In: 2015 11th international conference on network and service management (CNSM), pp 360–365
17. Olivier F, Carlos G, Florent N (2015) New security architecture for IoT network. Procedia Comput Sci 52:1028–1033

18. Balaram VVSSS, Mukundha C, Bhutada D (2017) Enhancement of network administration through software defined networks
19. Jain S, Kumar A, Mandal S, Ong J, Poutievski L, Singh A, Venkata S, Wanderer J, Zhou J, Zhu M, Zolla J, Hölzle U, Stuart S, Vahdat A (2013) B4: experience with a globally-deployed software defined wan. SIGCOMM Comput Commun Rev 43:3–14
20. Barakabitze AA, Ahmad A, Mijumbi R, Hines A (2020) 5G network slicing using SDN and NFV: a survey of taxonomy, architectures and future challenges. Comput Netw 167:106984
21. Omid Mahdi Ebadati E, Eshghi F, Zamani A (2020) A hybrid encryption algorithm for security enhancement of wireless sensor networks: a supervisory approach to pipelines. Comput Model Eng Sci 122:323–349

Service Quality of FiberVNN Prediction Using Deep Learning Approach

Bui Thanh Hung

Abstract Fiber optic technology, a commonly used technology nowadays, is also applied in FiberVNN to connect to the homes of end-users to provide high-speed Internet access and broadband-based services. FiberVNN is considered to be the most modern fiber optic Internet access service in Vietnam at this time which is provided by Vietnam Posts and Telecommunications Group (VNPT) with a transmission line entirely by optical cable from the connector to the user. With the aim to improve service quality, the challenge comes up is how to predict the quality of FiberVNN service in a timely manner. In this paper, we collect the dataset, do the preprocessing, then find the most important features and apply Long Short Term Memory (a deep learning approach) to predict the service quality of FiberVNN. Experimental results have shown that deep learning approach has achieved the best results for this task comparing with three machine learning algorithms: K-Nearest Neighbors, Logistic Regression and Support Vector Machine.

Keywords Deep learning · Long short term memory · FiberVNN · Service quality prediction

1 Introduction

FiberVNN service is a high-speed fiber optic service which is provided by VNPT. Using a transmission line entirely by optical cable from the connector to the user, FiberVNN can be seen as the most modern fiber optic Internet access service in the local market as of now [1].

FiberVNN uses fiber optic technology to connect to the end-users' homes to provide high-speed Internet access and broadband-based services. With FTTH technology, FiberVNN service can provide download speed up to 10 Gigabit/s, 200 times faster than ADSL + 2. Fiber optic cable is capable of transmitting data running from

B. T. Hung (✉)
Faculty of Information Technology, Ton Duc Thang University, 19 Nguyen Huu Tho Street, Tan Phong Ward, District 7, Ho Chi Minh City, Vietnam
e-mail: buithanhhung@tdtu.edu.vn

© The Author(s), under exclusive license to Springer Nature Switzerland AG 2022
N. H. T. Dang et al. (eds.), *Artificial Intelligence in Data and Big Data Processing*,
Lecture Notes on Data Engineering and Communications Technologies 124,
https://doi.org/10.1007/978-3-030-97610-1_55

the network hub straight to the end-user's home for stable signal quality without signal loss due to electromagnetic interference, weather or cable length. FTTH allows for a balanced transmission rate, which can serve hundreds of computers at the same time with extremely high security. FTTH is especially effective with the following services: Private server hosting, VPN (Virtual Private Network), data transmission, online games, IPTV (Interactive television), WoD (watching movies on demand), Video conference, IP camera with the advantage of high data transmission bandwidth, unaffected by electrical interference, magnetic field … [1, 2].

The industrial revolution 4.0 has created more-than-ever favorable conditions for the businesses if they could observe and seize the opportunity. Products and services could be delivered to customers in a faster way; meanwhile, there are also a variety of choices for modern sales channels. Therefore, the competition on price, distribution channel is no longer as important as before; the critical competitive factor today is the quality of products and services as well as the customers' experience.

With a broad vision, VNPT has realized this trend and without any delay, they have invested from the hay day in network infrastructure, equipment, digital transformation in operation and management, etc. which has proved to achieve many positive results. In which, Fiber Optic Internet service (FiberVNN) plays a pivotal role in the current competition. To win in this competitive market, it requires to figure out a key solution to improve service quality in a timely manner. Therefore, building an application to predict the quality of service of FiberVNN has come up as a critical task today.

In this research, we propose a solution to predict the service quality of FiberVNN at Trang Bang, Tay Ninh province by deep learning approach. Based on the data collected from the VNPT system, this research has conducted the following steps:

- Collecting the data and preprocessing.
- Analyzing features which affect the service quality of FiberVNN at Trang Bang, Tay Ninh province.
- From the features affecting the service quality of FiberVNN, we have conducted service quality prediction by deep learning method: Long Short Term Memory. We also compare the result of deep learning method with three machine learning algorithms such as: K-Nearest Neighbors, Logistic Regression and Support Vector Machine.

The paper consists of five sections with introduction, related work, the proposed model, experiment and conclusion. Section 1 introduces about the problem including research method and practical scientific significance and research purpose of the paper. Section 2 presents related works related to this paper. Section 3 presents the proposed model and features of this model. The experiences are presented in Section 4 with the data collection, preprocessing and the result of the proposed model. Section 5 concludes the paper as well as gives future research direction.

2 Related Works

There are not many researches focusing on the topic: service quality of Fiber prediction. This problem still remains as a huge challenge. The first problem is how to detect the features which affect the service quality automatically. At VNPT, they used factors which were advised by their experts to achieve high results and easy to apply in practice. And the second problem is how to predict the service quality. In this session we review some papers related to our research. Kozdrowski et al. used supervised machine learning approach with Random Forest, Decision Trees, Support Vector Machines, XGBoost and Logistic Regression algorithms to predict transmission quality in fiber optic networks. They evaluated by AUC score and got the score from 0.86 to 0.89. By their experiments, Random Forest algorithm gave the best results [3]. The achievement of this study is to predict the fault points on the entire fiber optic cable. However, due to the small data set, the use of machine learning has not achieved high accuracy.

Ahmad et al. predicted customers switching networks in telecommunications using machine learning technology. The author uses Logistic regression and KNN with big data to predict the volatility of customers in the telecommunications sector. Research results show that the accuracy of large and small datasets of Logistic Regression performed much better than the accuracy of KNN. From this study, data analysis with machine learning techniques has been proven to be accurate and effective for predicting customer churn in the future [4].

Deep learning is applied in many applications in the optical domain such as intrusion detection, fault prediction, impairment-aware routing, physical flow, traffic-aware capacity reconfigurations and security low-margin design. Deep learning can be divided to three main categories: deep supervised learning, deep unsupervised learning and deep reinforcement learning [5–7].

Karanov et al. proposed end-to-end deep learning of optical fiber communications [8]. Prakash et al. used multi-model algorithms for early predication of osteoporosis [9]. Ankur Gupta et al. proposed machine learning and deep learning in steganography and steganalysis [10]. Hung et al. proposed hybrid deep learning approach in many researches of text and image [11–13]. Our research is different with the others mentioned above. We analyze the factors affecting to the service quality of FiberVNN and propose these features for deep learning model-Long Short Term Memory to predict service quality. The effectiveness of our proposal is illustrated in the following sections.

3 Methodology

3.1 The Proposed Model

The proposed model for predicting service quality of FiberVNN consists of 4 components as follows:

- Collecting the raw data
- Preprocessing the raw data
- Extracting features
- Building the model for prediction.

The proposed model is shown in Fig. 1, we will describe in details in the next part.

Fig. 1 The proposed model

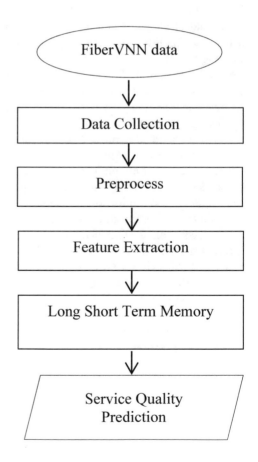

3.2 Data Collection

We collected the data at VNPT Trang Bang Telecommunication Center, Tay Ninh province. The data is collected automatically by VNPT Management Application. The data consists of all FiberVNN user of VNPT Trang Bang, Tay Ninh.

3.3 Preprocessing

Handling outliers or unusual data is one of the most widely used terms in the data science. Identifying and eliminating outliers is an extremely important step in the data processing process. The processing of outliers will help to increase the accuracy of predictive models. We handle outliers of FiberVNN data first; after the data is rendered, we then remove unnecessary fields, error lines due to empty data; finally we double check the results by manual work.

3.4 Feature Extraction

From the previous study about the factors affecting to the service quality of FiberVNN [2–4], in order to measure these factors through the collected data, we explorered the features which are presented in detail as follows.

Time

Time is selected at various times, including night. Determining the time feature provides a clear picture of how this factor affects to service quality.

Weather

The weather feature taken for research is specific to the Southeast region with two seasons: rainy and sunny, so this feature only includes two values: "rain" and "sun".

Transceiver power feature at OLT

The optical power transceiver is installed at the telecommunication station; it provides extremely fast bandwidth, meets full service access and many scenario requirements, and provides QoS guarantee and security at the station. This factor has an impact on customers when the level of reception and transmission is weak or out of threshold, increasing the overall loss.

Power feature of receiving and transmitting at the terminal

The receiving and transmitting terminal is installed at the customer's house, used to receive the signal from the device in the incoming station and return the signal after processing. Figure 2 shows that the received signal is −30.46 dBm and the

Serial Number:	VNPT00C3BF08
Laser Bias Current:	17054 uA
Optics Module Voltage:	3246 mV
Optics Module Temperature:	39 oC
Rx Optics Signal Level at 1490 nm:	-30.46 dBm
Tx Optics Signal Level at 1310 nm:	2.36 dBm
Optics Threshold - Lower:	-8 dBm
Optics Threshold - Upper:	-27 dBm

Fig. 2 Illustration of transceiver element at the terminal

Table 1 Illustration of upload/download direction loss features

Parameter	Download	Upload
Emission level (dBm)	4.1	2.6
Collection rates (dBm)	−25.7	−25.9
Attenuation (dB)	29.8	28.5

transmitted signal is 2.36 dBm.

Attenuation feature for upload and download direction

The upload/download value is used to manage the upload/download bandwidth of data up and down. When the attenuation value for each factor increases or decreases, it will give us different service quality. Table 1 shows that the upload interpolation is 28.5 dB and the download is 29.8 dB.

Subscriber house distance feature

The subscriber's house distance is the length of the cable line calculated from the equipment installed at the station (OLT) to the customer's terminal. The line depends on many factors such as: bend, connection ability, insect bites.

After exploring the factors affecting the service quality of FiberVNN are described as above, in order to measure these features through the collected data, we chose 8 key features: Weather, Transmit power OLT, Collection capacity OLT, Terminal output power, Terminal receiving power, Attenuation download, Attenuation upload and Subscriber distance which are described in Table 2.

To evaluate the impact of the above features on the service quality of FiberVNN, we used the Pearson correlation coefficient. The correlation coefficients are presented in the following section:

Pearson correlation coefficient

Pearson correlation coefficient measures the degree of linear correlation between two variables. In principle, Pearson correlation will find a line that best fits the linear

Table 2 The feature of service quality prediction

Feature	Description
W	Weather with 2 values: "rain" and "sun"
TPO	Transmit power OLT
CCO	Collection capacity OLT
TOP	Terminal output power
TRP	Terminal receiving power
AD	Attenuation download
AU	Attenuation upload
SD	Subscriber distance

relationship of two variables. Pearson correlation coefficient (r) will take the value from +1 to −1. The condition for meaningful correlation is the sig <0.05. The Pearson fomula is caculated as follows [14]:

$$r = \frac{\sum (x - \overline{x})(y - \overline{y})}{\sqrt{\sum (x - \overline{x})^2 \sum (y - \overline{y})^2}} \tag{1}$$

where x, y are two sets, and $\overline{x}$ and $\overline{y}$ are mean of each set.

- r < 0 indicates a negative correlation between the two variables, that is, if the value of one increases, the value of the other will decrease.
- r = 0 shows no correlation.
- r > 0 indicates a positive correlation between two variables, that is, if the value of one increases, the value of the other will increase.

3.5 Long Short Term Memory

Long Short Term Memory networks, commonly known as LSTMs—are a special type of RNN that is capable of learning distant dependencies. LSTM was introduced by Hochreiter and Schmidhuber [15], and has since then been refined and popularized by many people in the industry. They work extremely effectively on many different problems, so they have gradually become as popular as they are today.

The main point in the network architecture of the LSTM is the memory cells with ports that allow the storage or retrieval of information. These ports allow for over-writing (input gates), redundancy removal (forget gates) and access (output gates) of information stored inside memory cells.

LSTM is designed to avoid the problem of long-term dependency. Remembering information for a long time is their default property, we don't need to train it to be able to remember it. That is, its internals can already be memorized without any intervention [11–13, 15].

Fig. 3 RNN architecture

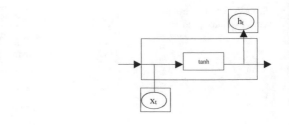

Fig. 4 LSTM architecture

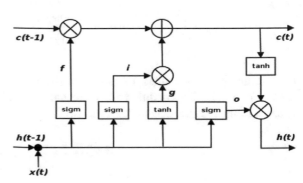

Every regression network takes the form of a sequence of repeating modules of the neural network. With standard RNNs, these modules have a very simple structure, usually a tanh layer where x_t is the input data at time t, and h_t is the output data at time t. Figure 3 shows architecture of RNN.

LSTM also has the same chain architecture, but the modules in it have a different structure from the standard RNN. Instead of just one neural network layer, they have up to four layers that interact with each other in a very special way. In addition to using two inputs, current input (x_t) and hidden state (h_{t-1}) like a normal RNN, LSTM also uses one more input, cell state (c_{t-1}). Here we can consider the cell state as the 'memory' of the LSTM network and it can store the information of the first timesteps. The LSTM architecture is shown in Fig. 4.

3.6 Service Quality Prediction

We divide the service quality manually to three values: "Good", "Pass", "Poor" or 3, 2, 1 and named as SQ field. Predicting service quality of FiberVNN is done through the five steps:

Step 1: Divide the dataset to Training and Testing.
Step 2: Put the features extraction into Long Short Term Memory.
Step 3: Train the model.
Step 4: Do activation function.
Step 5: The service quality prediction results.

4　Experiments

4.1　Dataset

Data is collected automatically from the internal production management software of VNPT Tay Ninh. The data includes 18,952 samples with 18,952 rows and 09 columns. We divided the dataset to Training and Testing separately. To know how the size of dataset affects to the result of the model, we chose two ratios with Train/Test are 6/4 and 8/2. We also evaluate by cross validation to analysis how random Training/Testing dataset affects to the result, by this experiment we choose k-fold equals 10.

4.2　The Result

We used Sklearn, Tensorflow [16] and Keras [17] library to train the model. Parameters used in the LSTM model are: Number of hidden nodes: 128, Drop out: 0.2, Activate function: Sigmoid, Epochs: 100, Batch size: 128, Adam Optimization, Loss function: Categorical cross entropy. We evaluated the result by Accuracy and F1 score calculated as follows:

$$Accuracy = \frac{Number\ of\ True\ Predictions}{Number\ of\ Predictions} \tag{2}$$

$$F1 - score = 2\frac{Precion * Recall}{Precision + Recall} \tag{3}$$

The result of features affecting the model by Pearson scores is shown in Fig. 5. Table 3 shows descending order features affecting the service quality of FiberVNN.

The analysis results of Pearson correlation coefficients shows that the selected features have a certain influence on the service quality of FiberVNN. On that basis, we decide to choose these factors for service quality prediction. We compared the result of our proposed model with 3 machine learning models: K-Nearest Neighbors (KNN), Logistic Regression (LR) and Support Vector Machine (SVM). The result of different ratio Train/Test dataset is shown in Tables 4 and 5 shows the result of cross validation K-fold = 10.

The results of Train/Test ratio 60/40, 80/20 and K-fold = 10 show that LSTM model got the highest score in both Accuracy and F1-score. From the above analysis results, it is easy to see that the Attenuation download (AD) and the Terminal Receiving Power (TRP) at the customer terminal have an impact on the most quality of FiberVNN service at Trang Bang, Tay Ninh. This proves that when the download direction loss is high and the reception level at the customer terminal is weak, it will lead to poor service quality. Through this result, we need to pay attention to the signal reception level at the customer terminal, which has not been taken into account at

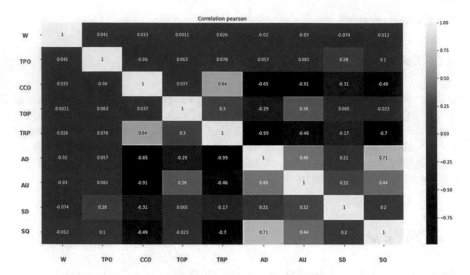

Fig. 5 Pearson correlation result of features affecting the service quality

Table 3 Descending order features affecting the service quality

No	Feature	Score
1	AD	**0.712840**
2	AU	0.441050
3	SD	0.200989
4	TPO	0.099613
5	W	−0.011708
6	TOP	−0.023331
7	CCO	−0.487797
8	TRP	**−0.698357**

Table 4 The result of different ratio Train/Test dataset

Method	Train/Test (60/40)		Train/Test (80/20)	
	Accuracy	F1-Score	Accuracy	F1-Score
LR	0.7456	0.4862	0.7581	0.4954
SVM	0.6123	0.3728	0.6343	0.4168
KNN	0.8425	0.6546	0.8733	0.6952
LSTM	**0.9447**	**0.8946**	**0.9530**	**0.9165**

Table 5 The result of cross validation K-fold = 10

Method	Accuracy	F1-Score
LR	0.7281	0.4846
SVM	0.6081	0.3924
KNN	0.8542	0.6723
LSTM	**0.9470**	**0.8987**

present. This will also be a direction for future research for FiberVNN service to contribute to improving the quality of service to achieve excellent rating.

5 Conclusion

In this paper, we proposed the deep learning model for prediction of service quality of FiberVNN. We analyzed the factors affecting the service quality of FiberVNN using Pearson correlation coefficients and used these features for deep learning model-Long Short Term Memory to predict service quality. We evaluated the proposed model by our collected data at VNPT Trang Bang, Tay Ninh with accuracy and F1-scores. We applied different Train/Test and cross validation on the model and compared the results of our proposed model with the result of three machine learning algorithms: Logistic Regression, SVM and KNN. Our experiments determined that the LSTM model gave the best results in all experiments. In the future, we will explore other features and methods to improve the result of service quality prediction.

References

1. Mukherjee B (2006) Optical WDM networks. Springer Science & Business Media
2. DeCusatis C (2014) Optical interconnect networks for data communications. J Lightwave Technol 32(4):544–552
3. Kozdrowski S, Cichosz P, Paziewski P, Sujecki S (2021) Machine learning algorithms for prediction of the quality of transmission in optical networks. Entropy 23:7. https://doi.org/10.3390/e23010007
4. Ahmad AK, Jafar A, Aljoumaa K (2019) Customer churn prediction in telecom using machine learning in big data platform. J Big Data 6:28. https://doi.org/10.1186/s40537-019-0191-0191-6
5. Musumeci F et al (2019) An overview on application of machine learning techniques in optical networks. IEEE Commun Surv Tutorials 21(2):1383–1408. https://doi.org/10.1109/COMST.2018.2880039
6. Mata J, De Miguel I, Duran RJ, Merayo N, Singh SK, Jukan A, Chamania M (2018) Artificial intelligence methods in optical networks: a comprehensive survey. Opt Switch Netw 28:43–57
7. Rafique D, Velasco L (2018) Machine learning for network automation: overview, architecture, and applications. J Opt Commun Netw 10(10):D126–D143
8. Karanov B, Chagnon M, Thouin F, Eriksson TA, Bülow H, Lavery D, Schmalen L (2018) End-to-end deep learning of optical fiber communications. J Lightwave Technol 36(20):4843–4855

9. Prakash UM, Cengiz KKK, Kose U, Hung BT (2021) 4x-expert systems for early predication of osteoporosis using multi-model algorithms. Measurement 2021.https://doi.org/10.1016/j.measurement.2021.109543
10. Gupta A, Pramanik S, Bui HT, Ibenu NM (2021) Machine learning and deep learning in steganography and steganalysis. In: Multidisciplinary approach to modern digital steganography.https://doi.org/10.4018/978-1-7998-7160-6.ch004
11. Hung BT, Tien LM (2021) Facial expression recognition with CNN-LSTM. In: Research in intelligent and computing in engineering. Springer series in advances in intelligent systems and computing. https://doi.org/10.1007/978-981-15-7527-3_52
12. Hung BT, Semwal VB, Gaud N, Bijalwan V (2021) Hybrid deep learning approach for aspect detection on reviews. In: Proceedings of integrated intelligence enable networks and computing. Springer series in algorithms for intelligent systems. https://doi.org/10.1007/978-981-33-6307-6_100
13. Hung BT, Semwal VB, Gaud N, Bijalwan V (2021) Violent video detection by pre-trained model and CNN-LSTM approach. In: Proceedings of integrated intelligence enable networks and computing. Springer series in algorithms for intelligent systems. https://doi.org/10.1007/978-981-33-6307-6_99
14. Kirch W (2008) Pearson's correlation coefficient. Encyclopedia of Public Health, Springer, Dordrecht. https://doi.org/10.1007/978-1-4020-5614-7_2569
15. Hochreiter S, Schmidhuber J (1997) Long short-term memory. Neural Comput 9:1735–1780. https://doi.org/10.1162/neco.1997.9.8.1735
16. François Chollet, Keras (2015). https://github.com/fchollet/keras
17. Abadi M, Barham P, Chen J, Chen Z, Davis A, Dean J, Devin M, Ghemawat S, Irving G, Isard M, Kudlur M, Levenberg J, Monga R, Moore S, Murray DG, Steiner B, Tucker P, Vasudevan V, Warden P, Wicke M, Yu Y, Zheng X (2016) Tensorflow: a system for large-scale machine learning. In: Proceedings of the 12th USENIX conference on operating systems design and implementation, OSDI'16, pp 265–283. https://doi.org/10.1007/s10107-012-0572-5

Simulation of Digital Control Systems by Nonlinear Objects

E. V. Larkin, V. S. Nguyen, and A. N. Privalov

Abstract The principle of simulation the digital control system by multi-contour objects, which include actuators with significant nonlinearities, is proposed. The simulation method intends analytical description of object under control with use transfer functions matrix, description of control law in the form of transfer function matrix, estimation of time delays, contributed by controller, introducing delays into the controller transfer function matrix, description of actuator with nonlinearities by non-linear differential equation in Cauchy form. For estimation of time delays semi-Markov model of control algorithm is worked out. The method of transformation the ergodic semi-Markov model to non-ergodic one for estimation of wandering time from-point-to-point is proposed. the proposed method is confirmed by simulation of double-contour object with dry friction on actuators.

Keywords Object under control · Analogue controller · Digital controller · Actuator · Significant nonlinearity · Control algorithm · Runtime · Time delay

1 Introduction

Digital control by complex multi-circuit objects assumes the usage of Von Neumann controllers for computation signals supplied to actuators at the base of digital signals coming from the sensor subsystem [1]. Due to the fact, that control algorithms are implemented in the form of a program interpreted by the controller sequentially, command-by-command, codes, supplied to actuators, have a time delay relative to codes received from sensors [2], which significantly disturb such control system performance characteristics, as regulation time and overshoot [3–6]. Another factor

E. V. Larkin
Tula State University, Tula, Russia

V. S. Nguyen (✉)
Military Weapon Institute, Hanoi, Vietnam

A. N. Privalov
Tula State Lev Tolstoy Pedagogical University, Tula, Russia

influencing on the transient process in the system is the presence of significant nonlinearities at the joints of the actuator electric drive with actuator mechanism (valves, rudders, flaps, etc.), such as backlash, dry friction, various kinds of restrictions, dead zones, etc. [7]. To take into account both factors when controller design it is necessary to have an adequate model of the system, which describes object under control, the digital controller and the nonlinearities of actuators. The methods of forming such models are insufficiently developed, which explains the necessity and urgency of research in this area [8–11].

2 Object Under Control

Flowcharts of control systems with analogue (AC) and digital (DC) controllers are shown in Fig. 1. On the Fig. 1 object under control is represented by matrix impulse response $w(t) = w_{k,l}(t)$ and characterized by the state vector $x(t) = [x_1(t), \ldots, x_k(t), \ldots, x_K(t)]$, where t—is the physical time. Analogue controller is represented by matrix impulse response $w_0(t) = [w_{0,k,l}(t)]$, which describes, f.e. PID-regulator with cross-links, which converts state vector $x(t)$ onto feedback vector $x_0(t) = [x_{0,1}(t), \ldots, x_{0,k}(t), \ldots, x_{0,K}(t)]$. Feedback vector $x_0(t)$ is compared with the states setting vector $f(t) = [f_1(t), \ldots, f_k(t), \ldots, f_K(t)]$, as a result of which the vector of the control signal is formed $u(t) = [u_1(t), \ldots, u_k(t), \ldots, u_K(t)]$. Signals $u(t)$, supplied to actuators A inputs put into action executive mechanisms, such as valves, rudders, flaps, etc., forming vector $v(t) = [v_1(t), \ldots, v_k(t), \ldots, v_K(t)]$, whose elements, in fact, affect the object under control $w(t)$, changing its state.

In digital control system state vector $x(t)$ is transformed at the interface between the object and the digital controller (DC) into state codes vector $x_c(n\tau) = [x_{c,1}(n\tau), \ldots, x_{c,k}(n\tau), \ldots, x_{c,K}(n\tau)$, where n is the discrete controller's time; τ is timing pulse repetition period, which code-by-code is inputted into DC. Also code-by-code digital states setting vector $f_c(n\tau) = [f_{c,1}(n\tau), \ldots, f_{c,k}(n\tau), \ldots, f_{c,K}(n\tau)]$, is formed. DC executes just the same functions, as AC, but on the output control codes vector $u_c(n\tau) =$

Fig. 1 Analogue **a** and digital **b** control systems (AC*t*—analogue controller; DC*t*—digital controller)

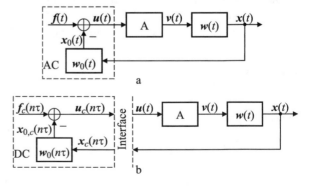

$[u_{c,1}(n\tau), \ldots, u_{c,k}(n\tau), \ldots, u_{c,K}(n\tau)]$ is formed, which at the interface code-by-code is transformed into analogue vector signal $u(t)$.

The object under control is described by the following matrix convolution:

$$x(t) = v(t) * w(t) \tag{1}$$

where matrix convolution is executed according to the rules of vector-to-matrix multiplication, in which the multiplying operation is replaced with the convolution.

Vector $u(t)$, supplied to drive inputs is defined as follows:

$$u(t) = f(t) - x(t) * w_0(t). \tag{2}$$

Applying to both parts of (1) and (2) the Laplace transform the following equation may be obtained:

$$U(s) = F(s) - V(s).W(s).W_0(s), \tag{3}$$

where $U(s) = L[u(t)]; F(s) = L[f(t)]; V(s) = L[v(t)]; W(s) = L[w(t)]; W_0(s) = L[w_0(t)]; L[\ldots]$ is the direct Laplace transform; $s = \sigma + i\omega$ is the Laplace variable; σ is the abscissa of absolute convergence; ω is athe circular frequency.

Transfer function matrices $W(s)$ and $W_0(s)$ would be represented as,

$$W(s) = \begin{bmatrix} W_{1,1}(s) \ldots W_{1,l}(s) \ldots W_{1,K}(s) \\ \cdots \\ W_{k,1}(s) \ldots W_{k,l}(s) \ldots W_{k,K}(s) \\ \cdots \\ W_{K,1}(s) \ldots W_{K,l}(s) \ldots W_{K,K}(s) \end{bmatrix}; \tag{4}$$

$$W_0(s) = \begin{bmatrix} W_{0,1,1}(s) \ldots W_{0,1,l}(s) \ldots W_{0,1,K}(s) \\ \cdots \\ W_{0,k,1}(s) \ldots W_{0,k,l}(s) \ldots W_{0,k,K}(s) \\ \cdots \\ W_{0,K,1}(s) \ldots W_{0,K,l}(s) \ldots W_{0,K,K}(s) \end{bmatrix}, \tag{5}$$

where $W_{k,l}(s)$ and $W_{0,k,l}(s)$ are elements of matrices (5) and (6), correspondingly,

$$W_{k,l}(s) = \frac{B_{k,l}(s)}{C_{k,l}(s)}, \quad W_{0,k,l}(s) = \frac{B_{0,k,l}(s)}{C_{0,k,l}(s)},$$

$$C_{k,l}(s) = \sum_{j(k,l)=0}^{J(k,l)} c_{j(k,l)} s^{j(k,l)},$$

$$B_{k,l}(s) = \sum_{m(k,l)=0}^{M(k,l)} b_{m(k,l)} s^{m(k,l)},$$

$$C_{0,k,l}(s) = \sum_{j(0,k,l)=0}^{J(0,k,l)} c_{j(0,k,l)} s^{j(0,k,l)},$$

$$B_{0,k,l}(s) = \sum_{m(0,k,l)=0}^{M(0,k,l)} b_{m(0,k,l)} s^{m(0,k,l)}.$$

In polynomials $C_{k,l}(s)$, $B_{k,l}(s)$, $C_{0,k,l}(s)$, $B_{0,k,l}(s)$ $1(k,l) \le m(k,l) \le M(k,l)$, $1(k,l) \le j(k,l) \le J(k,l)$, $1(0,k,l) \le m(0,k,l) \le M(0,k,l)$, $1(0,k,l) \le j(0,k,l) \le J(0,k,l)$ are index functions, introduced to simplify the expressions description; $c_{j(k,l)}$, $b_{m(k,l)}$, $c_{j(0,k,l)}$, $b_{m(0,k,l)}$ are constants or slowly changing coefficients; $M(k,l)$, $J(k,l)$, $M(0,k,l)$, $J(0,k,l)$ are proper polynomials orders; $M(k,l) < J(k,l); M(0,k,l) < J(0,k,l)$.

Matrices (4) and (5) product is as follows:

$$\tilde{W}(s) = W(s).W_0(s) = \left[\tilde{W}_{k,l}(s)\right],\tag{6}$$

where

$$\tilde{W}_{k,l}(s) = \sum_{\xi=1}^{K} \frac{\sum_{m(k,\xi)=0}^{M(k,\xi)} b_{m(k,\xi)} s^{m(k,\xi)}}{\sum_{j(k,\xi)=0}^{J(k,\xi)} c_{j(k,\xi)} s^{j(k,\xi)}} \cdot \frac{\sum_{m(0,\xi,l)=0}^{M(0,\xi,l)} b_{m(0,\xi,l)} s^{m(0,\xi,l)}}{\sum_{j(0,\xi,l)=0}^{J(0,\xi,l)} c_{j(0,\xi,l)} s^{j(0,\xi,l)}}.\tag{7}$$

Vector $u(t)$ is defined as inverse Laplace transform

$$u(t) = L^{-1}\left[F(s) - V(s).\tilde{W}(s)\right],\tag{8}$$

where $L^{-1}[\ldots]$ is the notation of the inverse Laplace transform.

3   Nonlinearity

In real actuators A significant nonlinearities such as backlash, dry friction, dead zone, limitation, etc. are always present in the places where the mechanical energy source (engine) docks with the actuator mechanism. On practice the most common type of nonlinearity is a dry friction on moving parts of named assembly. The k-th drive with control by the position of the actuator mechanism may be simulated by the following system of nonlinear differential equations in the Cauchy form:

$$\begin{cases} {}^1\dot{v}_k(t) = {}^2v_k(t); \\ {}^2\dot{v}_k(t) = -\alpha_{k,1}{}^2v_k(t) - \alpha_{k,2}.\text{sgn}[{}^2v_k(t)] + \alpha_{k,3}.[u_k(t) - {}^1v_k(t)], \end{cases} \quad (9)$$

where ${}^1v_k(t)$, ${}^2v_k(t)$ are executive mechanism position and its speed, correspondingly; $\alpha_{k,1}$, $\alpha_{k,2}$, $\alpha_{k,3}$ are actuator parameters; ${}^1\dot{v}_k(t) = \frac{dv_k(t)}{dt}$; ${}^2\dot{v}_k(t) = \frac{d^2\dot{v}_k(t)}{dt}$.

Expression (9) closes mathematical description of control system with AC.

4 Control System with DC

In the system with DC, shown on the Fig. 1b, DC processes digital signals $x_c(n\tau)$ and $f_c(n\tau)$, performed as a sequences of codes. Computation of vector $u_c(n\tau)$ elements is executed according to the dependence:

$$u_c(n\tau) = f_c(n\tau) - x_c(n\tau)\bar{*}w_0(n\tau), \quad (10)$$

where $\bar{*}$ is the notation of the digital convolution.

When practical realization of digital control, for access to inputting elements of $f_c(n\tau)$ and $x_c(n\tau)$, and outputting elements of $u_c(n\tau)$ in DC a polling procedure is organized, unfolding in real physical time. In turn, polling born time delays between:

- Forming analogue vectors $x(t)$, $f(t)$ and a beginning of $x_c(n\tau)$, and $f_c(n\tau)$ processing;
- Starting of $x_c(n\tau)$ and $f_c(n\tau)$ processing and finishing of $u_c(n\tau)$ computation;
- Finishing of $u_c(n\tau)$ computation and arriving elements of $u(t)$ on physical inputs of actuators.

Time lags of $f_c(n\tau)$ elements forming born data skew. Besides, signal $f_c(n\tau)$ may be realized by software in DC directly, so, it is no matter when $f_c(n\tau)$ elements will be processed. Due to the fact, below only delays of passing $x(t)$ vector signals, when computation $u(t)$ elements are investigated.

When transmitting data through the interface following delays take place

$$[x_{c,1}(n\tau), \ldots, x_{c,k}(n\tau), \ldots, x_{c,K}(n\tau)]$$
$$= [x_1(t - \tau_{x,1}), \ldots, x_k(t - \tau_{x,k}), \ldots, x_K(t - \tau_{x,K})]; \quad (11)$$

$$[u_{c,1}(n\tau - \tau_{u,1}), \ldots, u_{c,k}(n\tau - \tau_{u,1}), \ldots, u_{c,K}(n\tau - \tau_{u,1})]$$
$$= [u_1(t), \ldots, u_k(t), \ldots, u_K(t)], \quad (12)$$

where $\tau_{x,k}$ is the lag when the signal $x_k(t)$ processing; $\tau_{u,k}$ is the lag when the signal $u_{c,k}(n\tau)$ processing, $1 \le k \le K$.

Application direct Laplace transform to discrete function of argument $n\tau$ gives so called D-transform, while exception from D-transform the timing pulse repetition period τ gives the Z-transform [12]. When sampling period $\tau \to 0$, then substitution Z-transform by ordinary Laplace transform is possible [13]. In this case $x_c(n\tau) \to x_c(t)$, $u_c(n\tau) \to u_c(t)$ and (11), (12) became as follows:

$$X_c(s) = X(s).D_x(s); \tag{13}$$

$$U(s) = U_c(s).D_u(s), \tag{14}$$

where $D_x(s) = \left[D_{x,k,l}(s)\right]$ and $D_u(s) = \left[D_{u,k,l}(s)\right]$ are delay matrices in which

$$D_{x,k,l}(s) = \begin{cases} \exp(-s\tau_{x,k}), & \text{when } k = l \\ 0, & \text{when } k \neq l \end{cases} \tag{15}$$

$$D_{u,k,l}(s) = \begin{cases} \exp(-s\tau_{u,k}), & \text{when } k = l \\ 0, & \text{when } k \neq l \end{cases} \tag{16}$$

With taking delays into account, value $u(t)$ may be expressed according dependence, quite similar to (8):

$$u(t) = L^{-1}\left[\tilde{F}(s) - V(s).W(s).\tilde{W}_c(s)\right], \tag{17}$$

where $\tilde{F}(s) = F(s).D_u(s)$; $\tilde{W}_c(s) = D_x(s).W_c(s).D_u(s)$;

$$\tilde{W}_c(s) = \begin{bmatrix} W_{1,1}(s)e^{-s\tau_{1(x),1(u)}} \dots W_{1,l}(s)e^{-s\tau_{1(x),l(u)}} \dots W_{1,K}(s)e^{-s\tau_{1(x),K(u)}} \\ \dots \\ W_{k,1}(s)e^{-s\tau_{k(x),1(u)}} \dots W_{k,l}(s)e^{-s\tau_{k(x),l(u)}} \dots W_{k,K}(s)e^{-s\tau_{k(x),K(u)}} \\ \dots \\ W_{K,1}(s)e^{-s\tau_{K(x),1(u)}} \dots W_{K,l}(s)e^{-s\tau_{K(x),l(u)}} \dots W_{K,K}(s)e^{-s\tau_{K(x),K(u)}} \end{bmatrix} \tag{18}$$

$$\tau_{k(x),l(u)} = \tau_{x,k} + \tau_{u,l}. \tag{19}$$

Description out-of-controller part of the system is performed by (1), (9), so model of control system with DC is closed too.

To estimate lags $\tau_{x,k}$ and $\tau_{u,l}$ one should to analyze time complexity of control polling algorithm, which generates quests to sensors and actuators. For simplicity propose, that in polling algorithm there is an only operator $k(x)$, which generates quests to sensor x_k, $1 \leq k \leq K$, and an only operator $l(u)$, which generates quests to actuator u_l, $1 \leq l \leq K$ (Fig. 2). Also in algorithm abstract analogue only quest operators are included, and time, spent for computation control actions, is included

Fig. 2 Structure of polling
control algorithm **a**,
semi-Markov process for
delays estimation **b** and
simplest polling procedure **c**

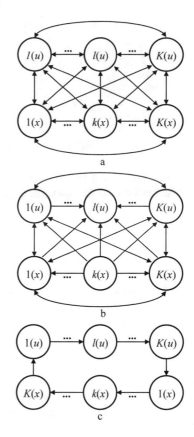

to proper input–output operators. In common case, quantity nodes in the graph, performed polling control algorithm, is $2K$. Graph is full, but without loops, due to the restriction, that consistent quests to the same peripheral are excluded.

Taking into account the fact, that for external observer, operator execution times and switches at branching points are random ones, polling procedure may be considered as semi-Markov process [14, 15].

$$\boldsymbol{h}(t) = \left[h_{\eta,\zeta}(t)\right] = \left[g_{\eta,\zeta}(t)\right] \otimes \left[p_{\eta,\zeta}(t)\right], \tag{20}$$

where $\boldsymbol{h}(t)$ is the $2K \times 2K$ semi-Markov matrix; $p_{\eta,\zeta}(t)$ is the probability of direct switching from the state η to the state ζ; $g_{\eta,\zeta}(t)$ is the time of sojourn at the state η, when a priori is known, that next switching will be to the state ζ; $\otimes$ is the direct multiplying sign; $\eta \in \{1(x), \ldots, k(x), \ldots, K(x), 1(u), \ldots, l(u), \ldots, K(u)\}$, $\zeta \in \{1(x), \ldots, k(x), \ldots, K(x), 1(u), \ldots, l(u), \ldots, K(u)\}$.

Due to quests periodicity and features of a polling procedure, semi-Markov process (20) is the ergodic one and does not contain absorbing or semi-absorbing states, i.e.

$$0 < T(\eta, \zeta, \min) \le \arg\left[g_{\eta,\zeta}(t)\right] \le T(\eta, \zeta, \max) < \infty, 1 \le \eta, \zeta \le \varsigma;$$

$$\sum_{\zeta=1}^{\varsigma} p_{\eta,\zeta} = 1;$$

$$\int_{T(\eta,\zeta,\min)}^{T(\eta,\zeta,\max)} g_{\eta,\zeta}(t)dt = 1,$$

where $T(\eta, \zeta, \min)$ and $T(\eta, \zeta, \max)$ are lower and upper boundaries of $g_{\eta,\zeta}(t)$ domain.

Contribute to time intervals between inputting of x_k and outputting of u_l may both direct switching, and wandering through the semi-Markov process from $k(x)$ to $l(u)$. For estimation of time delay the semi-Markov matrix (20) should be transformed as follows (Fig. 2b):

$$\boldsymbol{h}(t) \to \boldsymbol{h}'(t) = \left[g_{\eta,\zeta}(t).p'_{\eta,\zeta}\right]. \tag{21}$$

In the matrix $\boldsymbol{h}'(t)$ $[\eta = k(x)]$-th column and $[\zeta = K(x) + l(u)]$-th row are zeroed and probabilities $p_{\eta,\zeta}$ in all other rows and columns are recalculated according dependence

$$p'_{\eta,\xi} = \frac{p_{\eta,\xi}}{1 - p_{\eta,}}. \tag{22}$$

After recalculation, semi-absorbing states, emerging after $[\eta = k(x)]$-th column zeroing become non-absorbing, time density of wandering may be defined as follows:

$$\tilde{g}_{\eta,\zeta}(t) = \boldsymbol{I}_{R,\eta} \cdot L^{-1}\left[\sum_{\kappa=1}^{\infty}\{L[\boldsymbol{h}'(t)]\}^{\kappa}\right] \cdot \boldsymbol{I}_{C,\zeta}, \tag{23}$$

where $\boldsymbol{I}_{R,\eta}$ is the row vector of size $2\,K$, η th element of which is equal to 1, and all other elements are equal to zero; $\boldsymbol{I}_{C,\zeta}$ is the row vector of size $2\,K$, ζ th element of which is equal to 1, and all other elements are equal to zero.

In simplest case of polling, shown on the Fig. 2c, $\tilde{g}_{k(x),l(u)}(t)$ after sequential inputing of $\boldsymbol{x}(t)$ follows element-by-element outputting of vector $\boldsymbol{u}(t)$. In this case (23) is estimated as

$$\tilde{g}_{k(x),l(u)}(t) = L^{-1}\left[\prod_{\eta=k(x)}^{K(x)+l(u)+1} L\left[g_{\eta,\eta+1}(t)\right]\right]. \tag{24}$$

For time density $\tilde{g}_{k(x),l(u)}(t)$ according to «three sigma rule» [16–19] may be obtained estimations of time delays, which influence on the system performance:

$$\tau_{k(x),l(u)} = \tilde{T}_{k(x),l(u)} + 3\sqrt{\tilde{D}_{k(x),l(u)}}, \tag{25}$$

where

$$\tilde{T}_{k(x),l(u)} = \int_0^\infty \tilde{g}_{k(x),l(u)}(t)dt;$$

$$\tilde{D}_{k(x),l(u)} = \int_0^\infty \left(t - \tilde{T}_{k(x),l(u)}\right)^2 \tilde{g}_{k(x),l(u)}(t)dt.$$

5 Method

From described above, the proposed method of simulation of control system with digital controller and non linear actuators is presented below:

(1) Analytical description of abject under control in the form (1) and formation of its linear part transfer functions matrix $W(s)$.
(2) Synthesis of DC's control law in the form of $W_c(s)$ matrix.
(3) Working out control algorithm for its realization on DC, evaluation of its operators time complexities and probabilities of switching at branching points.
(4) Working out the semi-Markov ergodic model of algorithm (20) and transform it according (21).
(5) Estimation of time delays according (23) and (25) for every pair of signals $x_k(t)$, $u_l(t)$, $1 \le k, l \le K$.
(6) Description actuators by means of nonlinear differential equations in the Cauchy form.
(7) Integration created model by any convenient way.

6 Implementation in Digital Control System

Confirm the method with example of imitational simulation of system, shown on the Fig. 3.

Fig. 3 Double-contour control system with DC

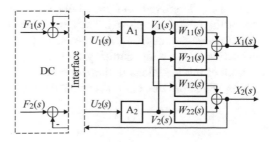

Fig. 4 Transient performance of the system

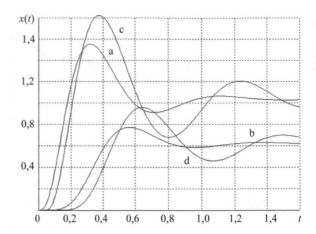

Transfer functions, which describe dynamics of object under control, are the next: $W_{11}(s) = W_{21}(s) = \frac{3}{0,5s+1}$; $W_{12}(s) = W_{22}(s) = \frac{1,5}{0,5s+1}$. In the system proportional control law is realized. $F_1(s) = F_2(s) = L[1(t)]$, where $1(t)$ is the Heaviside function. Actuators are described with Eq. (9) in which $\alpha_{1,1} = \alpha_{2,1} = 400$, $\alpha_{1,2} = \alpha_{2,2} = 1$, $\alpha_{1,3} = \alpha_{2,3} = 29$. Lags are the follows: $\tau_{1(x),1(u)} = \tau_{2(x),2(u)} = 0,04$ s; $\tau_{1(x),2(u)} = \tau_{2(x),1(u)} = 0$ s (absence cross links in DC).

Figure 4 shows system performance. Graphs a and b show transient processes in the system with AC. In transient process, shown on graphs c and d time delays are taken into account. Graphs a and c show $x_1(t)$, while graphs b and d show $x_2(t)$. As it follows from the example time delays at control contours increase overshooting and time of transient process. Due to it fact runtime of control programs should be taken into account when designing multicontour systems with DC, realized on Von Neumann computer.

7 Conclusion

As a result, the method of digital control systems simulation is worked out, in which considered not only nonlinearities of actuator, but also real time delays, which born Von Neumann type controller, organizing polling procedure of quest sensors and actuators. It is proved, that time delays may be estimated for control algorithms practically of any complexity. Theoretical results, confirmed by imitational simulation of two-contour system, may be used when engineering synthesis both hardware and software of digital control system.

Further investigations in domain may be directed on working out genetic algorithms, which provide optimal control process by linear objects with nonlinearities.

References

1. Landau ID, Zito G (2006) In: Digital control systems, design, identification and implementation. Springer, pp 484
2. Larkin EV, Ivutin AN (2014) Estimation of latency in embedded real-time systems. In: 3-rd Meditteranean conference on embedded computing (MECO-2014). pp 236–239
3. Fridman E, Shaked U (2002) A descriptor system approach to H_∞ control of linear time-delay systems. IEEE Trans Autom Control 4(2):253–270
4. Sanz R, García P, Albertos P, Fridman E (2020) Robust predictive extended state observer for a class of nonlinear systems with time-varying input delay. Int J Control 93(2):217–225
5. Wu M, He Y, She JH, Liu GP (2004) Delay-dependent criteria for robust stability of time-varying delay systems. Automatica 40(8):1435–1439
6. Zhang XM, Min WU, Yong HE (2004) Delay dependent robust control for linear systems with multiple time-varying delays and uncertainties. Control Decision 19(5):496–500
7. Wu R, Fan D, Iu HH-C, Fernando T (2019) Adaptive fuzzy dynamic surface control for uncertain discrete-time non-linear pure-feedback mimo systems with network-induced time-delay based on state observer. Int J Control 92(7):1707–1719
8. Fadali MS, Visioli A (2013) Digital control engineering: analysis and design. Elsevier Inc 239–272
9. Arnold KA (2008) Timing analysis in embedded systems. In Ganssler J, Arnold K et al. (eds) Embedded hardware MA. 01803 USA. Elsevier Inc. pp 239–272
10. Hamann A, Racu R, Ernst R (2007) Multi-dimensional robustness optimization in heterogeneous distributed embedded systems. In: Proceedings of the 13th IEEE real time and embedded technology and applications symposium, RTAS '07, IEEE Computer Society, Washington, DC, USA, pp 269–280
11. Briat C (2018) Stability and performance analysis of linear positive systems with delays using input-output methods. Int J Control 91(7):1669–1692
12. Pospišil M (2017) Representation of solutions of delayed difference equations with linear parts given by pairwise permutable matrices via Z-transform. Appl Mathem Comput 294:180–194
13. Schiff JL (1991) In: The laplace transform: theory and applications USA, NY: Springer, pp 233
14. Howard RA (2012) Dynamic probabilistic systems. Vol. 1: Markov Models. Vol. II: Semi-Markov and Decision Processes. Courier Corporation
15. Larkin EV, Bogomolov AV, Privalov AN (2017) A method for estimating the time intervals between transactions in speech-compression algorithms. Autom Document Mathem Linguistics 51(5):214–219
16. Kobayashi H, Marl BL, Turin W (2012) In: Probability, random processes and statistical analysis: Cambridge University Press, pp 812
17. Pukelsheim F (1994) The three sigma rule. American Stat 48(2):88–91
18. Pavlov AV (2016) About the equality of the transform of laplace to the transform of fourier. Issues of Anal 5(23)(4(76)):21–30
19. Li J, Farquharson CG, Hu X (2015) Three effective inverse Laplace transform algorithms for computing time-domain electromagnetic responses. Geophys 81(2):E75–E90

EEG-Based Biometric Close-Set Identification Using CNN-ECOC-SVM

Chi Qin Lai⑩, Haidi Ibrahim⑩, Mohd Zaid Abdullah⑩, and Shahrel Azmin Suandi⑩

Abstract As the growth of technology hikes, biometrics turns into an important field that identifies the individual's identity to access their sensitive pieces of information and asset. In the last few years, electroencephalography (EEG)-based biometrics have been explored by researchers as EEG is unique for each individual. Literature has shown that a convolutional neural network (CNN), which is one of the deep learning approaches, can skip feature extraction and feature selection stages. In this paper, a CNN architecture, which is incorporated with a majority voting ensemble of error-correcting output codes with a support vector machine (ECOC-SVM), is proposed in this study for biometrics identification. Open-source biometric EEG datasets used in this paper include both resting-state eyes open (REO) and resting-state eyes close (REC) EEG. It is shown that the proposed architecture achieved an identification accuracy of 98.49%. On the other hand, it is suggested that the proposed architecture can avoid complex feature extraction that is commonly seen in machine learning models.

Keywords Biometrics · Electroencephalography · Identification

C. Q. Lai
Peninsula College Malaysia, The Ship Campus, No. 1, Education Boulevard Batu Kawan
Industrial Park, 14110 Kampung Batu Kawan, Penang, Malaysia
e-mail: laichiqin@peninsulacollege.edu.my

H. Ibrahim (✉) · M. Z. Abdullah · S. A. Suandi
School of Electrical and Electronic Engineering, Engineering Campus, Universiti Sains Malaysia,
14300 Nibong Tebal, Penang, Malaysia
e-mail: haidi_ibrahim@ieee.org

M. Z. Abdullah
e-mail: mza@usm.my

S. A. Suandi
e-mail: shahrel@usm.my

© The Author(s), under exclusive license to Springer Nature Switzerland AG 2022   723
N. H. T. Dang et al. (eds.), *Artificial Intelligence in Data and Big Data Processing*,
Lecture Notes on Data Engineering and Communications Technologies 124,
https://doi.org/10.1007/978-3-030-97610-1_57

1 Introduction

In the era of swift technologies turnovers, personal information has been broadly required for all kinds of aspects. Thus, it is critical to keep them secure. Authentication is required to limit access to personal information [1]. There are three approaches to authentication. The first approach is based on knowledge. The second approach is based on the token. The third approach is based on biometrics [2]. Knowledge-based procedures rely on information, which is defined by humans, such as a textual password or personal identification number (PIN). The approach that uses the token makes use of the belongings of users, such as smart cards and identity cards. Both approaches are the most common. However, they carry a disadvantage as the identifiers can be misplaced or stolen [3]. Loss of the identifiers can cause crucial problems, causing leakage of sensitive information, leading to financial loss.

A biometric-based approach is a better alternative. The biometric-based approach utilized the user's behavioral and physiological characteristics as features to authenticate a person. Voices and gait are examples of behavioral characteristic, whereas physiological characteristic refers to DNA, fingerprints, iris scans, and palm print. Biometrics can be grouped into two classes, which are conventional biometrics [2] and cognitive biometrics [4]. Conventional biometrics extract the physiological and behavioral traits of the user. On the other hand, cognitive biometrics perform authentication-based features extracted from the signals from the human brain, either in a cognitive or emotional brain state.

Studies have shown that electrical signals from the human brain are unique to every individual [5–7]. Thus, electroencephalography (EEG) is suitable for person authentication [8, 9]. Implementing EEG for biometrics offers a high confidential advantage as it is generated based on the human brain corresponding to the task performed during the recording. It is difficult to imitate and copy EEG signals for illegal purposes. Among all task-related EEG acquisition, the combination of resting-state eyes close (REC) and resting-state eyes open (REO) EEG recording is the best option. Subjects are only required to close (i.e., REC) and open their eyes in resting state (i.e., REO) during the EEG acquisition, making resting-state EEG suitable for all individuals, including those that are severely injured and bedridden. Existing works have shown that using REC and REO EEG effectively represent the identity of individuals [10–13].

2 Related Works

In the literature, several existing works make use of REC and REO EEG for biometrics authentication. In work by Choi et al. [10], both REC and REO are used as the traits for biometrics. Alpha activities are extracted as features because the alpha's power of the recording gets stronger in the REC state [14–16]. To determine the identification accuracy, leave-one-out with cross-correlation (CC) was performed. It was found that the spatio-spectral patterns of the changes in alpha signals are distinct between persons and can be used to achieve authentication effectively.

Thomas and Vinod [11] extracted sample entropy from REC and REO at several EEG bands. The features are fed into Mahalanobis distance-based classifier for individual authentication. Their results show that beta-band entropy gave the highest intersubject variability. The performance of their work is further improved by concatenating the entropy traits with the power spectral density (PSD).

In work by Suppiah and Vinod [12], PSD is taken from REC and REO EEG to train a classifier known as Fischer linear discriminant, to do authentication. Similarly, Lee et al. [13] have used spectral power, maximum power, and frequency of maximum power in the alpha band from the REC EEG for their proposed biometric authentication system and were fed into a linear discriminant analysis for authentication. By implementing eigenvector centrality in resting-state brain networks, Fraschini et al. [17] designed a biometric system that uses REC REO EEG as input. Their result shows that the resting-state functional brain network gives a better performance as compared to a normal functional brain network. Their work strengthens the usability of REO or REC EEG as a biometrics input.

From the literature, it can be seen that the implementation of REC and REO in biometrics is getting popular. In the general design of a resting-state biometric system, EEG recordings have to undergo pre-processing to eliminate unwanted elements such as noises and artifacts. Subsequently, feature extraction and selection are made to select the important features. Both processes are time-consuming and extensive. Extracting features from resting-state EEG can be even more difficult as it contains less information than task-related EEG. It is optimum for the machines to automate the process, specifically in implementing resting-state EEG. To avoid the complicated feature extraction process, CNN is potentially applicable.

CNN is one of the deep learning approaches which can automatically perform feature extraction. The basic architecture of a CNN is built up of multilayer perception (MLP), consisting of many hidden layers. The hidden layers consist of convolution layers, pooling layers, and the conventional backpropagation neural network dense layer. The first convolutional layer computes the feature map from the input via the weighted learnable kernels [18]. Each input to the convolutional layers will output a feature map. When patterns of interest are detected in the input, the convolutional layers stimulate the feature maps. Next, the first pooling layer pools the traits collectively and feed-forward to the second convolutional layer. This layer extracts the features from the output of the first pooling layer, resulting in a feature map. This feature map is then will be directed to the second pooling layer. The second pooling layer pool the features. Before passing the resulting feature map into the fully connected layer, it will be flattened. The fully connected layer, which is similar to the conventional artificial neural network (ANN) [19] will perform classification and categorize the input into their labeled classes. Each layer in the hidden layers carries learnable parameters, which need to go through multiple iterations of learning and validation to empirically find the optimum value [20].

This paper presents a REC REO EEG-based biometric close-set identification model. Using EEG, the model can identify specific individuals, proposed as a biometric identification approach for brain situations to prevent personal information illegal access. CNN utilizing the error-correcting output codes with support vector machine (ECOC-SVM) is selected as the deep learning approach to overcome the complex stages of feature extraction and selection. The CNN-ECOC-SVM has shown good performance in classifying traumatic brain injury EEG signals [21]. Therefore, in this paper, we would like to investigate whether the same approach applies to EEG biometric identification.

This paper is divided into six sections. Section 1 gives an introduction, followed by Sect. 2 discussing on related works. It then continues with data acquisition in Sect. 3, which explains the dataset that is used for this research and also the methodology of preparing the EEG signal into compatible input for the proposed architecture. Then, Sect. 4 gives an overview of the proposed architecture for identification. Next, Sect. 5 presents the performance of the proposed architecture as well as the comparison to existing works. Lastly, Sect. 6 concludes this paper.

3 Dataset

The dataset used in this study is obtained from an open-source dataset available online [22]. This dataset has been utilized in the research by Ma et al. [23]. Both the REO and REC EEG are combined into one dataset for this study, as the study has shown that the combination of REC+REO presents better accuracy than only using either REC or REO dataset itself [24]. The EEGs from the dataset are acquired at a sampling rate 160 Hz, utilizing a 64-channel BCI2000 system. A total of 109 subjects have contributed to REO and REC EEGs. The first 60 s of the EEG data are utilized for each situation. This is due to the appearance of the more discriminative characteristic of EEG is located nearer to the starting point of the acquired EEG data [25]. Then, each signal is separated into 60 subsets of one second each. The dataset of a combination of REC + REO is separated into the training set, testing set, and validation set. Twenty-five subsets will be utilized for training, five will be utilized for validation, and 30 will be utilized for testing. After the division, there will be 5450 training samples, 1090 validation samples, and 6540 testing samples for a total of 109 subjects. In this paper, persons that are included in the dataset will be used for classification. As the research is restricted to a close-set identification, additional impostor persons will not be considered.

The EEG forms a matrix of the amplitude of the channel versus sampling point as the input to the CNN [26]. The example for matrix M is presented in Fig. 1. The order of the channels is based on the default order in the dataset [22].

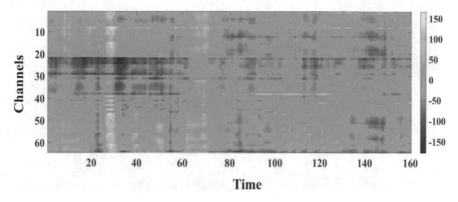

Fig. 1 Matrix M

The matrix size of the EEG will be $N \times F_s$, where N is the number of channels and F_s is the sampling frequency. In this paper, the size of the matrix is 64×160. This is due to the sampling rate of the EEG 160 Hz, 64 channels, and partitioned into one-second segments. The elements in the matrix from the EEG data points is constructed by using the formula:

$$M(i, t) = x_i(t) \tag{1}$$

where i is the EEG channel, t is index of the sampling point and $x_i(t)$ is the EEG amplitude at channel i, at sampling point t.

4 Overview of Proposed Architecture

The proposed biometric identification architecture is divided into two sections. The section performs feature extraction, while the second section performs classification, which is shown in Fig. 2. The activations from each fully connected layer are used to train three ECOC-SVM classifiers, respectively (i.e., feature vector from Layer 3, 4, 5). Error-correcting output codes (ECOC) is a combination of SVM to perform multi-class classification, as SVM alone is a binary classifier.

For this study, the input to the CNN is a matrix of size 64×160. There are six 3×3 filters, sliding using stride length of one in the convolution layer. The activation function used for the convolutional layer is a rectified linear unit (ReLU). Batch normalization is placed after the convolutional layer. The convolution layer's output will be the same as the input matrix, as zero paddings are performed before the convolution. Six 64×160 feature maps are produced and directed to a 2×2 average pooling layer with a stride length of 1. The average pooling layer downsampled the feature maps, resulting in six 32×80 feature maps. These six feature maps are flattened into a long feature vector of size 9300 (i.e., $6 \times 31 \times 50$) and then directed

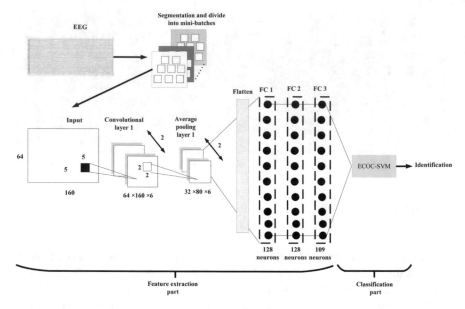

Fig. 2 Proposed architecture

Table 1 Parameters and values

Parameter	Setting
Batch normalization	L_2 normalization
Optimizer	ADAM
Mini batch size	16
L_2 regularization	0.0005
Learning rate	0.001
Training repetitions per epoch	30

into the fully connected (FC) layers. By referring to Fig. 2, FC 3 has a feature vector with 109 activations. Features from FC 3 are used to train an ECOC-SVM classifier.

Six parameters chosen for the proposed architecture are shown in Table 1. The learning rate of 0.001 is used and kept constant throughout the training of CNN. L_2 normalization is utilized to do batch normalization of the convolutional layer. For every iteration, the mini-batch size is set at 16. The training iteration per epoch is fixed at 30. L_2 regularization, with a factor of 0.0005, is utilized to avoid overfitting. The optimizer utilized for the back-propagation for CNN training is adaptive moment estimation (ADAM).

5 Results and Discussions

The performance of the proposed CNN is measured using identification accuracy in terms of percentage. The identification accuracy is acquired from three-fold cross-validation.

The accuracy of identification is calculated by using the following formula:

$$\text{Accuracy} = \frac{\sum C_P}{\sum N_P} \tag{2}$$

where C_P is the correct prediction and N_P is the number of prediction done.

To ensure the proposed architecture presents better results, a comparison has been made with three other existing methods. The first comparison model is the CNN architecture proposed by Ma et al. [23]. The second model is an artificial neural network (ANN) proposed by Hema et al. [27] trained using power spectral density (PSD) extracted from the beta frequency band of the EEG. The third model is a support vector machine (SVM) classifier developed by Ferreira et al. [28] that is trained using spectral power extracted from the gamma frequency band.

For a fair comparison, the same dataset is used for each of the models. To determine the performance of each model, identification accuracy is computed using three-fold cross-validation. The identification accuracy is calculated using Eq. (2). The performance of the proposed architecture and three comparison models are shown in Table 2. The best identification accuracy is given by the proposed architecture with 98.49%. In the second position is the shallow CNN proposed by Ma et al. [23], followed by the ANN [27] and SVM [28].

The proposed architecture uses more filters in the convolution layer compared to the shallow CNN by Ma et al. [23]. Therefore, the features that are extracted are finer and contain more distinct information, which benefits the training of the fully connected layers. ANN proposed by Hema et al. [27] shows identification accuracy of 31.81%, which is relatively low. PSD is extracted using the Welch algorithm from the beta band of EEG signals. The PSD features are not sufficient to represent the unique property of each individual. In their work, high identification accuracy was claimed. However, their EEG dataset is recorded using evoked potential (EP) with the task. It can be seen that EV EEG plotted locations in the signal, which makes PSD a useful feature for the training of the classifier. The datasets that are used in this study are

Table 2 Identification accuracy of proposed architecture and existing methods

Method	Identification accuracy (%)
Shallow CNN [23]	79.07
ANN [27]	31.81
SVM [28]	21.10
Proposed architecture	98.49

REC and REO, which do not contain any EP. Therefore, PSD itself is not sufficient to represent the properties of each individual. The proposed architecture can extract features even though there is no EP in the EEG. Complex feature extraction and selection can be avoided.

A similar situation exists in the comparison for the SVM model proposed by Ferreira et al. [28]. In their work, the spectral power is extracted from the gamma frequency band as a feature to train an SVM. By using their EEG dataset made up of visually evoked potential (VEP), high identification accuracy is presented. Nonetheless, when the resting-state EEG dataset is used, the identification accuracy is low as 21.10%. This is because more features have to be extracted when there are no labeled locations in the signals to represent the solitary property of each individual. Again, this scenario manifests the advantage of the proposed architecture. The pre-selected convolution layers can extract features that are beneficial to the training of fully connected layers without any VEP.

In addition, important features can be extracted from the alpha frequency band of the EEG signal, as it is more pronounced in a relaxed and awake subject with closed eyes [29]. Hema et al. [27] and Ferreira et al. [28] extract features from the beta band and gamma-band respectively. This can be the reason both of the methods do not perform well on resting-state EEG. To archive comparable performance by implementing their methods on resting-state EEG, feature extraction has to be restructured, which can be a complicated and time-consuming task. The proposed method does not require a pre-defined frequency band to perform feature extraction. Important features can be selected automatically by the convolution layers using the weighted learnable kernels [18].

6 Conclusion

When existing approaches were compared to the suggested design, it was discovered that the proposed method outperformed in terms of accuracy. Furthermore, it is demonstrated that the suggested architecture outperforms previous models by obviating the need for complicated feature extraction. Feature extraction is critical in the development of a reliable biometric system. Explicit features must be defined to achieve practical classifier training. In this investigation, the CNN feature extraction component succeeded in feature extraction and had a high accuracy rate.

References

1. Jain AK, Uludag U (2003) Hiding biometric data. IEEE Trans Pattern Anal Mach Intell 25(11):1494–1498
2. Jain AK, Ross A, Prabhakar S (2004) An introduction to biometric recognition. IEEE Trans Circ Syst Video Technol 14(1):4–20

3. Ratha NK, Connell JH, Bolle RM (2001) An analysis of minutiae matching strength. In: Bigun J, Smeraldi F (eds) Audio- and video-based biometric person authentication. Springer, Berlin, pp 223–228
4. Revett K (2012) Cognitive biometrics: a novel approach to continuous person authentication. Int J Cogn Biometrics 1:1–9
5. Poulos M, Rangoussi M, Chrissikopoulos V, Evangelou A (1999) Person identification based on parametric processing of the EEG. In: ICECS'99. Proceedings of ICECS'99. 6th IEEE International conference on electronics, circuits and systems (Cat. No.99EX357)
6. Palaniappan R, Mandic DP (2007) Biometrics from brain electrical activity: a machine learning approach. IEEE Trans Pattern Anal Mach Intell 29(4):738–742
7. Marcel S, Millan JDR (2007) Person authentication using brainwaves (EEG) and maximum a posteriori model adaptation. IEEE Trans Pattern Anal Mach Intell 29(4):743–752
8. Campisi P, Rocca DL (2014) Brain waves for automatic biometric-based user recognition. IEEE Trans Inf Forensics Secur 9(5):782–800
9. Maiorana E, Campisi P (2018) Longitudinal evaluation of EEG-based biometric recognition. IEEE Trans Inf Forensics Secur 13(5):1123–1138
10. Choi G, Choi S, Hwang H (2018) Individual identification based on resting-state EEG. In: 2018 6th International conference on brain-computer interface (BCI), Jan 2018, pp 1–4
11. Thomas KP, Vinod AP (2016) Biometric identification of persons using sample entropy features of EEG during rest state. In: 2016 IEEE International conference on systems, man, and cybernetics (SMC), Oct 2016, pp 003-487–003-492
12. Suppiah R, Vinod AP (2018) Biometric identification using single channel EEG during relaxed resting state. IET Biometrics 7(4):342–348
13. Lee HJ, Kim HS, Park KS (2013) A study on the reproducibility of biometric authentication based on electroencephalogram (EEG). In: 2013 6th International IEEE/EMBS conference on neural engineering (NER), Nov 2013, pp 13–16
14. Barry RJ, Clarke AR, Johnstone SJ, Magee CA, Rushby JA (2007) EEG differences between eyes-closed and eyes-open resting conditions. Clin Neurophysiol 118(12):2765 – 2773. [Online]. Available: http://www.sciencedirect.com/science/article/pii/S1388245707004002
15. Travis T, Kondo C, Knott J (1974) Parameters of eyes-closed alpha enhancement. Psychophysiology 11:674–681
16. Klimesch W, Doppelmayr M, Pachinger T, Russegger H (1997) Event-related desynchronization in the alpha band and the processing of semantic information. Cognitive Brain Res 6(2):83–94. [Online]. Available: http://www.sciencedirect.com/science/article/pii/S0926641097000189
17. Fraschini M, Hillebrand A, Demuru M, Didaci L, Marcialis GL (2015) An EEG-based biometric system using eigenvector centrality in resting state brain networks. IEEE Signal Process Lett 22(6):666–670
18. Sehgal A, Kehtarnavaz N (2018) A convolutional neural network smartphone app for real-time voice activity detection. IEEE Access 6:9017–9026
19. Vapnik VN (1999) An overview of statistical learning theory. IEEE Trans Neural Networks 10(5):988–999
20. Bengio Y (2009) Learning deep architectures for AI. Found Trends Mach Learn 2(1):1–127
21. Lai CQ, Ibrahim H, Abdullah JM, Azman A, Abdullah MZ (2021) Convolutional neural network utilizing error-correcting output codes support vector machine for classification of non-severe traumatic brain injury from electroencephalogram signal. IEEE Access 9:24-946–24-964
22. Goldberger AL, Amaral LAN, Glass L, Hausdorff JM, Ivanov PC, Mark RG, Mietus JE, Moody GB, Peng C-K, Stanley HE (2000) PhysioBank, PhysioToolkit, and PhysioNet: components of a new research resource for complex physiologic signals. Circulation 101(23):e215–e220. http://circ.ahajournals.org/content/101/23/e215.fullPMID:1085218; https://doi.org/10.1161/01.CIR.101.23.e215
23. Ma L, Minett JW, Blu T, Wang WSY (2015) Resting state EEG-based biometrics for individual identification using convolutional neural networks. In: 2015 37th Annual international

conference of the IEEE engineering in medicine and biology society (EMBC), Aug 2015, pp 2848–2851

24. Lai CQ, Ibrahim H, Abdullah MZ, Abdullah JM, Suandi SA, Azman A (2019) Arrangements of resting state electroencephalography as the input to convolutional neural network for biometric identification. Comput Intell Neurosci 2019:10 p (Article ID 7895924)

25. Maiorana E, Campisi P (2018) Longitudinal evaluation of EEG-based biometric recognition. IEEE Trans Inf Forensics Secur 13(5):1123–1138

26. Lai CQ, Ibrahim H, Abdullah MZ, Abdullah JM, Suandi SA, Azman A (2019) A literature review on data conversion methods on EEG for convolution neural network applications. In: Zawawi MAM, Teoh SS, Abdullah NB, Mohd Sazali MIS (eds) 10th International conference on robotics, vision, signal processing and power applications. Springer, Singapore, pp 521–527

27. Hema CR, Paulraj MP, Kaur H (2008) Brain signatures: a modality for biometric authentication. In: 2008 International conference on electronic design, Dec 2008, pp 1–4

28. Ferreira A, Almeida C, Georgieva P, Tomé A, Silva F (2010) Advances in EEG-based biometry. In: Campilho A, Kamel M (eds) Image analysis and recognition. Springer, Berlin, pp 287–295

29. Mitra S (2011) Digital signal processing: a computer-based approach. McGraw-Hill Series in Electrical and computer engineering. McGraw-Hill

Author Index

Printed in the United States
by Baker & Taylor Publisher Services